Plant Biotechnology

BIOTECHNOLOGY

JULIAN E. DAVIES, *Editor*
Pasteur Institute
Paris, France

BIOTECHNOLOGY SERIES

1. R. Saliwanchik	*Legal Protection for Microbiological and Genetic Engineering Inventions*
2. L. Vining (editor)	*Biochemistry and Genetic Regulation of Commercially Important Antibiotics*
3. K. Herrmann and R. Somerville (editors)	*Amino Acids: Biosynthesis and Genetic Regulation*
4. D. Wise (editor)	*Organic Chemicals from Biomass*
5. A. Laskin (editor)	*Enzymes and Immobilized Cells in Biotechnology*
6. A. Demain and N. Solomon (editors)	*Biology of Industrial Microorganisms*
7. Z. Vaněk and Z. Hošťálek (editors)	*Overproduction of Microbial Metabolites: Strain Improvement and Process Control Strategies*
8. W. Reznikoff and L. Gold (editors)	*Maximizing Gene Expression*
9. W. Thilly (editor)	*Mammalian Cell Technology*
10. R. Rodriguez and D. Denhardt (editors)	*Vectors: A Survey of Molecular Cloning Vectors and Their Uses*
11. S.-D. Kung and C. Arntzen (editors)	*Plant Biotechnology*
12. D. Wise (editor)	*Applied Biosensors*
13. P. Barr, A. Brake, and P. Valenzuela (editors)	*Yeast Genetic Engineering*

PLANT BIOTECHNOLOGY

Edited by

Shain-dow Kung
Center for Agricultural Biotechnology
Maryland Biotechnology Institute
University of Maryland
College Park, Maryland

Charles J. Arntzen
Office of the Deputy Chancellor for Agriculture
The Texas A&M University System
College Station, Texas

Butterworths
Boston London Singapore Sydney Toronto Wellington

Library of Congress Cataloging-in-Publication Data
Plant biotechnology/edited by Shain-dow Kung, Charles J. Arntzen.
p. cm.—(Biotechnology ; 12)
Includes bibliographies and index.
ISBN 0-409-90068-0
1. Plant genetic engineering. 2. Plant biotechnology. I. Kung, Shain-dow. II. Arntzen, Charles J. III. Series.
QK981.5.P54 1988
660′.6—dc19 88-10595
CIP

British Library Cataloguing in Publication Data
Plant biotechnology.—(Biotechnology series).
1. Plants. Biotechnology
I. Kung, Shain-dow II. Arntzen, Charles J.
III. Series
660′.6

ISBN 0-409-90068-0

Butterworth Publishers
80 Montvale Avenue
Stoneham, MA 02180

10 9 8 7 6 5 4 3 2 1

Printed in the United States of America

CONTRIBUTORS

Chong-maw Chen
Biomedical Research Institute
University of Wisconsin-Parkside
Kenosha, Wisconsin

J.H. Choi
School of Applied Biology
Georgia Institute of Technology
Atlanta, Georgia

Keith Elliston
Waksman Institute
Rutgers, The State University of New Jersey
Piscataway, New Jersey

David A. Evans
DNA Plant Technology Corporation
Cinnaminson, New Jersey

Robert T. Fraley
Monsanto Company
St. Louis, Missouri

A.A. Gatenby
Central Research and Development Department
E. I. duPont de Nemours and Company
Wilmington, Delaware

John C. Gray
Botany School
University of Cambridge
Cambridge, United Kingdom

R.J. Griesbach
Florist and Nursery Crops Laboratory
USDA, ARS
Beltsville, Maryland

John E. Hearst
Chemical Biodynamics Division
Lawrence Berkeley Laboratory
University of California
Berkeley, California

Tuan-hua David Ho
Department of Biology
Washington University
St. Louis, Missouri

Shain-dow Kung
Center for Agricultural Biotechnology, Maryland Biotechnology Institute and Department of Biology
University of Maryland
College Park, Maryland

Joachim Messing
Waksman Institute
Rutgers, The State University of New Jersey
Piscataway, New Jersey

Harry Roy
Plant Science Group
Biology Department
Rensselaer Polytechnic Institute
Troy, New York

Masahiro Sugiura
Center for Gene Research
Nagoya University
Chikusa, Nagoya, Japan

Zinmay R. Sung
Department of Plant Pathology and Genetics
University of California
Berkeley, California

Athanasios Theologis
Plant Gene Expression Center
ARS, USDA
Albany, California
and
Department of Molecular Plant Biology
University of California
Berkeley, California

William F. Thompson
Department of Botany and Genetics
North Carolina State University
Raleigh, North Carolina

John C. Watson
Department of Botany and Center for Agricultural Biotechnology of the Maryland Biotechnology Institute
University of Maryland
College Park, Maryland

Frank F. White
Department of Plant Pathology
Kansas State University
Manhattan, Kansas

Madeline Wu
Department of Biological Sciences
University of Maryland Baltimore County
Catonsville, Maryland

Ray Wu
Section of Biochemistry, Molecular and Cell Biology
Cornell University
Ithaca, New York

Yu Sheng Zhu
Sungene Technologies Corporation
San Jose, California

CONTENTS

PREFACE

Since the advent of biotechnology, we have been witnessing a golden age in biology that brings with it unprecedented opportunities to study and understand fundamental life processes and to modify and regulate those processes precisely for the benefit of humankind. *Biotechnology* as defined in the introduction by William F. Thompson is the manipulation of biological systems to solve practical problems in industry or agriculture. In a broad sense, the manipulation of biological systems can be achieved only after a basic understanding of how plants work has been attained. In a historical sense, biotechnology is an extension and refinement of ancient techniques for genetic manipulation that date back to when our ancestors recognized that both animals and plants could be selected and crossed to reproduce and upgrade species. Thus, biotechnology has its roots in agriculture. As horticulturist and physiologist employing biotechnology in our research, we succumbed to the temptation to record and promote its advances and applications and invited researchers working in biotechnology to share their knowledge and excitement in a book. The focus of the book is the review of current topics relevant to plant biotechnology.

There are four parts to this book. Part I contains five chapters that review the basic techniques essential to plant biotechnology. In Part II, there are seven chapters that describe the various aspects of the regulation of gene expression in plants. Part III includes five chapters devoted to the chloroplast genome. Part IV presents the practical examples of the importance and uses of biotechnology in crop improvement today. Although not every author points out the future direction or potential application of the techniques under review, it is understood that the potential is enormous and can be quickly realized and amplified. The invited contributors are well-known experts and active scientists in their respective fields. They provide an overall review and a general analysis of the state of biotechnology in their fields and above all, insight information that can be offered only by the scientist who did the work. It would be difficult to find better authorities in the fields covered in this book.

In Part I, Frank F. White provides a detailed description, critical evaluation, and clear prospects for the use of both tumor-inducing and root-inducing plasmids as vectors. The practical applications of the techniques he introduces will be valuable to many researchers. Ray Wu reviews many of the currently used techniques for introducing genes into plant cells. Some of the methods are new and less popular. By listing all of them together he provides the reader with needed flexibility for achieving the desired genetic transformation. David A. Evans summarizes many practical aspects of the tissue culture techniques that are essential to the regeneration of transformed cells into whole plants. He gives a critical appraisal and clear description of sources for many of the current techniques that will guide workers in choosing the appropriate approaches to plant regeneration. R.J. Griesbach describes the techniques for the isolation and introduction of chromosomes into plant cells. This topic is seldom covered in other books. It is a unique contribution and will prove to be a valuable reference for many researchers. A.A. Gatenby contributes the final chapter of Part I, presenting a thorough review and detailed analysis of the methods for inducing, and the results of, plant gene expression in heterologous prokaryotic and eukaryotic systems. This is a useful and timely chapter, since these methods will be widely applied as the study of plant gene expression in a defined and simplier system becomes an important tool in genetic research.

In Part II, Keith Elliston and Joachim Messing lead us through a review of the molecular architecture of plant genes. The knowledge of other systems helps plant scientists to understand the sequences and important elements in the regulated expression of plant genes. J.H. Choi and Zinmay R. Sung's chapter gives detailed information on the background of and problems involved in the isolation of developmentally regulated genes that are expressed at the early stage of embryo development. The carrot system serves as a useful model of increasing importance for the study of gene expression in early embryogenesis. John C. Watson gives an extensive account of the state of the art of light-regulated gene expression in plants. He synthesizes and summarizes the up-to-date information to paint an overall picture of this field at this time. This chapter is followed by three devoted to the hormonal regulation of plant gene expression as we understand it today. Tuan-hua David Ho uses the barley aleurone system to detail the molecular mechanism of gene regulation by gibberellic acid and abscisic acid. The elucidation of this mechanism and that of the stress-regulation of gene expression through heat shock constitutes a valuable contribution to the field of plant biotechnology. Athanasios Theologis presents us with first-hand information on auxin-regulated gene expression in pea. He gives detailed results of experiments and primary data that will be valuable to the workers in this field. Chong-maw Chen summarizes our current knowledge of cytokinin-mediated gene expression. He presents the state of the art of a field that is notably in its infancy. The need for more concerted effort in research into the mechanism of cytokinin action is obvious. The last chapter in this part is the contribution by Yu Sheng Zhu and John E. Hearst. They provide a comprehensive overview of the organization and expression of photosynthetic genes in bacteria, offering a valuable comparison with higher plants.

Because of the great excitement generated and rapid advances in the field of molecular biology of chloroplast genomes, the entirety of Part III is devoted to this subject. We would be less than candid if we did not confess that the study of chloroplast genomes is one of our favorite subjects. In 1986, by taking advantage of biotechnology, the complete nucleotide sequence of the chloroplast genome from tobacco, the *Escherichia coli* equivalent in the plant kingdom, was determined by Masahiro Sugiura's group. This is the first chloroplast genome from a higher plant ever sequenced. Therefore, Sugiura leads Part III by contributing a chapter on the organization and expression of the *Nicotiana* chloroplast genome. He presents a concise and precise description of the structure and organization of the chloroplast genome at the base-pair level. This is the key to understanding organelle genetics. With this knowledge in hand, John C. Gray presents us with an in-depth prognosis for its genetic manipulation. He first carefully analyzes and synthesizes the results of current attempts, and then critically discusses the problems involved at each step of chloroplast genome manipulation. As he points out, many problems have to be resolved before chloroplast genome transformation can be achieved. To bring the techniques for genetic manipulation of the chloroplast genome into focus, Harry Roy selects the gene coding for the large subunit of ribulose bisphosphate carboxylase/oxygenase as a subject. He reviews the potential applications of biotechnology to such manipulation and also voices the potential difficulties involved. He points out the shortcomings of various approaches to improving photosynthesis and offers logical suggestions for overcoming them. Madeline Wu's chapter describes electron microscopy techniques that are valuable, but not widely applied, in the study of plant genomes. She demonstates the practical value of those techniques with the hope that electron microscopy may be applied more in the future. Since most of the examples cited in her chapter have to do with chloroplast or mitochondrial genomes, the chapter is included in Part III. In order to broaden the scope of this book, a chapter on the molecular evolution of the *Nicotiana* chloroplast genome is included. This should be of interest to evolutionary biologists and particularly to systematists. The chloroplast genome has become a valuable and popular genetic tool with which to study evolution and taxonomy.

In Part IV the practical applications of biotechnology to crop improvement are summarized by the leader of the pioneering group. Robert T. Fraley describes the current exciting results of genetic engineering of disease resistance and herbicide and insect tolerance in transgenic plants. He details the ways by which plants are protected from viruses or insects by expressing virus coat proteins or *Bacillus thuringiensis* toxin proteins and how herbicide-resistant plants were engineered. These accomplishments signify the beginning of an unprecedented era of biotechnology and demonstrate the unmatched value of biotechnology in achieving crop improvement.

There is no intention on our part to cover all aspects of biotechnology in a single book; it is simply impossible. The aim of this book is to present a solid case to promote biotechnology and its application in plant systems. The articles assembled in this book by no means convey the current state of biotechnology; they simply reflect the state of the technology at the time they were prepared. The authors are to be commended for their valuable contributions. Their generosity in accepting the invitation, their efforts in completing the articles, their patience in allowing us to place deadline after deadline on their busy schedules are deeply appreciated. Without their contributions there would be no book. It is the quality of their contributions that makes this book valuable. We are extremely grateful to them all.

We want to thank the editorial staff of Butterworths Publishers, Helen Phillips and Christine Modes for their help in making this book possible.

Shain-dow Kung
Charles J. Arntzen

INTRODUCTION

The collection of disciplines and new techniques loosely described as biotechnology has created a miniature renaissance in plant biology over the last several years. Excitement generated by the advent of techniques permitting directed gene transfer and regeneration of "transgenic" plants has stimulated renewed interest in plant biology as a discipline and as a set of technologies with potential practical importance for agriculture. The relative ease with which transgenic plants may be produced has given new respectability to plants as experimental objects, while projects such as those dealing with herbicide-resistant plants and plants resistant to diseases have attracted the attention of numerous commercial enterprises.

It is appropriate to ask, at the beginning of a book such as this, just what *biotechnology* is and how it relates to conventional plant biology. The word is easy to use and difficult to define; lacking an authoritative definition, I distinguish at least two senses currently in common use. In its narrowest sense, biotechnology refers to techniques for manipulating biological systems to solve practical problems in industry or agriculture. This definition includes established technologies such as plant breeding in addition to recently developed technologies for gene isolation and transfer. Progress is made by modifying organisms or adapting their culture conditions to increase their yield of useful end products. In this sense biotechnology is almost an engineering discipline; the main goal is not to advance our understanding of biology (although it may happen that such advances do occur), but rather to put our present understanding to work doing something useful. Examples of this type of biotechnology might include the transfer of known genes for insect resistance to economically important crop plants, or efforts to develop plants resistant to herbicides. Research related to these themes utilizes a broad spectrum of techniques and subdisciplines of plant biology such as biochemistry, plant physiology, tissue culture, and plant breeding. Integrating these subdisciplines in research aimed at producing profitable products is a major challenge, and experience in such efforts will undoubtedly be beneficial to our basic understanding of plants.

In the long run, however, viable plant-improvement programs must also include efforts to extend the range of modifications that can be made. In most instances, progress of this type will come from research whose main motivation is simply to achieve a better understanding of how plants work. Even now, as the technology for gene transfer and expression is just beginning to be developed, our ambitions are frequently frustrated because we lack fundamental knowledge about frost hardiness, say, or lipid synthesis, or virus movement. Expanding our knowledge base in such areas—and hundreds of others—is vital. Otherwise we will soon run out of promising targets for genetic engineers to attack.

The importance of basic knowledge in many disciplines leads me to prefer a broader definition of biotechnology than that given above, that is, one that includes all of the many techniques relevant to understanding the genetic and biochemical functions of organisms as well as those for making directed genetic modifications. This definition is designed to include basic research as a full partner to applied science and technology. It is one that I hope will be considered carefully by those in a position to influence funding of plant biotechnology research.

Historically, plant biology in the United States has developed largely from the leavings of applied agricultural research and medical enterprises. Plant biology currently receives about five percent of the total competitive grant funding for biological and medical research. Even this abysmally low level of funding includes several quite-recent increases. The competitive grants progam developed over the last several years at the US Department of Agriculture may be considered a sign of progress, but much more needs to be done. Both the number of grants and the level of funding typically available in this program are far too low to support a critical mass of vigorous research programs around the country, and basic research within the department itself falls far short of making up the deficiency. Put bluntly, basic plant biology research in the United States has been grossly underfunded for several decades. If plant biotechnology is to be successful, this situation must change. We can no longer afford to rely on medically oriented researchers to provide us with the knowledge we need; tools such as gene cloning and antibody techniques have already been provided. The knowledge we need now pertains to the biology of plants. It is not available second hand. We must produce it ourselves.

Luckily, we will be aided in our efforts by the power of the technology itself. Investigators in biochemistry and physiology can now isolate individual genes, assay individual messenger RNAs and proteins, and transfer isolated genes back into plants. Genes of unknown function can be analyzed without knowledge of their encoded protein, and monoclonal antibodies to previously unknown proteins can be used to isolate both protein and gene. Gene expression can be monitored by sensitive assays for individual messenger RNAs or proteins, either biochemically or by one of a variety of emerging in situ localization techniques. Deliberate genetic alterations can be made in order to sort out cause-and-effect relationships between genes and phenotypes. Molecular markers such as restriction fragment length polymorphisms will be valuable in studies of complex quantitative characters. Transposon mutagenesis can provide an efficient way of identifying and subsequently isolating genes specifying phenotypic characters not yet understood biochemically—and so on, and on. Readers of this book will soon realize that the new technology creates opportunities for major advances in virtually all areas of experimental plant biology.

William F. Thompson
Raleigh, North Carolina

Plant Biotechnology

PART I

Basic Techniques in Plant Biotechnology

CHAPTER

1

Vectors for Gene Transfer in Higher Plants

Frank F. White

Progress in the development of plant gene transfer vectors can be attributed to developments in genetic and recombinant DNA techniques, plant tissue culture techniques and the design of selection methods for transgenic plant cells or tissue. In terms of vector development and use, the greatest progress has involved the *Agrobacterium* system. Although this chapter emphasizes the *Agrobacterium* system of DNA transfer, methods of direct DNA transfer are well developed and in many important circumstances are the clearest choice for achieving gene transfer. Viral vectors also present interesting possibilities for plant transformation and may have advantages in certain applications.

This chapter will not delve into the development of plant cell manipulation and regeneration methods. However, progress in plant cell transformation, particularly by direct DNA transfer methods, has been largely dictated by cell culture techniques. Thus, one of the critical factors in the transformation of a particular plant species is the ability to manipulate and regenerate whole plants from explants or protoplasts. Fortunately, many important crop species are amenable to plant regeneration. On the other

The work was supported by a research grant, RF84066, Allocation No. 3, from the Rockefeller Foundation.

hand, other important crop species—notably soybeans, corn, and wheat—have been more recalcitrant toward regeneration schemes. In general, the ability to introduce DNA into the genomes of these species is not the limiting factor. Stable gene transfer has been demonstrated in most cases, and progress in genetic engineering depends on developments in techniques for inducing tissue differentiation. These impediments are being removed in some species such as rice (Fujimura et al. 1985; Abdullah et al. 1986; Coulibaly and Demarly 1986; Yamada et al. 1986) and corn (Rhodes et al. 1988) and are likely to be overcome in others in the near future. Most recently, successful transformation and regeneration has been reported for rice (Torigama et al. 1988), corn (Rhodes et al. 1988), and soybean (Hinchee et al. 1988; McCabe et al.) by electroporation (rice and corn), microprojectiles (soybean), and agrobacterium (soybean).

1.1 *AGROBACTERIUM*-MEDIATED TRANSFORMATION

The most-well-developed system for higher plant transformation is derived from the tumor-inducing mechanism of the soil bacterium *Agrobacterium tumefaciens* (for reviews see Zambryski et al. 1983; Nester et al. 1984; White and Sinkar 1987). *A. tumefaciens* is the causal agent in crown gall disease. A closely related disease is hairy root disease, which is caused by *A. rhizogenes*. The utility of these bacteria has developed from an understanding of the molecular basis of the diseases, namely, the transfer of DNA from the bacterium to the plant nuclear genome. *Agrobacterium* was early recognized to be the causative agent in crown gall disease and was, in fact, one of the earliest phytopathogenic bacteria to be characterized. Hairy root disease was recognized as a variant of crown gall disease, but because of cultural and pathological differences between the *A. tumefaciens* and hairy root disease strains that were studied at the time, separate species were defined. The distinction is now kept largely for convenience, since it is known that relatively few genes control the differences between the ability to induce crown gall tumors and the ability to induce hairy root tumors. From a practical perspective, the *A. rhizogenes* strains commonly used in gene transfer experiments, namely A4 and 15834, are what are now known as biotype 2 strains (Keane et al. 1970). These strains generally require biotin (2 μg/1) for growth on minimal media. All of the *A. tumefaciens* strains that have been used for transfer experiments have been biotype 1 strains (Keane et al. 1970). These strains do not have any special nutritional requirements. For protocols for the use of *Agrobacterium* the reader is referred to two other reviews (Lichtenstein and Draper 1985; Rogers et al. 1986).

The aberrant plant growths caused by either type of *Agrobacterium* strain are due largely to genes on large plasmids within the bacteria (Van Larabeke et al. 1974; Watson et al. 1975; White and Nester 1980). Only a relatively small discrete portion of the plasmid is transferred to and maintained in the plant cells during the tumor-formation process, and this region is now familiarly known as the transferred DNA (T-DNA). The types of morphological effects observed in the transformed plant tissues are controlled by the complement of genes that is transferred. Plasmids that confer the tumor-inducing ability on *Agrobacterium* strains have become known as tumor-inducing (Ti) plasmids. The plasmids controlling the hairy root morphological features are commonly called the root-inducing (Ri) plasmids. Three T-DNA genes of the Ti plasmids are now known to control the biosynthesis of two phytohormones, indole-3-acetic acid (an auxin) and isopentyladenosine (a cytokinin precursor). The genes encode the enzymes for phytohormone synthesis (Akiyoshi et al. 1984; Barry et al. 1984; Schröder et al. 1984; Thomashow et al. 1984; Thomashow et al. 1986), and incorporation of the genes into the plant genome leads to the constitutive synthesis of the phytohormones. The constitutive synthesis of phytohormones in turn alters the developmental program of the transformed plant tissue.

A second characteristic trait of *Agrobacterium*-induced tumor tissue was later identified by Menagé and Morel (1964) when they discovered that tumor cells synthesized unusual amino acid compounds that are given the general name *opines*. Opines are derived from a variety of compounds occurring in the plant cell (for review see Tempé and Goldmann 1982; Smith and Neilands 1987). The first two discovered were called octopine and nopaline and are derived from arginine and pyruvate or arginine and α-ketoglutarate, respectively. The type of opine that is produced by the plant cells is dependent on the type of inciting bacterial strain. The discovery that *Agrobacterium* infection results in the production of specific opines has led to the so-called opine hypothesis, which states that bacteria attempt to create a niche in the plant tumor and environs by genetically diverting plant compounds into substances that are preferentially utilized by the bacteria. Many organisms including the plant cannot effectively utilize opines, but genes for the utilization of a particular opine reside in the bacteria on particular types of Ti plasmids. Opine production is controlled by T-DNA genes that encode the enzymes for opine synthesis (Schröder et al. 1981; Murai and Kemp 1982). The genes function in the plant genome, and the presence of opine synthesis can be used as a genetic marker. The gene for nopaline synthetase has been the most widely used opine gene and was derived from the nopaline plasmid pTiT37 (Depicker et al. 1982; Bevan et al. 1983a).

Ti and Ri plasmids from several strains of *Agrobacterium* have been well characterized by a variety of physical and genetic techniques. The findings indicate that the plasmids share the same basic architecture and contain many genes that are of marginal interest in regard to vector development, including genes involved in replication, conjugal transfer between bacteria, and utilization of the various opines that are synthesized in the tumors. Two critical regions for gene transfer are also found on the plasmids, although the exact organization and content of these regions may vary. The first region of the plasmid is known as the virulence or *vir* region. This region contains genes that are essential for gene transfer but are not themselves transferred to the plant genome. The virulence region from strain A6 covers approximately 35 kb and includes seven genetic loci that can affect virulence as indicated by symptoms on test plants (Klee et al. 1983; Stachel and Nester 1986). The second important region is the T-DNA. The length and number of T-DNA regions vary in Ti and Ri plasmids. The nopaline-type Ti plasmid has a single 23-kb T-DNA region (Engler et al. 1981), while the octopine-type Ti plasmid has two closely adjoining regions of 13 kb and 8 kb (Barker et al. 1983). The phytohormone and octopine synthesis genes are located in the left (T_L-DNA) region of the octopine-type Ti plasmid. More recently, several plasmids in addition to the octopine-type plasmid have been demonstrated to possess multiple T-DNA regions. The agropine-type Ri plasmids have two T-DNA regions that are separated by approximately 16 kb (Huffman et al. 1984; Jouanin 1984). Two T-DNA regions were also found in the so-called limited-host-range and supervirulent Ti plasmids (Hood et al. 1984; Yanofsky et al. 1985b; Komari et al. 1986). A single T-DNA region has been identified in the mannopine-type Ri plasmids (Koplow et al. 1984). Multiple T-DNA regions may turn out to be the rule as more wild-type Ti and Ri plasmids are characterized, but all have similar essential features. Introduction of an individual T-DNA region from one plasmid into a strain containing another type of plasmid results in the transfer of the newly acquired T-DNA segment (Hoekema et al. 1984; Hooykaas-Van Slogteren et al. 1984; White et al. 1985). Therefore, the T-DNA regions from various Ti and Ri plasmids appear to be mobilized to plant cells by a conserved mechanism.

The T-DNA regions of the octopine-type Ti plasmid provide a good model for demonstrating the basic features of all T-DNA regions (Figure 1–1). The genes that are located in the T-DNA region contain the structural features necessary for expression by the plant. The messenger RNA (mRNA) is polyadenylated and contains a 7-methylguanosine cap at the 5′ end. A total of eight mRNAs have been detected in plant tissue containing the T_L-DNA of the octopine-type Ti plasmid (Willmitzer et al. 1983). Five mRNA species have been mapped to the right (T_R-DNA) region (Salomon et al. 1984). None of the transcriptional units contains introns. Two T_L-DNA genes control the synthesis of indole-3-acetic acid from tryptophan. Transcript 1 (using the nomenclature of Willmitzer et al. 1983) encodes a tryptophan mono-oxygenase that converts tryptophan to indole acetamide (Schröder et al. 1984; Thomashow et al. 1984), and transcript 2 encodes a hydrolase that converts the indole acetamide to indole-3-acetic acid (Thomashow et al. 1986). The transcript 4 product has been identified as an isopentyltransferase that catalyzes the formation of isopentyladenosine from adenosine monophosphate and isopentylpyrophosphate (Akiyoshi et al. 1984; Barry et al. 1984). With the exception of the octopine synthase gene, the functions of other T_L-DNA loci are not understood. In the case of the octopine-type T-DNA, mutations at the *tml* locus (which corresponds to

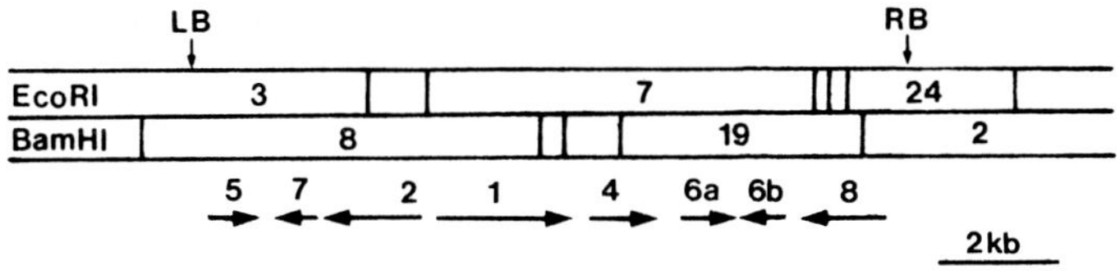

Transcript	Enzyme Encoded
1	Tryptophan Mono-oxygenase
2	Indoleacetamide Hydrolase
4	Dimethyallylpyrophosphate Transferase
8	Octopine Synthase

FIGURE 1–1 Left T-DNA region of octopine-type Ti plasmid pTIA6. Arrows beneath the map represent recognized transcripts from the T-DNA. The functions of the encoded proteins for several of the genes are listed. Vertical arrows designate the positions of the left (LB) and right (RB) borders.

transcript 6b) result in larger tumors on some plants (Garfinkel et al. 1981). The biochemical basis for this response is not known, nor is the phenotype observed when the homologous region is mutated in the nopaline-type T-DNA (Joos et al. 1983). Evidence regarding transcript 6a of the nopaline plasmid suggests that the gene product allows leakage of opines from the plant cells, and therefore may encode a permease (Messens et al. 1985). The genes of the T_R-DNA appear to be involved predominantly with the synthesis of agropine-type opines (Salomon et al. 1984). While no vectors have been designed from the T_R-DNA, several of the agropine synthesis genes have been the source of promoters for chimeric genes (Velten et al. 1984).

The key aspect of the T-DNA structure in regard to gene transfer is that none of the T-DNA genes are involved in the transfer process (Garfinkel et al. 1981; Leemans et al. 1982a; Ream et al. 1983). Any or all of the genes can be removed or mutated, and the T-DNA region can still be transferred to the plant genome. The key elements of the T-DNA are the 25-bp imperfect direct repeats present at the boundaries of the T-DNA (Yadav et al. 1982; Caplan et al. 1983). Most plant/T-DNA junction sequences that were isolated from transformed cells occur within or near the border sequences (Simpson et al. 1982; Yadav et al. 1982; Zambryski et al. 1982). As will be discussed further below, recent experiments demonstrate that the interaction of *vir* genes with the border sequences initiates the transfer events. As represented by standard convention, the right border is critical for T-DNA transfer and tumorigenesis, while removal of the left border appears not to affect tumor formation (Ooms et al. 1982; Shaw et al. 1984; Wang et al.

1984; Peralta and Ream 1985). A sequence, termed *overdrive*, that is required for maximal transfer efficiency has been identified immediately adjacent to the right border (Peralta et al. 1986). All Ti and Ri plasmids that have been sufficiently analyzed appear to contain related border sequences (Slightom et al. 1986). Also, the regions occupied by the *vir* genes appear to be closely related for the most part among the different plasmids, as indicated by DNA hybridization and sequencing experiments (Huffman et al. 1984; Hirayama et al. 1988).

T-DNA transfer begins with the introduction of the bacteria into a plant wound. Wounding is a necessary event in the process and may, at least in part, be required for the synthesis by the plant of compounds that induce the expression of the *vir* genes (Bolton et al. 1986; Janssens et al. 1986; Stachel et al. 1986a). A group of compounds have been identified in tobacco (*Nicotiana tabacum*) wound exudate that when added to *Agrobacterium* cultures cause *vir* gene induction (Stachel et al. 1985). Two of the most active substances identified are acetosyringone and β-hydroxyacetosyringone. Two *vir* genes, *vir*A and *vir*G are constitutively expressed and are involved in mediating the induction of the other genes (Stachel and Zambryski 1986a). On the basis of amino acid sequence homology, *vir*G belongs to a class of positive regulatory genes in bacteria that link gene induction to different external stimuli (Nixon et al. 1986; Winans et al. 1986). *Vir*A, in turn, belongs to a second group of genes that in some way leads to the activation of the positive regulatory gene proteins. *Vir*A is envisioned to be a membrane-associated protein that monitors the presence of the induction compounds. The presence of the inducer is then signaled to the positive regulatory protein in some unknown manner and induction of the other *vir* genes occurs (Stachel and Zambryski 1986b). The homologous products in other systems have been shown to involve protein phosphorylation (Ninfa and Magasanik 1986; Hess et al. 1987).

A particularly intriguing area of research involves deciphering the functions of the inducible *vir* genes in the transfer process. Circular T-DNA molecules can be isolated from the bacteria after induction with acetosyringone or plant-wound extracts (Koukolikova-Nicola et al. 1985; Machida et al. 1986). Originally thought to represent transformation intermediates, the circles appear to result from an excision and subsequent recombination or ligation events at the right and left border sequences. The covalently closed circular T-DNA molecules may in fact be artifacts or inadvertent products of the transfer process. More recently, induction has been shown to result in the appearance of single-strand nicks within the bottom strand of the 25-bp border (Albright et al. 1987; Wang et al. 1987). The nick site occurs predominantly after the third nucleotide from the left end of the right border (T-DNA side). The DNA nicking is followed by the generation of a single-strand copy (called the T-strand) of the T-DNA region that is thought to be the intermediate in plant cell transformation (Stachel et al. 1986b; Albright et al. 1987). *Vir*D has been shown to be required for the nicking activity and single-strand synthesis and likely encodes an endonuclease (Yanofsky et al. 1986; Yamamoto et al. 1987). Despite the presence of single-strand molecules, the form of the transfer intermediate is unknown. Double-strand molecules have also been detected.

The recent findings of T-DNA processing and *vir* gene function suggest a model with striking similarity to the model for bacterial conjugal transfer mechanisms (Willetts and Wilkins 1984; Stachel and Zambryski 1986b). In fact, a recent report indicates that bacterial origins of transfer (*ori*T) and the corresponding gene for DNA cleavage (*mob*) can mediate plant cell transformation (Buchanan-Wollaston et al. 1987). DNA transfer during conjugation occurs by the generation of a single-strand copy of the plasmid. The *ori*T sequence contains the site where strand cleavage occurs and DNA synthesis begins. Both *ori*T and the T-DNA border direct transfer in a polar manner. A hypothetical model for T-DNA processing is presented in Figure 1–2. Presumably the left T-DNA border provides a termination point for single-strand synthesis but is not required. If the T-DNA border is placed on a relatively small plasmid, the orientation of the border sequence is less critical: in this situation genes are transferred in either orientation of the border, since there is a high probability that the entire plasmid will be transferred to the plant cells (Horsch and Klee 1986; Jen and Chilton 1986).

The role that the other *vir* genes (*vir*B, *vir*C, and *vir*E) play in the tumorigenesis process is unknown. The analogy with the conjugation system suggests involvement in functions thought to be involved in conjugal transfer. Therefore, *vir* genes may function in DNA binding, replication, covalent modification of terminals or other form of packaging of the transfer intermediate, and/or transport of the T-DNA across the cell envelope (Thompson et al. 1988). Some of the genes are apparently not absolutely required for T-DNA transfer, but may affect the efficiency of transfer or integration into host genomic DNA or both (Klee et al. 1983). T-DNA transfer has been detected even in the presence of *vir*E and *vir*C mutations (Yanofsky et al. 1985a; Gardner and Knauf 1986; Horsch et al. 1986). Again, this phenomenon is similar to the function of some of the bacterial genes involved in conjugation that lower the transfer frequency when mutated but do not eliminate transfer completely (Willetts and Wilkins 1984). Recently, *vir*E has been shown to behave as a single-strand DNA binding protein (Das 1988; Gietly et al. 1987).

1.2 *AGROBACTERIUM* VECTORS

The *Agrobacterium*/Ti plasmid system exploits the elements of the *Agrobacterium* transformation mechanism. While a considerable number of plant transformation vectors now exist, all consist of modifications of two elements (Bevan 1984; An et al. 1985; Klee et al. 1985; Matzke and Matzke 1986; van den Elzen et al. 1985a; Spena and Schell 1987). The first element consists of a plasmid vector containing a DNA segment that is linked to a border sequence such that the region functions as a T-DNA region. The

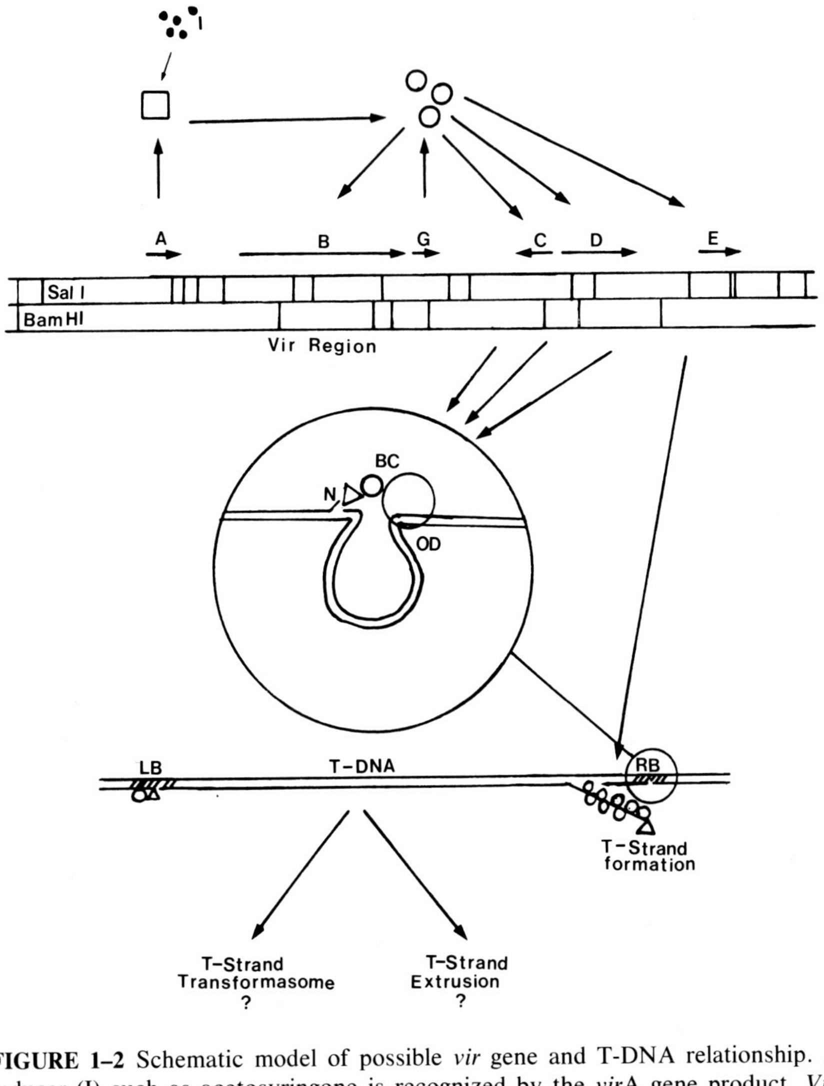

FIGURE 1–2 Schematic model of possible *vir* gene and T-DNA relationship. An inducer (I) such as acetosyringone is recognized by the *vir*A gene product. *Vir*A and *vir*G protein interaction results in the transcription of other *vir* genes. One *vir*D protein causes a nick (N) at the right border (RB). Other proteins may be involved in the border complex (BC) that directs nicking in part by recognition of the overdrive sequence (OD) and/or T-strand formation. The *vir*E protein is involved as a single-strand binding protein. Other possible *vir* functions involve covalent modification of 5′ and 3′ T-strand ends, bacterial membrane modifications for plant–cell interaction and delivery mechanism of T-DNA transfer to plant cell. LB, left border.

vector provides a shuttle that allows the recombinant DNA manipulations to be performed in *Escherichia coli* hosts before subsequent transfer to *Agrobacterium*. Secondly, transfer to the plant is mediated by a strain of *Agrobacterium* that contains a full complement of *vir* genes. Basically, two types of *Agrobacterium* vectors are currently in use. Vectors that recombine via DNA homology into a resident Ti plasmid are often referred to as integrative or intermediate vectors (Leemans et al. 1981; Matzke and Chilton 1981; Fraley et al. 1983). The second type of vectors need not recombine into the resident Ti plasmid and are referred to as binary or autonomous vectors (de Framond et al. 1983; Hoekema et al. 1983; Bevan 1984; An et al. 1985; Klee et al. 1985). Both types provide three features essential for their utility: convenient (unique) endonuclease restriction sites for cloning a target gene or DNA fragment within the T-DNA region; a plant-expressed gene marker for selection and/or screening of transgenic plant cells; and a bacterium-expressed gene marker for selection of the vector in *Agrobacterium* and *E. coli*. The development of the binary vectors came after the discovery that a T-DNA region would function in trans to the *vir* genes when present on a separate replicon (de Framond et al. 1983; Hoekema et al. 1983).

The strategy for the construction of integrative vectors was based on earlier techniques for site-directed mutagenesis (Ruvkun and Ausubel 1981). In simplest form an integrative vector requires only a plasmid that shares homology with the T-DNA. Homologous recombination between the plasmid and the T-DNA will produce a hybrid structure that can then be mobilized to the plant cell. However, in practice a number of important modifications have been incorporated into the design of integrative and binary vectors, and some variations in different vectors exist. Since the wild-type T-DNA contains oncogenic genes, these genes must be removed if normal and stable plant development is desired (Zambryski et al. 1983; Fraley et al. 1985). Without the selection for tumorous growth, alternative selection schemes have been devised. The most widely used selection marker for transformed plant cells has been the bacterially derived gene for resistance to kanamycin (or the related aminoglycoside antibiotic G418), which is fused to a plant promoter (Bevan et al. 1983b; Fraley et al. 1983; Herrera-Estrella et al. 1983). A variety of other selection techniques have been devised or envisioned (Table 1–1). However, not all of these possibilities have been put into practice. In addition, some vectors include a second marker for screening putative transformants after the initial selection (see Table 1–1). A second marker provides a convenient check for cells that have escaped the selection regimen. The second marker may be a selectable-type gene or, more commonly, an opine synthesis gene such as the nopaline synthesis gene.

The integrative vector is well illustrated by the pMON200 vector (Figure 1–3) (Rogers et al. 1986). In this particular vector the right T-DNA border from pTiT37 has been utilized. The critical feature is that the vector by

TABLE 1–1 Plant Transformation Markers

	Reference
Selection traits[1]	
Kanamycin or G418 resistance	Bevan et al. 1983b Fraley et al. 1983 Herrera-Estrella et al. 1983
Methotrexate resistance	Herrera-Estrella et al. 1983 Brisson et al. 1984
Hygromycin resistance	van den Elzen et al. 1985b Waldron et al. 1985
Bleomycin resistance	Hille et al. 1986
Aminoethylcysteine resistance	Dahl and Tempé 1983
Glyphosate resistance[2]	Comai et al. 1985 Shah et al. 1986
Streptomycin resistance	Jones et al. 1987
Sulfonylurea resistance[2]	Haughn and Somerville 1986
Phosphinothricin resistance[2]	De Block et al. 1987
Gentamicin resistance	Hayford et al. 1988
Indole acetamide utilization or sensitivity[3]	Inzé et al. 1984 Depicker et al. 1988
Hairy root formation	Stougaard et al. 1986
Screens and assays	
Opine synthesis	Otten and Schilperoort 1978
Neomycin phosphotransferase	Reiss et al. 1984
Chloramphenicol acetyltransferase	Gorman et al. 1982
β-Galactosidase	Helmer et al. 1984
β-Glucuronidase[4]	Jefferson et al. 1986
Firefly luciferase	Ow et al. 1986
Bacterial luciferase	Koncz et al. 1987

[1] These traits constitute selectable phenotypes for plants or plant cells that are transformed. The list does not include the tumorous phenotype of phytohormone autonomy.

[2] Herbicide resistance traits have not been reported in use as selection traits but obvious potential for their use exists.

[3] Cells can be selected on the basis of use of indole acetamide as an auxin, either at nontoxic or at toxic levels.

[4] Toxic, or "suicide," substrates may be developed for use of β-glucuronidase as a selective agent.

itself will not replicate in *Agrobacterium* without first integrating into the resident Ti plasmid. In this case the origin of replication is derived from pBR322, which has a ColEl-type origin. Replication of the plasmid is limited to *E. coli* and several related enteric bacteria. The plasmid also contains the *bom* site, which allows mobilization to another bacterium if other transfer functions are supplied in trans. Integrative vectors are mobilized into

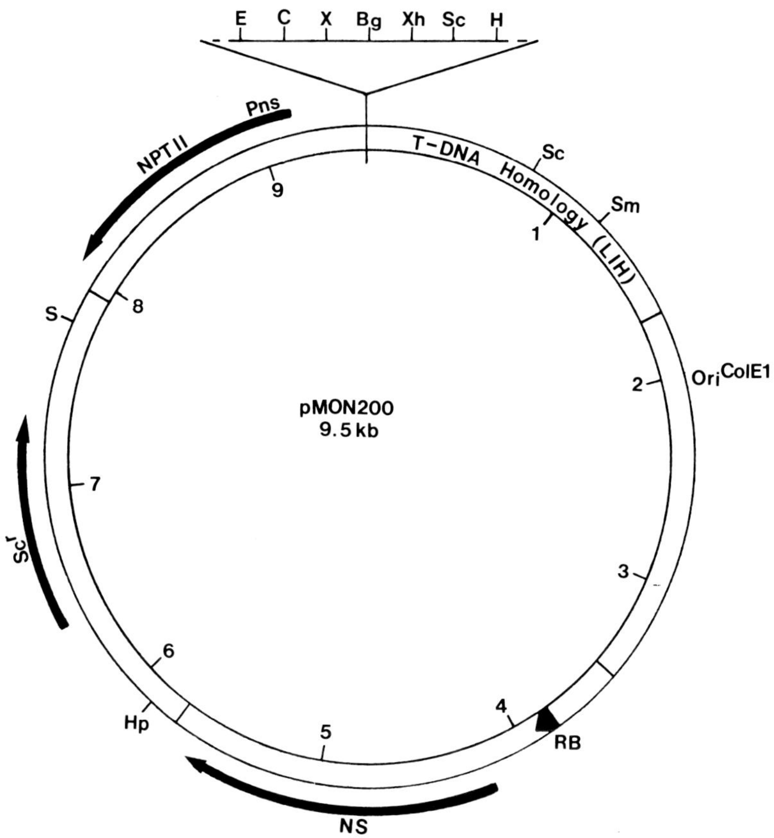

FIGURE 1–3 Integrative, or intermediate, type T-DNA vector. The vector pMON200 contains two plant-expressed genes: the chimeric kanamycin-resistance gene (NPTII) driven by the nopaline synthase promoter (Pns) and the nopaline synthase (NS) gene. The T-DNA homology region is a sequence that allows homologous recombination between the vector and the resistant Ti plasmid (B6 or A6 octopine Ti T-DNA). Sc^r, bacterially expressed gene for spectinomycin resistance; Ori^{ColEI}, origin of replication from ColEI (pBR322)-type plasmid; RB, right T-DNA border; Bg, *Bgl*II; C, *Cla*I; E, *Eco*RI; H, *Hin*dIII; Hp, *Hpa*I; S, *Stu*I; Sc, *Sac*I; Sm, *Sma*I; X, *Xba*I; Xh, *Xho*I; LIH, limited internal homology.

Agrobacterium by triparental conjugation systems (Ditta et al. 1981). In triparental conjugations the functions necessary for plasmid transfer are supplied on another helper plasmid that is often in a separate bacterial strain. A mating, therefore, consists of three participants: the recipient *Agrobacterium*, the helper strain, and the strain harboring the plant transformation vector. The helper transfer genes have also been placed on the

E. coli chromosome (Simon et al. 1983). The *Agrobacterium* transconjugant is then isolated on the basis of selection for stable antibiotic resistance conferred by the bacterially expressed resistance gene on the plant transformation vector, which in pMON200 is the gene for spectinomycin resistance. Routinely, a minimal media is used and is generally adequate to counterselect against the auxotrophic *E. coli* cells in the mating mixture. Stabilization of the vector in *Agrobacterium* requires integration of the vector into a resident Ti plasmid via a region of homology shared between the resident plasmid and vector. In pMON200 the integration site is a short segment of the Ti plasmid (limited internal homology, or LIH, site; Fraley et al. 1985). The exact requirements for efficient recombination have not been extensively studied. However, evidence from site-directed mutagenesis suggests that a DNA sequence of 500 bp or greater is adequate (Ruvkun and Ausubel 1981). In theory, a large homologous region creates a large duplication in the T-DNA that could result in instability due to reversal of the integration process. However, given continual antibiotic resistance selection, the rate of loss has not prevented the use of this approach. An alternative approach is to select for a double recombination event that would eliminate sequence duplication, provided that flanking sites of homology are present in the vector.

Integrative vectors can be designed in several fashions. The T-DNA borders and the plant-selectable marker gene can reside on the disarmed Ti plasmid (Zambryski et al. 1983) or these elements can be introduced with the shuttle vector (Fraley et al. 1985). The Ti plasmid pGV3850, one of the earliest disarmed resident Ti plasmids for use with integrative vectors, is constructed such that the oncogenic genes of the T-DNA region are replaced with pBR322 sequences (Zambryski et al. 1983). The T-DNA borders and the nopaline synthetase genes remain intact. The shuttle vector portion contains a plant-selectable marker (in later uses, Herrera-Estrella et al. 1983), a homologous recombination site, and a bacterially expressed antibiotic resistance gene for selection and maintenance in bacteria. The construction of the pMON200 vector involves a slightly different approach. The shuttle vector contains the plant-expressed kanamycin-resistance gene, the nopaline synthetase gene, the spectinomycin-resistance gene, and the T-DNA border. The pMON200 plasmid has been referred to as a split end vector, since one border is supplied by the shuttle vector and the other border resides on the Ti plasmid (Fraley et al. 1985).

The binary, or autonomous, vectors have been gaining in popularity. Binary vectors are designed like integrative vectors, with the exception that the components are assembled on a replicon that is stably maintained in *Agrobacterium*. The vector pGA482 is illustrative of the binary vector (Figure 1–4) (An 1986). The vector contains a plant-selectable marker [neomycin phosphotransferase (NPTII) coding sequence fused to the nopaline synthetase promoter] and a multiple restriction endonuclease cloning site that are situated between right and left borders. In addition, pGA482 con-

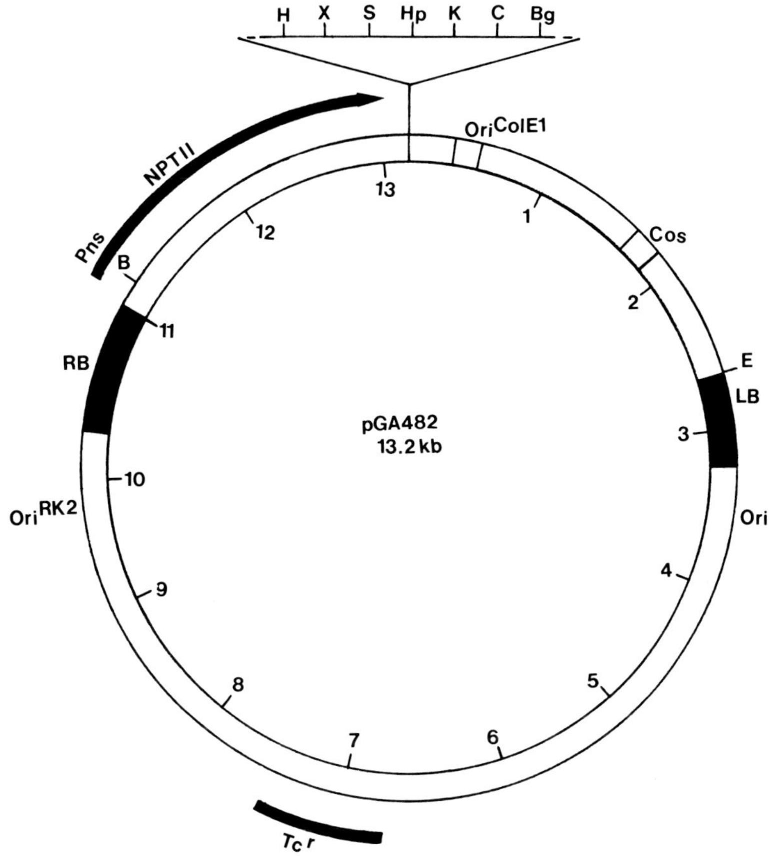

FIGURE 1–4 Autonomous, or binary, type T-DNA vector. The vector pGA482 contains a single plant-expressed chimeric kanamycin-resistance gene (NPTII) driven by the nopaline synthase promoter (Pns). The plasmid has two origins of replication: the pBR322 origin (Ori^{ColE1}) and the pRK2 origin (Ori^{RK2}), which allows replication in *Escherichia coli* and *Agrobacterium*. RB, right T-DNA border; LB, left T-DNA border; Cos, lambda phage *cos* site; OriT, pRK2 origin of transfer; Tc^r, bacterially expressed tetracycline-resistance gene; C, *Cla*I; B, *Bam*HI; Bg, *Bgl*II; H, *Hin*dIII; Hp, *Hpa*I; K, *Kpn*I; S, *Sst*I; X, *Xba*I; E, *Eco*RI.

tains a gene for resistance to tetracycline for selection of bacteria containing the vector and a *cos* site that allows utilization of the lambda bacteriophage packaging system. The replication functions of pGA482 are derived from the RK2 broad host range plasmid (Ditta et al. 1980). Theoretically, other bacterial plasmids that can be used for the basis of a binary vector include the RSF1010-related plasmids (Bagdasarian et al. 1981), pSa (incompatibility group W) plasmids (Leemans et al. 1982b; Tait et al. 1983), and endogenous *Agrobacterium* plasmids that are compatible with resident Ti or Ri plasmids (Jouanin et al. 1985). The discovery that *mob* and *ori*T (*bom*) functions of other plasmids can replace the T-DNA border suggests that an even wider variety of plasmids can not only serve as transformation vector replicons but supply the T-DNA border sequence as well (Buchanan-Wollaston 1987).

Binary vectors, like integrative vectors, can be used with either disarmed or wild-type Ti plasmid–containing strains. For most general gene transfer studies involving regenerated plants, disarmed strains are preferred. LBA4404 is representative of a disarmed strain (Hoekema et al. 1983). The strain is derived from an octopine-type Ti plasmid strain and the *vir* genes are intact. The complete T_R-DNA and T_L-DNA segments have been removed by in vivo deletion techniques (Ooms et al. 1982). This strain can mediate transfer if a T-DNA region is supplied in trans.

More sophisticated disarmed strains are available. Koncz and Schell (1986) have introduced the trans-acting RK2 replication and stability functions into the *Agrobacterium* chromosome and the Ti plasmid. These strains allow smaller RK2 replicons to be constructed and also provide a safety feature since the cloning vector is less likely to be maintained in other bacteria. The RK2 trans-acting mobilization functions have also been engineered into the *Agrobacterium* chromosome to facilitate transfer of the binary vector back to *E. coli* (Koncz and Schell 1986). A disarmed Ti plasmid has been created from the supervirulent Bo542 strain which is thought to mobilize T-DNA at a higher frequency than other Ti plasmids (Hood et al. 1986). The region of the supervirulence *vir* region has been identified (Jin et al. 1987) and can be introduced into other strains in order to promote more efficient transfer (Pythoud et al. 1987).

If the undesirable effects of the T-DNA oncogenes are not important, the cotransformation approach, in which a binary vector is introduced with a resident wild-type Ti or Ri plasmid, can be useful. Independent integration of T-DNA from a vector and a coresident Ti or Ri plasmid occurs frequently enough so that cells can generally be selected and screened for inheritance of the vector T-DNA alone (Depicker et al. 1985). This approach has been exploited most often with the Ri plasmid strains. Shahin et al. (1986b) introduced pARC8 (a standard binary vector conferring plant resistance to kanamycin; Simpson et al. 1986) into *A. rhizogenes* strain A4. Upon infection of punctured tomato cotyledons, no Ri T-DNA was found in plants when transformants were initially selected for kanamycin resistance via callus induction. However, the outcome of regenerated plants can vary with the method of selection. If roots were selected from inoculated inverted cotyledons virtually all the transformants contained vector and Ri T-DNA. Instead of selecting for cells that have not received the Ri T-DNA, Stougaard and colleagues (1986) exploited the propensity of Ri-transformed cells to differentiate into roots in order to transform the leguminous plant *Lotus corniculatus*. Transformation and regeneration of *L. corniculatus* is apparently facilitated by the effects of the Ri plasmid T_L-DNA. Whether the Ri

T_L-DNA improves regeneration is not clear. However, roots containing the Ri T_L-DNA do appear to form more lateral branches. A Ri T_L-DNA integrative vector was designed so as not to disturb the T-DNA genes affecting root formation (Stougaard et al. 1987). A drawback to this system is that the T_L-DNA genes also alter the developmental behavior of other plant organs, particularly the leaves (Tepfer 1984; Taylor et al. 1985). This precludes the use of the cotransformation system for general gene transfer with plants other than species that are recalcitrant to regeneration by other means. On the other hand, Ri-transformed roots offer potential for the production of plant metabolites (for review see Hamill et al. 1987).

Methods for inoculation of plant tissue vary depending on the plant species and the *Agrobacterium* delivery system. Initial in vitro infections were performed with regenerating protoplasts (Wullems et al. 1981). This approach has been largely replaced with the leaf disc procedure (Horsch et al. 1985). In practice, the procedure can be performed with any tissue explant that provides a good source for initiation of whole plant differentiation. In this regard, newly emerged cotyledons have provided useful material that can be prepared fairly rapidly (Shahin et al. 1986b). In the disc, or explant, method, a tissue segment is excised and inoculated with bacteria. The infected tissue is allowed to incubate for one to three days, depending on the extent of bacterial growth. Some plants appear to permit more rapid bacterial growth, and these explants should be rescued before overgrowth of the bacteria occurs. Nurse tissue may be included during this step to help promote division and growth of the infected cells (Horsch et al. 1985). Nurse tissue may also help in *vir* gene induction. Tissue explants are then moved onto a selective medium that eliminates the bacteria. The medium is also designed for the selection of transformed plant cells and, therefore, contains the appropriate antibiotic, such as kanamycin or hygromycin. The appropriate levels of phytohormones are also included in the medium for the type of tissue development desired. Generally, callus and shoot formation has been induced. However, in the case of Ri plasmid–mediated transformation, root development is often desired and usually occurs without any addition of phytohormones. By inoculating epidermal segments from flowering stems of *N. plumbaginifolia*, transgenic seeds can be obtained rapidly (8 weeks) in culture (Trinh et al. 1987). A radical new method of inoculation that has recently been reported is the treatment of steeping *Arabidopsis thaliana* seeds with bacteria (Feldmann and Marks 1987). Subsequent germination of the progeny of the treated seeds results in a low percentage of transformed plants. Examination of the T-DNA structure of these plants indicates that transformation occurs through the normal T-DNA delivery process. With some improvements in efficiency, this process could yield large numbers of transformed plants without the investment of time and equipment required by other procedures.

The last issue regarding T-DNA transformation is the nature of the integration events and the stability of integrated sequences. T-DNA copy number can vary considerably in transformed plants (Thomashow et al.

1980; Zambryski et al. 1982; Spielmann and Simpson 1986; Jones et al. 1987), and two unlinked T-DNA elements can integrate into the same plant cell (Depicker et al. 1985; de Framond et al. 1986; Petit et al. 1986). T-DNA junctures can vary from tandem inverted repeats, to direct repeats, to single copy insertions (Zambryski et al. 1982; Spielmann and Simpson 1986; Jorgensen et al. 1987). Recently, the rearrangement of target plant sequences at the site of insertion has also been documented (Gheysen et al. 1987). The type of insertion event that predominates may vary depending on several factors, such as the particular vector employed. Considerable differences in the percentage of inverted tandem insertions between the LBA4404/pKLBG2 (pARC8 derivative) and the C58/pGV3850KanR systems (approximately 10% and 70%, respectively) have been observed (Jorgensen et al. 1987). The presence of both multiple insertion sites and inverted tandem structures suggests some differences in the mode of DNA uptake from direct DNA transformation of mammalian and plant cells (De Jonge and Bootsma 1984; Potrykus et al. 1985a; Roth et al. 1985). Direct DNA transfer and incorporation in cells often results in large direct repeated arrays of introduced DNA. Presumably these differences reflect the efficiency and mode of transfer of the T-DNA.

The type of insertion probably has some bearing on the stability of the insertion, and large tandem arrays are likely to be less stable than single insertions (Potrykus et al. 1985b; Müller et al. 1987). Several plants with multiple insertions were noted to produce untransformed progeny at a higher frequency than plants containing simple insertions (Chyi et al. 1986; Feldmann and Marks 1987). However, with the possible exception of plants containing large tandem inserts, single-copy and low-copy-number T-DNA insertions appear to be very stable (Müller et al. 1987). Frequencies of T-DNA loss in transgenic tobacco plants with low-copy-number inserts were measured at 1.3 and 5.6 per 10^4 gametes. Studies suggest that the majority of T-DNA insertions by *Agrobacterium*-mediated transfer are stably inherited in normal Mendelian fashion (Budar et al. 1986; Chyi et al. 1986; Wallroth et al. 1986; Feldmann and Marks 1987; Müller et al. 1987). On the other hand, not all T-DNA copies in a given line are expressed, and increased copy number does not always correlate with higher levels of gene expression (Feldmann and Marks 1987; Jones et al. 1987). Several studies have indicated that suppression of T-DNA gene expression is associated with DNA methylation (Hepburn et al. 1983; Amasino et al. 1984).

1.3 USES OF *AGROBACTERIUM* VECTORS

The greatest use of T-DNA gene transfer has been in the analysis of plant gene expression. Although an in-depth review of plant gene expression is not intended, a number of basic observations will be discussed. In general, gene transfer as a tool to study plant gene expression has exceeded most expectations. In most cases, transferred gene function is similar to that of endogenous copy. Thus, regulatory elements from plants that are very diffciult to work with in culture can be studied by isolation, modification, and reintroduction of those elements to a manageable species. For example, light-regulated genes from wheat also respond to light induction in tobacco (Lamppa et al. 1985). In one case, the maize alcohol dehydrogenase gene was not functional after transfer to tobacco. However, the expression elements were not completely defective in tobacco, and upon introduction of an octopine synthesis promoter in the *Adh* promoter region, anaerobic regulated expression was observed (Ellis et al. 1987). On the other hand, genes from more widely divergent organisms such as mammals and yeast have not been found to be expressed efficiently in plant cells (Barton et al. 1983; Caplan et al. 1983). Expression can be analyzed after stable incorporation and callus growth or plant regeneration, or it can be measured transiently soon after bacterial infection (Horsch et al. 1985; An 1986). Low-level expression of otherwise developmentally regulated genes has been noted for callus tissue and primary regenerated tissue that was assayed after gene transfer. Expression can revert to the normal pattern after passage through seed (An 1987). Many studies involving examples of gene expression analysis are now in progress, and one can expect a sizable increase in the amount of information regarding promoter control of plant gene expression.

T-DNA transfer has numerous other applications in addition to studies designed to characterize gene regulatory elements. The system affords an ever-increasing array of interspecific and modified gene transfer possibilities. Perhaps the greatest attention has been given to the prospects of improved plant protection by the engineering of herbicide resistance (Comai et al. 1985; Haughn and Somerville 1986; Shah et al. 1986; Fillatti et al. 1987a), insect resistance (Fischhoff et al. 1987; Vaeck et al. 1987), and plant viral resistance (Abel et al. 1986; Loesch-Fries et al. 1987; Tumer et al. 1987). Although built on the same principle as the integrative and binary vectors, a number of vectors are available that have multiple cloning sites situated between a constitutive promoter and an RNA polyadenylation site [RNA(polyA)] in order to facilitate cloning and expression of heterologous genes (Velten and Schell 1985; Rothstein et al. 1987). In one case, the chloramphenicol acetyltransferase gene (CAT) was also incorporated between the cloning site and the polyadenylation site to facilitate expression assays (An 1986).

In addition to the great potential for new crop protection strategies, T-DNA gene transfer mechanisms have already provided important tools for the study of virus biology in general. By making infectious clones of viruses and viroids—as has been done for geminiviruses, maize streak virus (Grimsley et al. 1987), tomato golden mosaic virus (Rogers et al. 1986), and potato spindle tuber viroid (Gardner and Knauf 1986)—specific mutations can be engineered in the viral genome and tested for the effects on viral replication and disease development. The approach is equally applicable to single viral genes, such as that encoding the coat protein (Bevan et al. 1985; Abel et al. 1986; Loesch-Fries et al. 1987; Tumer et al. 1987) and cell-to-cell movement genes (Deom et al. 1987), and different components of viral genomes, such as the satellite RNA of cucumber mosaic virus (Baulcombe et al. 1986).

The study of organelle genes is also assisted by the T-DNA transfer system. Chloroplast and mitochondrial protein synthesis is directed by the chloroplast genome and synthetic machinery or by nuclear and cytoplasmic machinery imported into the chloroplast. One early report indicated that some T-DNA insertions occur in the chloroplast genome (De Block et al. 1984). This finding opened the way for the study of organelle-encoded proteins. However, no confirmation of the chloroplast genome transformation has ensued, and no other organelle transformants have been obtained, suggesting that they are very rare. Progress has been made regarding the elucidation of the mechanism of protein transport into the chloroplast or mitochondria. Nuclear genes whose products are destined for organelles function normally when reintroduced into plasmid cells via the T-DNA (Broglie et al. 1984). When reporter genes (NPTII and CAT) were properly fused to the coding sequences of nuclear chloroplast and mitochondrial genes, these genes were also transported (Van de Broeck et al. 1985; Kuntz et al. 1986; Boutry et al. 1987). Therefore, T-DNA-mediated transformation can at least be used to target proteins to organelles. One can envision that in the future some chloroplast genes might be functional when properly engineered and expressed as nuclear genes. The transfer of the quinone–binding membrane protein gene (p*sb A*) from atrazine–resistant biotype of *Amaranthus hybridus* suggests that conversion of a chloroplast gene to a nuclear gene is feasible (Cheung et al. 1988).

T-DNA transfer has also been used to introduce transposable elements from heterologous species, such as the transfer of the *Ac* element from maize to tobacco (Baker et al. 1986). The transposon can then in turn be used for insertional mutations (transposon tagging; Freeling 1984). Even the T-DNA itself can be used as a mutagen; an interesting variation of this type of approach has been to fuse a promoterless NPTII gene near the T-DNA border (André et al. 1986; Teeri et al. 1986). The absence of a promoter allows selection for plant cells that can grow in the presence of kanamycin because the T-DNA insertion has occurred behind an endogenous promoter element. While aberrant plants have been obtained by T-DNA insertion, the utility of T-DNA-mediated insertional mutagenesis awaits further characterization of putative mutants.

A hypothetical application of T-DNA-mediated transfer involves the shotgun approach, whereby an entire genomic or complementary DNA (cDNA) library is transferred to plant cells en masse (van den Elzen et al. 1985a; Simeons et al. 1986). The shotgun approach requires a good selection scheme for the relatively rare transformants containing the desired gene. A model ten-member library was transferred to tobacco en masse (Prosen and Simpson 1987). Out of 66 plants, at least 2 that contained each of the library members were obtained. One plant contained at least 4 members, while the average plant contained 1.3 different T-DNA regions. Without further improvements, shotgun experiments would require the manipulation of extremely large numbers of plants to ensure a 95% probability of transferring a given gene or fragment. Traits that can be selected in tissue culture appear to be the best candidates for shotgun transfer at this time.

1.4 HOST RANGE OF *AGROBACTERIUM*

Initially, the *Agrobacterium* system appeared to be limited to the transformation of susceptible plants, and while the list of susceptible plants was extensive (De Cleene and De Ley 1976), most of them were dicotyledonous. However, susceptibility was always judged on the basis of tumor growths or swellings, not DNA transfer. The lack of tumor formation on many species was taken as evidence that monocotyledons were not susceptible. In terms of prolific tumor tissue formation, this may indeed be correct. However, the limiting factor in tumor formation is not necessarily DNA transfer, and for genetic engineering purposes, a distinction between T-DNA transfer and the appearance of disease symptoms (galls) must be made. The evidence clearly demonstrates that T-DNA transfer does occur with some monocotyledon species. Opine assays were the first to suggest that T-DNA enters monocotyledon cells when opines were detected in maize embryos (Graves and Goldman 1986), *Chlorophytum capense* (Hooykaas-Van Slogteren et al. 1984), *Narcissus* cv. Paperwhite (Hooykaas-Van Slogteren et al. 1984), and *Asparagus officinalis* (Hernalsteens et al. 1984) after exposure to *Agrobacterium*. Opine synthesis is always accepted as evidence of DNA transfer with caution in the absence of T-DNA analysis, particularly in light of a report that some plants can synthesize opines without the T-DNA genes (Christou et al. 1986). However, in the reports showing opine synthesis in infected tissue, appropriate controls seem to indicate that opine synthesis is a consequence of T-DNA transfer. More recently, infectious genomes of maize viruses have been introduced into the T-DNA (Grimsley et al. 1987). The appearance of disease symptoms is a sensitive indicator of T-DNA transfer into the monocotyledon cells. Disease symptoms were observed in a high percentage of plants when virus was transmitted by *Agrobacterium*, indicating that T-DNA transfer is not the limiting step (Grimsley et al. 1987). Recently, actual T-DNA transfer data demonstrated DNA integration in *Dioscorea bulbifera* (Schafer et al. 1987) and *A. officinalis* (Bytebier et al. 1987). Thus, monocotyledon transformation may prove less intransigent than initially perceived, and the utility of T-DNA transfer may expand even further. However, only dicotyledonous plants have been transformed and regenerated to date (Table 1–2). The

TABLE 1–2 Transgenic Plant Species Produced by *Agrobacterium* Transformation

Species[1]	*Reference*
Nicotiana tabacum	De Block et al. 1984
Nicotiana plumbaginifolia	André et al. 1986
Nicotiana glauca	Taylor et al. 1985
Petunia hybrida	Horsch et al. 1985
Solanum tuberosum	Ooms et al. 1985b
Lycopersicon esculentum	Sukhapinda et al. 1987
	McCormick et al. 1986
Helianthus annuus	Everett et al. 1987
Daucus carota	Chilton et al. 1982
Brassica oleracea L. var *Botrytis*	David and Tempé 1988
Brassica napus	Ooms et al. 1985a
	Guerche et al. 1987
	Pua et al. 1987
Apium graveolens	Catlin et al. 1988
Cucumis sativus	Trulson et al. 1986
Populus species	Fillatti et al. 1987b
Arabidopsis thaliana	Lloyd et al. 1986
	Feldmann and Marks 1987
Convolvulus arvensis	Tepfer 1984
Medicago sativa	Shahin et al. 1986a
Lotus corniculatus	Stougaard et al. 1986
Stylosanthes humilis	Manners 1988

[1] The list includes only species that have been regenerated to whole plants. Root cultures of numerous species have been established (Hamill et al. 1987; Mugnier 1988).

evidence clearly suggests that transformation of monocotyledonous cells capable of regeneration is the limiting step in *Agrobacterium* or other plant transformation schemes.

1.5 ALTERNATIVE DNA TRANSFER METHODS

Agrobacterium delivery of genetic material has no precedence in other eukaryotic organisms. However, other gene transformation methods for mammalian and yeast cells, which were developed at the same time as *Agrobacterium*-mediated transfer in plants, have been adapted to plant cells. These methods have been overshadowed by the *Agrobacterium* method, but some techniques will undoubtedly gain more prominence as they improve.

Alternative methods can be divided into two types: direct physical introduction of DNA and transmission of genetic material by modified plant viruses. Viral gene transfer can also involve physical transmission to the plant (e.g., rub inoculation). The difference lies in that the viral approach attempts to exploit the replicative and, in some cases, the expressive properties of the viral genome. Alternative methods have been suggested as ways to circumvent the apparent inability of *Agrobacterium* to deliver DNA efficiently to some species. Several of the techniques also offer the potential of greatly simplifying gene transfer procedures and eliminating the need for extensive tissue culture facilities. However, some of the alternatives have been reported only once and have not been developed to the point of routine use in gene transfer.

An interesting variety of approaches to the direct incorporation of DNA, or free DNA delivery, as these protocols are generally termed, have been explored. The most prominent method by far for the direct introduction of DNA into plant cells is the transformation of plant cell protoplasts (reviewed in Potrykus et al. 1985b). The approach necessarily requires the ability to isolate viable protoplasts. Whole plant studies further require the ability to regenerate plants from the protoplasts. Initial experiments involved addition of the DNA to protoplasts in the presence of poly-L-ornithine (Davey et al. 1980). This compound has now largely been replaced with $Ca^{2+}PO_4$/polyethylene glycol (Krens et al. 1982) or Mg^{2+} and polyethylene glycol (Negrutiu et al. 1987). Recently, electroporation has emerged as the predominant method for protoplast transformation (Neumann et al. 1982; Fromm et al. 1986, Morikawa et al. 1986; Okada et al. 1986). The protoplasts are exposed to a strong electric field and briefly shocked. This treatment is believed to create small depolarized pores in the protoplast membrane that allow passage of small molecules, including DNA or RNA. Originally developed for mammalian cell transformation, the electroporation method has been adapted for use in protoplasts from both monocotyledon and dicotyledon species. Where regeneration techniques exist, stable transformed calli have been obtained. The method is very useful for transient expression assays (Ou-Lee et al. 1986; Ow et al. 1986).

A number of transformation protocols have been developed with the goal of avoiding the isolation of protoplasts and the inherent difficulties of callus and plant regeneration. Microinjection is one such method (Crossway et al. 1986). However, so far only protoplasts have proven to be manageable for injection because of the difficulty of monitoring cell penetration and injection. Toward the same end, DNA has been absorbed to magnesium sulfate crystals and literally blasted into the cells by detonating a gunpowder charge behind a load of crystals (Klein et al. 1987). Again, the idea is to introduce DNA directly into cells without protoplast isolation. However, the technique works with protoplasts and, more recently, intact whole cells (Klein et al. 1988). Pollen transformation has been explored as a means of introducing genes, but the technique has not been convincingly demonstrated (Ohta 1986).

One technique, referred to here as macroinjection, has been used to transform rye plants (de la Pená et al. 1987). DNA was injected into the immature tillers of 135 plants and was incorporated into germ-line (archesporial) cells. Previous work had established that the archesporial cells were sensitive to macromolecules of caffeine and colchicine 14 days before the cells' first meiotic division. Two transformed plants were obtained from 4,023 seedlings. The transformation of the plants was confirmed by Southern blot analysis. This technique has great potential for wide practical application, although further demonstrations of this technique have not been reported. The frequency of transformation may be too low for mass experiments such as shotgun-type cloning or screening experiments.

The techniques of free-DNA transformation have not been as involved as the use of complex vectors in the *Agrobacterium* system, since DNA uptake and incorporation rely on the plant physiological processes. Most of the DNA used for transformation involves an *E. coli* replicon (often pBR322 or pUC derivative) containing a plant-expressed marker. Monocotyledon tissue culture cell lines have been transformed using either the nopaline synthase or the cauliflower mosaic virus (CaMV) 35S promoter and selected for kanamycin resistance (Lörz et al. 1985; Fromm et al. 1986). In the case of the rye macroinjection procedure, the nopaline synthase promoter fused to the NPTII gene was used (de la Pená et al. 1987). The actual process of DNA incorporation is not well understood, and refinement of vectors may be possible once the process is better understood. One possible improvement of the technique might involve designing a vector that utilizes the activity of a transposon to improve the efficiency of DNA integration. Such vectors would be similar to the P element transformation vectors in drosophila (Rubin and Spradling 1982).

Both plant DNA and RNA viruses offer possibilities as plant transformation vectors. Spurred by efforts to understand viral biology further and to develop vectors, both DNA and RNA viruses have been demonstrated to mediate expression of foreign genes in plant cells (Brisson et al. 1984; French et al. 1986; Takamatsu et al. 1987). Theoretically, viral vectors could have some advantages over vectors involving nonviral DNA in certain circumstances beyond studies of virus function. Advantages of such vectors might involve improvements in the ease of gene introduction. Once the proper construct is obtained, the plant could simply be rub inoculated. If high levels of the transgenic product are desired, the high titer that is often achieved by plant viruses could facilitate this. The timing of gene expression could also be controlled by the timing of inoculation.

On the other hand, viral vectors present unique problems. The presence of undesirable disease symptoms may interfere with the intended use of the vector. Viral vectors may also involve the release of infectious agents into the environment and therefore may be difficult to control in terms of their effects on other plants. Finally, the problem of stable transformation, if desired, must be overcome. Only the DNA viruses appear to offer the possibility of integration into the host genome. While integration has not been observed, techniques enabling integration could expand the use of viral vectors for transformation of species that are difficult to regenerate in tissue culture.

DNA viruses would seem to offer the best system in terms of a classical vector for inserting foreign genes, and hence DNA viruses have received considerable attention as possible gene transfer agents. Two types of plant DNA viruses are known to exist; the CaMV (caulimovirus) group and the geminivirus group. DNA clones of both types of virus are infectious (Howell et al. 1980; Gronenborn et al. 1981; Hamilton et al. 1983). CaMV is a circular double-stranded virus that replicates via an RNA intermediate. Several open reading frames (ORF II and ORF VII) on the virus genome are dispensable for infectious DNA, and these regions can be substituted with alternate DNA sequences. A bacterium-derived gene for methotrexate resistance has been inserted into CaMV, and the engineered virus is able to infect and confer resistance on turnip plants (Brisson et al. 1984). Problems with CaMV are the current limit to the size of gene that can be delivered and the problem of DNA integration (Brisson and Hohn 1986). One could possibly engineer a specific plant to be a complementation host for defective CaMV vectors that are capable of encoding a much larger insert. Whether copies of CaMV become inserted into the plant genome at a relatively low frequency has been difficult to assess. With marker genes inserted in the genome it should be possible to determine if viral DNA does become inserted into germ line cells by screening large numbers of seeds from infected plants for drug resistance.

RNA viruses present special problems when considering the construction of gene vectors. The development of these vectors is based on concepts different from the ones underlying the *Agrobacterium* T-DNA vectors or DNA transformation. The primary consideration is designing a vector that produces infectious transcripts and, therefore, preserves the terminal structure of the RNA in order for the transcript to be a good substrate for the viral replicase (Ahlquist 1986). The bacterophage promoters have proven ideal for directing the synthesis of RNA from DNA clones (Melton et al. 1984). Thus, the basic plant virus vector consists of a DNA plasmid with a bacteriophage promoter. The promoter is designed to initiate a transcript with the proper 5′ and 3′ terminals (and other portions of the plant virus that are desired). The gene of interest is cloned into the vector in a manner that does not disrupt the terminal structure. Transcripts are synthesized in vitro, processed by adding a 7-methylguanosine cap, and rub inoculated onto a plant. The first RNA virus expression vector, pPM1, utilized a modified lambda P_r promoter (Ahlquist and Janda 1981). This vector was used to synthesize infectious transcripts of each brome mosaic virus (BMV) genomic RNA from BMV cDNA clones. More recent vectors have been designed around the more highly efficient T7 RNA polymerase promoter (Langereis et al. 1986; Janda et al. 1987). Gene chimeras of BMV-CAT and tobacco mosaic virus–CAT have demonstrated that this approach can mediate the expression of foreign genes that are properly engineered into the viral genome (French et al. 1986; Takamatsu et al. 1987).

1.6 CONCLUSIONS

Great strides have been made in our ability to transfer genes to higher plants, and a variety of plant transformation vectors are currently available for gene transfer. One might even argue that our ability to manipulate plants genetically has improved faster than our understanding of basic plant biochemical and developmental processes. The question that is confronted in many situations is not whether a particular gene can be transferred, but which gene or genes need to be manipulated for a desired plant phenotype. Single genetic traits such as herbicide resistance, and possibly insect resistance as conferred by *Bacillus thuringiensis* toxin, will likely first reach the field as practical applications of the new technology. Transference of other important traits of agronomic interest such as cold hardiness, drought and disease resistance, or salt tolerance awaits further characterization of the genes controlling the phenotypes. In addition to the immediate application of gene transfer technology to crop improvement, the techniques have already been demonstrated to provide important tools in the further understanding of plant molecular biology.

Given the array of plant transformation schemes, new and ingenious methods that might facilitate the transfer of genes to plants can be expected in the near future. The *Agrobacterium* system has steadily expanded to include both dicotyledonous and monocotyledonous species. This progress may be even further accelerated as a better understanding of the T-DNA transfer process is attained. Current work on the factors controlling the host range of *Agrobacterium* infections may also contribute to a wider application of T-DNA-mediated gene transfer. In addition, inroads are continually being made in regard to plant regeneration. The concurrent development of plant transformation technology and analyses of plant molecular biology marks an exciting period in plant science.

REFERENCES

Abel, P.P., Nelson, R.S., De, B., et al. (1986) *Science* 232, 738–743.

Abdullah, R., Cocking, E.C., and Thompson, J.A. (1986) *Bio/Technol.* 4, 1087–1090.

Ahlquist, P. (1986) *Methods Enzymol.* 118, 704–716.

Ahlquist, P., and Janda, M. (1981) *Mol. Cell. Biol.* 4, 2876–2882.

Akiyoshi, D.E., Klee, H., Amasino, R.M., Nester, E.W., and Gordon, M.P. (1984) *Proc. Natl. Acad. Sci. USA* 81, 5994–5998.

Albright, L.M., Yanofsky, M.F., Leroux, B., et al. (1987) *J. Bacteriol.* 169, 1046–1055.

Amasino, R.M., Powell, A.L.T., and Gordon, M.P. (1984) *Mol. Gen. Genet.* 197, 437–446.

An, G. (1986) *Plant Physiol.* 81, 86–91.

An, G. (1987) *Mol. Gen. Genet.* 207, 210–216.

An, G., Watson, B.D., Stachel, S., Gordon, M.P., and Nester, E.W. (1985) *EMBO J.* 4, 277–284.

André, D., Colau, D., Schell, J., Van Montagu, M., and Hernalsteens, J.-P. (1986) *Mol. Gen. Genet.* 204, 512–518.

Bagdasarian, M., Lurz, R., Ruckert, B., et al. (1981) *Gene* 16, 237–247.

Baker, B., Schell, J., Lörz, H., and Fedoroff, N. (1986) *Proc. Natl. Acad. Sci. USA* 83, 4844–4848.

Barker, R.F., Idler, K.B., Thompson, D.V., and Kemp, J.D. (1983) *Plant Mol. Biol.* 2, 335–350.

Barry, G.F., Rogers, S.G., Fraley, R.T., and Brand, L. (1984) *Proc. Natl. Acad. Sci. USA* 81, 4776–4780.

Barton, K.A., Binns, A.N., Matzke, A.J.M., and Chilton, M.-D. (1983) *Cell* 32, 1033–1043.

Baulcombe, D.C., Saunders, G.R., Bevan, M.W., Mayo, M.A., and Harrison, B.D. (1986) *Nature* 321, 446–449.

Bevan, M., Barnes, W.M., and Chilton, M.-D. (1983a) *Nucleic Acids Res.* 11, 369–379.

Bevan, M., Flavell, R., and Chilton, M.-D (1983b) *Nature* 304, 184–187.

Bevan, M. W. (1984) *Nucleic Acids Res.* 12, 8711–8721.

Bevan, M.W., Mason, S.E., and Goelett, P. (1985) *EMBO J.* 4, 1921–1926.

Bolton, G.W., Nester, E.W., and Gordon, M.P. (1986) *Science* 232, 983–985.

Boutry, M., Nagy, F., Poulsen, C., Aoyagi, K., and Chua, N.-H. (1987) *Nature* 328, 340–342.

Brisson, N., and Hohn, T. (1986) *Methods Enzymol.* 118, 659–684.

Brisson, N., Paskowsky, J., Penswick, J.R., et al. (1984) *Nature* 310, 511–514.

Broglie, R., Coruzzi, G., Fraley, R.T., et al. (1984) *Science* 224, 838–843.

Buchanan-Wollaston, V., Passiatore, J.E., and Cannon, F. (1987) *Nature* 328, 172–175.

Budar, F., Thia-Toong, L., Van Montagu, M., and Hernalsteens, J.-P. (1986) *Genetics* 114, 303–313.

Bytebier, B., Debroeck, F., DeGreve, H., et al. (1987) *Proc. Natl. Acad. Sci. USA* 85, 5345–5349.

Caplan, A., Herrera-Estrella, L., Inzé, D., et al. (1983) *Science* 222, 815–821.

Catlin, D., Ochoa, O., McCormick, S., and Quiros, C.F. (1988) *Plant Cell Reports* 7, 100–103.

Cheung, A.Y., Bogorad, L., Van Montagu, M., and Schell, J. (1988) *Proc. Natl. Acad. Sci. USA* 85, 391–395.

Chilton, M.-D., Tepfer, D.A., Petit, A., et al. (1982) *Nature* 295, 432–434.

Christou, P., Platt, S.G., and Ackerman, M.C. (1986) *Plant Physiol.* 82, 218–221.

Chyi, Y.S., Jorgensen, R.A., Goldstein, D., Tanksley, S.D., Loaiza-Figueroa, F. (1986) *Mol. Gen. Genet.* 204, 64–69.

Coulibaly, M.Y., and Demarly, Y. (1986) *Z. Pflanzenzuchtg.* 96, 79–81.

Comai, L., Facciotti, D., Hiatt, W.R., et al. (1985) *Nature* 317, 741–744.
Crossway, A., Oakes, J., Irvine, J., et al. (1986) *Mol. Gen. Genet.* 202, 179–185.
Dahl, G.A., and Tempé, J. (1983) *Theor. Appl. Genet.* 66, 233.
Das, A. (1988) *Proc. Natl. Acad. Sci. USA* 85, 2909–2913.
Davey, M.R., Cocking, E.C., Freeman, J., Pearce, N., and Tudor, I. (1980) *Plant Sci. Lett.* 18, 307–313.
David, C., and Tempé, J. (1988) *Plant Cell Reports* 7, 88–91.
De Block, M., Botterman, J., Vandewiele, M., et al. (1987) *EMBO J.* 6, 2513–2518.
De Block, M., Herrera-Estrella, L., Van Montagu, M., Schell, J., and Zambryski, P. (1984) *EMBO J.* 3, 1681–1689.
De Cleene, M.D., and De Ley, J. (1976) *Bot. Rev.* 42, 389–466.
de Framond, A.J., Back, E.W., Chilton, W.S., Kayes, L., Chilton, M.-D. (1986) *Mol. Gen. Genet.* 202, 125–131.
de Framond, A., Barton, K.A., and Chilton, M.-D. (1983) *Bio/Technol.* 1, 262–269.
de la Pená, A., Lörz, H., and Schell, J. (1987) *Nature* 325, 274–276.
Deom, C.M., Oliver, M.J., and Beachy, R.N. (1987) *Science* 237, 389–394.
Depicker, A., Herman, L., Jacobs, A., Schell, J., and Van Montagu, M. (1985) *Mol. Gen. Genet.* 201, 477–484.
Depicker, A., Jacobs, A.M., and Van Montagu, M.C. (1988) *Plant Cell Reports* 7, 63–66.
Depicker, A., Stachel, S., Dhaese, P., Zambryski, P., and Goodman, H. (1982) *J. Mol. Appl. Genet.* 1, 561–573.
Ditta, G., Stanfield, S., Corbin, D., and Helinski, D.R. (1980) *Proc. Natl. Acad. Sci. USA* 77, 7347–7351.
Ellis, J.G., Llewellyn, D.J., Dennis, E.S., and Peacock, W.J. (1987) *EMBO J.* 6, 11–16.
Engler, G., Depicker, A., Maenhaut, R., et al. (1981) *J. Mol. Biol.* 152, 183–208.
Everett, N.P., Robinson, K.E.P., and Mascarenhas, D. (1987) *Bio/Technol.* 5, 1201–1204.
Feldmann, K.A., and Marks, M.D. (1987) *Mol. Gen. Genet.* 208, 1–9.
Fillatti, J.J., Kiser, J., Rose, R., and Comai, L. (1987a) *Bio/Technol.* 5, 726–730.
Fillatti, J.J., Sellmer, J., McCown, B., Haissig, B., and Comai, L. (1987b) *Mol. Gen. Genet.* 206, 192–199.
Fischhoff, D.A., Bowdish, K.S., Perlak, F.J., et al. (1987) *Bio/Technol.* 5, 807–813.
Fraley, R.T., Rogers, S.G., Horsch, R.B., et al. (1985) *Bio/Technol.* 3, 629–635.
Fraley, R.T., Rogers, S.G., Horsch, R.B., et al. (1983) *Proc. Natl. Acad. Sci. USA* 80, 4803–4807.
Freeling, M. (1984) *Annu. Rev. Plant Physiol.* 35, 277–298.
French, R., Janda, M., and Ahlquist, P. (1986) *Science* 231, 1294–1297.
Fromm, M.E., Taylor, L.P., and Walbot, V. (1986) *Nature* 319, 791–793.
Fujimura, T., Sakurai, M., Agaki, H., et al. (1985) *Plant Tissue Culture Lett.* 2, 74–75.
Gardner, R.C., and Knauf, V.C. (1986) *Science* 23, 725–727.
Garfinkel, D.J., Simpson, R.B., Ream, L.W., et al. (1981) *Cell* 27, 143–153.
Gheysen, G., Van Montagu, M., and Zambryski, P. (1987) *Proc. Natl. Acad. Sci. USA* 84, 6169–6173.
Gietly, C., and Hohn, B. (1987) *Proc. Natl. Acad. Sci. USA* 84, 9006–9010.
Gorman, C.M., Moffat, L.F., and Howard, B.H. (1982) *Mol. Cell Biol.* 2, 1044–1051.

Graves, A.C.F., and Goldman, S.L. (1986) *Plant Mol. Biol.* 7, 43–50.
Grimsley, N., Hohn, T., Davis, J.W., and Hohn, B. (1987) *Nature* 325, 177–179.
Gronenborn, B., Gardner, R.C., Schaefer, S., and Shepherd, R.J. (1981) *Nature* 294, 773.
Guerche, P., Jouanin, L., Tepfer, D., and Pelletier, G. (1987) *Mol. Gen. Genet.* 206, 382–386.
Hamill, J.D., Parr, A.J., Rhodes, M.J.C., Robins, R.J., and Walton, N.J. (1987) *Bio/Technol.* 5, 800–804.
Hamilton, W.D.O., Bisaro, D.M., Coutts, R.H.A., and Buck, K.W. (1983) *Nucleic Acids Res.* 11, 7387–7396.
Haughn, G.W., and Somerville, C. (1986) *Mol. Gen. Genet.* 204, 430–434.
Hayford, M.B., Medford, J.I., Hoffmann, N.L., et al. (1988) *Plant Physiol.* 86, 1216–1222.
Helmer, G., Casadaban, M., Bevan, M., Kayes, L., and Chilton, M.-D. (1984) *Bio/Technol.* 2, 520–527.
Hepburn, A.G., Clarke, L.E., Blundy, K.S., and White, J. (1983) *J. Mol. Appl. Genet.* 2, 211–224.
Hernalsteens, J.-P., Thia-Toong, L., Schell, J., and Montagu, M.V. (1984) *EMBO J.* 3, 3039–3042.
Hess, J.F., Oosawa, K., Matsumura, P., and Simon, M.I. (1987) *Proc. Natl. Acad. Sci. USA* 84, 7609–7613.
Herrera-Estrella, L., De Block, M., Messens, E., et al. (1983) *EMBO J.* 2, 987–995.
Hille, J., Verheggen, F., Roelvink, P., et al. (1986) *Plant Mol. Biol.* 7, 171–176.
Hinchee, M.A., Connor-Ward, D.V., Newell, C.A., et al. (1988) *Bio/Technol.* 6, 915–922.
Hoekema, A., Hirsch, P.R., Hooykaas, P.J., and Schilperoort, R.A. (1983) *Nature* 303, 179–181.
Hoekema, A., Hooykaas, P.J., and Schilperoort, R.A. (1984) *J. Bacteriol.* 158, 383–385.
Hood, E.E., Helmer, G.L., Fraley, R.T., and Chilton, M.-D. (1986) *J. Bacteriol.* 168, 1291–1301.
Hood, E.E., Jen, G., Kayes, L., et al. (1984) *Bio/Technol.* 2, 702–708.
Hooykaas-Van Slogteren, G.M.S., Hooykaas, P.J.J., and Schilperoort, R.A. (1984) *Nature* 311, 763–764.
Horsch, R.B., Fry, J.E., Hoffmann, N.L., et al. (1985) *Science* 233, 1229–1231.
Horsch, R.B., and Klee, H.J. (1986) *Proc. Natl. Acad. Sci. USA* 83, 4428–4432.
Horsch, R.B., Klee, H.J., Stachel, S., et al. (1986) *Proc. Natl. Acad. Sci. USA* 83, 2571–2575.
Howell, S.H., Walker, L.L., and Dudley, R.K. (1980) *Science* 208, 1265.
Hirayama, T., Muranaka, T., Ohkawa, H., and Oka, A. (1988) *Mol. Gen. Genet.* (in press).
Huffman, G.A., White, F.F., Gordon, M.P., and Nester, E.W. (1984) *J. Bacteriol.* 157, 269–276.
Inzé, D., Follin, A., Van Lijsebettins, M., et al. (1984) *Mol. Gen. Genet.* 194, 265–274.
Janda, M., French, R., and Ahlquist, P. (1987) *Virology* 158, 259–262.
Janssens, A., Genetello, C., Van Montagu, M., and Zambryski, P. (1986) *Plant Sci.* 47, 185–193.

Jefferson, R.A., Burgess, S.M., and Hirsh, D. (1986) *Proc. Natl. Acad. Sci. USA* 83, 8447–8451.

Jen, G.C., and Chilton, M.-D. (1986) *Proc. Natl. Acad. Sci. USA* 83, 3895–3899.

Jin, S., Komari, T., Gordon, M.P., and Nester, E.W. (1987) *J. Bacteriol.* 169, 4417–4425.

Jones, J.D.G., Gilbert, D.E., Grady, K.L., and Jorgensen, R.A. (1987) *Mol. Gen. Genet.* 207, 478–485.

Jones, J.D.G., Svab, Z., Harper, E.C., et al. (1987) *Mol. Gen. Genet.* 210, 86–91.

Joos, H., Inzé, D., Caplan, A., et al. (1983) *Cell* 32, 1057–1067.

Jorgensen, R., Snyder, C., and Jones, J.D.G. (1987) *Mol. Gen. Genet.* 207, 471–477.

Jouanin, L. (1984) *Plasmid* 12, 91–102.

Keane, P.J., Kerr, A., and New, P.B. (1970) *Aust. J. Biol. Sci.* 23, 585–595.

Klee, H.J., White, F.F., Iyer, V.N., Gordon, M.P., and Nester, E.W. (1983) *J. Bacteriol.* 153, 878–883.

Klee, H.J., Yanofsky, M.F., and Nester, E.W. (1985) *Bio/Technol.* 3, 637–642.

Klein, T.M., Wolf, E.D., Wu, R., and Sanford, J.C. (1987) *Nature* 327, 70–73.

Klein, T.M., Fromm, M., Weissinger, A., et al. (1988) *Proc. Natl. Acad. Sci. USA* 85, 4305–4309.

Komari, T., Halperin, W., and Nester, E.W. (1986) *J. Bacteriol.* 166, 88–94.

Koncz, C., Olsson, O., Langridge, W.H.R., Schell, J., and Szalay, A.A. (1987) *Proc. Natl. Acad. Sci. USA* 84, 131.

Koncz, C., and Schell, J. (1986) *Mol. Gen. Genet.* 204, 383–396.

Koplow, J., Byrne, M., Jen, G., Tempé, J., and Chilton, M.-D. (1984) *Plasmid* 11, 17–27.

Koukolikova-Nicola, Z., Shillito, R., Hohn, B., et al. (1985) *Nature* 313, 191–196.

Krens, F.A., Molendijk, L., Wullems, G.J., and Schilperoort, R.A. (1982) *Nature* 296, 72–74.

Kuntz, M., Simons, A., Schell, J., and Schreier, P.H. (1986) *Mol. Gen. Genet.* 205, 454–460.

Lamppa, G., Nagy, F., and Chua, N.-H. (1985) *Nature* 316, 750–752.

Langereis, K., Neeleman, L., and Bol, J.F. (1986) *Plant Mol. Biol.* 6, 281–288.

Leemans, J., Deblaere, R., Willmitzer, L., et al. (1982a) *EMBO J.* 1, 147–152.

Leemans, J., Langenakens, J., DeGreve, H., et al. (1982b) *Gene* 19, 361–364.

Leemans, J., Shaw, C., Deblaere, R., et al. (1981) *J. Mol. Appl. Genet.* 1:149–164.

Lichtenstein, C., and Draper, J. (1985) in *DNA Cloning, a Practical Approach* (Glover, D.M., ed.), pp. 67–118, IRL Press, Oxford.

Lloyd, A.M., Barnason, A.R., Rogers, S.G., et al. (1986) *Science* 234, 464–466.

Loesch-Fries, L.S., Merlo, D., Zinnen, T., et al. (1987) *EMBO J.* 6, 1845–1851.

Lörz, H., Baker, B., and Schell, J. (1985) *Mol. Gen. Genet.* 199, 178–182.

McCabe, D.E., Swain, W.F., Martinell, B.J., and Christov, P. (1988) *Bio/Technol.* 6, 923–926.

McCormick, S., Niedermeyer, J., Fry, J., et al. (1986) *Plant Cell Rep.* 5, 81–84.

Machida, Y., Usami, S., Yamamoto, A., Niwa, Y., and Takebe, I. (1986) *Mol. Gen. Genet.* 204, 374–382.

Manners, J.M. (1988) *Plant Science* 55, 61–68.

Matzke, A., and Chilton, M.-D. (1981) *J. Mol. Appl. Genet.* 1, 39–49.

Matzke, A.J.M., and Matzke, M.A. (1986) *Plant Mol. Biol.* 7, 357–365.

Melton, D.A., Krieg, P.A., Rebagliati, M.R., et al. (1984) *Nucleic Acids Res.* 12, 7035–7056.

Menagé, A., and Morel, G. (1964) *C. R. Acad. Sci.* 259, 4795–4796.

Messens, E., Lenaerts, A., Van Montagu, M., and Hedges, R.W. (1985) *Mol. Gen. Genet.* 199, 344–348.

Morikawa, H., Iida, A., Matsui, C., Ikegami, M., and Yamada, Y. (1986) *Gene* 41, 121–124.

Mugnier, J. (1988) *Plant Cell Reports* 7, 9–12.

Müller, A.J., Mendel, R.R., Schiemann, J., Simoens, C., and Inzé, D. (1987) *Mol. Gen. Genet.* 207, 171–175.

Murai, N., and Kemp, J.D. (1982) *Proc. Natl. Acad. Sci. USA* 79, 86–90.

Negrutiu, I., Shillito, R., Potrykus, I., Biasini, G., and Sala, F. (1987) *Plant Mol. Biol.* 8, 363–373.

Nester, E.W., Gordon, M.P., Amasino, R.M., and Yanofsky, M.F. (1984) *Annu. Rev. Plant Physiol.* 35, 387–413.

Neumann, E., Schaeffer-Ridder, M., Wang, Y., and Hofschneider, P.H. (1982) *EMBO J.* 1, 841–845.

Ninfa, A.J., and Magasanik, B. (1986) *Proc. Natl. Acad. Sci. USA* 83, 5909–5913.

Nixon, B.T., Ronson, C.W., and Ausubel, F.M. (1986) *Proc. Natl. Acad. Sci. USA* 83, 7850–7854.

Ohta, Y. (1986) *Proc. Natl. Acad. Sci. USA* 83, 715–719.

Okada, K., Takebe, I., and Nagata, T. (1986) *Mol. Gen. Genet.* 205, 398–403.

Ooms, G., Bains, A., Burrell, M., Twell, D., and Wilcox, E. (1985a) *Theor. Appl. Genet.* 721, 325–329.

Ooms, G., Hooykaas, P.J.J., Van Veen, R.J.M., et al. (1982) *Plasmid* 7, 15–29.

Ooms, G., Karp, A., Burrell, M.M., Twell, D., and Roberts, J. (1985b) *Theor. Appl. Genet.* 70, 440–446.

Otten, L.A., and Schilperoort, R.A. (1978) *Biochim. Biophys. Acta* 527, 497–500.

Ou-Lee, T.-M., Turgeon, R., and Wu, R. (1986) *Proc. Natl. Acad. Sci. USA* 83, 6815–6819.

Ow, D.W., Wood, K.V., DeLuca, M., et al. (1986) *Science* 234, 856–859.

Peralta, E.G., Hellmiss, R., and Ream, W. (1986) *EMBO J.* 5, 1137–1142.

Peralta, E.G., and Ream, L.W. (1985) *Proc. Natl. Acad. Sci. USA* 82, 5112–5116.

Petit, A., Berkaloff, A., and Tempé, J. (1986) *Mol. Gen. Genet.* 202, 388–393.

Potrykus, I., Saul, M.W., Petruska, J., Paszkowski, J., and Shillito, R.D. (1985a) *Mol. Gen. Genet.* 199, 183–188.

Potrykus, I., Shillito, R.D., Saul, M.W., and Paszkowski, J. (1985b) *Plant Mol. Biol. Reptr.* 3, 117–128.

Prosen, D.E., and Simpson, R.B. (1987) *Bio/Technol.* 5, 966–971.

Pua, E.-C., Mehra-Palta, A., Nagy, F., and Chua, N.-H. (1987) *Bio/Technol.* 5, 815–817.

Pythoud, F., Sinkar, V.P., Nester, E.W., and Gordon, M.P. (1987) *Bio/Technol.* 5, 1323–1327.

Ream, L.W., Gordon, M.P., and Nester, E.W. (1983) *Proc. Natl. Acad. Sci. USA* 80, 1660–1664.

Reiss, B., Sprengel, R., Will, H., and Schaller, H. (1984) *Gene* 30, 211–218.

Rhodes, C.A., Lowe, K.S., and Ruby, K.L, (1988) *Bio/Technol.* 6, 56–60.

Rhodes, C.A., Pierce, D.A., Mettler, I.J., Mascarenhas, D., and Detmer, J.J. (1988) *Science* 240, 204–207.

Rogers, S.G., Horsch, R.B., and Fraley, R.T. (1986) *Methods Enzymol.* 118, 627–641.

Roth, D.B., Porter, T.N., and Wilson, J.H. (1985) *Mol. Cell. Biol.* 5, 2599–2607.

Rothstein, S.J., Lahners, K.N., Lotstein, R.J., et al. (1987) *Gene* 53, 153–161.

Rubin, G., and Spradling, A. (1982) *Science* 218, 348–353.

Ruvkun, G.B., and Asubel, F.M. (1981) *Nature* 289, 85–88.

Salomon, F., Deblaere, R., Leemans, J., et al. (1984) *EMBO J.* 3, 141.

Schafer, W., Gorz, A., and Kahl, G. (1987) *Nature* 327, 529–532.

Schröder, G., Waffenschmidt, S., Weiler, E.W., and Schröder, J. (1984) *Eur. J. Biochem.* 138, 387–391.

Schröder, J., Schröder, G., Huisman, H., Schilperoort, R.A., and Schell, J. (1981) *FEBS Lett.* 129, 166–168.

Shah, D.M., Horsch, R.B., Klee, H.J., et al. (1986) *Science* 233, 478–481.

Shahin, E.A., Spielmann, A., Sukhapinda, K., Yashar, M., and Simpson, R.B. (1986a) *Crop Sci.* 26, 1235–1239.

Shahin, E.A., Sukhapinda, K., Simpson, R.B., and Spivey, R. (1986b) *Theor. Appl. Genet.* 72, 770–777.

Shaw, C., Watson, M., Carter, G., and Shaw, C. (1984) *Nucleic Acids Res.* 12, 6031–6041.

Simeons, C., Alliotte, T., Mendel, R., et al. (1986) *Nucleic Acids Res.* 14:8073–8090.

Simon, R., Preifer, U., and Pühler, A. (1983) *Bio/Technol.* 1, 784–791.

Simpson, R., O'Hara, P., Lichtenstein, C., et al. (1982) *Cell* 29, 1005–1014.

Simpson, R.B., Spielmann, A., Margossian, L., and McKnight, T.D. (1986) *Plant Mol. Biol.* 6, 403–415.

Slightom, J.L., Durand-Tardif, M., Jouanin, L., and Tepfer, D. (1986) *J. Biol. Chem.* 261, 108–121.

Smith, M.J., and Neilands, J.B. (1987) in *Molecular Strategies for Crop Protection* (Arntzen, C.J., and Ryan, C., eds.), pp. 157–168, Alan R. Liss, New York.

Spena, A., and Schell, J. (1987) *Mol. Gen. Genet.* 206, 436–440.

Spielmann, A., and Simpson, R.B. (1986) *Mol. Gen. Genet.* 205, 34–41.

Stachel, S.E., Messens, E., Van Montagu, M., and Zambryski, P. (1985) *Nature* 318, 624–629.

Stachel, S.E., and Nester, E.W. (1986) *EMBO J.* 5, 1445–1454.

Stachel, S.E., Nester, E.W., and Zambryski, P.C. (1986a) *Proc. Natl. Acad. Sci. USA* 83, 379–383.

Stachel, S.E., Timmerman, B., and Zambryski, P.C. (1986b) *Nature* 322, 706–712.

Stachel, S.E., and Zambryski, P.C. (1986a) *Cell* 46, 325–333.

Stachel, S.E., and Zambryski, P.C. (1986b) *Cell* 47, 155–157.

Stougaard, J., Abildsten, D., and Marcker, K.A. (1987) *Mol. Gen. Genet.* 207, 251–255.

Stougaard, J., Marcker, K.A., Otten, L., and Schell, J. (1986) *Nature* 321, 667–674.

Sukhapinda, K., Spivey, R., Simpson, R.B., and Shahin, E.A. (1987) *Mol. Gen. Genet.* 206, 491–497.

Tait, R.C., Close, T.J., Lundquist, R.C., et al. (1983) *Bio/Technol.* 1, 269–274.

Takamatsu, N., Ishikawa, M., Meshi, T., and Okada, Y. (1987) *EMBO J.* 6, 307–311.

Taylor, B.H., Amasino, R.M., White, F.F., Nester, E.W., and Gordon, M.P. (1985) *Mol. Gen. Genet.* 201, 554–557.

Teeri, T., Herrera-Estrella, L., Depicker, A., Van Montagu, M., and Palva, E. (1986) *EMBO J.* 5, 1755–1760.

Tempé, J., and Goldmann, A. (1982) in *Molecular Biology of Plant Tumors* (Kahl, G., and Schell, J., eds.), pp. 427–449, Academic Press, Orlando, FL.

Tepfer, D. (1984) *Cell* 37, 959–967.

Thomashow, L.S., Reeves, S., and Thomashow, M.F. (1984) *Proc. Natl. Acad. Sci. USA* 81, 5071–5075.

Thomashow, M.F., Hugly, S., Buchholz, W., and Thomashow, L.S. (1986) *Science* 231, 616–618.

Thomashow, M.F., Nutter, R., Montoya, A.L., Gordon, M.P., and Nester, E.W. (1980) *Cell* 19, 729–739.

Toriyama, K., Arimoto, Y., Uchimiya, H., and Hinata, K. (1988) *Bio/Technol.* 6, 1072–1076.

Trinh, T.H., Mante, S., Pua, E.-C., and Chua, N.-H. (1987) *Bio/Technol.* 5, 1081–1084.

Trulson, A.J., Simpson, R.B., and Shahin, E.A. (1986) *Theor. Appl. Genet.* 73, 11–15.

Tumer, N.E., O'Connell, K.M., Nelson, R.S., et al. (1987) *EMBO J.* 6, 1181–1188.

Vaeck, M., Reynaerts, A., Hofte, H., et al. (1987) *Nature* 328, 33–37.

Van den Broeck, G., Timko, M.P., Kausch, A.P., et al. (1985) *Nature* 313, 358–363.

van den Elzen, P., Lee, K.Y., Townsend, J., and Bedbrook, J. (1985a) *Plant Mol. Biol.* 5, 149–154.

van den Elzen, P., Townsend, J., Lee, K.Y., and Bedbrook, J. (1985b) *Plant Mol. Biol.* 5, 299–302.

Van Larebeke, N., Engler, G., Holsters, M., et al. (1974) *Nature* 252, 169–170.

Velten, J., and Schell, J. (1985) *Nucleic Acids Res.* 13, 6981–6998.

Velten, J., Velten, L., Hain, R., and Schell, J. (1984) *EMBO J.* 3, 2723–2730.

Waldron, C., Murphy, E., Roberts J., et al. (1985) *Plant Mol. Biol.* 5, 103.

Wallroth, M., Gerats, A.G.M., Rogers, S.G., Fraley, R.T., and Horsch, R.B. (1986) *Mol. Gen. Genet.* 202, 6–15.

Wang, K., Herrera-Estrella, L., Van Montagu, M., and Zambryski, P. (1984) *Cell* 38, 455–462.

Wang, K., Stachel, S.E., Timmerman, B., Van Montagu, M., and Zambryski, P.C. (1987) *Science* 235, 587–591.

Watson, B., Currier, T.C., Gordon, M.P., Chilton, M.-D., and Nester, E.W. (1975) *J. Bacteriol.* 123, 255–254.

White, F.F., and Nester, E.W. (1980) *J. Bacteriol.* 141, 1134–1141.

White, F.F., and Sinkar, V. (1987) in *Advances in Plant Gene Research* (Hohn, T., ed.), pp. 149–178, Springer-Verlag, New York.

White, F.F., Taylor, B.H., Huffman, G.A., Gordon, M.P., and Nester, E.W. (1985) *J. Bacteriol.* 164, 33–44.

Willetts, N., and Wilkins, B. (1984) *Microbiol. Rev.* 48, 24–41.

Willmitzer, L., Dhaese, P., Schreier, P., et al. (1983) *Cell* 32, 1045–1056.

Winans, S.C., Ebert, P.R., Stachel, S.E., Gordon, M.P., and Nester, E.W. (1986) *Proc. Natl. Acad. Sci. USA* 83, 8278–8282.

Wullems, G.J., Molendijk, L., Ooms, G., and Schilperoort, R.A. (1981) *Cell* 24, 719–727.

Yadav, N.S., Vanderleyden, J., Bennett, D.R., Barnes, W.M., and Chilton, M.-D. (1982) *Proc. Natl. Acad. Sci. USA* 79, 6322–6326.

Yamada, M., Yang, Z., Tang, D. (1986) *Plant Cell Rep.* 5, 85–88.

Yamamoto, A., Iwahashi, M., Yanofsky, M.F., et al. (1987) *Mol. Gen. Genet.* 206, 174–177.

Yanofsky, M., Lowe, B., Montoya, A., et al. (1985a) *Mol. Gen. Genet.* 201, 237–246.

Yanofsky, M., Montoya, A., Knauf, V., et al. (1985b) *J. Bacteriol.* 163, 341–348.

Yanofsky, M.F., Porter, S.G., Young, C., et al. (1986) *Cell* 47, 471–477.

Zambryski, P., Depicker, A., Kruger, K., and Goodman, H. (1982) *J. Mol. Appl. Genet.* 1, 363–370.

Zambryski, P., Joos, H., Genetello, C., et al. (1983) *EMBO J.* 2, 2143–2150.

CHAPTER

2

Methods for Transforming Plant Cells

Ray Wu

Unprecedented opportunities for the manipulation of plant genes have arisen from recent advances in recombinant DNA research and research into the transformation of protoplasts or cells and the regeneration of several varieties of plants. These developments are revolutionizing our approaches to solving complex biological problems and allowing scientists to enter a new era in plant biotechnology. The availability of reliable genetic transformation systems for plants should allow a firm elucidation of the principles of gene expression and regulation, DNA replication, and recombination in plants. The transfer of novel or beneficial foreign genes into agronomically important plants is also feasible (Fraley et al. 1986; Kuhlemeier et al. 1987; Goodman et al. 1987).

The main focus of this chapter is the review of several general or promising methods for transforming plant cells. Emphasis will be placed on the transformation of monocotyledenous plants, because the transformation of dicotyledenous plants, using the *Agrobacterium tumefaciens*–Ti plasmid system or microinjection, has been well established and supported by many excellent papers and review articles.

2.1 TRANSFORMATION OF DICOTYLEDENOUS PLANTS

2.1.1 Methods Based on the *A. tumefaciens*–Ti Plasmid System

In nature, *A. tumefaciens* infects most dicots and some monocots by entry through wound sites. The infection of dicots with virulent strains of the bacterium *A. tumefaciens* results in the production of crown gall tumors. It has been shown that the bacterium carries large Ti plasmids. The crown gall disease occurs as a result of the transfer and integration of a small DNA segment (T-DNA) from the Ti plasmid into plant nuclear DNA (for review see Bevan and Chilton 1982; Gheysen et al. 1985). A foreign gene can be joined to the T-DNA and carried into plant cells. Both native genes and foreign genes introduced into plant cells are expressed (Willmitzer et al. 1982; Herrera-Estrella et al. 1983). Thus, the *A. tumefaciens*–Ti plasmid system is a convenient way to introduce foreign genes into dicots. This system has been simplified by the development of two procedures: (1) an in vitro transformation procedure in which regenerating tobacco protoplasts are cultured directly with *A. tumefaciens* cells (Marton et al. 1979; Wullems et al. 1981; Fraley et al. 1984), and (2) the use of the leaf disc technique, which bypasses the requirement to prepare protoplasts, thereby increasing the efficiency and speed of plant cell transformation (Horsch et al. 1985).

Successful transformation of plants by the Ti plasmid system usually depends on the use of essential portions of the Ti plasmid and a convenient reporter gene or a selectable marker gene. Two regions of the Ti plasmid are essential for the mobilization and integration of the T-DNA into plant cells: (1) the T-DNA border sequences, which consist of direct repeats of 25 base pairs, that flank the T-DNA (Wang et al. 1984), and (2) the *vir* region, which is required for the excision, transfer, and integration of the T-DNA from the Ti plasmid into the plant genome (Ooms et al. 1980; Klee et al. 1983). The Ti plasmid–based vectors all include these two regions, in addition to a suitable plant promoter sequence joined to a reporter gene (Figure 2–1), essential for the introduction of foreign genes into plant cells. There are a number of useful vectors that have been reported by Rogers et al. (1987), Deblaere et al. (1987), and An (1987). More information about vectors can be found in Chapter 1 by Frank F. White. The uses of these and other vectors in transforming plant protoplasts or cells, selection procedures, and regeneration of plants are expounded in articles listed in the references and will not be discussed in this chapter.

Recently, it was reported that the mobilization function on the naturally occurring wide-host-range plasmid RSF1010 can mediate the transfer of plasmids from *Agrobacterium* into plant cells as well as the transfer between bacteria. This suggests that plants have access to the gene pool of Gram-negative bacteria (Buchanan-Wollaston et al. 1987). It remains to be shown whether the DNA transferred by the *ori*T/*mob* plasmids is integrated at the *ori*T site in the same way as the T-DNA is integrated at the right border.

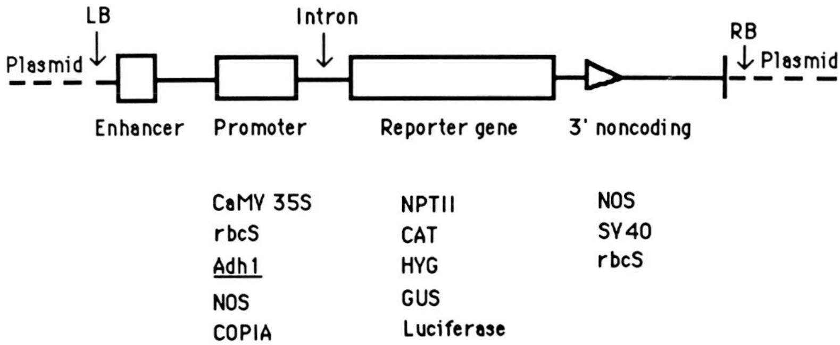

FIGURE 2–1 General structure of a vector for monitoring transformation of plant protoplasts or cells. The three basic components of a vector include a promoter, a reporter gene, and a 3′ noncoding region. The reporter gene may be a selectable marker gene (such as a bacterial antibiotic resistance gene: NPTII, CAT, or HYG) or a gene whose protein product can be readily detected (such as Gus or luciferase). The promoter is usually of plant origin (such as CaMV 35S, rbcS, *Adhl*, or NOS promoter). The 3′ noncoding region contains the poly(A) signal and perhaps the transcriptional terminator as well. The addition of an enhancer sequence (e.g., the duplication of CaMV 35S promoter sequence can serve as a strong enhancer for plant genes; Kay et al. 1987) or an intron of maize *Adhl* gene may increase the transcriptional activity of a promoter tenfold to one-hundred-fold (Callis et al. 1987). In principle, any combination of a promoter-reporter gene–3′ noncoding sequence may work for a given plant protoplast or cell system. However, in practice, one combination of the three basic components often gives higher expression than another in a specific plant system. In the case of a plasmid vector to be used with *Agrobacterium tumefaciens*, the left border (LB) and the right border (RB) of the T-DNA may be included. The location of the LB is usually on the left side of the promoter, and that of the RB is usually on the right side of the 3′ noncoding sequence. CaMV 35S, the promoter from the 35S RNA gene of cauliflower mosaic virus; rbcS, ribulose bisphosphate carboxylase small subunit gene; *Adhl*, alcohol dehydrogenase gene; NOS, nopaline synthase gene; NPTII [also known as APH(3′)II or neo], *Escherichia coli* Tn5 neomycin phosphotransferase gene; CAT, chloramphenicol acetyltransferase gene; HYG, hygramycin phosphotransferase gene; Gus, β-glucuronidase gene.

2.1.2 Methods Based on Microinjection

Microinjection is a precise technique for delivering DNA molecules into specific intracellular compartments of living cells. In principle, microinjection is not restricted by the host range of *A. tumefaciens* and has a potentially broad application. The method requires considerable skill to carry out, and so far it works mainly with protoplasts and not with intact cells.

Methods for the microinjection of plant protoplasts have been developed only over the last five years (Griesbach 1983; Steinbiss and Stabel 1983; Lawrence and Davies 1985; Morikawa and Yamada 1985; Crossway et al.

1986). Efficient transformation of alfalfa protoplasts by the intranuclear microinjection of Ti plasmids will be briefly described as an example (Reich et al. 1986). Microinjection of 2,500 alfalfa protoplasts yielded 70 transformed callus lines without selection. Analysis of several lines transformed with pTiC58 showed that integration did not occur by the T-DNA mechanism. The presence of a functional T-DNA right border or the coinjection of a functional *vir* region on a separate plasmid did not increase the transformation efficiencies.

2.2 TRANSFORMATION OF MONOCOTYLEDENOUS PLANTS

Although the *A. tumefaciens*–Ti plasmid system is the most widely utilized system for introducing DNA molecules into dicots, most monocots are not susceptible to this bacterium. This limitation restricts the applicability of the Ti plasmid as a vector for transforming monocots, especially the important crop plants of the Gramineae family such as maize, rice, wheat, and sorghum. During the last few years, several alternative methods have been developed for transforming monocots. Well-established general methods for transforming protoplasts of monocots include the use of polyethylene glycol (PEG) and electroporation. The application and limitation of several other systems such as liposome fusion, sphaeroplast fusion, and the use of DNA viruses as gene vectors have been briefly discussed by Potrykus et al. (1985a) and will not be elucidated here. We will concentrate on discussing potentially useful methods for transforming intact cells or organs, thus bypassing the problem associated with the requirement of using protoplasts. These methods include DNA injection into the ovule, limited use of *A. tumefaciens*, and application of the particle bombardment process.

2.2.1 Methods for Transforming Protoplasts

2.2.1.1 PEG Method. PEG is a chemical used to promote protoplast fusion. Since PEG affects the membrane of protoplasts, it was tested for its ability to promote uptake of DNA molecules by protoplasts. This method was first worked out with tobacco because in this system it is relatively easy to isolate protoplasts and possible to regenerate plants from protoplasts. To monitor DNA uptake, Pazkowski et al. (1984) used a selectable hybrid plasmid that included the protein-coding region of the Tn5 aminoglycoside phosphotransferase type II gene, APH(3′)II or NPTII, under the control of CaMV promoter (similar to that in Figure 2–1). The results showed that a foreign gene could be stably integrated into plant genomic DNA and could be expressed in selected, kanamycin-resistant, protoplast-derived cell clones. Tobacco plants regenerated from protoplasts were phenotypically normal and fertile, and they maintained and expressed the foreign genes throughout the development of vegetative and generative organs.

A more efficient chemical technique for protoplast transformation has been reported recently (Negrutiu et al. 1987). It is based on the synergistic interaction of PEG and $MgCl_2$. The technique yielded up to 1,400 transformants per 3×10^5 treated *Nicotiana tabacum* protoplasts. However, using *Nicotiana plumbaginifolia*, the frequencies were tenfold lower, indicating that the "competence" for transformation has a species-specific component.

Direct Gene Transfer into Wheat Protoplasts. Protoplasts isolated from cultured cells of *Triticum monococcum* were incubated in the presence of PEG and a plasmid carrying a selectable marker gene in the procedure developed by Lorz et al. (1985). The selectable chimeric gene consisted of the protein-coding region of NPTII, the nopaline synthase promoter (pNOS), and the polyadenylation signal of the octopine synthase gene. Stably transformed wheat cells were selected for in medium containing kanamycin and subsequently identified by NPTII enzyme activity (Lorz et al. 1985). The procedure used for the incubation of protoplasts with plasmid DNA was based on the method described by Krens et al. (1982) for transforming dicots with Ti-plasmid DNA. *T. monococcum* protoplasts (1 ml) were mixed with 0.4 ml medium F-PEG containing the salts of medium F (Krens et al. 1982) and 40% (v/v) PEG1500, 10 μg pBL1103-4 plasmid DNA, and 50 μg carrier DNA.

Transformed protoplasts were cultured at 4×10^5/ml at 27°C in the dark. Five to seven days later, the medium was solidified by adding the same volume of agarose medium containing 1.2% agarose. For selection purposes, 75 mg/l kanamycin sulfate was added to the medium. After about two months in this bead-type culture system (Shillito et al. 1983), proliferating colonies were transferred onto agar-solidified B5-1 medium containing 100 mg/l kanamycin sulfate. The transformation experiments using 18×10^6 protoplasts yielded ten kanamycin-resistant colonies (Lorz et al. 1985). Regeneration of plants from protoplast-derived wheat colonies has not yet been achieved.

Direct Gene Transfer to Italian Ryegrass Protoplasts. Definitive evidence was presented by Potrykus et al. (1985b) for stable gene transfer to cultured cells in another plant of the Gramineae family, *Lolium multiflorum* (Italian ryegrass), using DNA transformation of protoplasts. A plasmid consisting of the 35S promoter region of CaMV and the NPTII gene was used for transformation using the PEG method just described. The only modification was that the protoplasts (2×10^6) were subjected to a heat shock at 45°C for five minutes followed by ten seconds at 0°C immediately prior to the transformation treatment. Two weeks after the DNA treatment, the cultures were sedimented and the cells taken up in the same medium but supplemented with 25 mg/l G-418 for selection. A total of 71 resistant colonies were recovered from the transformed lines. Of these, 36 reached a fresh weight of 10 g after ten weeks of growth.

2.2.1.2 Electroporation Method. The electroporation method is based on the use of short electrical impulses of high field strength that increase the permeability of protoplast membrane (Zimmermann and Vienken 1982). This process has been used to transfer genes into protoplasts. It was first successfully applied in the uptake of DNA by animal cells (Neumann et al. 1982; Potter et al. 1984). The next application was in the efficient transfer of DNA into tobacco protoplasts. On the order of 2% of all colonies recovered without selection were transformed. This value is one-thousand-fold higher than that previously reported for direct gene transfer (using PEG only) to plant protoplasts. The improvements upon the method include treatment of protoplasts with a high-voltage electric pulse (initial field strength of 1.25 kV/cm), optimization of the PEG concentration (13% w/v), addition of PEG after addition of DNA, and application of a heat shock (five minutes at 45°C) (Shillito et al. 1985).

Transfer of Genes into Maize Protoplasts. Two plasmids were constructed and used to monitor DNA transfer into maize protoplasts by electroporation by Fromm et al. (1985 and 1986). The pNOSCAT plasmid consisted of the promoter region of the nopaline synthase (NOS) gene fused to the CAT gene (Alton and Vapnek 1979). The NOS poly(A) signal was located at the 3′ side of the CAT gene. The pCaMVCAT plasmid was identical to pNOSCAT except that the NOS promoter region was replaced by the CaMV 35S promoter (Hohn et al. 1982). Protoplasts were isolated from rapidly growing black Mexican sweet (BMS) corn suspension cells by incubating with cellulase and hemicellulase.

For electroporation, each sample of 3×10^6 protoplasts was resuspended in 1 ml of the desired concentration of a phosphate-buffered saline (or of a Hepes-buffered saline) containing 0.2 M mannitol. The optimal DNA concentration was 40 μg/ml. Electroporation was carried out by using a model 494 power supply (ISCO, Inc., Lincoln, NB) at 200 V and a 100-millisecond pulse length. Forty-eight hours later, extracts were prepared from the electroporated protoplasts and were assayed for CAT activity. High levels of CAT activity were found in maize protoplasts electroporated with pCaMVCAT, but much lower activity was obtained with pNOSCAT (Fromm et al. 1985).

Stable integration of the plasmid pCaMVNEO in the transformed maize protoplasts was achieved by regenerating cells and calli from the protoplasts. After BMS maize protoplasts were electroporated in the presence of the plasmid, approximately 1% of the protoplasts reformed their cell walls and divided after one week. Two weeks after electroporation, fresh medium containing kanamycin sulfate (100 μg/ml) was added. Microcalli appeared after two more weeks, and eight weeks later, the calli reached a weight of 3–6 g. The highest frequency observed was 161 kanamycin-resistant colonies per 2×10^6 initial protoplasts.

The kanamycin-resistant calli were shown to contain a NPTII-like phosphotransferase activity. Southern blot analysis of DNA from these calli showed hybridization, with a 1-kb NPTII fragment used as the probe. With undigested genomic DNA, the hybridizing band was larger than 20 kb, which indicated integration of pCaMVNEO sequences into maize chromosomes (Fromm et al. 1986).

Transfer of Genes into Rice, Wheat, and Sorghum Protoplasts. Two plasmids were used by Ou-Lee et al. (1986) to monitor DNA transfer by electroporation into protoplasts from the three other important crop plants: rice, wheat, and sorghum. The p35S-CAT plasmid (Morelli et al. 1985) consisted of a CaMV promoter region, the CAT coding region as the reporter gene, and a 3′ noncoding region from the gene for ribulose bisphosphate carboxylase small subunit (rbcS) (see Figure 2–1). The pCopia-CAT plasmid (DiNocera and Dawid 1983) consisted of a *Drosophila* copia LTR promoter, a CAT coding region, and a 3′ noncoding region from simian virus 40 (SV40). Protoplasts were isolated from rapidly growing suspension cells from *Oryza sativa* (cv. Tai-nung 67) seeds and anther-derived rice culture, *T. monococcum* cell line R-TM 1066, and sorghum bicolor callus.

For electroporation, two different systems were compared. One system (FH-GCA) consisted of a pulsar pulse generator connected to a helical fusion chamber (GCA, Chicago). The reaction chamber contained two platinum electrodes, which were immersed in the sample, as in most systems. The other electroporation system was a Baekon 2000 Advanced Gene Transfer system (Baekon, Saratoga, CA). The latter system, which operated in the noncontact mode, was easier to use and was found to be superior to the FH-GCA system: both the percentage of protoplast survival and the CAT activity were higher. Isolated protoplasts were resuspended in electroporation buffer at 2×10^6/ml. They were mixed with the plasmid DNA (20 μg/ml) and carrier calf thymus DNA (50 μg/ml) in a total volume of 200 μl. It was found that 10 kV, 65.5-microsecond pulses repeated 20 times was the optimal condition for rice, wheat, and sorghum protoplasts. Forty hours after electroporation, high levels of CAT activity were found in protoplasts from all three monocots. Both the p35S-CAT plasmid and the pCopia-CAT plasmid were equally effective.

2.2.1.3 Stable Integration of Transferred Genes and Regeneration of Rice Plants

Stable Integration of a Transferred Gene in Rice. Protoplasts isolated from suspension cultures of rice cells could be transformed with a plasmid in the presence of PEG. The plasmid consisted of the nopaline synthase promoter, the APH(3′)II structural gene, and the terminator region from CaMV. Colonies capable of growing in medium containing 100 μg/ml kanamycin sulfate were selected, and they grew into a callus. A transformation frequency of approximately 2–3% was obtained. The enzyme Tn5 aminoglycoside phosphotransferase was found in the kanamycin-resistant callus. Southern blot hybridization of DNA prepared from transformed calli showed positive results using a portion of the APH(3′)II-containing plasmid as the probe (Uchimiya et al. 1986).

Regeneration of Rice Plants from Untransformed Protoplasts. Recently, successful regeneration of rice plants from protoplasts has been achieved by several groups: Fujimura et al. (1985), Coulibaly and Demarly (1986), Yamada et al. (1986), Abdullah et al. (1986), and Kyozuka et al. (1987). These works represent substantial progress after many years of experimentation. So far, rice is the only one of the four most-important crop plants (rice, wheat, corn, and sorghum) in which regeneration of plants from protoplasts has been achieved. However, even with rice (*O. sativa*), only a small number of *japonica* rice varieties can be regenerated. Furthermore, the efficiency of regeneration is still relatively low. More progress is needed to improve the efficiency of regeneration, to broaden the range of regenerable rice varieties, and to extend the techniques to most of the common cultivars.

2.2.2 Methods for Transforming Intact Cells or Organs

In principle, the PEG method and the electroporation method are applicable to gene transfer into the protoplasts of a wide range of plants. However, the regeneration of plants from protoplasts remains a difficult problem. Even in rice, in which regeneration has been achieved, it takes approximately eight months and includes the use of rather difficult tissue culture techniques to regenerate mature plants from transformed protoplasts. It would be simpler and faster to regenerate plants from intact cells or organs instead of protoplasts. Recently, several methods have been developed that allow direct gene transfer into intact cells or organs of monocots, thus bypassing the requirement of protoplasts and the associated problems.

2.2.2.1 Injection Methods. A method has been reported by Zhou et al. (1983) for the introduction of exogenous DNA into cotton embryos. DNA was injected into the axial placenta from the top of the ovary about a day after self-pollination. The DNA appeared to have transformed the embryos, probably by entering the ovule and following the path along which the pollen tube grows. The ovule then developed into a mature seed that carried certain morphological characteristics of the donor plant.

Injecting DNA into the Rice Ovule. DNA of the donor plant *Spartina anglica* was injected into the rice ovule two to four days after self-pollination in a procedure developed by Duan and Chen (1985). The frequency of transformed seeds was 17%. However, only about 1% developed into plants that carried distinct morphological characteristics of the donor plant. This method was difficult to use on rice because the ovule is very small compared with that of cotton.

An alternative method was then developed that was simpler and gave higher transformation frequency. In this method, the pistil of the rice floret was cut off two to three hours after self-pollination and a drop of DNA solution containing donor DNA from purple rice was placed at the site of excision. Part of the DNA solution entered the ovule through or from the outside of the pollen tube. One hundred sixty-eight florets were treated and 53 seeds were obtained. The seeds were germinated, and morphological changes of the plants and their seeds were recorded. Of the 53 seeds, 46 grew into mature plants, 19 of which had purple glume and appeared in the first generation (F_1). More than 80% of the F_2, F_3, and F_4 plants showed purple-colored glume. Although phenotypic markers are not as reliable as molecular markers, the results suggested that foreign DNA could be transferred and possibly integrated into the genome of the recipient plants.

Injecting DNA into Rye Floral Tillers. Previous work on the development of the male germ line of rye has shown that at around 14 days before the first meiotic metaphase the archesporial cells are highly sensitive to colchicine injected into the developing floral tillers. Based on this observation, de la Pena et al. (1987) injected DNA carrying a dominant selectable marker gene into rye plants. The plasmid pLGVneo1103 carrying the APH(3′)II gene with the NOS promoter was injected into developing floral tillers of the diploid rye cultivar JNK. The seeds set on the injected tillers were screened for kanamycin resistance. From the 3,023 seeds derived by cross-pollinating 98 plants injected with the plasmid, seven seedlings remained green after ten days of growth on 10 μg/ml kanamycin-containing medium. Control seeds from untreated rye plants germinated in the presence of kanamycin gave almost totally white seedlings after ten days.

The seedlings that appeared to be kanamycin resistant were assayed for the presence of APH(3′)II enzymatic activity. Two of the seven plants showed APH(3′)II activity. To confirm that the activity was due to the presence of the introduced pLGVneo1103 DNA in the plant genome, DNA/DNA hybridization was carried out. The result showed that the two APH(3′)II-positive plants gave a 3-kb hybridizing band, corresponding to the expected DNA fragment from the plasmid. Thus, it is possible to introduce new genetic information into rye germ cells and to recover seeds capable of growing into normal plants in which the foreign gene is present and expressed.

2.2.2.2 Methods Based on the *A. tumefaciens*–Ti Plasmid System

Wounding of Seedlings of Zea mays. Seedlings of *Zea mays* were wounded by several incisions in the area that extends from the base of the scutellar node through the mesocotyl in a method described by Graves and Gôldman (1986). A total of 10^8 *A. tumefaciens* cells (strain B6) were dripped into the wound sites on the surface of the germinating seedling. After receiving an inoculum, the seedlings were incubated on agar at 27°C for 7 to 14 days. Octopine production was demonstrated in cell-free extracts of B6-transformed material but not in its buffer-treated control. This result was confirmed by showing that opine synthase activities in infected seedlings were T-DNA transfer dependent. Strain JK195, which carries a mutation in the critical *vir* region and therefore cannot convey the necessary portion of the Ti plasmid to its hosts yielded no opine in the transformed material. Thus, a commercially important crop plant, *Z. mays*, can be directly transformed by *A. tumefaciens*.

Injecting Agrobacterium *into Stems of Maize Plants*. A series of plasmids containing a tandemly repeated dimer of maize streak virus (MSV) DNA was constructed and mobilized into *Agrobacterium* by triparental mating. Ten-day-old seedlings of *Z. mays* were used as the recipients of 20 μl of a suitable *A. tumefaciens* suspension that was injected into the stems (Grimsley et al. 1987). It was observed that whole maize plants developed symptoms of viral infection if inoculated with strains of *Agrobacterium* carrying tandemly repeated copies of MSV genomes in their T-DNA. Mutant *Agrobacterium* strains defective in the ability to transfer T-DNA and naked DNA of this geminivirus failed to transmit MSV DNA using this method. Thus, suitably cloned MSV DNA in *Agrobacterium* can transfer DNA into maize plants, as detected by this very sensitive assay system.

2.2.2.3 Particle Bombardment Method. A novel method of introducing new genetic material into living plant cells has been developed by Sanford et al. (1987) using a particle gun that can accelerate microscopic tungsten projectiles to initial velocities of about 1,400 ft/sec (Figure 2–2). At this speed, the tungsten projectiles can penetrate the cell wall of intact plant cells and can carry RNA or DNA into the cells (Klein et al. 1987). The advantage of this method over DNA injection into floral organs or microinjection into protoplasts or cells is that with each bombardment, thousands of tungsten particles are accelerated at the same time, thereby delivering DNA into many cells or organs simultaneously. This method delivers DNA into intact cells, thus circumventing the difficulties associated with working with protoplasts. Another advantage of the particle bombardment method is the universality of its application. Neither cell type, size, shape, nor presence or absence of cell walls should significantly alter its effectiveness.

Delivery of Nucleic Acids into Onion Cells. A particle gun was used to accelerate tungsten microprojectiles (spherical particles 4 μm in diameter on average) into intact epidermal cells of *Allium cepa* (onion) by Klein et al. (1987). About 90% of the cells in a 1-cm^2 area of *A. cepa* epidermal tissue (with about 2,000 cells of around 100 μm in diameter) sustained microprojectiles. *A. cepa* survived well, with penetration of up to five microprojectiles per cell. To demonstrate that nucleic acid can be delivered into cells by this method, RNA isolated from tobacco mosaic virus was adsorbed to the surface of 4-μm tungsten particles prior to their acceleration into *A. cepa* cells. Expression of the viral RNA was indicated by the presence of viral inclusion bodies (crystallized virus particles) in 30–40% of the cells three days after bombardment.

Using the particle gun, DNA can be delivered into intact onion cells, resulting in transient expression of a foreign gene (Klein et al. 1987). Tungsten microprojectiles were coated with the plasmid p35S-CAT containing a gene that encodes CAT. Two days after bombardment, extracts from onion epidermal tissue showed very high levels of CAT activity. Thus, the particle gun method can deliver foreign nucleic acids into plant cells. The process does not require cell culture or the pretreatment of the recipient tissue in any way. Onion cells were chosen as a model system in which to test the method because of their large size and the fact that they are found in a single layer in the epidermal layer, allowing them to be easily stripped from the underlying onion bulb tissue.

Delivery of DNA into Maize Tissue Culture Cells. Several new plasmids were constructed by Fromm et al. (Callis et al. 1987) that yielded higher levels of expression in plant cells, and thus served as more-sensitive tools in detecting low-frequency gene transfer. The plasmid pCaMVI$_1$CN included the CaMV 35S promoter, the *Adh1* intron 1, and the CAT gene as the reporter gene. It was previously found that the level of CAT gene expression in transformed maize protoplasts was increased over one-hundred-fold by adding an *Adh1* intron fragment (I$_1$).

The particle bombardment method was found to deliver DNA efficiently into small clumps of maize cells derived from cell cultures of the cultivar BMS and one of two embryogenic cell lines using pCaMVI$_1$CN as the vector by Klein et al. (1988). High levels of CAT activity in a transient assay were observed following the bombardment of the cell cultures with pCaMVI$_1$CN, but not when the same vector (pCaMVCN) lacking the intron 1 from alcohol dehydrogenase 1 gene of maize is used. CAT activity was detected in cell cultures bombarded with tungsten particles having an average diameter of 1.17 μm, but not in those bombarded with particles having a 0.6-μm or 2.4-μm average diameter. Bombarding the same sample three times was found to enhance CAT activity severalfold. The results demonstrate that the particle bombardment method can be used to deliver foreign DNA into intact cells of maize, an important monocot. Because it circumvents the difficulties

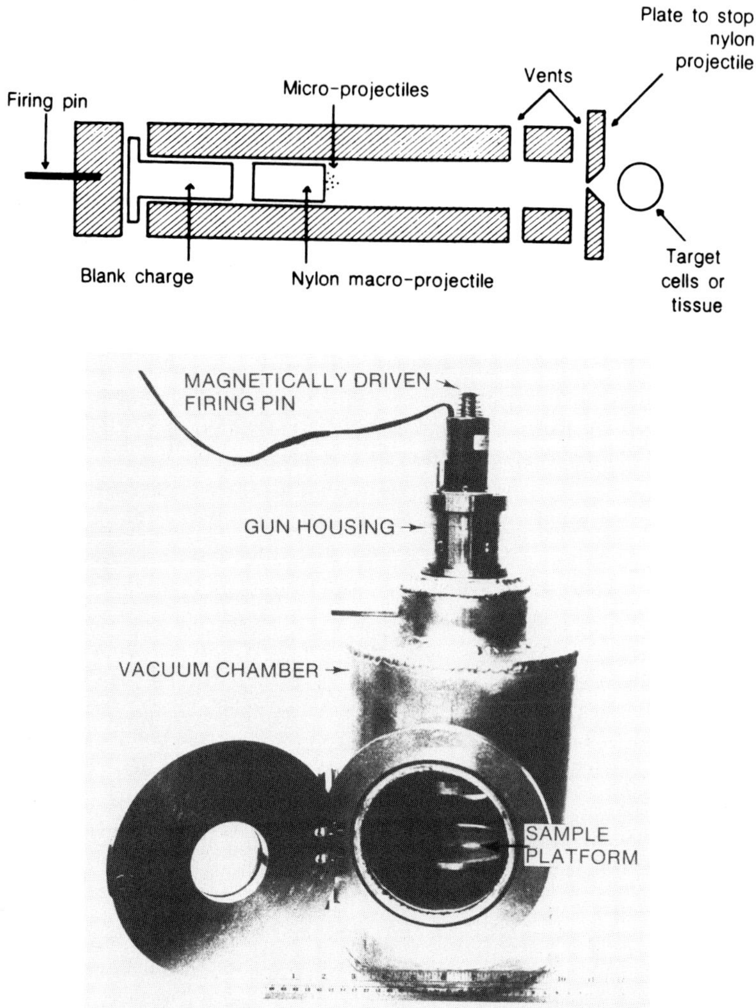

FIGURE 2–2 The particle gun for acceleration of microprojectiles. (Top panel) A schematic diagram of the particle gun. The particle gun consists of the following parts: (1) a gun barrel (about 12 cm in length) that allows a polyethylene macroprojectile to be accelerated, (2) a stopping plate with a small orifice in the center to stop the polyethylene macroprojectile while allowing the microprojectiles (tungsten powder) to continue to accelerate toward the target cells, and (3) a gun powder charge that can be fired by a firing pin. All of the above components of the particle gun are housed in a vacuum chamber to allow removal of air from the chamber. It is necessary to apply a partial vacuum (0.03–0.1 atm) to minimize the deceleration of the microprojectiles (especially those with an average diameter of 1.2 μm or smaller) due to air resistance. To carry out the particle bombardment, the following steps are necessary: (1) Approximately 1.25 mg of tungsten particles with an average diameter of 1.2 μm (GTE Sylvania, Precision Materials Group; Towanda, PA) are

associated with regenerating plants from protoplasts, the particle bombardment method may provide significant advantages over existing DNA delivery methods for the production of transgenic maize plants.

Delivery of DNA into Rice, Wheat, and Soybean Tissue Culture Cells. The plasmid pAI$_1$GusN (Klein et al. 1988) was used by Wang et al. (1988) to deliver DNA into rice, wheat, and soybean tissue culture cells. This plasmid included the promoter region of maize *Adh1* gene (A), the intron 1 (I$_1$) of maize *Adh1* gene, the β-glucuronidase gene (Gus) as the reporter gene, and the 3′ noncoding region of the NOS gene (N).

Using the plasmid pAI$_1$GusN and small clumps of cells derived from potentially embryogenic callus of rice and wheat, transient expression has been obtained following particle bombardment with 1.2-μm (average diameter) tungsten particles (Wang et al. 1988). Detection of the β-glucuronidase activity in the rice or wheat cells was carried out by an in situ staining method (Jefferson et al. 1986). Two days after bombardment by DNA-

mixed with 2.5 μg of plasmid DNA (carrying the reporter gene, etc.) in a volume of 25 μl in a 1.5 ml Eppendorf tube. After mixing, 25 μl of 2.5 M $CaCl_2$ is added, followed by 10 μl of 0.1 M spermidine. Following incubation at room temperature for 10 minutes, the mixture is centrifuged at 10,000 rpm for 1 minute and about 25 μl of the supernatant solution is removed and discarded. (2) Just before use, the tungsten particles with the adsorbed DNA at the bottom of the tube are mixed by briefly (about 1 second) touching the outside of the Eppendorf tube to the probe of a sonicator. About 2.5 μl is pipetted and placed in the center of the front surface of a high-density polyethylene macroprojectile (6 mm × 8 mm). The macroprojectile is then placed into the barrel of the particle gun. (3) A blank gun powder charge (#1 gray extra light, Speed Fasteners, Inc., St. Louis, MO) is loaded into the gun barrel behind the macroprojectile. (4) Tissue culture cells (about 100 mg fresh weight) are pipetted on top of a piece of #4 Whatman filter paper to cover an area of about 3-cm diameter evenly. The filter paper is then placed on top of three other pieces of filter paper in a 10-cm-diameter plastic Petri dish. (5) The Petri dish is placed inside the vacuum chamber on the sample platform. Vacuum is applied gradually so that a partial vacuum of 0.05–0.1 atmospheres is achieved in about 30–40 seconds. Partial vacuums (0.03–0.1 atm) do not appear to be detrimental to the plant tissues that we have studied thus far. (6) The firing pin device of the particle gun is activated by manually depressing it to cause the detonation of the gun powder charge, leading to the acceleration of the macroprojectile down the barrel of the device to the stopping plate. Upon impact, the microprojectiles continue to propel from the front surface of the macroprojectile toward the cells or tissues through a small aperture in the stopping plate. Following bombardment, the vacuum in the chamber is released (over 30 seconds) and the Petri dish removed from the apparatus. (Bottom panel) A photograph of the particle gun (particle acceleration device) currently in use. The gun housing is about 12 cm in length. The diameter of the sample platform is about 11 cm, and the outside dimension of the lower part of the vacuum chamber is about 18 cm.

microprojectiles, the staining solution for assaying β-glucuronidase activity was added to the cells or tissues: 100 mM sodium phosphate (pH 7.0), 5 mM potassium ferricyanide, 0.3% 5-bromo-4-chloro-3-indolyl-β-D-glucuronic acid (X-GLU), and 0.5% triton X-100. After incubation at 27°C for 24 to 48 hours, some blue spots were visible to the naked eye, but were more easily examined under a low-power (40 ×) microscope. When a cell suspension containing approximately 20 mg of cells (fresh weight) was spread over a 1.5-cm^2 area of filter paper, up to 200 blue cells (or clumps of cells) were observed. Expression of CAT gene transferred into intact cells of rice, wheat, and soybean was also obtained after using the particle bombardment method. Currently under study is the regeneration of rice and wheat cells transformed by the particle bombardment method (without treatment by triton and the X-GLU substrate, which kills the cells). These results demonstrate that this method can be used to deliver foreign DNA into intact cells of three agronomically important crops: rice, wheat, and soybean.

Delivery of DNA into Rice Embryo Sections. By introducing the plasmid pAI_1GusN into sections of rice embryo, transient expression in rice cells was detected by the in situ staining method for detecting β-glucuronidase activity (Jefferson et al. 1986). The experiments were carried out using embryos—from mature rice seeds that were first soaked in distilled water for 8 hours—that were cut into thin sections (about 300-μm thick). One hundred embryo sections containing shoot apex were placed on a piece of damp filter paper in the presence of the Murashige-skoog (MS) medium with 0.5% agar. After incubation at 27°C for 18 hours, they were all bombarded simultaneously with 1.2-μm (average diameter) tungsten particles carrying pAI_1GusN DNA. Two days later, the β-glucuronidase activity was assayed by the in situ staining method. Following incubation with the assaying solution at 27°C for 24 to 48 hours, blue spots on the embryo section were visible to the naked eye. When examined under a microscope, even more blue spots of smaller sizes were visible on almost every embryo section. Approximately 25% of the embryo sections showed one or more blue spots. Among 183 embryo sections that received the Gus gene, 60% of them showed one blue spot. However, some embryos showed up to ten spots. The size of the blue spots varied in diameter from around 15 μm (small spots, comparable to the size of rice meristematic cells) to 80 μm (large spots). The bombarded embryo sections (not treated with the staining solution) readily germinated. They formed shoots and roots and grew into rice plantlets within two weeks. So far, transformed cells and regions have been observed in rice shoots and plantlets (Cao et al. unpublished observations). Since the embryo section contains many cells, however, the foreign gene may not be present in many cells or in every resulting organ in the subsequent plantlets or mature plants. On the other hand, transgenic plants showing a mosaic pattern of transformed cells may be useful in studying the process of morphogenesis.

In summary, preliminary promising experiments show that the particle bombardment method is suitable for transforming sections of rice embryos. Transient expression of the introduced foreign gene has been detected, and the potential for stable expression is currently being studied.

2.3 CONCLUSIONS AND FUTURE PROSPECTS

By 1985, methods for transforming and regenerating model dicots such as tobacco and petunia had been well established by a number of laboratories. Yet, there was little or no progress for transforming and regenerating important crop plants such as maize, wheat, and rice. Now, in 1987, two effective methods (PEG and electroporation) have become generally applicable to a numer of protoplasts, including those from maize, wheat, rice, and sorghum. It is now possible to regenerate rice plants from protoplasts. Regeneration of plants from protoplasts, however, remains an obstacle. Several additional methods (such as DNA injection and particle bombardment) have shown promise as general methods for directly transforming intact plant cells or organs, thus bypassing the problem of regeneration of plants from protoplasts. With further developments, the particle bombardment method may prove to be both effective and generally applicable to all important crop plants. Moreover, these methods may be used to transform a wide variety of plant species, including cells from vegetables and trees, that cannot be readily transformed or regenerated today. Some of these methods may also be used to transform intact cells of other organisms such as certain species of plant-attacking pathogenic fungi.

In conclusion, the prospects for applying such methods to the transformation of many types of plant cells are bright, although some methods still need considerable improvement. The ability to transform and regenerate a wide variety of plants will have far-reaching benefits for both basic research and applied plant biotechnology.

ADDENDUM

Recently, transgenic rice plants were obtained in my laboratory by using two independent methods. First, protoplasts isolated from cultured rice cells were incubated with PEG and the plasmid pAl_1GusN. Out of 400 regenerated plants, 86 were transgenic as shown by the presence of the plasmid DNA and β-glucuronidase activity (Zhang and Wu, *Theor. Appl. Genet.*, in press 1988). Second, we have obtained transgenic rice plants following a modification of the pollen-tube-pathway technique of Duan and Chen (1985). We excised the pistil from 259 rice florets and applied to each floret a drop of DNA solution containing the plasmid p35S-NPTII-Hist3 to the excised style surface. Forty days later, 54 seeds were harvested. After germination, rice tissues from 10 plantlets gave NPTII-positive Southern hybridization signals, and four of these plantlets also showed NPTII enzyme activity (Luo and Wu, *Plant Mol. Biol. Reptr.*, July 1988).

REFERENCES

Abdullah, R., Cocking, E.C., and Thompson, J.A. (1986) *Bio/Technol.* 4, 1087–1090.

Alton, N.K., and Vapnek, D. (1979) *Nature* 282, 864–869.

An, G. (1987) *Methods Enzymol.* 153, 292–305.

Bevan, M., and Chilton, M.-D. (1982) *Annu. Rev. Genet.* 16, 357–384.

Buchanan-Wollaston, V., Passiatore, J.E., and Cannon, F. (1987) *Nature* 328, 172–175.

Callis, J., Fromm, M., and Walbot, V. (1987) *Genes Dev.* 1, 1183–1200.

Coulibaly, M.Y., and Demarly, Y. (1986) *Z. Pflanzenzuchtg.* 96, 79–81.

Crossway, A., Oakes, J.V., Irvine, J.M., et al. (1986) *Mol. Gen. Genet.* 202, 179–185.

Deblaere, R., Reynaerts, A., Hofte, H., et al. (1987) *Methods Enzymol.* 153, 277–292.

de la Pena, A., Lorz, H., and Schell, J. (1987) *Nature* 325, 274–276.

DiNocera, P.P., and Dawid, I.B. (1983) *Proc. Natl. Acad. Sci. USA* 80, 7095–7098.

Duan, X., and Chen, S. (1985) *China Agricult. Sci.* 3, 6–9.

Fraley, R.T., Horsch, R.B., Matzke, A., et al. (1984) *Plant Mol. Biol.* 3, 371–378.

Fraley, R.T., Rogers, S.G., and Horsch, R.B. (1986) *CRC Crit. Rev. Plant Sci.* 4, 1–46.

Fromm, M., Taylor, L.P., and Walbot, V. (1985) *Proc. Natl. Acad. Sci. USA* 82, 5824–5828.

Fromm, M.E., Taylor, L.P., and Walbot, V. (1986) *Nature* 319, 791–793.

Fujimura, T., Sakurai, M., Akagi, H., Negishi, T., and Hirose, A. (1985) *Plant Tissue Cult. Lett.* 2, 74–75.

Gheysen, G., Dhaese, P., Van Montagu, M., and Schell, J. (1985) in *Genetic Flux in Plants* (Hohn, B., and Dennis, E.S., eds.), pp. 12–27, Springer, New York.

Goodman, R.M., Hauptli, H., Crossway, A., and Knauf, V.C. (1987) *Science* 236, 48–54.

Graves, A.C.F., and Goldman, S.L. (1986) *Plant Mol. Biol.* 7, 43–50.

Griesbach, R.J. (1983) *Plant Mol. Biol. Reptr.* 1, 32–37.

Grimsley, N., Hohn, T., Davies, J.W., and Hohn, B. (1987) *Nature* 325, 177–179.

Herrera-Estrella, L., Depicker, A., Van Montagu, M., and Schell, J. (1983) *Nature* 303, 184–187.

Hohn, T., Richards, K., and Lebeurier, G. (1982) *Curr. Top. Microbiol. Immunol.* 96, 193–236.

Horsch, R., Fry, J., Hoffmann, N., et al. (1985) *Science* 227, 1229–1231.

Jefferson, R.A., Burgess, S.M., and Hirsh, D. (1986) *Proc. Natl. Acad. Sci. USA* 83, 8447–8451.

Kay, R., Chan, A., Daly, M., and McPherson, J. (1987) *Science* 236, 1299–1302.

Klee, H., White, F., Iyer, V., Gordon, M., and Nester, E. (1983) *J. Bacteriol.* 153, 878–883.

Klein, T.M., Gradziel, T., Fromm, M., and Sanford, J.C. (1988) *Bio/Technol.* 6, 559–563.

Klein, T.M., Wolf, E.D., Wu, R., and Sanford, J.C. (1987) *Nature* 327, 70–73.

Krens, F.H., Molendijk, L., Wullems, G.J., and Schiperoort, R.A. (1982) *Nature* 296, 72–74.
Kuhlemeier, C., Green, P.J., and Chua, N.-H. (1987) *Annu. Rev. Plant Physiol.* 38, 221–257.
Kyozuka, J., Hayashi, Y., and Shimamoto, K. (1987) *Mol. Gen. Genet.* 206, 408–413.
Lawrence, W.A., and Davies, D.R. (1985) *Plant Cell Rep.* 4, 33–35.
Lorz, H., Baker, B., and Schell, J. (1985) *Mol. Gen. Genet.* 199, 178–182.
Marton, L., Wullems, G., Molendijk, L., and Schilperoort, R. (1979) *Nature* 277, 129–131.
Morelli, G., Nagy, F., Fraley, R.T., Rogers, S.G., and Chua, N.-H. (1985) *Nature* 315, 200–204.
Morikawa, H., and Yamada, Y. (1985) *Plant Cell Physiol.* 26, 229–236.
Negrutiu, I., Shillito, R., Potrykus, I., Biasini, G., and Sala, F. (1987) *Plant Mol. Biol.* 8, 363–373.
Neumann, E., Schaefer-Ridder, M., Wang, Y., and Hofschneider, P.H. (1982) *EMBO J.* 1, 841–845.
Ooms, G., Klapwijk, P., Poulis, J., and Schilperoort, R. (1980) *J. Bacteriol.* 144, 82–91.
Ou-Lee, T.M., Turgeon, R., and Wu, R. (1986) *Proc. Natl. Acad. Sci. USA* 83, 6815–6819.
Paszkowski, J., Shillito, R.D., Saul, M., et al. (1984) *EMBO J.* 3, 2717–2722.
Potrykus, I., Saul, M., Petruska, J., Paszkowski, J., and Shillito, R.D. (1985b) *Mol. Gen. Genet.* 199, 183–188.
Potrykus, I., Shillito, R.D., Saul, M.W., and Paszkowski, J. (1985a) *Plant Mol. Biol. Reptr.* 3, 117–128.
Potter, H., Weir, L., and Leder, P. (1984) *Proc. Natl. Acad. Sci. USA* 81, 7161–7165.
Reich, T.J., Iyer, V.N., and Miki, B.L. (1986) *Bio/Technol.* 4, 1001–1004.
Rogers, S.G., Byrne, M., Horsch, R.B., and Fraley, R.T. (1987) *Methods Enzymol.* 153, 253–277.
Sanford, J.C., Klein, T.M., Wolf, E.D., and Allen, N. (1987) *Particle Sci. Technol.* 5, 27–37.
Shillito, R.D., Paszkowski, J., and Potrykus, I. (1983) *Plant Cell Rep.* 2, 244–247.
Shillito, R.D., Saul, M.W., Paszkowski, J., Muller, M., and Potrykus, I. (1985) *Biotechnol.* 3, 1099–1103.
Steinbiss, H.-H., and Stabel, P. (1983) *Protoplasma* 116, 223–227.
Uchimiya, H., Fushimi, T., Hashimoto, H., et al. (1986) *Mol. Gen. Genet.* 204, 204–207.
Wang, K., Herrera-Estrella, L., Van Montagu, M., and Zambryski, P. (1984) *Cell* 38, 455–462.
Wang, Y.-C., Klein, T., Fromm, M., Cao, J., Sanford, J. and Wu, R. (1988) *Plant Mol. Biol.* (in press).
Willmitzer, L., Simons, G., and Schell, J. (1982) *EMBO J.* 1, 139–146.
Wullems, G., Molendijk, L., Ooms, G., and Schilperoort, R. (1981) *Proc. Natl. Acad. Sci. USA* 78, 4344–4348.
Yamada, Y., Yang, Z.Q., and Tang, D.T. (1986) *Plant Cell Rep.* 5, 85–88.
Zhou, G.-Y., Weng, J., Zeng, Y., et al. (1983) *Methods Enzymol.* 101, 433–442.
Zimmermann, U., and Vienken, J. (1982) *J. Membr. Biol.* 67, 165–182.

CHAPTER 3

Techniques in Plant Cell and Tissue Culture

David A. Evans

Progress in refinement of techniques in the past several years has helped to establish tissue culture and cellular genetic manipulation as valuable tools for the development of new crops. This is an outgrowth of the first commercial use of tissue culture, which was the use of shoot tip propagation, as an alternative to conventional vegetative propagation. When applied to ornamental plants or to estate crops such as oil palm, tissue culture can be used for rapid, consistent reproduction of elite or difficult-to-propagate genotypes (Murashige 1978). Cloning techniques are, however, valuable in the genetic manipulation of only those cash crops or ornamentals of which each individual organism is precious. Hence, cloning would not be useful for the genetic manipulation of field crops such as corn or wheat.

However, several tissue culture techniques that could have commercial application to a wide range of crops have been perfected in recent years. These methods can be used to develop novel variation that has a genetic basis and can be transmitted through seed in a predictable manner. Somaclonal variants and gametoclonal variants may be used to introduce nonspecific variation into food crop species using standardized plant regeneration methods (Evans et al. 1984). While the full genetic nature of these variants has not been firmly established, new plant varieties have already been developed via somaclonal variation. Anther culture or gametoclonal variation can be used as an adjunct to conventional breeding for rapid gene transfer between breeding lines. The high-frequency of production of homozygous doubled-haploid breeding lines of crop species, particularly when anthers of F_1 hybrids are used, has led to development of new plant varieties. Development of haploid plants has already been exploited extensively in China. The combination of cell selection and protoplast fusion offers a method of introducing useful genetic changes, but requires additional basic research before widespread application is possible (Evans 1983).

In systems developed for the enhancement of secondary product synthesis, cell cultures themselves produce useful chemicals for the food industry. Using techniques of genetic manipulation, some cultures have already been identified that accumulate higher concentrations of chemicals than those reached in the intact plant. The application of all these newly developed technologies should result in significant new advances in the development of agricultural products.

3.1 CLONAL PROPAGATION

Innumerable plant species can be propagated in vitro (Murashige 1978). In most cases, clonal propagation is achieved by placing sterilized shoot tips or axillary buds onto a culture medium that is sufficient to induce formation of multiple buds. Shoot tip clonal propagation has already been used commercially to propagate a large number of marketable ornamentals (cf. Oglesby 1978). Murashige (1974) has described a three-stage procedure that normally requires alteration of culture medium or growth conditions between stages. Stage I concerns the establishment of tissue in vitro. Stage II is most important, as it involves the production of multiple shoots. Stage III concerns root formation and conditioning of propagules prior to transfer to the greenhouse. High light intensity is important in stage III (Murashige 1978). In many species, the media and culture conditions are not altered between stage I and stage II (Murashige 1978).

Advantages of cloning plants are the speed of plant multiplication and the quality and uniformity of plants produced. In many species, propagation in vitro is considerably faster than propagation by conventional means. Ferns, bromeliads, orchids, and ornamental trees can all be propagated more rapidly in vitro in controlled environments. For example, nearly all economically important orchids, except *Paphiopedilum*, are easily cloned in vitro (Murashige 1974). Orchids have traditionally been difficult to mass propagate using conventional methods.

Cell culture has already been demonstrated to be commercially important for the propagation of ornamentals, and it may also be useful for the propagation of certain crops. In plants with seed viability problems, breeding, or parent, lines can be regenerated in vitro. Somatic embryos produced in tissue culture can be directly planted into the field using systems already developed for pregerminated seed or transplants. Mutated plants, such as albino mutants or other genetic lines useful in cellular genetics, can be propagated in vitro (cf. Gamborg et al. 1979). Parents used in a breeding or hybrid seed program can also be propagated in vitro. Male sterile mutants have been used to produce commercial F_1 hybrids in a number of crops species. Normally, restorer lines must be used to maintain and propagate the male sterile. This procedure requires a series of sexual crosses with subsequent selection among all progeny for the male sterile types. Unless genetic markers are used, the selection must be completed on mature plants. This selection process is particularly time consuming when the male sterile parent line is under nuclear genetic control. In some species, it is possible to maintain male sterile parent lines and preserve the male sterile character using in vitro propagation to allow for consistent hybrid seed production. Vallos (1984) has used in vitro propagation to maintain trisomics and male sterile lines of pepper. Dore and Michel (1984) developed procedures to maintain male sterile lettuce plants for hybrid seed production. When advanced breeding lines have been identified, it is important to propagate as many seeds or plants as possible to permit variety release. At this point tissue culture may be used in conjunction with other asexual methods to propagate clonal lines for seed production. Such an approach was used in Ecuador to establish elite clones of pyrethrum (*Chrysanthemum cinerariaefolium*) for commercial production (Levy 1981). In some cases cultivated crops, particularly cash crops in which each individual plant is valuable, may be cloned using tissue culture. For example, varieties for greenhouse crop production or tropical estate crops can be propagated in vitro using shoot apexes. In addition, vegetatively propagated crops, such as the artichoke, can be multiplied for direct field planting (Pecaut et al. 1983).

In crops where male sterility is unavailable, production of hybrid seed is expensive. In crosses between divergent varieties, hybrid seed cannot be propagated sexually without genetic segregation in the F_2 population. The frequency of genetic segregation in the F_2 population is inversely correlated with relatedness of the parental lines. Hybrid plants can be propagated on a large-scale using in vitro techniques. Such hybrids may be cheaper to produce than hybrids originating from seed produced by hand pollination.

A business analysis of the economics of propagation of hybrids should consider the cost of transplanting regenerated hybrid plants to the field as opposed to the cost of seeding the field directly with F_1 hybrid seed. Unless plants can be regenerated very inexpensively or are of a species in which transplants are commonly used, F_1 hybrid in vitro propagation is not profitable.

In certain crops and ornamentals, it has also been economically useful to develop in vitro techniques to produce virus-eradicated plants. This can be accomplished by regenerating plants from a virus-free plant. Meristems are generally free of virus infection because they lack vascular differentiation that impairs intercellular movement of viruses and have mitotic cells whose active metabolism precludes viral infection. In order to take advantage of virus-free cells, the smallest-possible meristem explant should be cultured. Small explants do not, however, regenerate or produce roots as well as large meristems. These two factors must be balanced in identifying optimal explant size. Genetic stability as measured by phenotypic and chromosomal stability is usually preserved in plants recovered from cultured meristems. Moreover, novel tests such as highly sensitive enzyme-linked immunosorbent assays (ELISA) are being developed to monitor pathogen-free plants (Giles and Morgan 1987).

Heat treatment is commonly used to inactivate the virus prior to culturing meristems. In many cases, treatment with 34–36°C will result in inactivation of the virus without deleterious effects on the plant. About one half of the known plant viruses, particularly small spherical viruses, can be inactivated with thermotherapy (Hu and Wang 1983). Heat treatment during donor plant growth can result in successful use of larger meristem explants than those from unheated plants (Mullin et al. 1974). In some species, heat treatment is essential for successful virus elimination. In *Chrysanthemum*, 98.5% of progeny from heat-treated plants were virus-free, while only 15% of apexes from room-temperature-grown plants were virus-free (Hollings et al. 1972). Heat treatment can also be applied by incubating tissue cultures at elevated temperatures (Walkey and Cooper 1975).

Viruses have been eliminated using meristems from a number of economical species. Mottle virus has been eliminated from strawberry, potato virus X from potato, mosaic virus from cassava, and cauliflower mosaic virus from cauliflower (Quak 1977). Virus-free plants produced by meristem culture have resulted in significant increases in yield when compared with yield from virus-infected plants (Walkey and Cooper 1975). Results have been particularly favorable in potato (Gregorini and Lorenzi 1974), where yield of virus-free potatoes is 60% higher than that of infected plants. Virus-free plants, though, are not resistant to reinfection. As expected, field-grown plants often become reinfected with virus. Stock plants should be maintained to reproduce virus-free plants.

Plants regenerated from callus may also be free of viral infection. Callus tissue is similar to meristematic tissue in that cell mitosis is rapid and vascular differentiation is incomplete. In callus or cell suspension cultures, viral infection may be completely eliminated via repeated subculture. Plants regenerated from uninfected leaf tissue may also be used to recover virus-free plants. Murakishi and Carlson (1976) regenerated tobacco plants from dark-green uninfected islands of a tobacco mosaic virus–infected leaf. Approximately 50% of regenerated plants were virus-free.

3.2 SOMACLONAL VARIATION

In contrast to shoot tip propagation, which faithfully produces clones, evidence has accumulated suggesting that regeneration of plants from callus, leaf explants, or plant protoplasts results in recovery of somaclonal variants (Evans and Sharp 1986). Regeneration of plants from callus culture has been associated with recovery of aneuploid plants, sterile plants, and morphological variants. The observation that in both potato (Shepard 1982) and sugarcane (Larkin and Scowcroft 1981) somaclonal variants were resistant to diseases suggests that production of this variability in regenerated plants could prove agriculturally useful.

The phenotypic variability recovered in regenerated plants probably reflects both preexisting cellular genetic differences and tissue culture–induced changes. Skirvin and Janick (1976) systematically compared plants regenerated from callus of five cultivars of *Pelargonium* species. Plants obtained from geranium stem cuttings in vivo were uniform, whereas plants from in vivo root and petiole cuttings and plants regenerated from callus were quite variable. Changes were observed when regenerated plants were compared with parent plants in terms of plant and organ size, leaf and flower morphological characteristics, essential oil constituents, fasciation, pubescence, and anthocyanin pigmentation. Long-term cell cultures result in tissue culture–induced variability with regard to chromosome number that is expressed in both callus and plants regenerated from callus (D'Amato 1977). This results in regeneration of aneuploid plants that are commercially useless in sexually propagated species. On the other hand, aneuploidy may not interfere with productivity of asexually propagated crops such as sugarcane and potato.

Plants regenerated from callus of sugarcane have been examined in detail. A wider range of chromosome numbers (2N = 71–300) has been observed in plants regenerated from sugarcane callus (Heinz et al. 1977) than in parent lines. These lines of regenerated sugarcane were examined for disease resistance. Resistance to three diseases (eye-spot disease, Fiji disease, and downy mildew) was observed in some plants regenerated from a susceptible sugarcane clone (Heinz et al. 1977). One regenerated line was simultaneously resistant to both Fiji disease and downy mildew. These lines have been incorporated into a conventional breeding program.

Plants regenerated from mesophyll protoplasts of potato have also been analyzed. Although detailed genetic analysis has not been published, it is apparent that regenerated clones are highly variable. There is evidence that some of the variability is due to chromosomal rearrangements (Shepard 1982). As in sugarcane, lines have been identified that are resistant to diseases (late blight and early blight) to which the parent, Russet Burbank, is susceptible (Shepard 1982).

Plant regeneration has resulted in transient variation (epigenetic) as well as genetic variation. Epigenetic changes are usually induced by culture conditions and are not useful for crop improvement because they are not expressed in progeny of regenerated plants.

To determine accurately the value and genetic basis of somaclones, sexual progeny must be examined. This is accomplished as regenerated plants are raised to maturity in greenhouse conditions, self-fertilized, and seed collected. The first generation of self-fertilized regenerated plants (R_1) may be examined under greenhouse or field conditions. Detailed data on the transmission of variation to sexual progeny are necessary to facilitate the use of somaclonal variation for improvement of sexually propagated crops.

Genetic changes resembling single-gene mutations have now been detected in numerous crops. For example, recessive single-gene mutations are suspected if the variant does not appear in the R_0 plant and if self-fertilized R_1 progeny segregate in an expected 3 : 1 Mendelian ratio for a morphological trait. To confirm that the trait of interest is indeed a single-gene trait, progeny tests should be completed on selected single R_1 plants to identify segregators and nonsegregators and to ensure continued 3 : 1 transmission in R_2 segregating populations. Moreover, mutant plants should breed true in the R_2 generation. When possible, complementation tests should be completed on known mutants of the crop of interest. Detailed genetic analysis to confirm single-gene traits has been completed only for several tomato somaclones (Evans and Sharp 1983) and has been used to map somaclones to specific loci (Evans and Sharp 1986). Three-to-one ratios in R_1 plants, the first indication for this type of analysis, have been reported for maize (Edallo et al. 1981), *Nicotiana sylvestris* (Prat 1983), rice (Fukui 1983), and wheat (Larkin et al. 1984).

In addition to single nuclear gene changes, cytoplasmic genetic changes have also been detected by somaclonal variation. The most-detailed work on the evaluation of cytoplasmic genetic changes via somaclonal variation has been completed by Gengenbach et al. (1977) who evaluated plants for two cytoplasmic traits. Sensitivity to host-specific toxin of *Drechslera maydis* race T, the causative agent in southern corn leaf blight, is associated with all genotypes containing Texas male sterile (cms-T) ctyoplasm. In seed-derived plants these two characteristics are tightly associated and both controlled by mitochondrial DNA (mtDNA). Gengenbach et al. (1977) selected for resistance to toxin in vitro and regenerated resistant plants with the aim of recovering toxin-resistant, cms-T-recombinant breeding lines, but even among tissue culture regenerates resistance was associated with a concomitant reversion to male fertility. However, when the restriction endonuclease pattern of mtDNA was evaluated (Kemble et al. 1982), it was evident that significant changes had occurred in the mtDNA of plants derived from cell culture. This mutation to male fertility and toxin insensitivity has been shown to be a frameshift mutation in mtDNA (Wise et al. 1987).

Variations in mtDNA have also been detected among potato somaclones (Kemble and Shepard 1984). It is not surprising that organelle DNA variants are uncovered among regenerated plants because the number of organelles in a developing shoot apex is much smaller than that in a mature cell (Bendich and Gauriloff 1984). Hence, if a mutation occurs in organelle DNA, it is more likely to become the dominant organelle type during sorting out if the mutant occurs in one of ten organelles than if it occurs in one of one-hundred organelles.

Variant plants with altered chromosome number have been reported by several authors (D'Amato 1977). Polyploidy is the most frequently observed chromosomal abnormality, and the mechanisms for its origin have been discussed (D'Amato 1978). It has been suggested that in aneuploid species such as sugarcane and potato, when duplicate chromosomes are already present, the addition or loss of one chromosome is usually not lethal. Changes in chromosome number are commonly associated with reduced fertility and with altered genetic ratios in progeny of self-fertilized plants. In addition to changes in chromosome number, chromosome rearrangements have been detected by analyzing meiosis in regenerated plants. Translocations have been reported in ryegrass (Ahloowalia 1976), potato (Shepard 1982), and oat (McCoy et al. 1982). Published work also suggests that chromosome deletions, inversions, and other minor reciprocal and nonreciprocal rearrangements occur among regenerated plants. Depending on the ploidy of the crop plant used and the extent of chromosome modification, such regenerated plants could have reduced fertility or altered transmission of genetic characters.

Several authors have suggested that other cytogenetic changes are responsible for the range of variation detected in regenerated plants (Orton 1983). The most-frequently cited and well-documented changes are small chromosome rearrangements. By evaluating meiotic behavior of plants regenerated from cell culture, translocations, deletions, and inversions have all been detected (McCoy et al. 1982). While large changes in chromosome structure have been detected, it is also likely that less-dramatic structural changes that are not detected early occur frequently. Small changes in chromosome structure could alter expression and genetic transmission of specific genes. These could include deletion of one or more copies of a gene, duplications of one or more copies of a gene, or gene conversion during repair processes. In addition, recombination or chromosome breakage may occur in preferential regions or "hot spots" of particular chromosomes, thereby affecting some regions of the genome at a disproportionally higher frequency. Striking variation has also been detected in repetitive DNA (Cullis and Cleary 1986).

Mitotic crossing over could also account for some of the variation detected in regenerated plants. This could include both symmetrical and asymmetrical recombination. Mitotic crossing over may account for the recovery of homozygous recessive single-gene mutations in some regenerated plants (Evans and Sharp 1986). As breeders have previously had access only to variation that is normally transmitted through meiosis, the recovery of products of mitotic crossing over may constitute a unique source of new genetic variation.

In some cases use of specific genetically marked strains has aided the evaluation of plants regenerated from cell culture. Dulieu and Barbier (1982) and Lorz and Scowcroft (1983) have all regenerated plants from *Nicotiana tabacum* with specific chlorophyll-deficiency markers present in heterozygosity. Genetic changes could be detected by the appearance of albino or dark-green regenerated plants. Dulieu and Barbier (1982) reported a high frequency (9.6%) of variant regenerates at the a_1 and y loci and have ascribed these genetic changes to the combination of deletion and mitotic recombination. Lorz and Scowcroft (1983) detected genetic changes in morphogenetic cell colonies using the Su locus. When light-green regenerated plants (Su/su) were self-fertilized, up to 37% of the regenerated plants had segregation ratios that were distorted from the normal 1 : 2 : 1 ratio of dark-green–light-green–albino phenotypes obtained from seed-derived plants. These authors did not speculate on the genetic basis for the altered R_1 segregation ratios. The value of using genetically marked heterozygous donor material was evident in each of these studies, as the authors were able to demonstrate that much of the reported variation was proportional to the duration of culture.

In addition, reports of a molecular biological approach to analyzing somaclones have begun to appear. Brettell et al. (1986a) used a cloned gene of Adh to evaluate an electrophoretic variant in the Adh locus. They were able to demonstrate that the somaclonal variant was the result of an alteration of a single base pair resulting in change of a single amino acid in the polypeptide sequence. Landsmann and Uhrig (1985) used random cloned fragments of potato DNA to probe twelve regenerated potato plants. Two of the twelve plants had alterations resulting in deficiencies in ribosomal RNA (rRNA) genes, although they were not associated with a change in plant morphological characteristics. Such deficiencies in rRNA genes (deamplification) have also been reported for flax (Cullis and Cleary 1986). Brettell et al. (1986b) examined rRNA genes in families of plants regenerated from triticale cell cultures and detected a reduction in rDNA spacer sequences in one family of regenerated plants. These and other reports confirm that somaclonal variation causes genetic changes that range from simple single base pair changes (mutation) to more-complicated chromosomal changes such as deletions, translocations, and changes in chromosome number.

Since single-gene mutations and organelle gene mutations have been produced by somaclonal variation, one obvious strategy is to introduce the best-available varieties into cell culture to select for incremental improvements in existing varieties by somaclonal variation. Somaclonal variation could thus be used to uncover new variants that retain all the favorable qualities of an existing variety while adding one additional trait, such as disease resistance or herbicide resistance. Work with sugarcane and tomato has already suggested that this approach is feasible. A possible sample procedure is outlined in Figure 3–1. In this procedure, improved lines are identified based on selection for specific desirable traits from among R_1 plants. Once new R_1 variants are identified, they are field tested in replicated plots to ascertain stability of the somaclonal variants. For most crops it would also be appropriate to make single plant selections on R_1 plants to ensure stable inheritance. New promising breeding lines can be reintroduced into cell culture to add an additional character or to improve agronomic performance of a selected somaclonal variant. By using this approach, it is possible to produce new breeding lines with desirable traits in a short period of time. The value of somaclonal variation rests in the ability to recover in high frequency new genetic variants with desirable characteristics.

During the course of a somaclonal program, two important selection steps serve as sieves to permit recovery of a population of R_1 plants that are most suitable for a breeding program. (1) The culture medium and plant regeneration protocol provide a sieve for singling out cells from the original explant that possess genetic competence for plantlet regeneration; and (2) greenhouse selection permits identification of those regenerated plants that are capable of undergoing flower and fruit formation and that set seed. This selection eliminates R_0 plants with deleterious genetic information. Hence, the population that is examined in the field is most suitable for rapid variety development.

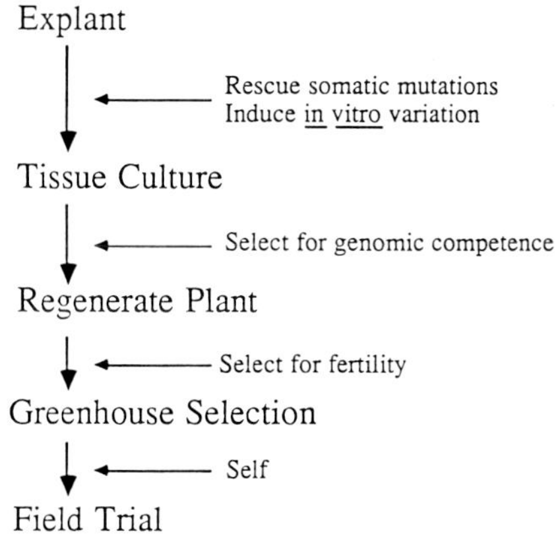

FIGURE 3–1 Application of somaclonal variation to crop improvement.

3.3 GAMETOCLONAL VARIATION

Haploid plants derived from male or female gametes have value in plant breeding because they greatly reduce the time required to obtain homozygous breeding lines. Doubled haploids, which contain two copies of every gene present, have been integrated into conventional breeding programs and have resulted in rapid development of new varieties. Kasha and Reinbergs (1980) have reported that use of doubled haploids reduced the time for release of a new variety of barley from twelve years to five years. Haploid plants can also be used for genetic analysis. Using an aurea mutant, Burk (1970) has demonstrated that single-gene heterozygous diploid plants produce a 1 : 1 ratio of the two haploid parental types following anther culture. In *Nicotiana*, haploids have been used to analyze the number of genes controlling disease resistance (Nakata and Kurihara 1972) and nicotine production (Nakamura et al. 1974), as well as to aid in analysis of mutants induced in vitro (Chaleff 1981) and of somatic hybrids (Schieder 1978).

Haploid plants have been obtained in crop species using conventional breeding methods, anther culture, immature ovule culture, or pollen culture. Because the haploid plants produced by each method may be equally desirable, the method that yields the highest frequency of haploid production should be used.

Haploid plants arise frequently following interspecific hybridization, and haploids produced this way have been integrated into breeding programs in barley, potato, and tobacco. In each of these species, haploids are produced following fertilization of a cultivated crop with a wild relative. In barley, haploids are produced following fertilization of *Hordeum vulgare* with *Hordeum bulbosum*; *H. bulbosum* chromosomes are preferentially eliminated. Recently, the *Hordeum* system has been extended to produce haploids of closely related cereals in which additional species have been used as the female parent. In potato, pollination of *Solanum tuberosum* with wild species pollen, particularly that of *Solanum phureja*, results in parthenogenic development of *S. tuberosum*. Potato doubled haploids and haploids can be used to aid in the introduction of wild species genes into the tetraploid cultivated potato. Haploid tobacco has been produced following pollination with *Nicotiana africana* (Burk et al. 1979). Each of these methods results in production of maternal haploids of the cultivated crop. On the other hand, the indeterminate gametophyte (ig) gene can be used in corn to produce haploids with unique mixtures of nuclear and cytoplasmic DNA. Both parental and maternal haploids can be produced using the ig gene in heterozygosity (Kermicle 1969). This system may be modified in the future to aid in transfer of cytoplasmic DNA between breeding lines without transfer of nuclear genes.

Haploid tissue and plants can be produced at a higher frequency in many species using anther culture. Haploid plants were first produced using this method in *Datura innoxia*. For each species examined, the best developmental stage of cultured anthers must be ascertained to optimize the frequency of haploid production. First pollen grain mitosis is optimal for tobacco and *Datura*, while an earlier stage is required for tomato and a later stage is required for *Brassica* (Sunderland 1974). In some species, the culture medium does not appear to be as important as the anther stage in determining the conditions for haploid production; the nutrient requirements of cultured anthers may be quite simple. On the other hand, in some species (e.g., rice varieties) the response of cultured anthers is medium dependent. Numerous other factors effect the degree of success in anther culture. The growth environment and age of the donor plant are important. Care should be taken to use the first floral buds that appear on the plant. Pretreatment of the anthers after removal from the plant, but prior to culturing in vitro, may be essential. Both cold shock (Nitsch 1974) and heat treatment (Keller and Armstrong 1979) have been proposed to increase the number of haploid plants produced per cultured anther.

Gametoclones are genetic variants from anther culture–derived plants. They are detected when haploids are derived by direct plant regeneration or if regeneration is preceded by callus formation. From a recent review (Sharp et al. 1984), it is evident that most species capable of haploid production proceed via direct plant formation (androgenesis). Taxonomic differences are apparent, though, between plants that undergo androgenesis and those that undergo callus-mediated regeneration. The majority of solanaceous species capable of producing haploid plants undergo direct androgenesis, while the majority of graminaceous species undergo callus-mediated haploid plant production. Commercially important crops for which haploid plants have been produced are listed in Table 3–1.

Since anthers are a mixed population of diploid and haploid cells, all of which are capable of cell reproduction, great effort has been directed toward isolated pollen culture. Pollen can be liberated by squashing anthers in

TABLE 3–1 Crops for Which Haploid Plants Have Been Produced

Alfalfa	Cucumber	Rye
Asparagus	Grape	Strawberry
Barley	Kale	Sugarcane
Broccoli	Peanut	Sunflower
Citrus	Pepper	Sweet potato
Clover	Potato	Tomato
Coffee	Rapeseed	Triticale
Corn	Rice	Wheat

See Sharp et al. (1984) for details.

liquid culture medium and subsequently collected via filtration. Isolated pollen can be cultured using a complex medium or nurse culture (reviewed by Nitsch 1971). Glutamine and serine are important additives to pollen culture medium.

Haploids are integrated into a breeding program as homozygous diploid lines. Diploidized haploid lines can be produced using a colchicine treatment or, alternatively, a leaf midrib culture. Selection from a population of doubled haploids, particularly an F_1 population, can result in rapid fixation of desired characteristics (Figure 3–2). Griffing (1975) has statistically demonstrated that use of doubled haploids, when rapidly obtained, is more efficient than conventional selection.

Doubled haploids have been incorporated into breeding programs to produce new cultivated varieties of barley (Kasha and Reinbergs 1980) and tobacco (Collins and Genovesi 1982). Although not used as extensively in other crop species, doubled haploids have also been used in *Asparagus, Brassica,* rice, corn, barley, wheat, rye, and potato (Nitzsche and Wenzel 1977). In *Asparagus*, anther culture of mature male (XY) plants can be used when chromosomes are doubled to produce supermales (YY). The supermales, when crossed with females (XX), have been very useful in establishing high-yielding, uniformly all-male (XY) lines of *Asparagus*. Doubled-haploid techniques are being utilized in *Brassica* to breed for low erucic

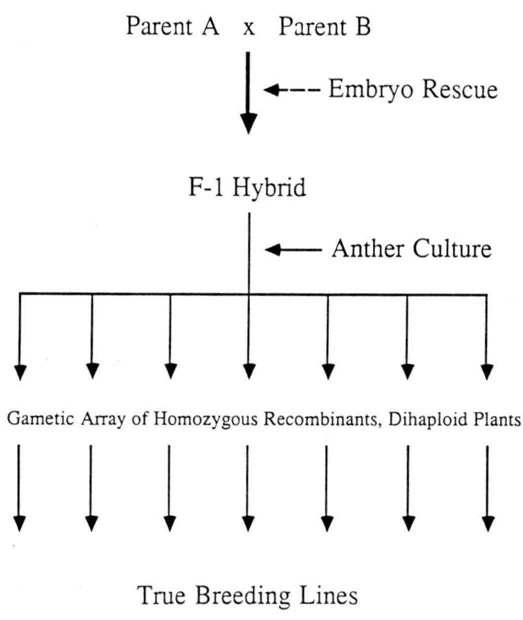

FIGURE 3–2 Hybrid sorting for crop improvement.

acid and glucosinolate content. Varietal development in rice and peppers also currently involves doubled haploids. In corn, doubled haploids are being utilized for analytical breeding. Because of the genetic uniformity of doubled haploids when compared with regular diploid lines, doubled haploids of inbred lines are used for evaluation of the breeding lines to be used as parents to produce F_1 hybrids. Following such analysis, F_1 hybrids are constructed with more confidence (Nitzsche and Wenzel 1977).

Doubled haploids can also be used in genetic analysis to study the inheritance of qualitative and quantitative traits. Because haploids and doubled haploids exhibit the genotype of the gamete, ratios obtained are less complex if doubled haploids rather than diploids are used for genetic analysis (Collins and Legg 1980). Doubled haploids have been utilized for the genetic analysis of several quantitative traits in tobacco. Haploids are also useful in cellular mutagenesis programs because recessive alleles are expressed in haploids while they may be undetected in diploids. Anther culture can also be used to manipulate chromosome number in intraspecific (Schieder 1978) and interspecific (Evans et al. 1983) somatic hybrids and to generate nullisomics useful for genetic analysis (Moore and Collins 1982).

3.4 IN VITRO CELL SELECTION—MUTANT ISOLATION

Cultured plant cells have been used for isolation of mutants in vitro with subsequent plant regeneration (Flick 1983). Mutants could be regenerated directly for incorporation into conventional breeding programs and mutated cell lines could also be useful in the establishment of biochemically stable and genetically marked cell lines for protoplast fusion or DNA uptake experiments. Variant cell lines with direct potential agricultural application include lines selected for pathotoxin (Sacristan 1986) or herbicide resistance (Chaleff 1986) as well as lines selected for salt or cold tolerance. Mutant lines used as cell genetic markers would ideally have genetic lesions that are recessive, to permit identification of genetically modified cells as corrected variants. It has been demonstrated that auxotrophic or temperature-sensitive mutants can be useful in cell genetic studies (Bourgin 1986). Dominant variants may be useful in selecting for genetic modifications that confer resistance to toxic chemicals. Certain plants resistant to amino acid analysis overproduce amino acids. Schaeffer and Sharpe (1987) have recently isolated rice plants with increased lysine in seed storage protein by selecting for resistance to lysine plus threonine and S-(2-aminoethyl) cysteine.

In addition, dominant mutants may be useful in basic cell genetics studies. Cell lines resistant to amino acid analogues may be quite useful in protoplast fusion because protoplasts are susceptible to very low concentrations of these chemicals, permitting discrimination between sensitive and resistant phenotypes and, hence, easy identification of heterokaryons. In lieu of well-characterized induced recessive variants, most researchers have used genetically marked cell lines with lesions in photosynthesis-related genes. Albino mutants occur naturally or may be induced by chemical treatment. Most albinos are recessive, and some albino mutants have been biochemically analyzed in great detail. While mutated and corrected cells following protoplast fusion or DNA uptake cannot normally be distinguished in vitro, corrected cells can subsequently be identified as green regenerating shoots. Isolation of more biochemically marked cell lines would permit analysis of genetic modification experiments.

3.5 PROTOPLAST FUSION

Protoplast fusion has often been suggested as a means to develop unique hybrid plants impossible to achieve via conventional sexual hybridization. While any two plant cells can be fused using a modified polyethylene glycol method, creation of unique somatic hybrid plants is limited by the difficulty of regenerating plants from isolated plant protoplasts (Evans 1983). Interspecies somatic hybrid plants have been produced in the following genera: *Datura, Daucus, Nicotiana, Petunia, Brassica, Medicago,* and *Solanum*. In most cases, these hybrids have been between sexually compatible species and have not been incorporated into conventional programs. Much of the emphasis of early fusion experiments was directed toward production of distant somatic hybrids not obtainable using conventional sexual hybridization.

Several intergeneric hybrid plants have been recovered. In each case, only aneuploid plants were recovered, and gross morphological abnormalities were reported for each combination. These intergeneric combinations included the following: *Solanum* + *Lycopersicon* (Melchers et al. 1978); *Atropa* + *Datura* (Krumbiegel and Schieder 1979); *Daucus* + *Aegopodium* (Dudits et al. 1980); *Arabidopsis* + *Brassica* (Gleba and Hoffman 1980); *Nicotiana* + *Atropa* (Gleba et al. 1982); *Brassica* + *Moricandia* (Toriyama et al. 1987); *Lycopersicon* + *Solanum* (Handley et al. 1986); and *Nicotiana* + *Petunia* (Pental et al. 1986). Unfortunately, none of these fusion products has resulted in crop improvement because the plants were morphologically and chromosomally abnormal. In most cases, a wide range of aneuploid plants was recovered that resulted in a wide range of morphological phenotypes. In the *Daucus* + *Aegopodium* experiment, all the *Aegopodium* chromosomes were lost, resulting in plants with the same chromosome number as *Daucus carota*. However, evidence was presented that a small amount of genetic information was transferred from *Aegopodium* into *Daucus* (Dudits et al. 1980).

Since it has been demonstrated that distant hybrids can be produced by protoplast fusion, attention has more recently been directed toward use of fusion products in crop improvement. Somatic hybridization should be critically analyzed for use as an aid to crop improvement. The primary limitation to application of protoplast technology is certainly the inability to regenerate plants from protoplasts (Evans and Bravo 1983). Numerous hybrids could be proposed to complement cereal and legume breeding programs, but very little success has been reported in plant regeneration of these important crops. Hence, the limitation of plant regeneration from protoplasts precludes application of this technology to cereals and legumes for now.

Techniques have been developed for the production of unique somatic hybrids in the genus *Nicotiana* between species that cannot normally be hybridized (Evans et al. 1981). Production of these somatic hybrids has been directed toward production of agriculturally useful hybrid germ plasm. Protoplasts from leaves of *Nicotiana nesophila* and *Nicotiana stocktonii*, wild species that are resistant to numerous diseases to which cultivated tobacco is susceptible, were fused with cell suspension culture protoplasts from an albino mutant (Su/Su) of cultivated *N. tabacum*. Hybrid plants produced by protoplast fusion were distinguished as light-green shoots, while *N. stocktonii* or *N. neosphila* shoots were dark-green and *N. tabacum* shoots were albino. The light-green plants recovered following fusion were verified as somatic hybrids using morphological, chromosomal, isoenzymatic, and genetic criteria. A number of clones with the amphiploid chromosome number (2N = 96) have been identified. Stable incorporation of disease resistance into somatic hybrids was tested using resistance to tobacco mosaic virus and to the insect tobacco hornworm. These somatic hybrids should be resistant to numerous other diseases of cultivated tobacco. Ability to sexually propagate these hybrids was variable between clones. At least one clone capable of self-fertilization has been identified and nearly all clones could be backcrossed to *N. tabacum*. Seed from backcrosses (BC_1 and BC_2) is being evaluated for disease resistance. Useful material will be incorporated into a breeding program to develop resistant tobacco varieties. Several other reports of transfer of useful traits via protoplast fusion have recently been published (Table 3–2).

Another use of protoplast fusion is in the transfer of cytoplasmic genetic information (Figure 3–3). During fertilization, the male cytoplasm is preferentially excluded from the developing embryo. Thus, in most crops, conventional sexual hybrids contain and express only maternally derived cytoplasmic genes. Because protoplast fusion combines both parent cytoplasms into a single protoplast, fusion can be used to produce unique nuclear-cytoplasmic combinations not possible using conventional breeding methods (Gleba and Evans 1983).

Evidence from a number of somatic hybrids suggests that although two types of cytoplasms are initially mixed during protoplast fusion, resulting in heteroplasmons, eventually one parent-type cytoplasm predominates, resulting in cytoplasmic segregation. Following segregation, which may not occur until after meiosis, genetic information from the second cytoplasm is irreversibly lost. This phenomenon was first reported by examination of the fate of the large subunit of fraction-1 protein (F-1-P) in *Nicotiana glauca* +

TABLE 3–2 Agriculturally Important Genetic Traits Transferred Via Protoplast Fusion

Trait	*Somatic Hybrid*	*Reference*
Tobacco mosaic virus resistance	*Nicotiana tabacum* + *Nicotiana nesophila*	Evans et al. 1981
Potato virus X resistance	*Solanum tuberosum* + *Solanum chacoense*	Butenko and Kuchko 1982
Tobacco hornworm resistance	*Nicotiana tabacum* + *Nicotiana nesophila*	Evans, unpublished observations
Cytoplasmic male sterility	*Nicotiana tabacum* + *Nicotiana sylvestris*	Aviv and Galun 1980
Atrazine (herbicide) resistance	*Solanum tuberosum* + *Solanum nigrum*	Binding et al. 1982
Nematode resistance	*Solanum melongena* + *Solanum sisymbrifolium*	Gleddie et al. 1985
Potato leaf roll virus resistance	*Solanum tuberosum* + *Solanum brevidens*	Austin et al. 1985

Nicotiana langsdorffii somatic hybrids (Chen et al. 1977). Random chloroplast segregation resulted in fixed *langsdorffii*- or *glauca*-type F-1-P in each clone prior to meiosis (ten *langsdorffii*-type and seven *glauca*-type), except for one clone in which segregation followed self-fertilization. Cytoplasmic segregation has been observed for other cytoplasmically controlled traits. Segregation of male sterility and male fertility has been reported in several cybrids (Gleba and Evans 1983). Cosegregation has been reported for several cytoplasmic traits (Aviv et al. 1980). F-1-P and tentoxin sensitivity have been found to cosegregate in *N. glauca* + *N. tabacum* somatic hybrids (Flick and Evans 1982). Cosegregation implies that following cytoplasmic segregation, the cytoplasm contains all characters from one parent or the other, and therefore, precludes widespread ctyoplasmic recombination. Evidence has been presented that mitochondria and chloroplasts sort independently in somatic hybrids (Galun 1982).

Despite reports of cosegregation of ctyoplasmic genes, biochemical evidence using restriction analysis of mtDNA of some cybrids has suggested that recombination occurs between parental mtDNAs prior to cytoplasmic segregation in somatic hybrids between different male sterile and fertile parent plants (Belliard et al. 1979). In this and other reports, mitochondrial recombination following protoplast fusion produced a wide range of morphological phenotypes controlling male sterility in the hybrid plants. Nagy et al. (1981) detected recombination in mtDNA of *Nicotiana tabacum* + *Nicotiana knightiana* somatic hybrids resulting in novel mtDNA restriction

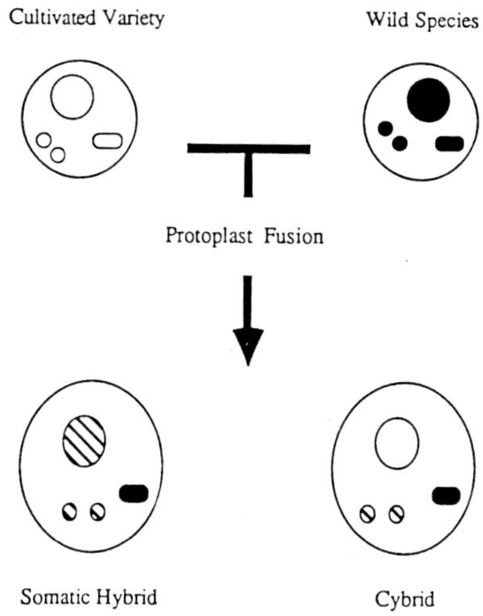

FIGURE 3–3 The production of unique gene combinations following protoplast fusion.

enzyme patterns. Recombination of chloroplast DNA has only recently been documented using products of protoplast fusion (Medgyesy et al. 1985). Most likely these intriguing results of ctyoplasmic segregation and recombination predate an era of extensive research on ctyoplasmic genetics with higher plants. Newly established cytoplasmic markers for disease resistance (Gengenbach et al. 1977) and antibiotic resistance (Maliga 1982) should facilitate research in this area. The production of unique nuclear-cytoplasmic combinations using protoplasts will aid in the development of novel germ plasm not obtainable using conventional methods.

3.6 SYNTHESIS OF SECONDARY PRODUCTS

Most proposed benefits of plant tissue culture require regeneration of a genetically modified or cloned plant to achieve successful application of tissue culture methodology. It has been suggested, though, that cultured plant cells may produce useful chemicals. While plant cells cannot compete with bacterial or yeast batch cell culture for cost of production, certain unique compounds synthesized exclusively by plant cells may have commercial value.

While there are some similarities between the large-scale culture of microbes and that of plant cells, the striking number of physiological differences between these cell types precludes the use of microbial fermentation systems for plant cell culture (Prenosil and Pedersen 1983). Plant cells are generally much larger than microbial cells and they possess rigid cell walls. These features render plant cells much more susceptible to the shear forces that develop in conventional microbial blade fermentors. Plant cells also grow much slower than microbial cells, and therefore, stringent aseptic conditions need to be observed. Because of the tendency for plant cells to aggregate, settling becomes a problem in large-scale cultures in terms of oxygen transfer. Most importantly, plant cells generally do not excrete their secondary products, but retain them inside the cell vacuole. Therefore, a destructive harvest would be necessary to release intercellular secondary compounds. This would preclude long-term recyclable cell cultures and increase the overall cost of production. Studies on plant cell excretion have suggested that accumulation and excretion of secondary products may be conveniently modulated by strict control of external pH (Renaudin 1981). Precise definition of specific excretion and accumulation values for individual cell lines would undoubtedly prove invaluable in designing cost-effective bioreactors for scale-up of producing cell cultures (Fowler 1986). Nondestructive induction of permeability for scale-up of producing cells by treatment with organic chemicals has been reported (Brodelius and Nilsson 1983). Moreover, treatments with specific compounds, such as vanadylsulphate (Smith et al. 1987) or fungal elicitors (DiCosmo and Misawa 1985), have been proposed to stimulate product synthesis.

Clearly the major hurdle to the cost-effective production of secondary compounds by plant cells is the isolation of high-producing cell lines. While cell cultures have been shown to produce a number of valuable secondary metabolites (Table 3–3), production levels are generally low and sometimes quite variable. The use of somaclonal variation technology in conjunction with cell aggregate cloning methods developed by Yamada and Fujita (1983) has already produced dramatic results in the induction and selection of high-yielding cell lines. Cell aggregate cloning has successfully been used to select for photoautotrophic cells (Yasuda et al. 1980), vitamin high-producing cells (Watanabe and Yamada, 1982) and high-pigment cells (Yamamoto et al. 1982) in various plant species. In some instances, cell cultures have been selected that produce a significantly higher concentration of a particular secondary metabolite than is found in the intact plant. An illustrative example of cell aggregate cloning is the selection of anthocyanin high-producing cell lines from cultured cells of *Euphorbia millii* by Yamamoto et al. (1982). In these experiments, cell aggregates were plated on solid medium and cultured. The resulting calli were then divided: one half for continued growth and the other for analysis of anthocyanin content. The highest-pigmented calli were continually subcultured in this manner for 24 selections. In the end, the amount of pigment found in the highest-producing lines was seven times greater than that produced by the original cells.

The most economically successful example of plant cell culture production of valuable chemicals is shikonin, a naphthoquinone used as both a

TABLE 3–3 Secondary Metabolites with Commercial Value That Can Be Produced in Cell Culture

Product	*Use*	*Species*
Anthraquinones	Laxative, dyes	*Morinda citrifolia*
Ajmalicine	Tranquilizer	*Catharanthus roseus*
Camptothecin	Antileukemic, antitumor agent	*Camptotheca acuminata*
Cephalotaxine	Antitumor agent (esters of)	*Cephalotaxus harringtonia*
Diosgenin	Raw material for hormone synthesis	*Dioscorea deltoides*
Rosmarinic acid	Spice, antioxidant	*Coleus blumei*
Shikonin	Dye, pharmaceutical	*Lithospermum erythrorhizon*
Stevioside	Nonnutritive sweetener	*Stevia rebaudiana*
Thebaine	Source of codeine	*Papaver bracteatum*
Ubiquinone-10	Cardiovascular agent	*Nicotiana tabacum*

See Zenk (1978) and Nickell (1980) for details.

dye and a pharmaceutical (Fujita et al. 1981; Curtin 1983). The development of high-producing cell cultures was based on the selection of variant cell aggregates that produced shikonin, followed by detailed investigations of proper nutritional and cultural techniques that further increased production thirteenfold.

Cell aggregate cloning is one tissue culture method that permits the identification of cells with increased biosynthetic capacities for a particular secondary product (Yamada and Hashimoto 1982). The genetic variation that is the basis for the observed chemical production is the result of spontaneous genetic variation already present in the original explant material or a variation that was induced during the process of tissue culture initiation and subculture, that is, somaclonal variation. Therefore, the same principle that has been successfully applied to agricultural crop plants to identify and select agronomically improved varieties is also widely applicable to the isolation of cell cultures that produce high levels of secondary metabolites. The only difference in the two applications is the material being evaluated for genetic alterations. Somaclonal variation technology, as it is applied to crop plants, relies on the regeneration of whole plants, their self-fertilization, and the analysis of the resulting R_1 progeny, whereas the application of somaclonal variation for secondary product synthesis relies on the generation of cell aggregates for secondary product synthesis and on the generation of cell aggregates that can be individually selected and evaluated for specific chemical production.

A detailed financial analysis is necessary before secondary product synthesis by plants is undertaken. It is important to consider cost of the final product per kilogram. Unless the cost of the chemical is high, it will be impossible to use cell cultures to produce a commercially competitive product. Cost evaluation should also consider demand of the product, opportunities for and value of producing a product with improved purity, and government regulations on distribution. Finally, it should be emphasized that if there is any possibility that a microbe can make the product, plant cell biosynthesis will not be economical. The production of only expensive drug compounds and selected, valuable flavor and fragrance chemicals are currently commercially feasible (Zenk 1978; Whitaker and Hashimoto 1985).

3.7 CONCLUDING REMARKS

The potential agricultural application of plant cell culture to crop improvement is astonishing. Tissue culture represents a loose collection of techniques that when carefully integrated into an existing breeding program can result in the development of unique germ plasm. In addition, as techniques of cellular genetics continue to develop it may be possible to alter the nature of future crop species. Even if only the less-technical methods of meristem culture and plant regeneration for cloning and genetic variability are utilized, plant tissue culture should have a tremendous impact on crop improvement in the near future.

REFERENCES

Ahloowalia, B.S. (1976) in *Current Chromosome Research* (Jones, K., and Brandham, P.E., eds.), pp. 115–122, Elsevier, Amsterdam.

Austin, S., Baer, M.A., and Helgeson, J.P. (1985) *Plant Sci.* 39, 75–82.

Aviv, D., Fluhr, R., Edelman, M., and Galun, E. (1980) *Theor. Appl. Genet.* 56, 145–150.

Aviv, D., and Galun, E. (1980) *Theor. Appl. Genet.* 58, 121–127.

Belliard, G., Vedel, F., and Pelletier, G. (1979) *Nature* 281, 401–403.

Bendich, A.J., and Gauriloff, L.P. (1984). *Protoplasma* 119, 1–7.

Binding, H., Jain, S., Finger, J., et al. (1982) *Theor. Appl. Genet.* 63, 273–277.

Bourgin, J.P. (1986) in *Cell Culture and Somatic Cell Genetics of Plants*, vol. 3 (Vasil, I.K., ed.), pp. 475–498, Academic Press, Orlando, FL.

Brettell, R.I.S., Dennis, E.S., Scowcroft, W.R., and Peacock, W.J. (1986a) *Mol. Gen. Genet.* 202, 235–239.

Brettell, R.I.S., Pallotta, M.A., Gustafson, J.P., and Appels, R. (1986b) *Theor. Appl. Genet.* 71, 637–643.

Brodelius, P., and Nilsson, K. (1983) *Eur. J. Appl. Microbiol. Biotechnol.* 17, 275–228.

Burk, L.G. (1970) *J. Hered.* 61, 279.

Burk, L.G., Gerstel, D.U., and Wernsman, E.A. (1979) *Science* 206, 585.

Butenko, R.G., and Kuchko, A.A. (1982) in *Advances in Protoplast Research* (Ferenczy, L., and Farkas, G.L., eds.), pp. 293–300, Pergamon Press, Elmsford, NY.

Chaleff, R.S. (1981) *Genetics of Higher Plants—Applications of Cell Culture*, pp. 93–95, Cambridge University Press, New York.

Chaleff, R.S. (1986) in *Cell Culture and Somatic Cell Genetics of Plants*, vol. 3 (Vasil, I.K., ed.), pp. 499–512, Academic Press, Orlando, FL.

Chen, K., Wildman, S.G., and Smith, H.H. (1977) *Proc. Natl. Acad. Sci. USA* 74, 5109–5112.

Collins, G.B., and Genovesi, A.D. (1982) in *Application of Plant Cell and Tissue Culture to Agriculture and Industry* (Tomes, D.T., Ellis, B.E., Harney, P.M., Kasha, K.J., and Peterson, R.J., eds.), pp. 1–24, University of Guelph Press, Guelph, Canada.

Collins, G.B., and Legg, P.D. (1980) in *The Plant Genome* (Davies, D.R., and Hopwood, D.A., eds.), pp. 197–213, The Johns Innes Charity, Norwich, England.

Cullis, C.A., and Cleary, W. (1986) *Can. J. Genet. Cytol.* 28, 247–251.

Curtin, M.E. (1983) *Bio/Technol.* 1, 649–657.

D'Amato, F. (1977) in *Plant Cell, Tissue, and Organ Culture* (Reinert, J., and Bajaj, Y.P.S., eds.), pp. 343–357, Springer-Verlag, New York.

D'Amato, F. (1978) in *Frontiers of Plant Tissue Culture* (Thorpe, T.A., ed.), pp. 287–295, IAPTC/University of Calgary Press, Calgary, Canada.

DiCosmo, F., and Misawa, M. (1985) *Trends Biotechnol.* 3, 318–322.

Dore, C., and Michel, H. (1984) in *Reunion Eucarpia "Legumes a Feuilles,"* pp. 10–15, INRA, Versailles, France.

Dudits, D., Fejer, O., Hadlaczky, G., et al. (1980) *Mol. Gen. Genet.* 179, 283–288.

Dulieu, H., and Barbier, M. (1982) in *Variability in Plants Regenerated from Tissue Culture* (Earle, L., and Demarly, Y., eds.), pp. 211–229, Praeger Press, New York.

Edallo, S., Zuccinalli, C., Perenzin, M., and Salamini, F. (1981) *Maydica* 26, 29–56.

Evans, D.A. (1983) in *Handbook of Plant Cell Culture*, vol. 1 (Evans, D.A., Sharp, W.R., Ammirato, P.V., and Yamada, Y., eds.), pp. 291–321, Macmillan, New York.

Evans, D.A., and Bravo, J.E. (1983) in *Handbook of Plant Cell Culture*, vol. 1 (Evans, D.A., Sharp, W.R., Ammirato, P.V., and Yamada, Y., eds.), pp. 124–176, Macmillan, New York.

Evans, D.A., Flick, C.E., Bravo, J.E., and Kut, S.A. (1983) *Theor. Appl. Genet.* 65, 93–101.

Evans, D.A., Flick, C.E., and Jensen, R.A. (1981) *Science* 219, 907–909.

Evans, D.A., and Sharp, W.R. (1983) *Science* 221, 949–951.

Evans, D.A., and Sharp, W.R. (1986) *Bio/Technol.* 4, 528–532.

Evans, D.A., Sharp, W.R., and Medina-Filho, H.P. (1984). *Am. J. Bot.* 71, 759–774.

Flick, C.E. (1983) in *Handbook of Plant Cell Culture*, vol. 1 (Evans, D.A., Sharp, W.R., Ammirato, P.V., and Yamada, Y., eds.), pp. 393–441, Macmillan, New York.

Flick, C.E., and Evans, D.A. (1982) *J. Hered.* 73, 264–266.

Fowler, M.W. (1986) *Trends Biotechnol.* 4, 214–219.

Fujita, Y., Hera, Y., Ogino, Y., and Suga, L. (1981) *Plant Cell Rep.* 1, 59–60.

Fukui, K. (1983) *Theor. Appl. Genet.* 65, 225–230.

Galun, E. (1982) in *Plant Improvement and Somatic Cell Genetics* (Vasil, I.K., Scowcroft, W.R., and Frey, K.J., eds.), pp. 205–220, Academic Press, Orlando, FL.

Gamborg, O.L., Shyluk, J.P., Fowke, L.C., Wetter, L.R., and Evans, D.A. (1979) *Z. Pflanzenphysiol.* 95, 255–264.

Gengenbach, B.G., Green, C.E., and Donovan, C.M. (1977) *Proc. Natl. Acad. Sci. USA* 74, 5113–5117.

Giles, K.L., and Morgan, W.M. (1987) *Trends Biotechnol.* 5, 35–39.

Gleba, Y.Y., and Evans, D.A. (1983) in *Handbook of Plant Cell Culture*, vol. 1 (Evans, D.A., Sharp, W.R., Ammirato, P.V., and Yamada, Y., eds.), pp. 322–357, Macmillan, New York.

Gleba, Y.Y., and Hoffmann, F. (1980) *Planta* 149, 112–117.

Gleba, Y.Y., Momot, V.P., Cherep, N.N., and Skarzynskaya, M.V. (1982) *Theor. Appl. Genet.* 62, 75–79.

Gleddie, S., Fassuliotis, G., Keller, W.A., and Setterfield, G. (1985) *Z. Pflanzenzuchtg.* 94, 352–355.

Gregorini, G., and Lorenzi, R. (1974) *Potato Res.* 17, 24–25.

Griffing, B. (1975) *Theor. Appl. Genet.* 46, 367–385.

Handley, L.W., Nickels, R.L., Cameron, M.W., Moore, P.P., and Sink, K.C. (1986) *Theor. Appl. Genet.* 71, 691–697.

Heinz, D.J., Krishnamurathi, M., Nickell, L.G., and Maretzki, A. (1977) in *Plant Cell, Tissue, and Organ Culture* (Reinert, J., and Bajaj, Y.P.S., eds.), pp. 3–17, Springer-Verlag, New York.

Hollings, M., Stone, O.M., and Dale, W.T. (1972) *Plant Pathol. (London)* 21, 46–47.

Hu, C.Y., and Wang, P.J. (1983) in *Handbook of Plant Cell Culture*, vol. 1 (Evans, D.A., Sharp, W.R., Ammirato, P.V., and Yamada, Y., eds.), pp. 177–227, Macmillan, New York.

Kasha, K.J., and Reinbergs, E. (1980) in *The Plant Genome* (Davies, D.R., and Hopwood, D.A., eds.), pp. 215–230, The Johns Innes Charity, Norwich, England.

Keller, W.A., and Armstrong, K.C. (1979) *Theor. Appl. Genet.* 55, 65–67.

Kemble, R.J., Brettell, R.I.S., and Flavell, R.B. (1982) *Theor. Appl. Genet.* 62, 213–217.

Kemble, R.J., and Shepard, J.F. (1984) *Theor. Appl. Genet.* 69, 211–216.

Kermicle, J.L. (1969) *Science* 166, 1422–1424.

Krumbiegel, G., and Schieder, O. (1979) *Planta* 145, 371–375.

Landsmann, J., and Uhrig, H. (1985) *Theor. Appl. Genet.* 71, 500–505.

Larkin, P.J., Ryan, S.A., Brettell, R.I.S., and Scowcroft, W.R. (1984) *Theor. Appl. Genet.* 67, 443–455.

Larkin, P.J., and Scowcroft, W.R. (1981) *Theor. Appl. Genet.* 60, 197–214.
Levy, L.W. (1981) *Environ. Exp. Bot.* 21, 389–395.
Lorz, H., and Scowcroft, W.R. (1983) *Theor. Appl. Genet.* 66, 67–75.
Maliga, P. (1982) *Plant Mol., Biol. News.* 3, 88–94.
McCoy, T.J., Phillips, R.L., and Rines, H.W. (1982) *Can. J. Genet. Cytol.* 24, 37–50.
Medgyesy, P., Fejes, E., and Maliga, P. (1985) *Proc. Natl. Acad. Sci. USA* 82, 6960–6964.
Melchers, G., Sacristan, M.D., and Holder, A.A. (1978) *Carlsburg Res. Commun.* 43, 203–218.
Moore, G.A., and Collins, G.B. (1982) *Biochem. Genet.* 20, 555–568.
Mullin, R.H., Smith, S.H., Frazier, N.W., Schlegel, D.E., and McCall, S.R. (1974) *Phytopathol* 64, 1425–1429.
Murakishi, H.H., and Carlson, P.S. (1976) *Phytopathol* 661, 931–932.
Murashige, T. (1974) *Annu. Rev. Plant Physiol* 25, 135–166.
Murashige, T. (1978) in *Frontiers of Plant Tissue Culture* (Thorpe, T.A., ed.), pp. 15–26, University of Calgary Press, Calgary, Canada.
Nagy, F., Torok, I., and Maliga, P. (1981) *Mol. Gen. Genet.* 183, 437–439.
Nakamura, A., Yamada, T., Kadotani, N., Itagaki, R., and Oka, M. (1974) *Sabrao J.* 6, 107–131.
Nakata, K., and Kurihara, T. (1972) *Jpn. J. Breed.* 22, 92–98.
Nickell, L.G. (1980) in *Plant Tissue Culture as a Source of Biochemicals* (Staba, E.J., ed.), pp. 236–256, CRC Press, Boca Raton, FL.
Nitsch, C. (1974) in *Haploids in Higher Plants—Advances and Potential* (Kasha, K.J., ed.), pp. 123–135, University of Guelph Press, Guelph, Canada.
Nitsch, J.P. (1971) *Colloque Internationaux C.N.R.S.* 193, 282–294.
Nitzsche, W., and Wenzel, G. (1977) *Haploids in Plant Breeding*, Parey Paul, Scientific Publications, New York.
Oglesby, R.P. (1978) in *Propagation of Higher Plants Through Tissue Culture* (Hughes, K.W., Henke, R., and Constantin, M., eds.), pp. 59–61, US Department of Energy, Washington, DC.
Orton, T.J. (1983) *Theor. Appl. Genet.* 67, 17–24.
Pecaut, P., Dumas de Vaulx, R., and Lot, H. (1983) *Acta Horticulturae* 131, 303–309.
Pental, D., Hamill, J.D., Pirrie, A., and Cocking, E.C. (1986) *Mol. Gen. Genet.* 202, 342–347.
Prat, D. (1983) *Theor. Appl. Genet.* 64, 223–230.
Prenosil, J.E., and Pedersen, J. (1983) *Enzyme Microb. Technol.* 5, 323–331.
Quak, F. (1977) in *Plant Cell, Tissue, and Organ Culture* (Reinert, J., and Bajaj, Y.P.S., eds.), pp. 598–615, Springer-Verlag, New York.
Renaudin, J.P. (1981) *Plant Sci. Lett.* 22, 59–69.
Sacristan, M.D. (1986) in *Cell Culture and Somatic Cell Genetics of Plants*, vol. 3 (Vasil, I.K., ed.), pp. 513–525, Academic Press, Orlando, FL.
Schaeffer, G.W., and Sharpe, F.T. (1987) *Plant Physiol.* 84, 509–515.
Schieder, O. (1978) *Planta* 141, 333–334.
Sharp, W.R., Reed, S.M., and Evans, D.A. (1984) in *Contemporary Bases For Crop Breeding* (Vose, P.B., and Blixt, S., eds.), pp. 347–381, Pergamon Press, Elmsford, NY.
Shepard, J. (1982) *Sci. Am.* 246, 154–166.

Skirvin, R.M., and Janick, J. (1976) *J. Am. Soc. Hort. Sci.* 101, 281–290.

Smith, J.L., Smart, N.J., Misawa, M., et al. (1987) *Plant Cell Rep.* 6, 142–145.

Sunderland, N. (1974) in *Haploids in Higher Plants—Advances and Potential* (Kasha, K.J., ed.), pp. 92–122, University of Guelph Press, Guelph, Canada.

Toriyama, K., Hinata, K., and Kameya, T. (1987) *Plant Sci.* 48, 123–128.

Vallos, C. (1984) Multiplication in vitro du piment (Capsicum annuum L.). Etude du microbouturage et de l'embruiogenese somatique, pp. 1–36. Memoire de D.E.A., U.S.T.L., Montepellier, France.

Walkey, D.G.A., and Cooper, V.C. (1975) *Ann. Appl. Biol.* 80, 185–190.

Watanabe, K., and Yamada, Y. (1982) *Phytochem.* 21, 513–516.

Whitaker, R.J., and Hashimoto, T. (1985) in *Handbook of Plant Cell Culture,* vol. 4 (Evans, D.A., Sharp, W.R., and Ammirato, P.V., eds.), pp. 264–286, Macmillan, New York.

Wise, R.P., Pring, D.R., and Gengenbach, B.G. (1987) *Proc. Natl. Acad. Sci. USA* 84, 2858–2862.

Yamada, Y., and Fujita, Y. (1983) in *Handbook of Plant Cell Culture,* vol. 1 (Evans, D.A., Sharp, W.R., Ammirato, P.V., and Yamada, Y., eds.), pp. 717–728, Macmillan, New York.

Yamada, Y., and Hashimoto, T. (1982) *Plant Cell Rep.* 1, 101–103.

Yamamoto, Y., Muzriquchi, R., and Yamada, Y. (1982) *Theor. Appl. Genet.* 61, 113–116.

Yasuda, T., Hashimoto, T., Sato, F., and Yamada, Y. (1980) *Plant Cell Physiol.* 21, 729–732.

Zenk, M.H. (1978) in *Frontiers of Plant Tissue Culture* (Thorpe, T.A., ed.), pp. 1–13, University of Calgary Press, Calgary, Canada.

CHAPTER
4

Selected Topics in the Genetic Manipulation of the Nuclear Genome

R. J. Griesbach

There are three different types of genome within a plant cell—nuclear, plastid, and mitochondrial. Each of these genomes is inherited in a different manner. Inheritance and control of expression have been most studied in the nuclear genome. It is by far the largest of the three genomes and contains the vast majority of economically important genes. Quite a few techniques have been developed for transferring nuclear genetic material between somatic cells. These techniques involve two distinct steps. The first step involves the isolation of the genetic material, and the second step is concerned with the physical transfer of the genetic material into a host cell.

There have been a number of reviews that describe many aspects of the manipulation of the nuclear genome. *Agrobacterium tumefaciens'* Ti plasmid is very efficient at transferring single genes into plant cells. There have been quite a few transgenic plants created with the help of the Ti plasmid (Goodman et al. 1987). The only limitation with the Ti plasmid is that only small one- or two-gene pieces of DNA can be transferred. Many economically important characteristics such as drought tolerance are encoded not by single genes but by blocks of polygenes. Other potential single gene vectors are the DNA plant viruses (Howell 1982) or transposable elements (Döring and Starlinger 1986).

A vector is not always needed to introduce small pieces of DNA into plant cells. Several recent reviews describe direct gene transfer (Bryant 1985; Moffat 1986). Electrical shock (electroporation) has also been used to induce DNA uptake (Neumann 1984). The use of synthetic genes has also been reviewed (Jaynes et al. 1986).

Polygene transfer can occur during somatic hybridization. Many excellent reviews describe results in this area (Vasil et al. 1979; Sybenga 1983; Pelletier 1986). In most instances too many genes are transferred. Many somatic hybrids contain the complete genomes of their two parent cell lines. This could be a problem if the goal is to introduce a single desirable characteristic from a weed species into a cultivated crop. How can all the economically undesirable traits of the weed species be eliminated in the hybrid? This chapter will discuss several selected techniques—preparation of chromosomes for transfer, preparation of microcells and their use in gene transfer, and microinjection—that have not recently been reviewed and that could allow polygene transfers.

4.1 CHROMOSOME TRANSFER

4.1.1 Preparation of Chromosomes

The first physiologically active chromosomes were isolated from cultured mammalian cells (Chorazy et al. 1963; Somers et al. 1963). Since then considerable progress has been made in isolating mammalian chromosomes. The procedure generally used involves exposing cultured cells to an agent that holds the cells in mitotic metaphase. Colchicine and its derivatives are the most widely used agents. Once sufficient quantities of mitotic cells are obtained, the cells are placed in a hypotonic solution, which makes them swell. This helps disperse the metaphase chromosomes. The cells are then lysed by adding a suitable buffer, and the cellular debris is removed via differential centrifugation. Finally, the chromosomes are pelleted.

The most critical feature of the isolation procedure is the composition of the lysis buffer (Hanson 1973) Table 4–1 lists the chemicals found in several of the more-successful buffers. Which buffer is selected depends upon how the chromosomes will be used after isolation. Certain properties are required of the chromosomes if they are to be used in transformation studies. First, the chromosomes must contain high-molecular-weight DNA that is unnicked or unmodified. Second, the acidic and basic proteins must be unmodified and correctly associated with the DNA. Finally, the chromatin and number of contaminating interphase nuclei should be kept to a minimum.

The method of cell lysis affects chromosome morphology. For example, prolonged incubation in hypotonic solutions prior to lysis or lysis in hypo-

TABLE 4–1 Composition of Chromosome Isolation Buffers

Mendelsohn et al. 1968	*Stubblefield and Wray 1978*	*Griesbach 1987*	*Hadlaczky et al. 1983*
1 mM $MgCl_2$	1 mM $CaCl_2$	15 mM Hepes (pH 7.0)	100 mM Glycine
1 mM $CaCl_2$	1 M Hexylene glycol	1 mM EDTA	100 mM Hexylene glycol
100 mM Sucrose	1 mM CAPS (pH 10.5)	15 mM DTT	75 mM Glucose
100 mM Sodium acetate	or 1 mM Hepes (pH 6.5)	80 mM KCl	$Ca(OH)_2$ (pH 8.5)
(pH 3.0)		20 mM NaCl	
		300 mM Sucrose	
		0.5 mM Spermine	
		500 mM Hexylene glycol	
		1% Triton X-100	

Hepes, N-2-hydroxyethylpiperazine-N-2-ethanesulfonic acid; CAPS, cyclohexylaminopropane sulfonic acid; EDTA, ethylenediaminetetraacetic acid; DTT, dithiothreitol.

tonic solutions can lead to irreversible chromosome expansion or disintegration (Mendelsohn et al. 1968). The use of nonionic detergents should also be avoided because they can cause nuclei to rupture as well as increase the chromosomes' sensitivity to shearing forces (Blumenthal et al. 1979). In lysing the cells, mechanical means, such as passage through a hypodermic needle, prove to disrupt chromosome morphology the least (Hanson 1973).

Chromosomes with a high-molecular-weight-DNA component (6.5×10^7 daltons) were isolated by Wray (1973). He found that a basic lysis buffer (pH around 10) could reduce most of the nuclease activity without producing some of the harmful side effects found when low pH buffers or metal chelators are used to reduce the enzyme activity. Low pH can reduce nuclease degradation; however, it can also lead to depurination of the DNA as well as remove histones and other basic chromosomal proteins. A high pH may prevent DNA degradation; however, it does not preserve the protein integrity of the chromosome. Up to 25% of the acidic chromosomal proteins can be removed with a high pH buffer (Hearst and Botchan 1970). Thus, basic and acidic pHs can preserve the integrity of the DNA, but they destroy the chromosomal protein composition. As a consequence, chromosomes must be isolated at near-physiological pH if one wishes to preserve their protein composition. It is possible at neutral pH to reduce the degradative effects nucleases have on DNA. For example, DNA that is bound to protein is protected from nuclease attack. The greater the quantity of protein bound, the less likely nucleases can bind and digest the DNA. By adding protein stabilizers, such as dithiothreitol (DTT) and hexylene glycol, it is possible to help preserve the protein composition of the chromosomes (Wray 1973). Metal chelators like ethylenediaminetetraacetic acid (EDTA) can also restrict nuclease activity by binding Mg^{2+} and Mn^{2+} cations. These cations are the cofactors that are needed for nuclease activity. Agents like colchicine and polyamines and conditions like cold temperature and high ionic strength can also reduce nuclease binding by helping to condense the chromosomes. Chromosomes with a high-molecular-weight DNA (2×10^8 daltons) were isolated from neutral buffers containing polyamines, chelators, and protein stabilizers (Blumenthal et al. 1979).

Isolated mammaliam chromosomes have been fractionated based upon their mass (Huberman and Attardi 1967), size (Hanson 1973) density (Stubblefield and Wray 1973), and electrical charge (Landel et al. 1972). In order to separate or fractionate chromosomes based upon their charge or density, the protein/DNA composition of the chromosomes must be selectively modified prior to fractionation. For example, all unmodified HeLa cell chromosomes band at the same density—1.31 g/ml (Huberman and Attardi 1967)—and have the same pK of about 4.0 (Landel et al. 1972). If however, the chromosomes are treated with trypsin before fractionation, it is possible to change the density and the electrical charge of the individual chromosomes selectively (Stubblefield and Wray 1973). One very serious problem with these methods of fractionation is that they permanently alter the composition of the chromosomes. Although separation based upon size via selective filtration does not damage the chromosomes' composition, it does have severe limitations. The major problem lies in the limited availability of filters that do not absorb chromosomes and that have a sufficient range of uniform pore sizes. As a result, selective filtration is of use only in very crude separations or in selecting chromosomes of unusually large or small size. The most widely used technique for fractionating unmodified chromosomes is based upon their mass and differential sedimentation through sucrose gradients. In this method, an unfractionated chromosome mixture is layered onto a preformed sucrose gradient usually at a concentration between 10 and 40% (w/v) and centrifuged for approximately 40 minutes at 1,000 *g*.

Differential sedimentation does not allow the complete separation of individual chromosomes. At best, chromosomes can be separated into a few distinct classes. High-resolution chromosome sorting can be accomplished through flow microfluorimetry (Lebo et al. 1984; Stubblefield and Wray 1978). Flow microfluorimetry sorts chromosomes based upon their DNA content. Chromsomes are first stained with a fluorescent dye and then singly passed through a laser beam. After passing through the laser beam, the flow is broken into droplets containing a single chromosome. As the droplets break free, a charge is applied based upon the fluorescent pulse. The charged droplets are deflected based upon their charge when they pass through two charged plates. The deflected droplets are then collected. Three-dimensional dual-laser flow microfluorimetry has been used to sort and identify 22 human chromosome fractions composed of 21 unique types (Lebo et al. 1984). Thirty-thousand chromosomes of each type were sorted in two days. The chromosomes were stained with two fluorescent dyes. The two laser-excited fluorescence signals were measured separately and corresponded to an x-axis deflection for the first fluorescent dye and to the y-axis deflection for the second fluorescent dye.

Only recently have chromosomes been isolated from higher plants (Malmberg and Griesbach 1980; Griesbach et al. 1982; Szabados et al. 1981; Hadlaczky et al. 1983). Table 4–2 describes the two most-commonly used procedures for isolating higher plant chromosomes.

The two chromosome isolation procedures differ in one major aspect—the pH of the chromosome buffer. My buffer has a pH of 7.0. At neutral pH nuclease activity is high; therefore this buffer has added ingredients to decrease nuclease-induced degradation. Hadlaczky's buffer has a pH of 8.5. At this pH nucleases have very low activity. Both buffers have disadvantages. The added ingredients in my buffer could affect chromosome structure. For example, the EDTA could remove Ca^{2+}, Mg^{2+}, and Mn^{2+} from the chromosomes. On the other hand, Hadlaczky's buffer could affect the chromosomal proteins. Acidic proteins that regulate chromosome function are structurally changed at basic pHs.

Plant chromosomes have been sorted through flow microfluorimetry (deLaat and Blaas 1984). Using my procedure *Haplopappus* (2N=4) chromosomes were sorted at a rate of 5×10^3 chromosomes per minute.

TABLE 4–2 Chromosome Isolation Procedures

Griesbach 1987	*Hadlaczky et al. 1983*
1. Induce mitotic synchrony in cultured cells—0.1% colchicine, 18 h.	1. Induce mitotic synchrony in cultured cells—0.1% colchicine, 12 h.
2. Isolate protoplasts—3-h enzyme incubation at 5°C—4% cellulase, 2% macerase, 10 mM $CaCl_2$, and 10% mannitol.	2. Isolate protoplasts—3-h enzyme incubation at 5°C—6% cellulase, 2% Rhozyme®, 2% pectinase, 5% driselase®, 170 mM mannitol, 250 mM glucose.
3. Lyse protoplasts in buffer (see Table 4–1)	3. Lyse protoplasts in buffer (see Table 4–1)
4. Pass suspension through 27-gauge hypodermic needle.	4. Add 0.1% triton X-100 for 5 min.
5. Spin at 1,000 *g* for 20 min.	5. Spin at 1,000 *g* for 20 min.
6. Resuspend in chromosome buffer.	6. Resuspend in chromosome buffer.
7. Layer on top of 10–40% (w/v) Nycodenz® gradient.	7. Add 1 mM phenylmethyl-sulfonylfluoride and 1% isopropyl alcohol.
8. Spin at 100,000 *g* for 16 h.	8. Spin at 1,000 *g* for 20 min.
9. Resuspend chromosome band in 10 mM $CaCl_2$ and 5 mM morholinoethane-sulfonic acid, pH 5.7.	9. Adjust chromosome-containing supernatant to 0.5 M sucrose and layer on top of 1 M sucrose.
	10. Spin at 1,000 *g* for 20 min.
	11. Resuspend chromosome pellet in buffer.

4.1.2 Use of Isolated Chromsomes in Gene Transfer

There are many reports in the literature of chromosome-mediated gene transfer between mammalian cells (Willecke 1978; Klobutcher and Ruddle 1981). The first demonstration of chromosomal gene transfer was reported in 1973 by McBride and Ozer (see reference), who transferred the Chinese hamster hypoxanthine guanine phosphoribosyltransferase (HGPRT) gene into a mouse cell. Mouse cells were incubated with poly-L-ornthine and chromosomes for two hours. Approximately 40 hamster chromosomes per mouse cell were used. The frequency of mouse cells expressing the hamster gene was 1×10^{-7}. Since then, improvements have been made. Transformation frequencies approaching 5×10^{-5} were obtained when mitotically synchronized cells were treated with calcium-precipitated chromosomes and dimethylsulfoxide (DMSO) (Klobutcher and Ruddle 1981). Almost all the cells took up at least a single chromosome. All of the chromosomes appeared to enter intracellular vacuoles that in a few cases became associated with the nucleus. Most of the vacuole-entrapped chromosomes were degraded (Willecke 1978). Once inside the nucleus, the foreign chromosomes were expressed. This expression was usually unstable and was lost at a frequency of up to 1.4×10^{-1} per cell generation. Almost all of the stable transformants contained a foreign-to-host chromosome translocation that could consist of up to 10^3 kb of DNA (Klobutcher and Ruddle 1979).

Chromosome transformation of plant cells has also been reported (Griesbach 1987). Szabados and others (1981) developed procedures for introducing isolated chromosomes into plant protoplasts. The host protoplasts and chromosome suspension were treated with 17% polyethylene glycol (MW 1,600 daltons) for five to ten minutes. The frequency of cells that contained a foreign chromosome was 5×10^{-5}. The effects of the polyethylene glycol concentration and the ratio of chromosomes to protoplasts on the frequency of uptake have also been tested (Griesbach et al. 1982). The frequency of uptake was highest (1×10^{-2}) when chromosomes were treated with 35% polyethylene glycol for 20 minutes at a ratio of ten chromosomes per protoplast.

Plant chromosomes have also been microinjected into protoplasts (Griesbach 1987). It appeared that the mechanisms involved in the stabilization of the foreign chromosomes in plant cells were similar to those in mammalian cells. Microinjection techniques are discussed below.

4.2 MICROCELL TRANSFER

Microcell transfer is a variation of somatic hybridization in which a host cell is fused with a microcell. A microcell consists of a single micronucleus surrounded by cytoplasm and an intact plasma membrane. The micronucleus usually contains a small fraction of the total genome.

4.2.1 Preparation of Microcells

Microcells are generally produced through a two-step process. The first step involves the production of cells containing micronuclei. The second step is enucleation. Enucleation of micronucleated cells results in the production of microcells and enucleated cells. This process was first reported for mammalian cells in 1974 by Ege and Ringertz (see reference). There are many slight procedural variations for obtaining different populations of mammalian microcells (Schor et al. 1975; Mullinger and Johnson 1976; Fournier and Ruddle 1977). A typical procedure can be outlined as follows: Cultured cells are overexposed to a mitotic arresting agent such as Colcemid at 20 μg/ml for 48 hours. This results in micronucleation of about 50% of the cells. The cells are then enucleated by centrifugation in 10 μg/ml cytochalasin B. The small microcells are purified from enucleated cells and normal cells through membrane filtration. These microcells can then be fused with normal host cells.

Using these procedures, Tourain and coworkers (1978) were able to transfer up to 11 human chromosomes into mouse cells. Treated mouse cells expressed the human HGPRT gene at a frequency of approximately 2×10^{-5} in the first report of gene transfer using microcells. In 1982 Fournier and Frelinger (see reference) constructed a collection of 20 hamster hybrid clones, each containing a different single mouse chromosome. These hybrids were stabilized by forcing the donor mouse chromosomes to complement gene deficiencies within the recipient hamster cells.

Within plants, microcell, or better, microprotoplast, transfer has also been reported. Procedures for inducing microprotoplasts are more complicated than those for inducing microcells. There are a number of reasons for this. Plant cells, unlike animal cells, contain a large central vacuole. This vacuole can occupy up to 90% of the cell's volume. The vacuole's presence within the cell complicates the enucleation step in microprotoplast formation. In addition, plant cells, unlike animal cells, have a cell wall that must be removed prior to enucleation. The cell wall is usually degraded by exposing cells to a cellulytic enzyme mixture. This incubation, if too long, can interfere with the micronucleation step.

Mitotic arresting agents can induce micronucleation. For example, after a five-day exposure to 10^{-4} M colchicine, many of the carrot cells in a liquid suspension culture contained large lobed nuclei. A few cells contained micronuclei (Okamura 1979). Another mitotic arresting agent, griseofulvin, can induce higher frequencies of micronucleation. When alfalfa liquid cell suspension cultures were exposed to 150 μg/ml griseofulvin for two days, approximately 60% of cells formed micronuclei after the drug was removed (Schiavo et al. 1980).

Griseofulvin treatment can also result in microcell formation without an enucleation step. Griseofulvin-induced micronucleated cells undergo a series of divisions that produce irregularly shaped cells containing intricate cross-wall partitions (Okamura 1979). The result of these abnormal divisions is the production of microcells. About 40% of alfalfa cells treated with griseofulvin contained a single micronucleus (Schiavo et al. 1980). The number of chromosomes within these microcells ranged from two to thirty. A diploid alfalfa cell contains 32 chromosomes.

Once cells that contain micronuclei are obtained, they must be converted to protoplasts. The cellulytic digestion must be performed under conditions that will allow the cell to retain the micronuclei. These conditions usually include incubation in high concentrations of cellulytic enzymes (e.g., 5% cellulysin and 2% macerase) for short periods of time (e.g., three hours) under refrigeration (4°C).

There are several methods for enucleating protoplasts (Bradley 1983). Plant cells, like animal cells, can be enucleated using cytochalasin B (Wallin et al. 1977; Bracha and Sher 1981). A general procedure is as follows: Protoplasts are layered onto a three-step gradient containing 45% sucrose, 33% sucrose, and 1 M mannitol and centrifuged for 20 minutes at 20,000 *g* (Bracha and Sher 1981). Cytochalasin B (50 μg/ml) and DMSO (0.5%) are included in the 1 M-mannitol step. Enucleated protoplasts are obtained at the 33–45% sucrose interface. This procedure results in about 60% enucleation of the protoplasts; more than 95% of protoplasts at the 33–45% sucrose interface are enucleated.

Protoplasts can also be enucleated by high-speed density gradient centrifugation (Archer et al. 1982; Griesbach and Sink 1983). The first procedures utilized 5–50% step gradients of Percoll®. After centrifugation at 30,000 *g* for 60 minutes, between 60 and 65% of the protoplasts were recovered as enucleated (Zorg et al. 1981). More-recent procedures for enucleation utilized a continuous 0–100% Percoll gradient. After 15 minutes of centrifugation at 150,000 *g*, cytoplasmic contents in most protoplasts aggregated into a portion of the cell. Further centrifugation forced the vacuole and a thin layer of peripheral cytoplasm to float to the top of the gradient and the remaining protoplast containing the nucleus and cytoplasm to move down to the bottom of the gradient (Griesbach and Sink 1983). Table 4–3 describes a procedure for obtaining microprotoplasts.

In plants, several different types of protoplasts can be produced during enucleation procedures. In the literature several terms are used—protoplast, subprotoplast, evacuolated protoplast, cytoplast, karyoplast, miniprotoplast, and microprotoplast. The definition of these structures is not always clear. Usually, three structures are produced (Burgess and Lawrence 1985). The first structure is a protoplast that lacks its vacuole but retains its nucleus and most of its cytoplasm. I prefer to call these structures evacuolated protoplasts. The second structure is a protoplast minus its vacuole and its nucleus. I prefer to call these structures cytoplasts. The third structure is the central vacuole surrounded by a thin layer of cytoplasm and a plasma

TABLE 4–3 Procedure for Obtaining Microprotoplasts

1. Treat culture cells with 150 μg/ml griseofulvin for 1 to 2 cell cycles.
2. Remove griseofulvin and wait 1 cell cycle.
3. Treat with 5% cellulase, 2.5% macerase, 1% pectinase, 10% mannitol, 100 mM $CaCl_2$, and 5 mM morholinoethanesulfonic acid.
4. Isolate protoplasts and resuspend in Percoll containing 100 mM $CaCl_2$, 5 mM hydroxyethyl piperazineethanesulfonic acid (pH 7.0), and 8% mannitol.
5. Centrifuge at 100,000 *g* (SW 56 rotor-Bechman) for 50 min at 23°C.
6. Collect microprotoplasts at the silica interface at the bottom of tube.

membrane. I prefer to call these structures vacuoleoplasts. By definition, an enucleated protoplast should contain a vacuole and cytoplasm but no nucleus. According to the mammalian definition a microprotoplast would be a protoplast that contains cytoplasm, a vacuole, and a subgenomic amount of chromatin within its nucleus.

4.2.2 Use of Microcells in Gene Transfer

There are many excellent reviews that describe a number of procedures for fusing plant cells (Evans et al. 1983; Butenko 1979; Vasil et al. 1979). Any of these procedures would be appropriate for microprotoplast-protoplast fusions. Even electricity is being used to fuse protoplasts (Zimmerman and Greyson 1983).

Evacuolated protoplasts have been used in cell fusion. Evacuolated protoplasts from a wild-type tobacco were fused with nitrate reductase mutant tobacco protoplasts (Wallin et al. 1979). The resulting hybrid cells were able to grow in nitrate-containing media. Evacuolated protoplasts and cytoplasts of onion have also been fused to normal onion protoplasts (Bracha and Sher 1981). Hybrid calli were not, however, produced.

There are no reports of microprotoplasts being used in fusing experiments. Most investigators have been trying to create cybrids, which result from fusions between cytoplasts and nucleoplasts. A nucleoplast contains a nucleus surrounded by very little cytoplasm and a plasma membrane. It appears that no one has attempted gene transfer in plants via microprotoplast fusions, even though the technology is readily available.

4.3 MICROINJECTION

Microinjection is a very old technique dating to the mid-1880s; however, it was not until the 1940s that this technique gained widespread use. At this time, several new micromanipulator designs were introduced. Three designs proved to be important—the pneumatic, lever, and rack-and-pinion designs. Very recently (1986) an electronically controlled manipulator design was introduced.

The design of the micromanipulator is very important to successful microinjection. Equally important is the design of the microscope. A microscope with a long working-distance lens is essential. The longer the working distance (distance between subject and lens), the more room for manipulation. Another important new development is the inverted microscope. In an inverted microscope the objective lens is under the subject and the condenser lens is above the subject. The benefit of this inversion comes from the fact that the working distance between the subject and condenser lens can be longer than the working distance between the objective lens and subject. A large number of inverted microscopes are currently available.

Once again, animal cells were microinjected before plant cells. The first important uses of microinjection were in amphibian nuclear transplantation, in which nuclei from differentiated cells were injected into enucleated egg cells (Gurdon 1977). Results from these experiments provided evidence for totipotency during development. Microinjection was also used to study the biochemistry of transcription and translation and the effects of various compounds on cellular metabolism.

It was not until recently that microinjection was used to modify the nuclear genome. In 1974 Jaenisch and Mintz (see reference) reported the first genetically transformed animals produced by microinjection. Mice embryos at the blastocyst stage were injected with SV40 DNA. The DNA was injected into the blastocyst cavity. Cells of about 40% of resulting animals contained SV40 sequences, but not all the cells in the animal were transformed. Such transformed animals were called mosaics. Many of the mosaic animals did not harbor the foreign genes within their germ cells and thus did not transmit the foreign genes to progeny.

If fertilized egg cells were microinjected, then fully transformed, and not mosaic, animals could be obtained. Fully transformed animals have been called transgenic. The procedure usually involves microinjecting cloned DNA directly into a nucleus. Approximately 1 cell in 500 was transformed by Capecchi (1980) when he injected 125 molecules of the thymidine kinase gene into mouse egg nuclei. No transformed cells were recovered when the DNA was injected into the cytoplasm. The first transgenic mice were produced in 1980 by Gordon and others (see reference). The production of transgenic mice via microinjection is now a routine laboratory practice (Palmiter and Brinster 1986). In many instances, the procedures are so common that they are not even described in research papers. In a recent *Science* paper (Goring et al. 1987), the procedure was described as follows: "Two hundred fifty-six fertilized eggs from CD-1 mice were microinjected with a 4.5 kb DNA fragment. . . . " In this report, 69 animals were produced; of these only 5 expressed the foreign gene.

In a recent review (Palmiter and Brinster 1986) it was stated that since 1980 several thousand transgenic mice and a few transgenic sheep, pigs, and rabbits have been produced through microinjection. In general, it appears that the foreign DNA is randomly inserted into the host genome and is stably transmitted for many mitotic, as well as meiotic, generations. The expression of the foreign genes has been modulated and controlled by the construction of hybrid genes in which different promoters, enhancers, and silencers are attached to the injected structural gene (Kollias et al. 1986). The level of expression of a foreign gene varies widely from one transgenic animal to another. This is mostly like the result of foreign gene integration. The chromosomal position of a gene is important in gene expression (Townes et al. 1985).

The microinjection of plant cells is far from routine. Plant cells are more difficult to inject than animal cells for two reasons. First, plant cells have a cell wall and a vacuole. The cell wall is composed of thick layers of lignin and cellulose. These compounds are very strong, making it almost impossible for a glass microneedle to penetrate the cell wall. Protoplasts, not cells, are better suited to injection. Second, the vacuole contains many hydrolases and toxic compounds. When injecting plant cells, one must not damage the vacuolar membrane. If the vacuolar contents are emptied into the cytoplasm, the cell will most likely die. In some instances, the cell's nucleus can be found embedded within the vacuole, making it impossible to inject. These problems can be overcome in a number of ways.

There are several methods for reducing the effects of the vacuole. In the first method, two micromanipulators are used. One micromanipulator holds the protoplast while the other holds the injection needle (Morikawa and Yamada 1985; Crossway et al. 1986). One can use the holding manipulator to select a protoplast with a nucleus peripherally exposed to the plasma membrane. The protoplast can then be manipulated such that the side of the protoplast containing the exposed nucleus is in line with the injection needle. A disadvantage of this system is that protoplasts that have a peripheral nucleus are less viable than those with a central nucleus. In addition, this procedure is very time consuming. Another approach is the use of evacuolated protoplasts (Griesbach 1987), which was discussed in the previous section. Like peripherally nucleated protoplasts, evacuolated protoplasts have a reduced viability.

If protoplasts are not held by a micromanipulator during injection, then some other method is required for immobilizing them. Animal cells naturally attach to surfaces. In addition, upon attachment the cells flatten, exposing their nuclei. This is not the case for plant protoplasts. Protoplasts can be attached to glass that has been coated with poly-L-lysine. Protoplasts attached by such a technique have been injected (Steinbiss and Stabel 1983; Reich et al. 1986a); however, the concentration of poly-L-lysine required for attachment is toxic to some species. The toxicity can be avoided by embedding the protoplasts in agarose (Griesbach 1987; Reich et al. 1986b; Lawrence and Davies 1985). Agarose has several disadvantages: It is difficult to move the injection needle within the gel and the gel reduces visibility and resolution.

The injection equipment can be quite simple. A micrometer attached to the plunger of a syringe has been used (Steinbiss and Stabel 1983; Reich et al. 1986a). A more-complicated system can also be used. Pressurized gas tanks attached to gas-burst regulators have been used (Griesbach 1987; Morikawa and Yamada 1985). The pressurized gas systems allow a more controlled and uniform injection rate. Dyes have been used to visualize the site of injection. The most common dye is lucifer yellow (Morikawa and Yamada 1985); however, Hoechst 33258, Diamidinophenylindole (DAPI), and rhodamine have also been used (Reich et al. 1986b). Such dyes could reduce the viability of injected protoplast. Table 4–4 compares several successful injection systems.

There are two reports of genetic transformation via microinjection. Tobacco protoplasts were injected with an *Agrobacterium* plasmid. Approximately 2,000 copies were injected into each protoplast. The frequency of transformation was 1.4×10^{-1} when nuclei were injected and 6×10^{-2} when the cytoplasm was injected (Crossway et al. 1986). Transformation was determined by in situ cDNA hybridization. In the other instance, *Petunia hybrida* protoplasts were injected with *Petunia alpicola* chromosomes. Approximately one chromosome was injected into each protoplast. The frequency of transformation was 4×10^{-2}. Transformation was determined by changes in the flavonoid biosynthetic pathway. Transgenic plants were obtained and the foreign genes were genetically inherited in a Mendelian manner (Griesbach 1987).

4.4 CONCLUSION

There is a wealth of technology available for modifying the nuclear genome. Many vectors are currently available for the transfer of single genes. The widespread practical application of these vectors is still in the future. Basic biology has not caught up with biotechnology. Very few economically important single genes have been identified, much less isolated and cloned. Single-gene vectors are currently being used to transfer herbicide resistance.

TABLE 4–4 Comparison of Different Microinjection Systems

Investigators	*Speed of Injection (no./h)*	*Viability (%)*	*Method of Injection*[1]	*Method of Protoplast Immobilization*[2]
Morikawa and Yamada 1985	10	—	P	H
Reich et al. 1986a	90	88	S	L
Crossway et al. 1986	—	86	S	H
Griesbach 1987	80	58	P	E
Steinbiss and Stabel 1983	20	30	S	L
Lawrence and Davies 1985	—	40	S	E

[1] P, pressurized gas; S, syringe.
[2] H, hold protoplast with a pipette; E, embed protoplasts in agarose; L, attach protoplasts to glass coated with poly-L-lysine.

There are several methods for transferring polygenic characteristics. Once again biotechnology is ahead of basic biology. The underlying biochemical basis for many polygenic characteristics is unknown. Without this information it is difficult to design a selection screen to identify transformed cells or plants.

In the future, transgenic plants will most likely be created through non–tissue culture procedures. Many species are almost impossible to regenerate from single cells. In addition, the regeneration process can create many genetic abnormalities. New technology is now being developed to introduce foreign genes into cells within intact plants. For example, preliminary studies have already shown that it is possible to obtain transgenic seedlings from pollinations with pollen that had been treated with foreign DNA.

REFERENCES

Archer, E., Landgren, C., and Bonnett, H. (1982) *Plant Sci. Lett.* 25, 175–185.

Blumenthal, A., Dieden J., Kapp, L., and Sedat, J. (1979) *J. Cell Biol.* 81, 255–259.

Bracha, M., and Sher, N. (1981) *Plant Sci. Lett.* 23, 95–101.

Bradley, P. (1983) *Plant Mol. Biol. Reptr.* 1, 117–123.

Bryant, J. (1985) *Trends Biotechnol.* 3, 90.

Burgess, J., and Lawrence, W. (1985) *Protoplasma* 126, 140–146.

Butenko, R. (1979) *Int. Rev. Cytol.* 59, 323–371.

Capecchi, M. (1980) *Cell* 22, 479–488.

Chorazy, M., Bendich, A., Borenfreund, E., Ittenson, O., and Hutchison, D. (1963) *J. Cell Biol.* 19, 71–77.

Crossway, A., Oakes, J., Irvine, J., et al. (1986) *Mol. Gen. Genet.* 202, 179–185.

deLaat, A., and Blaas, J. (1984) *Theor. Appl. Genet.* 67, 463–467.

Döring, H., and Starlinger, P. (1986) *Annu. Rev. Genet.* 20, 175–201.

Ege, T., and Ringertz, N. (1974) *Exp. Cell Res.* 87, 378–382.

Evans, D.A., Sharp, W.R., Ammirato, P.V., and Yamada, Y., eds. (1983) *Handbook of Plant Cell Culture,* Macmillan, New York.

Fournier, R., and Frelinger, J. (1982) *Mol. Cell. Biol.* 2, 526–534.

Fournier, R., and Ruddle, R. (1977) *Proc. Natl. Acad. Sci. USA* 74, 319–323.

Goodman, R., Hauptli, H., Crossway, A., and Knauf, V. (1987) *Science* 236, 48–54.

Gordon, J., Scangos, G., Plotkin, D., Barbosa, J., and Ruddle, F. (1980) *Proc. Natl. Acad. Sci. USA* 77, 7380–7384.

Goring, D., Rossant, J., Clapoff, S., Breitman, M., and Tsui, L. (1987) *Science* 23, 456–458.

Griesbach, R. (1987) *Plant Sci.* 50, 69–77.

Griesbach, R., Malmberg, R., and Carlson, P. (1982) *J. Hered.* 73, 151–152,

Griesbach, R., and Sink, K. (1983) *Plant Sci. Lett.* 30, 297–301.

Gurdon, J. (1977) *Methods Cell Biol.* 14, 125–139.

Hadlaczky, G., Bisztray, G., Praznovsky, T., and Dudits, D. (1983) *Planta* 157, 278–285.

Hanson, C. (1973) in *New Techniques in Biophysics and Cell Biology* (Pain, R., and Smith, B.J., eds.) pp. 43–83, Wiley, New York.

Hearst, J., and Botchan, M. (1970) *Annu. Rev. Biochem.* 39, 151–165.

Howell, S. (1982) *Annu. Rev. Plant Physiol.* 33, 609–650.

Huberman, J., and Attardi, G. (1967) *J. Mol. Biol.* 29, 487–491.

Jaenisch, R., and Mintz, B. (1974) *Proc. Natl. Acad. Sci. USA* 72, 1250–1254.

Jaynes, J., Yang, M., Espinoza, N., and Dodds, J. (1986) *Trends Biotechnol.* 4, 314–320.

Klobutcher, L., and Ruddle, F. (1979) *Nature* 280, 657–660.

Klobutcher, L., and Ruddle, F. (1981) *Annu. Rev. Biochem.* 50, 533–554.

Kollias, G., Wrighton, N., Hurst, J., and Grosveld, F. (1986) *Cell* 46, 89–94.

Landel, A., Aloni, Y., Raftery, M., and Attardi, G. (1972) *Biochemistry* 11, 1654–1658.

Lawrence, W., and Davies, D. (1985) *Plant Cell Rep.* 4, 33–35.

Lebo, R., Gorin, F., Fletterick, R., et al. (1984) *Science* 225, 57–59.

Malmberg, R., and Griesbach, R.J. (1980) *Plant Sci. Lett.* 17, 141–147.

Mendelsohn, J., Moore, D., and Salzmann, N. (1968) *J. Mol. Biol.* 32, 101–105.

McBride, O., and Ozer, H. (1973) *Proc. Natl. Acad. Sci. USA* 70, 1258–1262.

Moffat, A. (1986) *Mosaic* 17, 26–39.

Morikawa, H., and Yamada, Y. (1985) *Plant Cell Physiol.* 26, 229–236.

Mullinger, A., and Johnson, R. (1976) *J. Cell Sci.* 22, 243–285.

Neumann, J. (1984) *Bioelectrochem. Bioengin.* 13, 219–224.

Okamura, S. (1979) *Cell Struct. Funct.* 4, 11–22.

Palmiter, R., and Brinster, R. (1986) *Annu. Rev. Genet.* 20, 465–499.

Pelletier, G. (1986) *Oxford Surv. Plant Mol. Cell Biol.* 3, 96–121.

Reich, T., Iyer, V., Haffner, M., Holbrook, L., and Miki, B. (1986a) *Can. J. Bot.* 64, 1259–1267.

Reich, T., Iyer, V., Scobie, B., and Miki, B. (1986b) *Can. J. Bot.* 64, 1255–1258.

Schiavo, F., Ronchi, V., and Terzi, M. (1980) *Theor. Appl. Genet.* 58, 43–47.

Schor, S., Johnson, R., and Mullinger, A. (1975) *J. Cell Sci.* 19, 281–303.

Somers, C., Cole, A., and Hsu, T. (1963) *Exp. Cell Res.* 9, 220–226.

Steinbiss, H., and Stabel, P. (1983) *Protoplasma* 116, 223–227.

Stubblefield, E., and Wray, W. (1978) *Biochem. Biophys. Res. Commun.* 83, 1404–1414.

Sybenga, J. (1983) *Theor. Appl. Genet.* 66, 179–201.

Szabados, L., Hadlaczky, G., and Dudits, D. (1981) *Planta* 151, 141–145.

Tourian, A., Johnson, R., Burg, K., Nicholson, S., and Sperling, K. (1978) *J. Cell Sci.* 30, 193–209.

Townes, T., Lingrel, J., Chen, H., Brinster, R., and Palmiter, R. (1985) *EMBO J.* 4, 1715–1723.

Vasil, I., Ahuja, M., and Vasil, V. (1979) *Adv. Genet.* 20, 127–177.

Wallin, A., Glimelius, K., and Eriksson, T. (1977) *Z. Pflanzenphysiol.* 87, 333–340.

Wallin, A., Glimelius, K., and Eriksson, T. (1979) *Z. Pflanzenphysiol.* 91, 89–94.

Willecke, K. (1978) *Theor. Appl. Genet.* 52, 97–104.

Wray, W. (1973) *Methods Cell Biol.* 6, 307.

Zimmerman, V., and Greyson, J. (1983) *Biotechniques* 1, 5–10.

CHAPTER

5

Regulation and Expression of Plant Genes in Microorganisms

A. A. Gatenby

The expression of plant genes in microorganisms has been used to investigate a wide range of biological issues, including the production of plant proteins to be used either in research or commercially and the elucidation of the mechanisms of gene regulation in plants based on observations of the types of plant gene expression in yeast and bacteria. In addition, detection of the synthesis of various proteins, or parts thereof, by screening clones with antibodies raised against the desired gene product has proved to be a useful method for the identification of gene expression. In general, most of these experiments have been conducted using *Escherichia coli* or *Saccharomyces cerevisiae,* and the type of regulation employed has relied on elements of the host's various regulatory mechanisms. Since the general principles of the expression of foreign genes in bacteria and yeast have been

A number of the experiments described in this review were carried out during my stay at the Plant Breeding Institute, Cambridge, England, and I am greatly indebted to the many people who contributed directly and indirectly to this work. These include Dave Baulcombe, Martine Boccara, Doug Bradley, Judy Castleton, Rick Cuellar, Tristan Dyer, Dick Flavell, John Gray, Colin Lazarus, Steve Rothstein, Mike Saul, Saskia van der Vies, and Dave Willey. I also thank Jane Cramer and Les Hoffman of Agrigenetics Advanced Science Company and Ad Kool of the Vrije Universiteit for preprints of their work.

described elsewhere, I have not dealt with them in great detail in this review. Instead, I have concentrated on the specific reported examples of plant gene expression in microorganisms, and have tried to describe these examples in terms of their potential contribution to the understanding of certain properties of the genes or their products, and how these can interact with the microbial cell.

5.1 RECOGNITION OF PLANT SEQUENCES THAT FUNCTION AS PROMOTERS OF TRANSCRIPTION IN MICROORGANISMS

Chloroplast genes show considerable homology to *E. coli* genes for sequences involved in recognition of DNA by RNA polymerase and subsequent interaction of RNA with ribosomes (Whitfeld and Bottomley 1983; Gray and Doolittle 1982). In early experiments it was demonstrated that isolated chloroplast DNA could direct the synthesis of a chloroplast protein in vitro using an *E. coli* cell-free system (Bottomley and Whitfeld 1979), indicating that the bacterial RNA polymerase could bind and initiate transcription at sites on the chloroplast genome. These findings were supported by in vivo expression studies in which synthesis of chloroplast proteins could be detected in bacteria harboring plasmids that contained chloroplast genes. Chloroplast genes from a wide range of plant species that have been expressed in *E. coli* include genes for the large subunit of ribulose bisphosphate carboxylase (Gatenby et al. 1981; Fluhr et al. 1983; Bovenberg et al. 1984b; Zhu et al. 1984; Somerville et al. 1986), various CF_1 adenosine triphosphatase (ATPase) polypeptides (Fluhr et al. 1983; Bovenberg et al. 1984a; Gatenby and Rothstein 1986), and cytochrome *f* (Rothstein et al. 1985). The patterns of expression observed in a number of investigations suggested that transcription may initiate on the chloroplast DNA, rather than occur from readthrough transcription from vector sequences. In some cases, chloroplast DNA restriction fragments have been used to provide a sequence for *E. coli* RNA polymerase to recognize, resulting in expression of a downstream promoterless gene. This technique of trying to identify chloroplast promoters by expression in *E. coli* has made use of transcriptional fusions with the genes for β-galactosidase, galactokinase, neomycin phosphotransferase, and chloramphenicol acetyltransferase (Gatenby et al. 1981; Kong et al. 1984; Kool et al. 1986). Although such an approach may identify a promoter that is active in both the chloroplast and *E. coli*, a significant problem arises when sequences fortuitously function as promoters in bacteria but are not authentic chloroplast gene promoters. For example, both DNA sequence homology and promoter activity have been observed in *E. coli* with a maize *rbcL* restriction fragment that is known not to contain the authentic chloroplast transcription initiation site (Gatenby and Cuellar, unpublished observations).

The clearest evidence for authentic chloroplast promoter recognition by *E. coli* RNA polymerase has been obtained from experiments using high-resolution transcript mapping or mutagenesis or both. Thus, Shinozaki and Sugiura (1982) using tobacco *rbcL* and *atpB* genes, Tohdoh et al. (1981) using tobacco rRNA genes, Erion et al. (1983) using spinach *rbcL* genes, and Boyer and Mullet (1986) using pea *psbA* genes were able to demonstrate that purified *E. coli* RNA polymerase would initiate transcription in the same position on chloroplast DNA as would the corresponding chloroplast RNA polymerase. Fukuzawa et al. (1985) also observed recognition of a chloroplast promoter from a liverwort by the bacterial RNA polymerase. Using deletion and single-base substitution mutants Bradley and Gatenby (1985) were able to show that in vitro the chloroplast RNA polymerase and *E. coli* RNA polymerase respond to the promoter mutations in a quantitatively similar fashion. In addition, these authors demonstrated that the chloroplast promoter was sufficiently homologous to a bacterial promoter to enable *E. coli* to be used as a screen to select for chloroplast promoter mutants. Additional evidence for the homology has been presented by Gruissem and Zurawski (1985a and 1985b). It is, therefore, well established that a chloroplast promoter can function as such in *E. coli*. Chloroplast sequences may also function as promoters in other Gram-negative bacteria, since we have observed expression of the maize *rbcL* gene in both *Rhizobium* species and *Pseudomonas oxalaticus* in an orientation-independent manner (Gatenby and O'Gara, unpublished observations).

Although the prokaryotic-like chloroplast promoters can be recognized in *E. coli,* the transcription of nuclear genes in a microorganism has relied on the yeast *S. cerevisiae*. Langridge et al. (1984) were able to demonstrate the correct recognition of maize storage protein promoters in yeast. Genes for both the 19,000- and 21,000-dalton-size classes of zein were integrated into chromosome II of yeast. The promoter regions of both size classes were accurately recognized by yeast RNA polymerase, giving rise to initiation at the same nucleotide position in yeast and in maize endosperm. The 5′ ends of the mRNA for the smaller protein were heterogeneous in both yeast and maize endosperm, indicating that two transcription starts separated by only two bases were employed. Zein promoter recognition in yeast has also been described by Coraggio et al. (1986).

Expression of phaseolin genes in yeast under the control of plant sequences has also been reported (Cramer et al. 1985). Two classes of polyadenylated RNA were detected in the yeast cells, and both of these initiated and terminated in plant DNA sequences. However, unlike the zein promoter, the authentic phaseolin promoter was not recognized. Instead a heterogeneous series of 5′ termini were observed to initiate in two clusters upstream of the major bean initiation site. It was also suggested that the 3′ termini were either also heterogeneous or polyadenylated to different degrees to account for the difference between the major 1,400- and 1,800-base transcripts.

It has also been suggested that the soybean leghaemoglobin promoter is recognized by yeast (Jensen et al. 1986). Several initiation sites were detected in yeast and they were all close together. Although transcription in yeast did not initiate at the same nucleotide as that used in the soybean nodule, it was initiated in the same region, indicating that some features of the leghaemoglobin promoter were recognized.

5.2 INCREASING THE RATE OF TRANSCRIPTION OF CLONED GENES

In general, a crucial requirement of plant gene expression in a microorganism is that the production of the gene product be maximized to obtain sufficient protein for biochemical or other investigations. Although chloroplast promoters are recognized in *E. coli,* they do not appear to be particularly powerful, and so for bacterial expression studies the chloroplast promoters can be replaced, or supplemented, with more-efficient promoters. When studying plant nuclear gene expression in *E. coli,* both transcriptional and translational signals of prokaryotic origin are usually used to replace the eukaryotic sequences, often by gene fusions. A particular advantage of chloroplast genes is that the ribosome binding site is recognized by bacterial ribosomes, thus amplification of expression will often require an increase in only the rate of transcription. It is known, however, that an antiterminator protein is required for readthrough transcription of the maize *rbcL* gene from a bacteriophage promoter (Gatenby and Cuellar 1985).

Several different promoters have been used to attain efficient transcription of cloned plant genes. Information on the general features of these promoters and their mechanism of induction has been published in other reviews (Gatenby 1983; Timmis 1981; Bernard and Helinski 1980; Pouwels et al. 1985). The most-favored promoter used is the *lac* operon promoter; often the UV5 mutation is employed to prevent catabolite repression of activity. The *lac* promoter is frequently used in combination with the *lacZ* translational initiation site to allow synthesis of fusion proteins fused to a β-galactosidase N-terminus. Examples in which both transcriptional and translational signals from the *lac* operon have been used to synthesize a complete polypeptide with a fused N-terminus include wheat glutenin (Bartels et al. 1985), maize prolamins (Wang and Esen 1985), *Brassica S*-locus-specific glycoprotein (Nasrallah et al. 1985), spinach ribulose bisphosphate carboxylase activase (Werneke et al. 1988), and maize phosphoenolpyruvate carboxylase (Izui et al. 1986). Other examples in which transcription was initiated at a *lac* promoter, but translation occurred at a eukaryotic or another bacterial site, include thaumatin (Edens et al. 1982), the small subunit of ribulose bisphosphate carboxylase from pea (Van den Broeck et al. 1985), the large subunit of maize ribulose bisphosphate carboxylase (Somerville et al. 1986), potato patatin (Nakamura et al. 1986), and wheat α-amylase (Gatenby et al. 1986). Sometimes *lac* transcription is required simply to enable synthesis of part of a polypeptide that is used for clone identification by in situ reaction with an appropriate antibody (Wang and Esen 1985; Tittgen et al. 1986; Crawford et al. 1986).

Other promoters have also been used. The thermoinducible leftward promoter of lambda bacteriophage has been used to express plant genes from either phages (Gatenby et al. 1981) or plasmids (Gatenby and Castleton 1982; Gatenby 1984; Gatenby and Cuellar 1985; Gatenby and Rothstein 1986; Piatak et al. 1988). The *trp* promoter was used by Edens et al. (1982) to attain expression of a cDNA encoding the sweet-tasting plant protein thaumatin, and by Piatak et al. (1988) to express ricin and Dewey et al. (1988) to express a mitochondrial protein. Stubbs et al. (1986) relied on the outer membrane lipoprotein (*lpp*) promoter of *E. coli* to transcribe and express a cDNA clone for pea lectin. A combination of a *recA* promoter and a T7 ribosome binding site was used to express a maize glutathionine-*S*-transferase cDNA clone (Moore et al. 1986). Synthetic or hybrid promoters have also been used for synthesis of bean phytohemaglutinin (Hoffman and Donaldson 1987) and alfalfa glutamine synthetase (DasSarma et al. 1986). Wasmann et al. (1986) used an inducible *tac* promoter to express a fusion protein made up of the small subunit of ribulose bisphosphate carboxylase and neomycin phosphotransferase II. O'Hare et al. (1987) expressed castor bean ricin from a T5 promoter, and a synthetic acyl carrier protein gene was transcribed with a *trc* promoter (Beremand et al. 1987).

In yeast several different promoters have been used to express plant genes, even though some plant promoters can be recognized by yeast RNA polymerase (Langridge et al. 1984; Coraggio et al. 1986; Jensen et al. 1986). The yeast glyceraldehyde-3P-dehydrogenase promoter was used to transcribe a cDNA clone for thaumatin in both *S. cerevisiae* (Edens et al. 1984) and *Kluyveromyces lactis* (Edens and van der Wel 1985), and the highly efficient phosphoglycerate kinase promoter enabled transcription and translation of wheat α-amylase to be detected in cells of *S. cerevisiae* (Rothstein et al. 1984). Transcription of a maize cDNA clone for glutathione-S-transferase in yeast was accomplished by using the *GAL10* promoter (Wiegand et al. 1986). The level of expression, as measured by enzyme assay, was substantially increased following induction by 1% galactose. The yeast acid phosphatase promoter has been used to express bean phaseolin that has achieved levels of 3% of the total soluble cellular protein (Cramer et al. 1986, 1987), and also to express bean phytohemagglutenin (Tague and Chrispeels 1987). The cytochrome *c* promoter was used to express maize zein (Coraggio et al. 1986) and wheat α-gliadin (Neill et al. 1987). Expression of barley β-gluconase was obtained by using the alcohol dehydrogenase promoter (Jackson et al. 1986).

5.3 TRANSLATIONAL FEATURES OF PLANT GENE EXPRESSION AND REGULATION

The similarity between the chloroplast ribosome binding sites and those of bacterial genes is exemplified by the recognition and correct translation of chloroplast genes by bacterial ribosomes both in vitro (Bottomley and Whitfeld 1979) and in vivo (Gatenby et al. 1981). In vitro transcription and translation using *E. coli* extracts has become a standard method to identify chloroplast genes (Bottomley 1982). There are many examples of different chloroplast genes from various plant species that can be successfully expressed in *E. coli* by recognition of chloroplast Shine-Dalgarno sequences (Gatenby et al. 1981; Gatenby and Castleton 1982; Fluhr et al. 1983; Bovenberg et al. 1984a and 1984b; Zhu et al. 1984; Rothstein et al. 1985; Fukuzawa et al. 1985; Gatenby and Rothstein 1986). There probably are, however, other features that define a ribosome binding site than a Shine-Dalgarno sequence that is complementary to the 3′ end of 16S rRNA. A number of chloroplast genes show this complementarity, but others do not (Whitfeld and Bottomley 1983). An example is the spinach *atpB* gene (Zurawski et al. 1982), which is reported to possess no obvious ribosome binding structure (Whitfeld and Bottomley 1983). Nevertheless, we find that the spinach gene can be highly expressed in *E. coli,* indicating that translational initiation is efficient (Gatenby and Rothstein, unpublished observations). Many chloroplast genes could be expressed in *E. coli* by direct cloning, but the presence of intervening sequences in some genes (Koller et al. 1984) would make them refractory to expression.

For nuclear gene expression in bacteria, translational fusion of cDNA clones with the ribosome binding site and N-terminus of β-galactosidase has frequently been used to ensure successful initiation of translation of the eukaryotic gene (Bartels et al. 1985; Wang and Esen 1985; Nasrallah et al. 1985; DasSarma et al. 1986; Izui et al. 1986). The N-terminus of β-lactamase can also be used for translational fusions (Gatenby et al. 1986), as can the outer membrane lipoprotein leader peptide (Stubbs et al. 1986). For synthesis of thaumatin in *E. coli,* Edens et al. (1982) placed the cDNA initiation codon directly downstream from the ribosome binding site of the tryptophan leader peptide. This apparently resulted in translation initiating at the authentic preprothaumatin AUG codon. Synthetic ribosome binding sites have also been used (Wasmann et al. 1986).

It is apparent that bacterial ribosomes can also fortuitously recognize plant nuclear RNA sequences in the region of an initiation codon, resulting in correct translation of the plant protein. Such recognition of a eukaryotic site by bacterial ribosomes is perhaps unusual because of the basic differences in the translational mechanisms between bacteria and eukaryotes (Kozak 1983); however, there are three recorded examples of just such a recognition involving plant genes. Nakamura et al. (1986) used immunological screening of potato tuber cDNA clones to identify clones that expressed patatin, a major tuber protein. A more-detailed analysis revealed that a prepatatin protein of the correct size was synthesized in *E. coli,* indicating that translation was initiating correctly at the eukaryotic AUG codon. A similar result was also obtained when a full-length cDNA clone for α-amylase from wheat was transformed into *E. coli* (Gatenby et al. 1986). A pre-α-amylase protein of the correct size was synthesized at about 20% of the efficiency of synthesis of a β-lactamase/α-amylase fusion protein. Sequence analysis of the 5′ nontranslated region of the α-amylase gene (5′-CACA*GAGCUGA*AGACG[AUG]-3′) showed striking complementarity to the 3′ end of *E. coli's* 16S rRNA (5′*UCACCUC*CU*U*A-3′) that could give rise to a Shine-Dalgarno-type interaction. The third example of translational recognition of a plant AUG codon was the expression of bean phytohemagglutinin (Hoffman and Donaldson 1987). The intronless structural gene contained a possible Shine-Dalgarno ribosome binding site nine nucleotides upstream from the initiator methionine, which apparently led to the synthesis of phytohemagglutinin of the correct size.

Several plant nuclear cDNA clones have been successfully expressed in yeast by relying on the correct recognition of the plant initiator methionine by yeast ribosomes. The genes for thaumatin (Edens et al. 1984), phaseolin (Cramer et al. 1985), α-gliadin (Neill et al. 1987), phytohemagglutinin (Tague and Chrispeels 1987), and glutathione-*S*-transferase (Wiegand et al. 1986) have all been expressed in this manner. Gene fusions with the small subunit of *Chlamydomonas* ribulose bisphosphate carboxylase have also been expressed using the algal initiator methionine (Hurt et al. 1986). Alternatively, gene fusions with the N-terminal coding sequence of yeast genes have been constructed to make use of yeast translational initiation signals. Examples are fusions between phosphoglycerate kinase and α-amylase (Rothstein et al. 1984, 1987), acid phosphatase and phaseolin (Cramer et al. 1986), and cytochrome *c* and zein (Coraggio et al. 1986).

A particularly interesting feature of plant gene expression in *E. coli* and yeast has been the identification of translational regulation. Jensen et al. (1986), using soybean leghaemoglobin sequences for expression in yeast, were able to show that heme modulates the translation of chimeric mRNAs fused to the plant 5′ noncoding region through interaction with the noncoding region. They suggested that heme controls expression at a posttranscriptional level, probably by regulating the efficiencies of translation. From this, a mechanism for control of leghaemoglobin expression in the soybean nodule can be inferred. It has also been shown that the expression in *E. coli* of the overlapping *atpB* and *atpE* genes from chloroplasts is regulated very closely by translational coupling. Synthesis of the downstream ε subunit is dependent on synthesis of the upstream β subunit, and indicates that such a regulatory mechanism may operate in chloroplasts (Gatenby et al., submitted for publication).

5.4 RECOGNITION OF PLANT SIGNAL PEPTIDES

The N-terminal signal peptides that direct proteins through the membranes of eukaryotic and prokaryotic cells have a highly conserved structure. This feature enables the signal peptides from widely divergent species to be recognized when expressed in foreign cells (reviewed in Randall and Hardy 1984; Garoff 1985). It is clear, from the accumulating DNA sequence data on plant genes, that secreted proteins also contain these conserved signal peptides at their N-terminus. Several recent experiments have revealed that plant signal peptides can be recognized across species barriers when the appropriate proteins are synthesized in *E. coli* or yeast (reviewed by Gatenby 1987).

Pea cytochrome *f* is a chloroplast-encoded thylakoid membrane protein (Willey et al. 1983) with an N-terminal signal peptide (Willey et al. 1984) that is synthesized within the organelle. When this gene was expressed in *E. coli,* two cytochrome *f* polypeptides were detected that corresponded in size to pre–cytochrome *f* and the mature, processed form (Rothstein et al. 1985). For a more-detailed examination of membrane binding, various sequences from the 5′ end of the cytochrome *f* gene were fused in the correct reading frame with *lacZ,* and the cellular location of the hybrid polypeptides in *E. coli* were determined. It was found that the N-terminal parts of cytochrome *f* directed the β-galactosidase fusion proteins to the cytoplasmic membrane, and that membrane localization was most efficient when the intact signal peptide was present. Fusion within the signal peptide so that the processing site was absent reduced the efficiency of membrane binding. In addition it was found that membrane insertion of fusion proteins containing signal peptides was prevented in a temperature-sensitive *secA* strain at the nonpermissive temperature. Taken together these results revealed both that a chloroplast signal peptide is recognized in a specific fashion by the bacterial secretory pathway, and that the mechanism of protein secretion or insertion within the chloroplast shows homologies with the standard signal peptide–mediated model for protein translocation (Rothstein et al. 1985).

Plant nuclear-encoded signal peptides can also be recognized in *E. coli.* When a cDNA clone for wheat α-amylase was expressed in *E. coli* two different proteins were synthesized using two types of plasmid constructions (Gatenby et al. 1986). In one plasmid, a fusion protein between the N-terminus of β-lactamase and the α-amylase coding sequence was synthesized that contained two signal peptides in tandem. The β-lactamase signal, followed by the α-amylase enzyme, was secreted into the periplasm and both signal peptides were removed, indicating that the bacterial signal peptidase had recognized the plant cleavage site. In a second plasmid, a native, unfused, α-amylase was synthesized that contained only the wheat signal peptide at its N-terminus. This was also recognized in *E. coli,* resulting in secretion of the enzyme into the periplasmic space. However, this process of recognition was not efficient: The α-amylase polypeptide appeared to be inactive as an enzyme until the signal peptide was removed. Although most of the enzyme activity from both plasmid constructions was recovered from the periplasmic space, indicating secretion and cleavage, most of the actual α-amylase protein was retained in the cytoplasm as an unprocessed and presumably inactive polypeptide (Gatenby et al. 1986). This illustrates that although signal peptide recognition can occur between species, it may not be particularly efficient. Low efficiency of signal peptide recognition was also observed for preprothaumatin expression (Edens et al. 1984). There are indications that other nuclear-encoded plant signal peptides may be recognized. Both Hoffman and Donaldson (1987), using phytohemagglutinin genes, and Nakamura et al. (1986), using patatin genes, detected polypeptides whose size was similar to that of the mature proteins and suggested that cleavage of the respective signal peptides may occur.

The recognition of plant signal peptides, and their subsequent secretion, has also been examined in yeast. Edens et al. (1984), using a preprothaumatin clone, were able to show, from analysis of the amino-terminal amino acid sequence, that there was a complete and precise removal of the plant signal peptide. They were unable to obtain any data on the location of the processed thaumatin (Edens and van der Wel 1985). The wheat α-amylase signal peptide can also be recognized in yeast, resulting in strains capable of secreting the wheat enzyme and degrading starch (Rothstein et al. 1984, 1987). In one particular construction the α-amylase cDNA clone was fused to the N-terminus of yeast phosphoglycerate kinase, resulting in the location of the α-amylase signal peptide in an internal position 38 amino acids from the N-terminus. Even so, the signal peptide was recognized, resulting in cleavage of the protein and secretion of a processed form from the cell into the medium. The specificity of this recognition was demonstrated by introducing a deletion into the signal peptide that abolished secretion. The processing time for conversion of the pre-α-amylase to its mature form was exceptionally long, taking about 30 to 60 minutes. This slow rate was due to the internal location of the signal peptide. In recent work Rothstein and colleagues (1987) have repositioned the signal peptide at the N-terminus of the α-amylase protein and have been unable to detect the precursor form, indicating a very rapid recognition of the signal peptide.

Cramer et al. (1985) obtained data showing that bean phaseolin synthesized in yeast was glycosylated and had a molecular weight consistent with the removal of the signal peptide. The protein was not secreted, but its cellular location was unknown. The observed glycosylation would suggest that the protein had passed through the endoplasmic reticulum, and perhaps the Golgi region. Hurt et al. (1986) have also shown that the cleavable presequence of the small subunit of ribulose bisphosphate carboxylase is recognized in yeast and will direct fused proteins into the mitochondria. Phytohemagglutinin is transported to protein bodies in plant cells, but when expressed in yeast it is localized efficiently to the vacuole (Tague and

Chrispeels 1987). This indicates that vacuolar targeting information is recognized between two highly divergent species.

5.5 ASSEMBLY OF MULTISUBUNIT PLANT PROTEINS

There are a number of reasons why it may be desirable to synthesize and assemble the polypeptides of oligomeric enzymes from plants in *E. coli*. One reason is that cDNA that complements a bacterial mutation may be identified through its ability to assemble into an active enzyme. Also, the separation from its normal cell of a pathway leading from gene expression to subunit assembly, and its reconstitution in *E. coli,* facilitates a more-precise investigation of what is required for successful assembly by providing the opportunity to mutate, use different subunit genes, and otherwise modify the system. Another reason, particularly important where subunits are encoded by a multigene family, is that assembly from a cloned gene will provide an enzyme with identical, well-characterized subunits of a particular type, rather than a plant-derived enzyme that may be a heterogeneous mixture of subunits giving rise to an average value when used in enzyme assays. The opportunity for synthesizing hybrid enzymes with subunits from different species, or even *E. coli* subunits, is also provided.

An attempt was made to synthesize a hybrid enzyme using plant and bacterial subunits (Gatenby and Rothstein 1986). To see if chloroplast β-ATPase subunits could be assembled into an *E. coli* membrane-bound ATPase complex, the maize *atpB* gene was fused to *lacZ* and the fate of the resulting hybrid β-galactosidase polypeptides was examined. By using different hybrid proteins, sucrose gradient and differential centrifugation, membrane solubilization, and mutations in the bacterial F_o membrane receptor, it was demonstrated that the maize β-ATPase subunit did bind to the *E. coli* inner membrane in a manner that indicated specific recognition by the F_o receptor.

The assembly of an active hexadecameric ribulose bisphosphate carboxylase in *E. coli* has become a particularly active area of research, since the synthesis of the active enzyme in bacteria would provide an opportunity to modify the enzyme to improve net photosynthetic yield (reviewed in Ellis and Gatenby 1984). Early experiments were designed to at first express (Gatenby et al. 1981), and then amplify (Gatenby and Castleton 1982), the levels of synthesis of the large subunit from maize. Unfortunately, the insoluble nature of the large subunit (Gatenby 1984; Somerville et al. 1986) prevented its use in enzymatic or reconstitution assays. The obvious experiment of expressing both small subunit and large subunit genes from plants together also failed to result in assembly of an active holoenzyme (Bradley et al. 1986). Several groups then developed various expression systems based on cyanobacterial ribulose bisphosphate carboxylase genes that have led to the assembly of an active holoenzyme (Christeller et al.

1985; Gatenby et al. 1985; Gurevitz et al. 1985; Tabita and Small 1985). This assembly of the active 16-subunit enzyme now provides an opportunity for genetic modification and subunit exchange experiments in vivo using an enzyme that has a plantlike structure. Attempts at forming a hybrid cyanobacteria–higher plant enzyme have also been successful, resulting in an enzyme with wheat small subunits and cyanobacterial large subunits (van der Vies et al. 1986).

Glutamine synthetase is another multisubunit plant enzyme that has been assembled in *E. coli* following the expression of the gene from alfalfa (*Medicago sativa*) (DasSarma et al. 1986). In plants the enzyme is of central importance in nitrogen metabolism and has an octomeric structure. The glutamine synthetase cDNA clone was fused to bacterial transcription and translation signals, and synthesis of the polypeptide was observed. Electrophoresis of bacterial extracts on denaturing and nondenaturing gels revealed that the plant polypeptide was of the expected size and its mobility was consistent with an octomeric quaternary structure. It also appeared that essentially all of the soluble plant glutamine synthetase was in its native conformation, indicating that it was efficiently folded and assembled (DasSarma et al. 1986).

Subunit assembly also seems to occur when the phytohemagglutinin polypeptide from *Phaseolus vulgaris* (Hoffman and Donaldson 1987) and the phosphoenolpyruvate carboxylase polypeptide from maize (Izui et al. 1986) are synthesized in *E. coli*. The phytohemagglutinin from bacteria was purified by affinity chromatography and its subunit structure investigated by size-exclusion high-pressure liquid chromatography. It was found to elute in precisely the same way as the bean tetrameric protein, suggesting that the lectin self-assembled into a tetrameric structure. In the example of maize phosphoenolpyruvate carboxylase, assembly of an active tetrameric enzyme might have occurred in *E. coli,* although direct evidence of assembly was not provided (Izui et al. 1986). The phosphoenolpyruvate carboxylase from maize is usually active as a tetramer, and the demonstration of activity in *E. coli* using the cloned gene, together with the presence of a high–molecular weight form on nondenaturing gels, suggests that assembly probably occurs.

5.6 PRODUCTION OF PROTEINS FOR COMMERCIAL APPLICATIONS AND ANALYTICAL STUDIES

In many of the examples described so far, the expression of plant genes in microorganisms was used to study the interaction of various microbial components (RNA polymerase, ribosomes, secretory mechanisms) with structural features of the plant genes or their products. However, for many investigations, the synthesis of a plant protein in a microbial cell serves the purpose of providing either a product of commercial value or a protein that can be isolated for biochemical analyses.

One protein that is considered to have potential commercial applications is thaumatin, the intensely sweet-tasting protein from the African plant *Thaumatococcus daniellii.* This protein is a potential substitute for sugar in low-calorie foods, and has other properties, such as flavor enhancement (Edens and van der Wel 1985). In early work Edens et al. (1982) cloned a cDNA for thaumatin and expressed this in *E. coli* as a preprothaumatin polypeptide. The authors observed low yields of thaumatin in bacteria, which, combined with the possible unsuitability of *E. coli* for the biotechnological production of food-grade proteins, suggested that production in yeast should be considered (Edens and van der Wel 1985). The thaumatin cDNA clone was expressed in both *S. cerevisiae* (Edens et al. 1984) and *K. lactis* and was correctly processed. No information on yields has been provided, but if adequate and economic, these transformed organisms could presumably be an attractive source of thaumatin; the plant does not bear fruit outside its natural environment.

Another possible application of the production of plant proteins is in the brewing industry, where the development of amylolytic yeast strains may have some value in the production of both potable alcohol and fuel alcohol from starchy materials. The development of strains of yeast with a starch-degrading ability has been reported using the α-amylase genes from wheat (Rothstein et al. 1984, 1987) and mouse (Thomsen 1983; Filho et al. 1986) and the glucoamylase gene from *Aspergillus* (Innis et al. 1985). In all of these examples the enzymes are secreted into the culture medium, where they can degrade starch. In the potable alcohol industry the use of recombinant DNA to produce amylolytic yeast is likely to require a barley (or other cereal) α-amylase for both legislative and customer-preference reasons. The successful exploitation of such yeast strains will depend on the price of alternative enzyme supplies from the commerical enzyme business. Since starch-degrading enzymes have now moved from the specialty to the commodity markets (Layman 1986) with a corresponding decrease in price, the use of modified yeasts may appear less attractive. The production of barley β-glucanase by yeast may also be of value in fermentation (Jackson et al. 1986).

A wide range of biological phenomena are being investigated by synthesizing an appropriate plant protein in *E. coli* and yeast and subsequently analyzing it. These include the molecular basis for the glycopeptide specificity of legume lectins (Stubbs et al. 1986); the structure and properties of cereal storage proteins (Bartels et al. 1985; Wang and Esen 1985; Neill et al. 1987; Coraggio et al. 1986); the assembly (van der Vies et al. 1986; Bradley et al. 1986) and properties (Gatenby 1984; Somerville et al. 1986) of higher-plant ribulose bisphosphate carboxylase subunits; glycosylation sites of phaseolin (Cramer et al. 1985 and 1986); x-ray analysis of ricin crystals (Robertus et al. 1987); the synthesis of fusion proteins for targeting into higher-plant chloroplasts (Van den Broeck et al. 1985; Wasmann et al. 1986); the immunological features of photosystem II (PSII) cores (Nixon et al. 1986); cytoplasmic male sterility in maize (Manson et al. 1986; Dewey et al. 1988); the properties of glutathione-*S*-transferase (Moore et al. 1986); and studies of fatty acid metabolism with the acyl carrier protein (Beremand et al. 1987).

5.7 SYNTHESIS OF PLANT PROTEINS FOR SCREENING AND CLONE IDENTIFICATION

Sometimes the only purpose of engineering a bacterium so that it synthesizes a particular plant protein, or part of a particular protein, is to be able to screen clones with an antibody that can detect and isolate that protein. Tittgen et al. (1986) used this type of screening to isolate the cDNA clones for 14 intrinsic and peripheral thylakoid membrane proteins and one stromal protein from spinach. Nakamura et al. (1986) also used immunological identifications of patatin cDNA clones when screening a library made from poly(A)$^+$ RNA from potato tubers, and Crawford et al. (1986) used similar methods to isolate a cDNA clone for nitrate reductase from squash.

In contrast to screening clones directly on Petri dishes, clone confirmation by polypeptide synthesis in intact cells has been used in two instances. Nasrallah et al. (1985) identified a cDNA clone thought to encode a *Brassica oleracea* *S*-locus-specific glycoprotein in *E. coli,* and then used an antiserum raised against the authentic protein to confirm that they had indeed obtained the required clone. Identification by expression in yeast was also used by Wiegand et al. (1986) to confirm that a cDNA clone for maize glutathione-*S*-transferase had been successfully isolated.

5.8 COMPLEMENTATION OF BACTERIAL MUTATIONS

Genes from maize, soybean, and alfalfa have been demonstrated to complement appropriate *E. coli* mutations. Complementation was used to isolate a cDNA clone for maize phosphoenolpyruvate carboxylase (Izui et al. 1986). The clone was conveniently screened by virtue of its ability to complement the glutamate-requiring phenotype of a phosphoenolpyruvate carboxylase mutant of *E. coli*. Bacterial *ppc* mutant cells were transformed with cDNA clones and screened for their ability to grow, without glutamate supplement, following replica plating. One out of 8,400 transformants grew. The specific activity of the maize-encoded enzyme was about half of that of the *E. coli* *ppc* gene product encoded on a plasmid. The cDNA-encoded maize enzyme also possessed the allosteric properties characteristic of the enzyme isolated from maize.

For complementation with the alfalfa glutamine synthetase gene, a previously identified cDNA clone was used to construct an expression plasmid (DasSarma et al. 1986). Expression of glutamine synthetase was initially demonstrated by complementation of a deficient (Δ*glnA*) bacterial mutant. Although the bacterial mutant required glutamine for growth, the presence of the expression plasmid allowed growth in the absence of glutamine. This result was supported by enzyme assays for glutamine synthetase that revealed activity in the complemented mutant, but not in the uncomplemented mutant. Lytic complementation was used to identify the soybean gene encoding meso-diaminopimelate dehydrogenase (Wenko et al. 1985). A phage clone was isolated that corrected via complementation the succinyl diaminopimelate aminotransferase lesion in an *E. coli* strain.

5.9 PROTEIN STABILITY, SOLUBILITY, AND ACCUMULATION

The diverse range of plant proteins that have been synthesized in both *E. coli* and yeast displays an equally diverse range of properties when produced in these microorganisms. Sometimes these proteins possess altered properties when compared with the plant-derived material, indicating that some requirement for correct folding, assembly, or processing has not been provided in the microbial milieu.

There have been very few studies of the stability of plant proteins using pulse-chase experiments. We have observed in published (Rothstein et al. 1984) and unpublished work that wheat α-amylase is very stable in yeast. Little or no degradation of wild-type or fusion α-amylase proteins was observed after 90 to 120 minutes. Similarly, wheat α-amylase expressed in *E. coli* is completely stable for more than 90 minutes (Gatenby, unpublished observations). This stability is not due to the formation of insoluble, stable protein aggregates, because the α-amylase activity from both *E. coli* and yeast is found in the soluble fraction of disrupted cells. In contrast, the observed stability of the large subunit of maize ribulose bisphosphate carboxylase in *E. coli* is possibly the result of formation of insoluble aggregates (Gatenby 1984; Somerville et al. 1986). Pulse-chase experiments revealed a total absence of protein degradation for up to four hours (Gatenby 1984). It is interesting that when the small subunit of wheat ribulose bisphosphate carboxylase is synthesized in *E. coli* as the mature protein it is very unstable, with a half-life of less than 15 minutes (Gatenby et al. 1987). However, if the large subunit of maize and the small subunit of wheat are synthesized together in the same bacterial cell, the small subunit is stabilized and is not degraded after 120 minutes. It should also be noted that in this example small subunit synthesized alone is soluble, but when synthesized with the large subunit it becomes insoluble. Perhaps the two subunits together form an aggregate that protects the small subunit from proteolysis. There may be some variation in the solubility of large subunits, since we have observed that the wheat large subunit synthesized in *E. coli* is significantly more soluble than the maize large subunit (Gatenby 1984; Bradley et al. 1986). Another very stable protein synthesized in *E. coli* is the spinach β-ATPase subunit. Pulse-chase experiments have revealed that this protein is stable for at least 60 minutes following induction of synthesis (Gatenby and Rothstein, unpublished observations). A standard method for trying to stabilize a foreign protein synthesized in *E. coli* is to fuse the gene with *lacZ* to form a fusion protein. We suspect that this may have occurred with the expression of pea cytochrome *f* in *E. coli* (Rothstein et al. 1985). Using immunoblotting of membranes from cells synthesizing wild-type cytochrome *f*, we were unable to detect the protein. However, if cytochrome *f* was then fused to β-galactosidase, the fusion protein could easily be detected in the membrane fraction. It is possible that the wild-type protein was recognized as foreign in the membrane and degraded, but that fusion with β-galactosidase stabilized it.

Various solubility properties have been reported for a number of proteins expressed in *E. coli*. Soluble proteins that have been described in the literature include castor bean ricin (O'Hare et al. 1987; Piatak et al. 1988), alfalfa glutamine synthetase (DasSarma et al. 1986), spinach acyl carrier protein (Beremand et al. 1987), maize phosphoenolpyruvate carboxylase (Izui et al. 1986), wheat large subunit of ribulose bisphosphate carboxylase (Bradley et al. 1986) and wheat small subunit (Gatenby et al. 1987). Several insoluble proteins are known to be synthesized in *E. coli*, but these must be distinguished as those that become insoluble because of a possible problem with folding and those that become insoluble because of membrane recognition and attachment. The insolubility of the maize large subunit of ribulose bisphosphate carboxylase (Gatenby 1984; Somerville et al. 1986) and pea lectin (Stubbs et al. 1986) have been attributed to the former reason. The lectin example is interesting because the plant-produced lectin and its precursor are soluble proteins. The insolubility of lectin aggregates formed in *E. coli* may have been caused by overexpression or may have resulted from the presence of additional hydrophobic amino acids at the N-terminus (Stubbs et al. 1986). The observed insolubility of maize large subunits was not due to overexpression since it was observed within a range of production from 100 to 60,000 molecules per cell (Gatenby 1984). Insolubility due to membrane binding in *E. coli* has been observed for the chloroplast proteins cytochrome *f* and the β-ATPase subunit. In cytochrome *f*, the presence of a plant signal peptide on a cytochrome *f*–β-galactosidase fusion protein directed the protein into the insoluble membrane fraction of bacterial cells (Rothstein et al. 1985). Similarly, the recognition of a maize β-ATPase–β-galactosidase fusion protein by the F_0 ATPase receptor sites on a bacterial membrane also caused much of the fusion protein to associate with the *E. coli* membrane fraction (Gatenby and Rothstein 1986). It has been observed that the temperature at which cells are grown can affect the solubility of castor bean ricin (O'Hare et al. 1987; Piatak et al. 1988).

The final levels to which a plant protein will accumulate in a microbial cell will be a reflection of the rates of synthesis, based on the efficiency of transcriptional and translational initiation, and the stability of the final product. As observed with numerous examples of foreign gene expression in microbial cells, the levels of accumulation of the plant product are frequently less than expected. Even so, respectable final yields have been obtained for a number of plant proteins. The following levels of accumulation have been reported in *E. coli*: 1% of total protein as spinach acyl carrier protein (Beremand et al. 1987); 1% of total protein as alfalfa glutamine synthetase (DasSarma et al. 1986); 10–20% of insoluble protein as pea lectin (Stubbs et al. 1986); 5% of total protein as maize gamma-zein (Wang and Esen 1985); 1% of total protein as wheat glutenin-β-galactosidase (Bartels et al. 1985); 2% of total protein as maize ribulose bisphosphate carboxylase large subunit (Gatenby 1984); 6–8% of total protein as castor bean ricin (Piatak et al. 1988); and 500 molecules per cell of thaumatin (Edens and van der Wel 1985). The maximum levels of expression that have been attained were for a fusion protein between the maize β-ATPase subunit and β-galactosidase. Transcription originated from an induced P_L promoter and translation initiated at the chloroplast ribosome binding site giving rise to total cellular protein content in *E. coli* of 7–8% (Gatenby and Rothstein, unpublished observations). Very low levels of expression have also been encountered with fusions between pea cytochrome *f* and β-galactosidase. Levels of only 15 to 60 units of enzyme activity per milligram of protein were observed and these were not signficantly improved by using a P_L promoter and a transcription antiterminator (Rothstein et al. 1985). It was suspected in this case that a problem with translational initiation was occurring, since there is a potential secondary structure at the ribosome binding site that could interfere wth ribosome entry.

In yeast there are several examples of the quantitation of the levels of expression of and protein accumulation from plant genes. Cramer et al. (1985) were able to show that bean phaseolin was present at levels of 0.01–0.03% of the total cellular yeast protein when expressed using plant sequences. These levels were subsequently improved to about 3% when the yeast acid phosphatase promoter and regulatory regions were used to express the phaseolin gene (Cramer et al. 1986). With wheat α-amylase, the levels of enzyme present in yeast cells represented only about 0.1% of cellular protein, even when using the efficient phosphoglycerate kinase transcriptional and translational signals (Rothstein et al. 1984). An interesting observation on thaumatin expression in yeast indicated that the protein was not detectable unless the protein's signal peptide was also present. It was suggested that unless the protein was transported through a membrane with the signal peptide, it would be degraded by proteases (Edens et al. 1984). Maize zein can accumulate to levels of 5% (Coraggio et al. 1986) and wheat α-gliadin is reported to represent 0.1% of yeast cell protein (Neill et al. 1987).

5.10 EXPRESSION IN OTHER BACTERIAL SPECIES

As described in this review, plant genes have been expressed principally in *E. coli* and in yeast. There are some examples of the expression of chloroplast genes in other bacterial species. Zhu et al. (1984) demonstrated the synthesis of *Nicotiana otophora* and *Chlamydomonas reinhardtii* large subunit of ribulose bisphospahte carboxylase in the Gram-positive bacterium *Bacillus subtilis*. Synthesis was detected by in situ immunoprecipitation. The gene for the large subunit of maize ribulose bisphosphate carboxylase has also been expressed in the Gram-negative bacteria *P. oxalaticus* and *Rhizobium melitoti* (Gatenby and O'Gara, unpublished observations). In this experiment the maize gene was cloned in both orientations in a broad host-range plasmid and transferred by conjugation from *E. coli* into the recipient species. Expression of the gene product was detected by immunoblotting and the similar levels of expression in both orientations indicated that chloroplast sequences were probably being used for expression.

REFERENCES

Bartels, D., Thompson, R.D., and Rothstein, S. (1985) *Gene* 35, 159–167.

Beremand, P.D., Hannapel. D.J., Guerra, D.J., Kuhn, D.N., and Ohlrogge, J.B. (1987) *Arch. Biochem. Biophys.* 256, 90–100.

Bernard, H.U., and Helinski, D.R. (1980) in *Genetic Engineering,* vol. 2 (Setlow, J.K., and Hollaender, A., eds.), pp. 133–167, Plenum Press, New York.

Bottomley, W. (1982) in *Methods in Chloroplast Molecular Biology* (Edelman, M., Hallick, R.B., and Chua, N.-H., eds), pp. 573–580, Elsevier, Amsterdam.

Bottomley, W., and Whitfeld, P.R. (1979) *Eur. J. Biochem.* 93, 31–39.

Bovenberg, W.A., Howe, C.J., Kool A.J., and Nijkamp, H.J.J. (1984a) *Curr. Genet.* 8, 283–290.

Bovenberg, W.A., Koes, R.E., Kool, A.J., and Nijkamp, H.J.J. (1984b) *Curr. Genet.* 8, 231–241.

Boyer, S.K., and Mullet, J.E. (1986) *Plant Mol. Biol.* 6, 229–243.

Bradley, D., and Gatenby, A.A. (1985) *EMBO J.* 4, 3641–3648.

Bradley, D., van der Vies, S.M., and Gatenby, A.A. (1986) *Philos. Trans. R. Soc. Lond.* 313, 447–458.

Christeller, J.T., Terraghi, B.E., Hill, D.F., and Laing, W.A. (1985) *Plant Mol. Biol.* 5, 257–263.

Corraggio, I., Compagno, C., Martegani, E., et al. (1986) *EMBO J.* 5, 459–466.

Cramer, J.H., Lea, K., Drong, R.F., et al. (1986) in *Molecular Biology of Seed Storage Proteins and Lectins* (Shannon, L., and Chrispeels, M.J., eds.), pp. 183–191, Waverly Press, Baltimore, MD.

Cramer, J.H., Lea, K., Schaber, M.D., and Kramer, R.A. (1987) *Mol. Cell. Biol.* 7, 121–128.

Cramer, J.H., Lea, K., and Slightom, J.L. (1985) *Proc. Natl. Acad. Sci. USA* 82, 324–338.

Crawford, N.M., Campbell, W.H., and Davis, R.W. (1986) *Proc. Natl. Acad. Sci. USA* 83, 8073–8076.

DasSarma, S., Tischer, E., and Goodman, H.M. (1986) *Science* 232, 1242–1244.

Dewey, R.E., Siedow, J.N., Timothy, D.H., and Levings, C.S. (1988) *Science,* 239, 293–295.

Edens, L., Bom, I., Ledeboer, A.M., et al. (1984) *Cell* 37, 629–633.

Edens, L., Heslinga, L., Klok, R., et al. (1982) *Gene* 18, 1–12.

Edens, L., and van der Wel, H. (1985) *Trends Biotechnol.* 3(3), 1–4.

Ellis, R.J., and Gatenby, A.A. (1984) in *The Genetic Manipulation of Plants and Its Application to Agriculture* (Lea, P.J., and Stewart, G.R., eds.), pp. 41–60, Oxford University Press, New York.

Erion, J.L., Tarnowski, J., Peacock, S., et al. (1983) *Plant Mol. Biol.* 2, 279–290.

Filho, S.A., Galembeck, E.V., Faria, J.B., and Frascino, S. (1986) *Bio/Technol.* 4, 311–315.

Fluhr, R., Fromm, H., and Edelman, M. (1983) *Gene* 25, 271–280.

Fukuzawa, H., Uchida, Y., Yamano, Y., Ohyama, K., and Komano, T. (1985) *Agricult. Biol. Chem.* 49, 2725–2731.

Garoff, H. (1985) *Annu. Rev. Cell Biol.* 1, 403–445.

Gatenby, A.A. (1983) in *Plant Biotechnology* (Mantell, S.H., and Smith, H., eds), pp. 269–297, Cambridge University Press, New York.

Gatenby, A.A. (1984) *Eur. J. Biochem.* 144, 361–366.

Gatenby, A.A. (1987) in *Plant Membranes: Structure, Function, Biogenesis* (Leaver, C., and Sze, H., eds.), pp. 289–304. Alan R. Liss, New York.

Gatenby, A.A., Boccara, M., Baulcombe, D.C., and Rothstein, S.J. (1986) *Gene* 45, 11–18.

Gatenby, A.A., and Castleton, J.A. (1982) *Mol. Gen. Genet.* 185, 424–429.

Gatenby, A.A., Castleton, J.A., and Saul, M.W. (1981) *Nature* 291, 117–121.

Gatenby, A.A., and Cuellar, R.E. (1985) *Eur. J. Biochem.* 153, 355–359.

Gatenby, A.A., and Rothstein, S.J. (1986) *Gene* 41, 241–247.

Gatenby, A.A., van der Vies, S.M., and Bradley, D. (1985) *Nature* 314, 617–620.

Gatenby, A.A., van der Vies, S.M., and Rothstein, S.J. (1987) *Eur. J. Biochem.* 168, 227–231.

Gray, M.W., and Doolittle, W.F. (1982) *Microbiol. Rev.* 46, 1–42.

Gruissem, W., and Zurawski, G. (1985a) *EMBO J.* 4, 1637–1644.

Gruissem, W., and Zurawski, G. (1985b) *EMBO J.* 4, 3375–3383.

Gurevitz, M., Somerville, C.R., and McIntosh, L. (1985) *Proc. Natl. Acad. Sci. USA* 82, 6546–6550.

Hoffman, L.M., and Donaldson, D.D. (1987) *Bio/Technol.* 5, 157–160.

Hurt, E.C., Soltanifar, N., Goldschmidt-Clermont, M., Rochaix, J.D., and Schatz, G. (1986) *EMBO J.* 5, 1343–1350.

Innis, M.A., Holland, M.J., McCabe, P.C., et al. (1985) *Science* 228, 21–26.

Izui, K., Ishijima, S., Yamaguchi, Y., et al. (1986) *Nucleic Acids Res.* 14, 1615–1628.

Jackson, E.A., Ballance, G.M., and Thomsen, K.K. (1986) *Carlsberg Res. Commun.* 51, 445–458.

Jensen, E.Ø., Marcker, K.A., and Villadsen, I.S. (1986) *EMBO J.* 5, 843–847.

Koller, B., Gingrich, J.C., Stiegler, G.L., et al. (1984) *Cell* 36, 545–553.

Kong, X.F., Lovett, P.S., and Kung, S.D. (1984) *Gene* 31, 23–30.

Kool, A.J., de Haas, J.M., Haring, M.A., and van Grinsven, M.Q.J.M. (1986) in *Nuclear Techniques and In Vitro Culture for Plant Improvement,* pp. 463–470, International Atomic Energy Agency, Vienna.

Kozak, M. (1983) *Microbiol. Rev.* 47, 1–45.

Langridge, P., Eibel, H., Brown, J.W.S., and Felix, G. (1984) *EMBO J.* 3, 2467–2471.

Layman, P.L. (1986) *Chem. Eng News* Sept. 15, pp. 11–14.

Manson, J.C., Liddell, A.D., Leaver, C.J., and Murray, K. (1986) *EMBO J.* 5, 2775–2780.

Moore, R.E., Davies, M.S., O'Connell, K.M., et al. (1986) *Nucleic Acids Res.* 14, 7227–7235.

Nakamura, K., Hattori, T., and Asahi, T. (1986) *FEBS Lett.* 198, 16–20.

Nasrallah, J.B., Kao, T.-H., Goldberg, M.L., and Nasrallah, M.E. (1985) *Nature* 318, 263–267.

Neill, J.D., Litts, J.C., Anderson, O.D., Greene, F.C., and Stiles, J.I. (1987) *Gene* 55, 303–317.

Nixon, P.J., Dyer, T.A., Barber, J., and Hunter, C.N. (1986) *FEBS Lett.* 209, 83–86.

O'Hare, M., Roberts, L.M., Thorpe, P.E., et al. (1987) *FEBS Lett.* 216, 73–78.

Piatak, M., Lane, J.A., Laird, W., et al. (1988) *J. Biol. Chem.* 263, 4837–4843.

Pouwels, P.H., Enger-Valk, B.E., and Brammer, W.J. (1985) *Cloning Vectors,* Elsevier, Amsterdam.

Randall, L.L., and Hardy, S.J.S. (1984) *Microbiol. Rev.* 48, 290–298.

Robertus, J.D., Piatak, M., Ferris, R., and Houston, L.L. (1987) *J. Biol. Chem.* 262, 19–20.

Rothstein, S.J., Gatenby, A.A., Willey, D.L., and Gray, J.C. (1985) *Proc. Natl. Acad. Sci. USA* 82, 7955–7959.

Rothstein, S.J., Lahners, K.N., Lazarus, C.M., Baulcombe, D.C., and Gatenby, A.A. (1987) *Gene* 55, 353–356.

Rothstein, S.J., Lazarus, C.M., Smith, W.E., Baulcombe, D.C., and Gatenby, A.A. (1984) *Nature* 308, 662–665.

Shinozaki, K., and Sugiura, M. (1982) *Nucleic Acids Res.* 10, 4923–4934.

Somerville, C.R., McIntosh, L., Fitchen, J., and Gurevitz, M. (1986) in *Methods in Enzymology,* vol. 118 (Weissbach, A., and Weissbach, H., eds.), pp. 419–433, Academic Press, Orlando. FL.

Stubbs, M.E., Carver, J.P., and Dunn, R.J. (1986) *J. Biol. Chem.* 261, 6141–6144.

Tabita, F.R., and Small, C.L. (1985) *Proc. Natl. Acad. Sci. USA* 82, 6100–6103.

Tague, B.W., and Chrispeels, M.J. (1987) *J. Cell Biol.* 105, 1971–1979.

Thomsen, K.K. (1983) *Carlsberg Res. Commun.* 48, 545–555.

Timmis, K.N. (1981) in *Genetics as a Tool in Microbiology* (Glover, S.W., and Hopwood, D.A., eds.), pp. 49–109, Cambridge University Press, New York.

Tittgen, J., Hermans, J., Steppuhn, J., et al. (1986) *Mol. Gen. Genet.* 204, 258–265.

Tohdoh, N., Shinozaki, K., and Sugiura, M. (1981) *Nucleic Acids Res.* 9, 5399–5406.

Van den Broeck, G., Timko, M.P., Kausch, A.P., et al. (1985) *Nature* 313, 358–363.

van der Vies, S.M., Bradley, D., and Gatenby, A.A. (1986) *EMBO J.* 5, 2439–2444.

Wang, S-Z., and Esen, A. (1985) *Plant Sci.* 42, 49–54.

Wasmann, C.C., Reiss, B., Bartlett, S.G., and Bohnert, H.J. (1986) *Mol. Gen. Genet.* 205, 446–453.

Wenko, L.K., Treick, R.W., and Wilson, K.G. (1985) *Plant Mol. Biol.* 4, 197–204.

Werneke, J.M., Zielinski, R.E., and Ogren, W.L. (1988) *Proc. Natl. Acad. Sci. USA* 85, 787–791.

Whitfeld, P.R., and Bottomley, W. (1983) *Annu. Rev. Plant Physiol.* 34, 279–310.

Wiegand, R.C., Shah, D.M., Mozer, T.J., et al. (1986) *Plant Mol. Biol.* 7, 235–243.

Willey, D.L., Auffret, A.D., and Gray, J.C. (1984) *Cell* 36, 555–562.

Willey, D.L., Huttly, A.K., Phillips, A.L., and Gray, J.C. (1983) *Mol. Gen. Genet.* 189, 85–89.

Zhu, Y.S., Lovett, P.S., Williams, D.M., and Kung, S.D. (1984) *Theor. Appl. Genet.* 67, 333–336.

Zurawski, G., Bottomley, W., and Whitfeld, P.R. (1982) *Proc. Natl. Acad. Sci. USA* 79, 6260–6264.

PART

II

Regulation of Gene Expression in Plants

CHAPTER 6

The Molecular Architecture of Plant Genes and Their Regulation

Keith Elliston
Joachim Messing

Plant gene expression is a highly regulated activity that is dependant upon a multitude of factors. One element of the regulation of gene expression is the primary structure of the gene itself. Primary structure, composed of the linear sequence elements of the DNA and RNA, may be responsible for the binding of various cellular factors that influence the expression of any particular gene (von Hippel et al. 1984; McKnight and Tjian 1986). This influence may be at the DNA level and affect the transcription of the gene, or at the level of transcription itself and affect the processing and stability of the mRNA. Translation of the mRNA is another level at which genes are regulated, as in most cases a protein is the end product of gene expression. To understand the regulation of genes at each of these levels better, it is essential to study their primary structure first, in an attempt to relate that structure to the expression of the gene.

We would like to thank our colleagues for the many thoughtful discussions that coalesced into this article. The computer analysis was supported by a grant to J.M. from the U.S. Department of Energy (FG02-84ER13210).

This study of the architecture of genes concentrates on the basic sequence elements that contribute to the primary structure of genes. Certainly secondary and tertiary structures are important, but we will limit ourselves to the discussion of the primary structure of genes transcribed by RNA polymerase II.

6.1 THE ELEMENTS OF PRIMARY STRUCTURE

Conserved primary sequence elements such as the TATA box, the AUG translation start codon, and the AATAAA poly(A) signal compose the primary structure of a gene. The TATA box is the basic primary structural unit of the promoter and has been well described (Proudfoot 1979). The promoter controls the transcription of a gene and constitutes the first level of gene regulation. Early studies on plant gene structure have shown that the TATA box is very well conserved among plants (Messing et al. 1983). However it has recently been demonstrated that monocots and dicots tend to observe differing forms of the TATA consensus (Elliston and Messing 1988). Most notably, the monocots conserve a cytosine immediately prior to the TATAAA sequence, while the dicots have no such conservation.

The processing of mRNAs comprises the second level of gene regulation and is influenced by two units of primary structure: the mRNA splicing site and the poly(A) signal. Initial analysis of intron splicing signals suggested that plants and animals conserved similar splicing signals (Brown 1986). However, recent data have suggested that plants may conserve an intron splicing signal that is distinct from that observed in animal systems (Wiebauer et al. 1987). There may even be a distinction between monocot and dicot intron splicing signals, as many monocot mRNAs are not properly spliced in transgenic dicots (Keith and Chua 1986). The correct splicing of introns from the hnRNA is an important step in mRNA processing prior to export from the nucleus. The recognition of the 5′ cap structure appears to be a requirement for the correct splicing of the first intron in a HeLa cell splicing system, and may be the first step in the processing of hnRNA in general (Ohno et al. 1987). Polyadenylation is also a requirement for export of the mRNA to the cytoplasm and its subsequent translation, and is dependant upon the recognition of the poly(A) addition signal.

The poly(A) addition signal has been well studied in animal systems and consists of the consensus AATAAA, which precedes the poly(A) addition site by 10 to 30 nucleotides (Proudfoot and Brownlee 1976; McLaughlin et al. 1985). This signal has been shown to form a complex with a nuclear factor that is necessary for both cleavage and polyadenylation of the transcript (Zarkower and Wickens 1987). The poly(A) signal in plants has a unique structure. A consensus AATAAA is frequently found 5 to 30 bases downstream from the termination codon. However, this signal is rarely used as the poly(A) signal, since the poly(A) tail usually occurs 10 to 30 bases downstream of a more-divergent poly(A) signal. This signal conserves the AAT of the consensus, but diverges in the 3′ direction. It has been postulated that either secondary structure or other factors may allow for the recognition of the upstream poly(A) signal, followed by polyadenylation at the downstream site (Heidecker and Messing 1986).

The context of the AUG start codon is a third level of gene regulation and has been of particular interest because it has a direct impact on the translation of mRNA and hence on the expression of a gene. The mechanisms behind translational control of gene expression have come under intense scrutiny lately, and it is generally accepted that one of the first steps in translation is the recognition of the sequences immediately preceding the AUG start codon by the 40S preinitiation complex, either directly or through related factors (Moldavé 1985; Kozak 1984 and 1986). Sequence specificity has been shown to affect the efficiency of translation, with a putative consensus derived (Kozak 1986; Heidecker and Messing 1986; Lütcke et al. 1987). Recognition of the AUG codon by the preinitiation complex is a rate-limiting step in the translation of a mRNA, and thus a major influence on gene expression.

All of these units combined comprise the basic architecture of a gene and influence the expression of the gene on these three main levels: transcription, mRNA processing, and translation.

6.2 THE ANALYSIS OF PRIMARY STRUCTURE

The advent of gene transfer in plants has provided a powerful tool for the understanding of the regulation of gene expression and the effects that primary structures have upon this regulation. Gene transfer, or the incorporation of new genetic material into a target cell, has been accomplished by various means in plants. To consider these, it is necessary first to understand the two types of expression in the target cells. These are transient expression and stable expression.

Transient expression is simply the expression of a gene for a short period of time, with a decay of expression over time. This is due to the lack of the incorporation of the DNA into the cell's genome. Thus, the inserted DNA is not replicated and is usually rapidly degraded. When the DNA is not incorporated into the cell's chromatin, it is impossible to study regulation that results from the normal chromatin environment. But the analysis of cis-acting regulatory elements and factors that bind them, the analysis of strong promoter units, and the processing of subsequent messages can all be performed in transient expression systems. Transient expression assays usually take advantage of reporter gene systems such as chloramphenicol acetyltransferase, β-glucuronidase, or luciferase. Convenient and sensitive assays for these gene products presently exist.

Stable expression results from the stable incorporation of the gene into the cell's genome followed by maintained expression over a long period of time, from several cell generations to several plant generations. This technique is much better suited to the study of tissue-specific gene regulation, because expression can be studied on the organismal level. Ideally, the information obtained from transient expression studies can be verified and studied in more depth using stable expression. This would allow the definition of other related sequences or cellular factors necessary for proper gene expression. Stable transformation is an important prerequisite for analysis of the inserted gene by classical genetic experiments, such as segregation, recombination, and quantitative genetic analysis.

Gene transfer experiments allow the manipulation of DNA by such in vitro methods as chemical or site-directed mutagenesis, followed by analysis of the effects of those manipulations. This is a powerful tool, especially for the study of effects that primary structure contributes to the expression of genes. The conserved structures of the TATA box, the AUG initiation context, and the poly(A) signal can be manipulated and assayed for activity under actual cellular conditions, rather than in cell-free systems. Such in vitro alterations, followed by introduction into protoplasts via electroporation (the permeabilization of the target cell's membrane using a pulsed electric field), are a powerful means of analyzing the conservation of primary structure and its effect on expression on a general level. Both transient and stable transformation systems are useful in this respect, and such experiments will provide considerable information on the actual function of these conserved primary structures. Primary structures that are recognized by tissue-specific factors may also be manipulated by mutagenesis, but only a stable transformation system would allow the effects of such mutations, on both tissue-specific and developmental expression, to be studied.

Several means of transforming plant cells exist and have varying efficiencies of both transient and stable transformation. The most widely used technique is that of *Agrobacterium*-mediated transformation (Rogers et al. 1986). This method takes advantage of the transducing Ti plasmid of *Agrobacterium tumefaciens* and results in the mobilization of the T-DNA into the plant cell with subsequent incorporation into the nuclear genome. This results in a relatively high frequency of stable transformants. Transformed cells from *Agrobacterium*-mediated gene transfer can, depending upon the initial source of the cells, be regenerated into complete plants. This is a critical step in the analysis of tissue-specific expression in transformed plants, as well as genetic analysis of subsequent generations resulting from crosses. The transformed gene can also be moved into different "genetic environments" using standard breeding techniques to generate completely new varieties of important crop species. These techniques may have a dramatic influence on agriculture in years to come. But, unfortunately, many agriculturally important crop species are not transformable using the *Agrobacterium* system, because it seems able to function only in a variety of dicotyledenous systems.

Electroporation requires that the cells first be freed of their cell walls (Fromm et al. 1985). These protoplasts are then incubated with a mixture of the target DNA and electroporated. This results in a high frequency of transient expression, and a very low frequency of stable transformation. Electroporation is a valuable technique, as it is a rapid and efficient means of transforming plant cells. A third method, that of calcium phosphate–polyethylene glycol–mediated DNA uptake has been successful in producing both transient expression and stable transformation (Potrykus et al. 1985). This system has proven difficult to reproduce and often results in variable yields of transformed cells. The regeneration of complete plants from protoplasts is very difficult, and has not yet been accomplished for many important crop species. This limitation affects the same plant species that are not transformable using the *Agrobacterium* system and makes stable gene transfer into these species difficult. Although progress is being made to remove these barriers, the characteristics of transient expression systems continue to limit the use of these techniques effectively for any type of tissue-specific or developmental expression studies in some of the best genetically studied plants. The use of microprojectiles to deliver DNA into plant cells has recently been reported (Klein et al. 1988). This technique has been successfully used to transform embryogenic maize tissue culture cells, and may provide the long sought means for generating transformed plants of all species.

6.3 FUNCTIONAL ORGANIZATION OF GENES

The study of primary gene structure requires organization of the data. Several different classes of regulated gene expression lend themselves to study: developmentally induced expression, environmentally induced expression, and constitutive expression.

Developmentally induced gene expression, the expression of a specific set of genes in specific tissues at certain periods during development, has been a significant area of plant research. Developmental expression presumably requires tissue-specific factors that are present only at certain stages of development. One of the most extensively studied developmentally regulated groups of genes in plants is the storage protein genes, which are expressed early in seed development and are of great agronomic importance. Specifically, the endosperm storage proteins of the cereal crop species such as corn, wheat, and barley and the seed storage proteins of the legumes have received intense scrutiny.

Environmentally induced genes, the specific genes that are expressed upon stimulation by various environmental factors, constitute the most widely studied group of genes. Hormonal induction of gene expression has received much attention recently and has applicability to commercial interests in agriculture. This is due in part to the fact that many herbicides imitate plant hormones. The induction of genes by stress is an area of investigation that has received renewed attention. The heat-shock genes are relatively well studied in *Drosophila* and other animal species and have molecular relatives in plants (Ananthan et al. 1986). Another stress-induced gene is that encoding alcohol dehydrogenase, which is induced by anaerobic stress much like its animal counterparts. The major class of environmentally induced genes that are unique to plants, however, are the light-induced genes. These genes are most-often regulated by the phytochrome system, yet many also have independent regulatory mechanisms.

Constitutive expression is most-often associated with the housekeeping genes. These are genes that are expressed in most or all tissues and are necessary for the maintenance of the individual cell. Among these are the genes encoding microtubule, microfilament, and histone proteins. A common characteristic of these genes is that they must all be expressed in every cell, regardless of its role in the organism. Thus these housekeeping genes utilize the various systems operating in the developmentally diverse cell types of the organism.

6.4 THE USE OF COMPUTERS IN STRUCTURAL ANALYSES

The primary structure of genes is most-easily studied using computers. This is primarily due to the fact that manual analysis of the data presented by the sequence of a single gene is unmanageable when it comes to the recognition of base sequence patterns. When comparing tens or hundreds of genes, this task is completely impossible without the aid of a microcomputer and is much easier when undertaken using a minicomputer. Considerable software is available both commercially and through academic institutions and ranges in capability from the simple analysis of restriction patterns and derived protein sequences, to the calculation of secondary structures and the elucidation of complex sequence patterns both within a sequence and among sequences.

DNA sequence data on most of the published DNA sequences are available through several databases. These include Genbank (Bilofsky et al. 1986) and the EMBL (Hamm and Cameron 1986) nucleotide sequence databases. Databases are very important to molecular biology and the study of gene structure. The amount of data available on nucleotide sequences to date is enough to fill an average-sized set of encyclopedias. This volume has been growing exponentially at such a rate that printing of the complete set of nucleic acid sequence information will soon be impossible. Electronic databases make the printing of sequence information unnecessary while allowing wide access to the information. Databases such as Genbank and EMBL are central locations for the deposit of sequence information, which is then available to the scientific public at large. These databases are available to individual investigators either on-line or as recorded information as supplied in many of the sequence analysis software packages on the market, making this vast amount of data immediately available for analysis.

The analysis of DNA sequence data based upon the relationship that exists between the sources of the data is a relatively new concept in the study of plants. Only recently have sufficient data been available for such comparisons. When we organize the present plant DNA sequence information, based upon the phylogenetic relationships between the species, many differences are found to exist. Focusing on the main conserved structures of the TATA box, the context of the AUG initiator codon, the codon usage, and poly(A) signals demonstrates that significant variation exists in these areas. The most-significant areas of divergence appear to be in the contexts of the AUG initiator codon and the conservation of the TATA box.

The conservation of the AUG context is quite different between monocots and dicots. Monocots conserve the context CCGC(A/C)AUGGC, while dicots conserve the context AA(A/C)AAUGGC. Since the translation of messages requires the recognition of the AUG initiator codon by the 40S preinitiation complex, the conservation of divergent contexts suggests that divergent recognition mechanisms may exist between monocots and dicots.

The divergence noted at the TATA box is somewhat less distinct. Monocots conserve the sequence CTATAAA, while dicots conserve the sequence TATAAATA. Whether the extension in the 3′ direction that the dicots conserve is significant as compared with the conservation in the 5′ end that the monocots demonstrate has yet to be investigated. Yet, these differences in conservation are sufficient to suggest that genes from these two groups should be considered separately when studying the conservation of primary structure.

Conservation of gene structure based upon gene function is an interesting alternative to the phylogenetic perspective. Different regulatory sequence elements may be conserved within related species, but are they also conserved between evolutionarily distant species because of functional constraints? Certain elements of primary structure may be conserved because the regulatory factors that interact with them may have little tolerance for mutation. The combination of cis-required sequence elements and trans-acting regulatory factors makes up a powerful genetic control system that is sensitive to a variety of influences. To begin to understand the requirements of such macromolecular interactions, it is helpful first to look for potential cis-acting regulatory sequences.

Our interest has thus been directed to the analysis of the conservation of primary structures of functionally similar genes; that is, how the conserved cis sequence elements vary when one considers the genes with respect to their function. Therefore, we have compiled DNA sequence data on all the plant RNA polymerase II genes from the Genbank and EMBL nucleotide sequence databases and organized them with respect to their regulation. We then further subdivided these sequences by their specific class of origin, that is, monocot versus dicot. Thus, we have a set of genes that are grouped first by gene function and second by phylogenetic relationship. For the sake of brevity, we have confined ourselves to the analysis of the TATA region and the AUG initiator codon context. Since much of the sequence data available concentrate on the coding regions of genes, the characterization of other more-distantly located regulatory elements would be premature. We analyzed the data using a VAX 11/785 minicomputer (Digital Equipment Corporation, U.S.A.) with the software package distributed by the University of Wisconsin Genetics Computer Group (GCG; Deveraux et al. 1984).

6.5 STORAGE PROTEIN GENES

The storage protein genes represent the majority of the data available for developmentally expressed genes. Storage proteins can be divided into several classes based upon their physicochemical properties. The alcohol-soluble prolamines and the acid- or base-soluble glutelins represent the major classes of monocot storage proteins. These proteins are usually found in protein bodies, extensions of the rough endoplasmic reticulum in the endosperm tissue of the seed. The salt-soluble globulins are the main representatives of this class of genes among the dicots, and are found in a vacuolelike structure in the cotyledon. These two tissues, the endosperm and the cotyledon, are considerably different and thus these two groups of functionally similar genes must be considered separately.

The endosperm is the major storage tissue of many monocot seeds and provides most of the protein and energy necessary for the postgermination development of the embryo before photosynthesis commences. The endosperm storage protein genes are expressed in the endosperm tissues of their respective plants, beginning several days after pollination and continuing up to seed maturation. Analysis of the regulation of these genes is complicated by the fact that the endosperm is a terminally differentiated tissue and has a triploid genome. Two of the genome equivalents are maternal, supplied by the two polar nuclei located in the center of the embryo sac and resulting from maternal meiosis. The remaining component is supplied by one of the sperm nuclei, a result of paternal meiosis. Thus, the expression of the genes in this tissue follows variant rules with respect to the classical diploid genome model. This can be utilized for the analysis of dosage-dependant gene regulation and for the study of regulatory mechanisms that are inherited in a non-Mendelian fashion, as for instance, the presetting and erasure or the genomic imprinting of regulatory genes in maize endosperm (Das and Messing, unpublished observations; Kirihara and Messing, unpublished observations). The analysis of regulation based upon the primary structure of genes expressed in the endosperm must consider these varying aspects of endosperm development.

The promoter regions of the monocot storage protein genes faithfully conserve the monocot TATA consensus CTATAAATA. In addition, several upstream sequences are conserved. Approximately 25 bases 5′ of the TATA box, the sequence $A_{2-3}GA_{2-5}$ is less-stringently conserved (Figure 6–1) and is similar to an earlier-described AGGA box (Messing et al. 1983). This sequence is again repeated another 15 bases upstream in the form of $RRA_{2-3}GA_{2-4}TG$. Thus the general structure of the monocot storage protein

```
                        Upstream Sequences                                      TATA box
---------------------------------------------------------------------------------------------------------------
 Blyhorb   AAACATCCAAACACAATTGTTAAATTTGATGCAAAGAAGGAAAGAGATGAAGCCATGGCTA.CTATAAATA.GGCATGTAGTATAGAGATCAT
Taglulag   TTCTTAGGCTGAACTAACATCACCGTACACACAACCATTGTCATGAACCTTCACCACGTCC.CTATAAAAG.CACAACCAATCCCCACAAT
Tagluldg   CTTCTTAGGGCTGAACTAACCTCGTCGTGCACAACCATTGTCCTGAACCTTGACCACGTCC.CTATAAAAG.CCCAACCAATCTCCACAATTTCATCATCA
Whtgliaa   AGCATCCAAGCACTTTCCACACAAACAAATGCCAATTGTGAAAGAGATCATTCCATGGCAA.CTATAAATA.GCCCCATAGCATGAC
Zma27kza   AAGCTGCCTACAGCCGTCTCGGTGGCATAAGAACACAAGAAATTGTGTTAATTAATCAAAG.CTATAAATA.ACGCTCGCATGCCTGTGCACTTCTCC
 Zmapml1   CTATACCTTATCTGTTGTTTGGAAAAAAGACAAAATCCAAAAAAATATATATGAGTCTCCT.GTATAAGCA.CACAAATTTTGGTATGTATGTCCA
   Zmaz4   TACTAATATTCTTTTTGCAAAATCGAAAATTAATCTTGCACAAGCACAAGGACTGAGATGT.GTATAAATA.TCTCTTAGATTAGTAGATAATAT
  Zmaza1   CTATATCTTATCTATTGTTTGGAAAAAATACAAAATCCAAAAAAAATATTTGAGACCTCAC.CTATATAAA.TAGCTCCCATATCAGTAGTTAATCC
---------------------------------------------------------------------------------------------------------------
Consensus  -T-T--C-TAC---TT-CAT--AA--AAA---AAATA---AAA-AAAT-TT--CATCGC---CTATAAATA--CC-TC-AAT-TC-A-A-T-T--------
---------------------------------------------------------------------------------------------------------------
```

FIGURE 6–1 The promoter sequences of the monocot storage protein genes were aligned at the TATA consensus, and a consensus was derived at the 50% level. The sequences are as follows: Blyhorb, a barley hordein gene; Taglulag, the wheat glutelin A gene; Tagluldg, the wheat glutelin D gene; Whtgliaa, the wheat gliadin gene; Zma27kza, the maize 27-kd A zein gene; Zmapml1, the maize pml1 zein gene; Zmaz4, the maize zein 4 gene; and Zmaza1, the maize zein A1 gene. Dots indicate artifically introduced gaps.

gene promoters is 5′-RRA_{2-3}GAA_{2-4}TG—8—A_{2-3}GA_{2-5}—12—YCTATAAATA-3′. The conservation of the TATA box is very strict, with variation occurring only at a single position within the sequence for any gene. The conservation of the first AAGAA box is less stringent and quite variable in length, and in several cases the second AAGAA box is almost absent. These are the results of analyzing five zein sequences, two gliadin genes, two glutelin genes, and one hordein gene.

The zein genes of maize have been shown to conserve the AUG context RCAAYANYNUANCAACAAUGGC (Heidecker and Messing 1986). When this is compared with the sequences from barley and wheat, several structures become apparent (Figure 6–2). Proximal to the AUG initiator codon the sequence ANCAACAAUGGC is very highly conserved. Approximately 20 bases 5′ of the AUG codon, the sequence UAGCAACAA is well conserved in the zein genes, and more distantly followed in the other storage protein genes. Another 22 bases upstream the sequence AAACNAACAAA is followed, although rather loosely. Thus the monocot storage protein genes follow the general AUG contextual structure 3′-AAACNAACAAA—12—UAGCAACAA—4—ANCAACAAUGGC-5′. Each of the sequences analyzed—fourteen distinct zein genes, three wheat genes, and one barley gene—conserved some variant of this motif.

Many of the seed storage protein genes of dicots are expressed in the cotyledons, a 2N tissue. Dicots do have a triploid endosperm tissue similar to the monocots but this tissue is small and transient in nature, as it is absorbed early in seed development. The size of the endosperm tissue in dicots is also quite variable: tobacco has a rather large endosperm and petunia has a very small endosperm. The major class of seed storage proteins is represented by the lectins, glycinins, and related proteins. Like the monocot storage proteins, these proteins are utilized as a nitrogen source by the developing embryo during and following germination.

The dicot seed storage protein gene promoters conserve a promoter consensus similar to the monocot's consensus. The TATA box follows the form TYTCTATAAATAG (Figure 6–3). Upstream of the TATA consensus, fairly long stretches of poly(A) or poly(T) occur, with an interspersed cytosine or guanine. While no discrete consensus sequence seems to exist, the pattern of recurring adenine or thymine (and often both) seems to suggest some role for this structure. Recently, Jofuku et al. (1987) have found that the sequence AAATTTTAAA from the soybean lectin promoter is bound by a 60-kd embryo-specific protein. This protein also binds other embryo-specific promoters, but not promoters that are not expressed in the embryo. This further suggests that the poly(A)-poly(T) regions found in the dicot storage protein genes may be of significance.

Little conservation of the AUG initiator context can be found in the dicot storage protein genes. The AUG context itself is AYNACNAUGGCY and is quite variable (Figure 6–4). At approximately 25 and 34 bases upstream of the AUG context, two distinct but degenerate regions are apparent

```
                        5' Untranslated Leader              AUG
----------------------------------------------------------------------

  Blyhorb   CACAAGTATCAAAACCAAGCAAC.ACTCGTTAACACCAATCCACCATGAAG
  Tagliagl  CGCATCATTCAA.GCCAAGCAAGCAGTGGTCAATACAAATCCACCATGAAG
  Taglulag  CGAGCACCACAAAATAGAGATCAATTCACTAACAGTCCATCGA.AATGGCT
Whtgligba   GCACAAGCATCAAAACCAAGCAAAGTAGTT.AGCACCAATCCATCATGAAG
  Zma10kz   ........................GGAAGCAAGGACACCACCGCCATGGCA
  Zma15kz   AGCAAAGCGGCAGTGCGTAGAGAGGATCGTCGAACCAGAAC.AGCATGAAG
  Zma27kza  TTAGCTTTATCTACTCCAGAGCGCAGAAGAACCCGATCGAC.ACCATGAGG
   Zmaa20   AAGCAACAUAGAAAGUGGAAUCCAGUAGCAACAACAGAG.CAACAAUGGCG
   Zmagl2   ................................CCGATCGAC.ACCATGAGG
  Zmapcm1   .CAGAAACACACCAAGCGAAGCTACCTAGCAACGACTTAACAACAATGGCT
  Zmaz19a   .UUGAGACCAACUAGCAAUAUAGAAAGCACAAUAUUGUACCAAUAAUGGCA
 Zmaz19b1   .GACCAACTAGCAACATAGAAAGCACTAGCAATAGTGTACCAACAATGGCA
 Zmaz19c1   ....CAACAUAGAAAGUGGAAUCCAGUAGCAACAACAGAGCAACAAUGGCG
 Zmaz19d1   .........ACATAGTGAAGTACACTAGCAACATCTTAGCACCAATGGCA
  Zmaz22a   .CAAAAACACACCAAGCGAAGCGCACUAGCAACGACCUAACACCAAUGGCU
  Zmaz22c   .................................AACCUAACAACAAUGGCU
    Zmaz4   .TTGAGACCAACTAGCAACATAGAAAGCACAATATTGTACCAATAATGGCA
    Zmaz7   .CAGAAACACACAAAGGGAAGTGCACTAGCAACTTCCTAACAACAATGGCT
----------------------------------------------------------------------

Consensus   ----AA-C--A-AA-----A----A-TAGCAA-A-C--A-CAACAATGGC-
----------------------------------------------------------------------
```

FIGURE 6–2 The 5′ untranslated regions of the monocot storage protein genes were compiled and aligned at the AUG initiator codon. Sequences are represented as either DNA or RNA (T versus U), depending upon the source of the sequence data (EMBL represents cDNA sequences as RNA). A consensus is displayed and was derived at the 50% level. The sequences represented are as follows: Blyhorb, a barley hordein gene; Tagliag1, a wheat gliadin gene; Taglu1ag, a wheat glutelin gene; Whtgligba, a wheat gliadin gene; Zma10kz, a maize 10-kd zein gene; Zma15kz, a maize 15-kd zein gene; Zma27kza, the maize 27-kd A zein gene; Zmaa20, the maize 20-kd A20 zein gene; Zmag12, a maize glutelin gene; Zmapcm1, a maize zein gene; Zmaz19a, Zmaz19b1, Zmaz19c1, and Zmaz19d1, 19-kd maize zein genes; Zmaz22a and Zmaz22c, 22-kd maize zein genes; Zmaz4, a maize zein 4 gene; and Zmaz7, a maize zein 7 gene. Dots indicate artifically introduced gaps.

from our analysis. At −25 the sequence ANUUARNA occurs, while at −34 the sequence ANCCA is found. The dicot storage protein AUG context can be considered roughly to follow 5′-ANCCA—4—ANUUARNA—9—AY-NACNAUGGCY-3′. It is also interesting to note that the 5′ untranslated leader of dicot seed storage protein transcripts tends to be much smaller than that of the monocot transcripts. In fact, 8 of 16 dicot storage genes analyzed had a 5′ untranslated leader of less than 30 bases. These results are from the analysis of 16 distinct dicot storage protein gene sequences.

```
                     Upstream Sequences                                       TATA box
-------------------------------------------------------------------------------------------------------------
 Peaabnl   TTAATTATGTTTATCATATCGTTTTGTTTTTACATTTACCGTCAGCTATGGTAAGAAGACA..TCTCTATAAATTGC..TTCATACAAGAATGAAATAATGCAA
 Pealega   CAGAATGTCCAATTCTCAACATCCCACTTTCAATGACGTGTCCACCTTCACCACCCTCTCT..TCTCTATAAATTAC..CACTTCTCATTAAGGTTCTCCGC
Pvudleca   CTCAAATAATATTATAAAAGAAAAAAGAATCCATAAAGTAGAGAGAGAGAGTGCAGTTGTT..GTTGTATAAATAGA..GAAGAGAGTGATGGTTAAT
Pvudlecb   AGCTCCCGCCTCTTCACCCGTGTCTTTCTCTCATGTCGCAGTGCTGCGTGCCAACCTCTTC..TCTCTATAAATAGA..GAGAAGAGAATGATGGTTAAT
 Pvulect   CCCTCAAATAATATAATAAAAGAAAAAAAAATGCATAAAGTAGAGAGTTTTTGCATTTGTT..GTTGTATAAATAGA..GAAGAGAGTGATGGTTAATG
Pvuphasl   ATGCGTGTCATCCCATGCCCAAATCTCCATGCATGTTCCAACCACCTTCTCTCTTATATAA..TACCTATAAATACC..TCTAATATCACTCACTTCTTTCA
 Rcricin   ATTGATCAAGTATTTGTGGTTTTCTTTATATAAAAAAATGTATTAGTGTTTTTCTGTATTA..ATTTTATAAGTTCA..TCTTTATGAGAATGCTAATGTATTT
  Soylea   TTAAAAATTATAAAAGAAATATTAATTTTGATAAACAACTTTTGAAAAGTACCCAATAATG..CTAGTATAAATAGG..GGCATGACTCCCCATGCA
  Vfleb4   TGCATGCTGAAGAATGTCACACACGTTCTGTCACACGTGTTACTCTCTCACTGTTCTCCTC..TTCCTATAAATCAC..CGCGCCACAGCTTCTCCACTTCA
-------------------------------------------------------------------------------------------------------------
Consensus  -T-A---T--T--T---AA-A----TT-T--CA-ATA---T--A---T--CT-C--T--T---TTTCTATAAATAGA-----AT-AGAG-T-G-T-AT-T----
-------------------------------------------------------------------------------------------------------------
```

FIGURE 6–3 The promoter sequences of the dicot storage protein genes were aligned at the TATA consensus, and a consensus was derived at the 50% level. The sequences represented are as follows: Peaabn1, a pea albumin gene; Pealega, a pea legumin gene; Pvudleca, Pvudlecb, Pvulect, and Pvuphasl, *Phaseolus vulgaris* lectin genes; Rcricin, a *Ricinus communis* ricin gene; Soylea, a soybean lectin gene; and Vfleb4, a broad bean legumin gene. Dots indicate artifically introduced gaps.

```
                       5' Untranslated Leader            AUG
------------------------------------------------------------
  Bnasspb  ...................................CATACACGAATGGCG
  Cencona  GCAGCACTACTAGTGAAGTAGTGAATATCAATAGGTATACCACCATGGCC
  Peaabn1  ...........................TTTCTTTCTATCAAACAATGGCT
  Peaabn2  ..........GTCCTCGACGCCTACTAAACTTACTACACCTAAGATGACG
  Pealega  ..........CATCACAACCAACATTCACTTAGTATCTCTCTTCATGGCT
 Pvudleca  .............................TGCATGAATGCATACATGGCT
  Pvulect  ..................................GCATGAATGCATGATC
 Pvuphasl  TACTATAATACCCCAACCCAACTCATATTCAATACTACTCTACTATGATG
 Rcppricr  ..............AAACCGGGAGGAAAUACUAUUGUAAUAUGGAUGUAU
  Rcricin  .........CAGCCAATAAAATTCCAGAATTGCTGCAATCAAGGATGAAA
Soyglyala  AAACATTCTCTCCATTGGTCCTTAAACACTCATCAGTCATCACCATGGCC
Soyglyaab  .............AACTTAATTATTAACACTTCCTTAGTTCAATATGGGG
  Soyglyr  CACAACACAAACATTCTCTCCATTGTCTTTGAACACTCATCACCATGGCC
   Soylea  ..............ATCACAGTGCAATTTAGCTGAAGCAAAGCAATGGCT
   Vfleb4  CTTCACTCACAATCCTTCATTAGTTGTTTACTATCACAGTCACAATGTCC
------------------------------------------------------------
Consensus  ------------------------------------A---AA--ATGGC-
------------------------------------------------------------
```

FIGURE 6–4 The 5′ untranslated leader sequence from the dicot storage protein genes was aligned at the AUG start codon. Sequences are represented as either DNA or RNA (T versus U), depending upon the source of the sequence data (EMBL represents cDNA sequences as RNA). A consensus is displayed and was derived at the 50% level. The sequences represented are as follows: Bnasspb, a rapeseed napin gene; Cencona, a jack bean concanavalin gene; Peaabn1 and Peaabn2, pea albumin genes; Pealega, a pea legumin gene; Pvudleca and Pvulect, *P. vulgaris* lectin genes; Pvuphasl, a *P. vulgaris* phaseolin gene; Rcppricr and Rcricin, *Ricinus communis* ricin genes; Soyglyala, Soyglyaab, and Soyglyr, soybean glycinin genes; Soylea, a soybean lectin gene; and Vfleb4, a broad bean lectin gene. Dots indicate artifically introduced gaps.

6.6 LIGHT-INDUCED GENES

Stimulation by light is the main basis for development in plants. The development of chloroplasts as well as various plant organs is determined by both the intensity and the duration of light. Many of the light-regulated genes are controlled by the phytochrome system. Phytochrome exists in two stable spectral forms, Pr and Pfr. Pr is transformed into Pfr when it absorbs red light; conversely, Pfr is converted to Pr when exposed to far red light. Pfr also slowly converts to Pr in the absence of light. The nature of the transduction of the signal from the phytochrome to the regulated gene is unknown. However, it is known that the light response affects not only transcription, but also translation, mRNA stability, and protein turnover rates. The most widely studied of the phytochrome-regulated genes are the ribulose-1, 5-bisphosphate carboxylase small subunit genes and the chlorophyll a/b binding protein genes (Ca/b).

Only two monocot light-induced genes are represented in the Genbank version 50 and EMBL release 11, and only one of these has any promoter data. Therefore little can be said about these genes. However it is interesting to note that both the wheat Ca/b gene and the *Lemna gibba* rbcS gene share the AUG context of AGCGCNAUGGC. As sequence data accumulate, analysis of this sequence will be more valid.

The dicots contribute most of the known sequence information on light-induced genes. Several rbcS genes from pea, tobacco, and petunia have been sequenced, as well as several of the Ca/b genes of petunia and pea. A plastocyanin and a ferredoxin cDNA from *Silene pratensis* have also been sequenced. All of these genes are regulated by the phytochrome system.

Transcription of light-induced mRNAs is postulated to be influenced by light-regulatory elements. One such element is the GT motif described by Kuhlemeier et al. (1987). This element occurs between −168 and −110 in a pea rbcS gene and consists of the sequence 5′-GTGTGGTTAATATATG-3′. Kuhlemeier et al. have also derived a GT consensus from additional pea rbcS genes and rbcS genes from soybean, petunia, and tobacco. The consensus is 5′-RTGTGGYYA(T/A)TA(T/A)G-3′. This sequence is present in the published rbcS sequences, but it is not present in the other published light-induced gene sequences.

The main features of the light-inducible promoters of the dicots are the TATA box and an upstream AG or AGGA-like box (Figure 6–5). The TATA box takes the form of CATTATATATACNRTTT, with the internal TTATATATAC being very highly conserved. The upstream AG box has a rather variable form and is represented by the consensus AGGA$_{1-3}$TAAGRA$_{1-4}$NCT$_{1-3}$. This sequence ends six to ten nucleotides before the beginning of the TATA consensus sequence and bears little resemblance to the CAAT box of animal systems. The CAAT box most-often occurs 30 to 50 nucleotides upstream of the TATA and is a variation of the sequence CAAT. This AG sequence is somewhat similar to the earlier-described AGGA box (Messing et al. 1983), as that sequence consists of YA$_{2-5}$(G/T)NGA$_{2-4}$YY. However, the AG box occurs immediately upstream of the TATA, whereas the AGGA box occurs 36 to 59 base pairs upstream from the TATA consensus. Another AG box of similar composition to the first occurs approximately five to twenty base pairs upstream of the TATA proximal AG box. This AG box is more degenerate and is located in the consensus position described as the AGGA box by Messing et al. (1983).

The 5′ untranslated leader of the dicot light-induced mRNAs shows some interesting conservation. There appear to be two distinct structural units that show heterogeneity in both content and length. The first unit,

```
            Upstream AGGA-like box                       AGGA-like box                           TATA Consensus
-----------------------------------------------------------------------------------------------------------------
Peacab80   .AGAATTGCAAAGAAAA.....(21)............AGATAACACACGATAAGAG...............(6)......CATTATAAATACAGACTCA
Pearubps   .AGGCAGTGGCCAA........(13)......AAGACCATAATATTGGAAATA.GAAAAA............(6)......CATTATATAAAGCAAGTTT
 Pecab13   ..............................................AAGTTTATGAAA..............(5)......AGTTATAAGTATCTTTTCC
Pecab221   ..GAAACCCAATGAAA......(4).......AGATAGAGATATCATAAGATAAGAA..............(13).....CCCCTATATATACACTTAAG
Pecab22r   .AATTAGGAAATAGA.......(11)....................AAGGATAAGGA..............(17).....CATTTAAATAAACTTGTTGG
 Pecab25   .AATGAAGAAACCGA.......(12).....................AGGATAAGGA..............(17).....TATTTATATACACACATGGT
Pecab91r   .AATGAGAAAACAAGA......(11)....................AAGGATAAGGACAATGGAAG......(6).....GTCATATATATACAATTGGT
Tobrbpco   .AAGATGAGGGGG.........(19)....................AGATATA.GGAAATATGTAAAA....(5).....TCATTATATATAGGGTGGTG
-----------------------------------------------------------------------------------------------------------------
```

FIGURE 6–5 The promoter regions of the dicot light-induced genes were aligned at the TATA consensus. Gaps (dots) were introduced to align the upstream AGGA elements, and the bases removed are indicated in parentheses. The sequences analyzed are as follows: Peacab80, a pea chlorophyll a/b binding protein gene; Pearubps, a pea rbcS gene; Pecab13, Pecab221, Pecab22r, Pecab25, and Pecab91r, petunia chlorophyll a/b binding protein genes; and Tobrbpco, a tobacco rbcS gene.

```
         5'cap  upstream AAG     upstream CUU       Consensus AAG                                Consensus CUUC        AUG Context     AUG
------------------------------------------------------------------------------------------------------------------------------------------
Peacab80  (9)....AAGGA.....(10).......................AAAAGC........(7)...............TCTTTTCAATTTCATT....GCAATATAATACACG...ATGGCC
Pearubps  (6)....AAG...............CTTT...(13)..AAGAACTAAGAAAGTCAGAAAAAA......................................................ATGGCT
 Pecab13  .....................CTTTCCTTTTTTGTT..(13)...AAAGAA......................TTTTCCATCTTACTTCTATCATTCT..........ACA...ATGGCT
Pecab221  (26)..................TTTCTCTATTACTTC.(9)...AAAAGAA........................CTCTTTTCTCTTCTTATT............AAACC...ATGGCT
 Pecab25  (26)..................CTCTTCTTTCTGTTT.(13)..GAAAGAG.....................TTCTTCATCTTACTTCT..................ACA...ATGGCT
Pecab91r  (13)...AAGCAACAAATAAA.CTCTTCTTTCTTT......GCAGTAGGTG.....(10)......................TTCTTCCATTTTACTATT.......ACA...ATGGCT
  Sippcy  ...AACTAATAAACAACAAG....TCATTTCATC....(8)..AAACGTAAA.....(10)..............CCCCTTATTCCTTCC.................AAAA...ATGGCA
Tobrbpco  (31).GGAAGTT...........................AAAGGAAAAGAGAGAAAGAGAAA.....................TCTTTCTGTCT...AAGTGTAATTAACA...ATGGCT
------------------------------------------------------------------------------------------------------------------------------------------
```

FIGURE 6–6 The 5′ untranslated leader of the dicot light-induced genes were aligned at the AUG initiation codon, and gaps (dots) were introduced to align the upstream CUUC and AAG elements. Bases removed are indicated in parentheses. Sequences are represented as either DNA or RNA (T versus U), depending upon the source of the sequence data (EMBL represents cDNA sequences as RNA). The sequences analyzed are as follows: Peacab80, a pea chlorophyll a/b binding protein gene; Pearubps, a pea rbcS gene; Pecab13, Pecab221, Pecab25, and Pecab91r, petunia chlorophyll a/b binding protein genes; Sippcy, a *Silene pratensis* plastocyanin gene; and Tobrbpco, a tobacco rbcS gene.

most proximal to the AUG, is the CUUC box (Figure 6–6). This unit varies in length and content, but is basically a repeated permutation of the trinucleotide CUU. Upstream of this moiety is the AAG box. This region is generally purine-rich with a variable central core of the form AAGAA. A third region, which is most prominent in the petunia Ca/b genes, seems to be a repeat of the CUU box. Several sequences also show a repeat of the AAG box upstream of the consensus AAG box and the repeated upstream CUU box. The uniformity of these sequences tends to deteriorate with increased distance from the AUG start codon. These upstream structures are of interest, but no experimental data exist concerning the level at which these sequences might affect gene expression. However, it has been noted that light-regulated translational regulation of rbcS mRNA does take place in amaranth seedlings (Berry et al. 1986). It is also conceivable that these units might exert their effects at the DNA level rather than the RNA level, or that they have no significance as modifiers of the transcription or translation of these messages.

6.7 STRESS-INDUCED GENES

Stress-induced genes range from the heat-shock protein genes to the wound-induced protease inhibitor genes to the alcohol dehydrogenase genes. These genes all share the common feature of induction by some form of physiological stress. The Adh gene is induced by anaerobic stress, as are the heat-shock protein genes. Additionally, as their name suggests, the heat-shock genes are induced by heat stress. The protease inhibitor genes are usually induced by wounding or some other form of physical damage, or by a related viral infection.

Most of the dicot stress-induced promoter data originate from the soybean heat-shock genes (4/6). These promoters have no striking sequence motif, with only a generalized poly(T) region about 30 bases upstream of the TATA box and separated from it by a fairly pyrimidine-rich region. The TATA box itself varies from the core TATAAA, and follows the form TTTAAAT in the heat-shock genes (Figure 6–7). The *Arabidopsis* Adh soybean protease inhibitor promoters do not resemble the heat-shock promoters.

The 5′ untranslated leader of the dicot stress-induced genes has a characteristic primary structure. All of the stress-induced transcripts, from Adh to the heat-shock genes to the protease inhibitors, have a region very high in adenine residues that appears to have a central core of AAAANNNAAAA (Figure 6–8). This sequence varies in length from 12 to 20 residues, and in position from 78 nucleotides upstream to immediately proximal to the AUG start codon. There is little remarkable about the region immediately upstream of the AUG, as a very high amount of variability is noticed.

```
                   poly T box          Pyrimidine Rich Spacer     TATA box                  Txn Start
----------------------------------------------------------------------------------------------------
   Athadh      ..GAGTTTTGTAAATATCTAC.....TTCTTCCAATTACC.....AGCTGCTATATAAAT....CCCCTTCTCTGTTTCTCTTTTCTT..
 Soyciipi      ........................CTGTAATAAGCCAACCT.....AACCTCATCAAACAC....GGGTTCTTTCTCATCACACTGACGG
Soyhsp175      ..CGATTTTTCTGGAA.CGTAC.....ACGATTATCCTTTC.....ACTTACTTTAAATAC....ATCACGATTAGTCAGAAAAACGA...
Soyhsp176      ..ATATTTTTCTGGAA.CATAC.....AAGAGTATCCTTTC.....ACTTCCTTTAAATAC....CTCGAGTGTCCCCATTGACATCA...
  Soyhsp2      ..GGACTTT.CTCGAA.AGTAC....TATATTGCTCCTCTA.....CATCATTTTAAATAC....CCCATGTGTCCTTTGAAGAC......
 Soyhspgm      ..GGATTTTTCTGGAA.CATAC.....AAGATTATCCTTTC.....ACTTCCTTTAAATAC....CTCGCGTATCCCCTTCGTCCTCG...
----------------------------------------------------------------------------------------------------
```

FIGURE 6–7 The promoter regions of the dicot stress-induced genes were aligned at the TATA consensus, and gaps (dots) were introduced to highlight conserved sequences. The sequences analyzed are as follows: Athadh, an *Arabidopsis thaliana* alcohol dehydrogenase gene; Soyciipi, a soybean bowman birk protease inhibitor gene; and Soyhsp175, Soyhsp176, Soyhsp2, and Soyhspgm, soybean heat-shock protein genes.

```
                         poly A box                                          AUG context
------------------------------------------------------------------------------------------------
   Athadh      ............AAAACTAACAAAAGA...(27)................TCAAAAGCAAGTTCTTCACTGTTGATA..ATGTCT
   Ntprpr      <4>.........AAAAUAAAACA.......(14)...........................UUUCUCCUAUAGUC..AUGGGA
   Phochl      <3>........AAAAAAAAAAAA.......(0)........................................ATGGCA
   Soybbi      ...........AAAAACGTAAAGA......(71)...<51>............TTCTTTTCCTTGTGGGGGTTACTA..ATGCAC
 Soyciipi      ..........GAAAAACGTAAAGA......(16)..........................TGGGTTTGAAATAACA..ATGGTG
Soyhsp175      ........AAACGAAACGAAGAAAAGA...(74)...<50>............TCTCGCGAATAAATATATCAAAAG..ATGTCT
Soyhsp176      <4>..........AAACAAGAGAAGAG...(76)...<52>............TATTAGCTAAAGTAACATCAAAAG..ATGTCA
  Soyhsp2      ............GAAAGAAGTGAAGGCA..(78)...<54>............TTTACACAAGACTGATAAGAGAAA..ATGTCT
 Soyhspgm      <3>..........AAACGAAGAAAAAA...(65)...<41>............ATCTCAGCTAAGAAAAACCAAAAG..ATGTCT
 Stpiiirl      <5>.........AAAAUUAAAAGAAAAAGA(24)..................GGCAGUACUAAUUAAUUAUCCAUC..AUGGAU
  Tomwipi      <18>.......ACAACTAAAACAAA....(0)........................................ATGGAG
Tomwipii       <6>.........AAAATTAAAAGAAGACAA(24)..................GGTACTAGTAATCAATTATCCATC..ATGGCT
------------------------------------------------------------------------------------------------
```

FIGURE 6–8 The 5′ untranslated leader of the dicot stress-induced genes were aligned at the AUG initiation codon, and gaps (dots) introduced to highlight conserved sequence elements. Numbers in parentheses indicate distance in bases to the AUG start codon, and numbers in angled brackets indicate bases removed for alignment. Sequences are represented as either DNA or RNA (T versus U), depending upon the source of the sequence data (EMBL represents CDNA sequences as RNA). The sequences analyzed are as follows: Athadh, an *A. thaliana* alcohol dehydrogenase gene; Ntprpr, a tobacco pr protein gene; Phochl, a parsley chalcone synthase gene; Soybbi and Soyciipi, soybean Bowman Birk protease inhibitor genes; Soyhsp175, Soyhsp176, Soyhsp2, and Soyhspgm, soybean heat-shock protein genes; Stpiiirl, a potato Bowman Birk protease inhibitor gene; and Tomwipi and Tomwipii, tomato wound-induced protese inhibitor genes.

6.8 NODULATION GENES

Nitrogen fixation takes place in a specialized structure called the nodule. Nodulation takes place only in the legumes, a small group of dicots of the family Rosidae. A group of host genes, the nodulation genes, are induced by the infection of the legume's root hairs by a member of the bacterial genus *Rhizobium*. These bacteria induce the plant to form nodules for the reduction of nitrogen to ammonia. Many different genes are induced by this infection, and all are hypothesized to be involved either in the reduction of nitrogen or in the maintenance of the *Rhizobium*-plant symbiosis. Among these are the genes for glutamine synthase, xanthine dehydrogenase, and leghaemoglobin. The function of a number of the nodulation genes has not yet been definitively determined. Nodulation has not been demonstrated in the monocot species, and therefore the monocots will not be discussed in this respect.

Little conservation of structure can be found in the nodulation genes. This may be due, in part, to the lack of sequence information on similar genes from diverse species. Most of the information on these genes comes from various nodulation genes from soybean. Further information from other leguminous species will make analysis of these genes possible in the future.

6.9 HOUSEKEEPING GENES

Various gene products are necessary for the basic maintenance of all cells. These housekeeping genes share the common feature that their gene products are required in all of the various cell types that make up the organism, and as such are not restricted to specific tissues. This means that the primary structure of these genes must allow for their efficient expression even in cell types that utilize very specific means of regulating gene expression. The major class of sequenced plant housekeeping genes are the actins. Actin is an integral component of microfilaments. Microfilaments are involved in organelle movement, cytoskeletal structure, and other processes such as cytoplasmic streaming. Extensin is another filament protein and is a component of the extracellular matrix (Lloyd 1987).

Little information is available on the structure of the actin promoters, and analysis of this feature of the actin genes must await further data. The actin and extensin AUG context, however, are well characterized. Upstream of the AUG initiation codon is a region that is common to the actin genes of maize and soybean and to the carrot extensin gene. This region contains the sequence UUUG and is designated the UUUG box. The UUUG box takes the form $U_{3-4}G_{1-2}UU$ and often extends in the 3′ direction with an additional GUU (Figure 6–9). The sequences proximal to the AUG codon have the common feature of beginning with a uracil and con-

```
                          UUUG box                 AUG context
---------------------------------------------------------------
  Darext        .....TTTTGGTTGT....<2>.....TAAAC..ATGGGA
Soyact1g        ......TTTGTTGTTT...<18>...TAGAAG..ATGGCA
Soyact3g        .....TTTTGTTACTT...<42>...TAAAAG..ATGGCA
  Zmaact        ......TTTGGTCTGTT..<9>....TTGAGA..ATGGCT
---------------------------------------------------------------
```

FIGURE 6–9 The 5′ untranslated leader of the housekeeping genes for extensin and actin were aligned at the AUG initiation codon, and gaps (dots) introduced to highlight the conserved UUUG box. Numbers in angled brackets indicate bases removed for alignment. The sequences are shown as DNA, as they were reported in Genbank as such. The sequences analyzed are as follows: Darext, a carrot extensin gene; Soyact1g and Soyact3g, soybean actin genes; and Zmaact, a maize actin gene.

taining three to four adenines followed by a guanine or cytosine immediately adjacent to the AUG. The maize actin varies somewhat from this, but is similar.

6.10 CONCLUSION

The study of gene architecture from the viewpoint of the regulation of gene expression allows the elucidation of primary sequence elements that are required for the correct and coordinate expression of functionally related genes. This type of analysis takes advantage of the variability seen in DNA sequences that are not required for either the function or the expression of a gene. Mutations that cause a function to be dramatically changed or lost are disfavored, and often cause a decrease or lack of viability in the organism. This means that mutations or genetic variation will be noted only where it either doesn't affect or only slightly affects the expression or function of the gene.

If one assumes that regulatory systems act in a hierarchical manner, then it is expected that many more structural genes exist than regulatory genes. Since mutations in the small number of regulatory genes would be amplified by their interaction with the large number of receptor elements, it is unlikely that a large amount of variability is tolerated at this level. However, since the targets for this small number of regulatory elements are relatively widespread and have a lesser impact on the organism, these should tolerate a higher amount of variability. The functionality of such regulatory factors is, therefore, determined not by only one form of macromolecular interaction, but by a broad range of interactions. Thus, in a finely tuned system, one would expect that the conservation of the regulatory trans-acting factors would be much higher than that in the average structural gene.

The cis-acting elements that are required for the correct interaction with the regulatory factors must conserve a basic molecular structure to be able to interact successfully with these factors. These cis-acting elements can be seen as "footprints" of the trans-acting regulatory factors. The conservation noted among similar genes may then be the result of a conserved function, as determined by the cellular factors that are responsible for their expression. Divergence is then expected to occur in regions that are not involved in the macromolecular recognition of the sequence elements by their transacting regulatory factors. From this argument, one can then assume that conserved structures among functionally equivalent genes from diverse species are responsible in some manner for the specific regulation of those genes. Although these cis-required signals may be located in various positions with respect to the transcriptional or translational start sites or both, they certainly involve the sequences around the TATA box and the AUG initiator codon. Additional information can be obtained by studying the specificity of the trans-acting regulatory factors and their coordinate sequence interactions.

The data presented here focus on two of these functional domains, the promoter and the translation initiation site. These data were derived from diverse species, but analyzed with respect to similarity of regulatory response. This has resulted in the observation of some interesting features that are different from those noted when comparing genes that are quite differently regulated but derived from related species.

The promoter structures of the monocot and dicot storage protein genes have several features in common. Both the monocots and the dicots conserve the TATA consensus C$\underline{\text{TATAAA}}$TA and a similar upstream element. In monocots this element is the AG box, which has the structure of poly(A) separated by guanine, while in dicots the conserved structure is a region of poly(A) with poly(T) interspersed. McKnight and Tjian (1986) have noted that a sequence element that is termed the GC box is bound by the cell-type-specific protein spl and seems to be intrinsically involved in gene regulation. Chen et al. (1986) have noted that a GC-like box is required for maximal expression of the alpha subunit of the β-conglycinin gene in soybean. A corollary system has been implicated in the dicot storage protein genes where the sequence AAATTTTAAA is involved in the developmental regulation of a soybean lectin promoter. Jofuku et al. (1987) have shown that an embryo-specific protein binds this sequence preferentially and binds only to embryo-specific promoter regions in general. The variation seen in the monocot sequence as compared with the dicot sequence could then be viewed as the result of the divergence of such a factor, to control expression in the endosperm as compared with the embryo.

Monocot storage protein mRNAs exhibit a very highly structured 5′ untranslated leader sequence. This sequence is found in one form or another in all of the analyzed monocot endosperm storage protein genes. Dicot storage protein genes, which are translated in the cotyledon as opposed to the endosperm, have a very short leader sequence with no observable conservation of structure. This difference may be in part due to the different subcellular locations of these two classes of proteins. The dicot storage proteins are most-often found in vacuolar structures, while the endosperm storage proteins are found in protein bodies that are extensions of the rough endoplasmic reticulum.

The promoter structures of the light-induced genes conserve the upstream AG sequence $AGGA_{1-3}TAAGRA_{1-4}NCT_{1-3}$. Therefore, this structure might be involved in the light-regulated expression of these genes. The TATA box also shows a conserved variation in the form of CATTATATA-TACNRTTT. The core TTATATATA is slightly, yet possibly significantly, different from the overall dicot TATA consensus of TATAAATA. The AUG context of the light-induced transcripts also has some very interesting features. The presence of repeating AAG and CUUC units suggests some sort of binding factors that recognize this alternating structure. Even the two monocot light-induced genes, wheat Ca/b and *L. gibba* rbcS, have the repeated AAG sequence. This is a further indication that this sequence may indeed be involved in the regulation of these genes.

The stress-induced genes also have conserved structures. The heat-shock genes conserve the TATA consensus of TTTAAAT, which is a variant of the TATAAA sequence usually noted for dicot promoters. Perhaps more interesting is the conservation found in the 5′ untranslated leader of the mRNAs of these genes. All of the stress-induced transcripts analyzed conserve the sequence AAAANNNAAAA. This sequence may be involved in a specific cellular translation recognition that would target certain transcripts for translation during periods of stress.

The most-interesting conservation may be that noted for the housekeeping genes. These genes must be expressed in every cell for the organism to survive; no functional mutants of these genes are tolerated by the cell. This necessitates strict control over the expression of these genes in the presence of a variety of different regulatory machinery. Data on the promoters for these genes are incomplete, but some sequence information exists for the 5′ untranslated leader of the mRNAs for actin and extensin. These transcripts all contain the GUU homology. This homology is not limited to either the dicots or the monocots, but exists in both systems and is very high among the various plant species. This also suggests a role for the specific recognition of transcripts that are necessary to the cell at all stages of its development, regardless of induction by outside stimuli.

Unfortunately this analysis has been limited by the amount of sequence data available on specific genes from the diverse species that make up the plant kingdom. The main current in plant molecular biology has been to analyze gene families within a species, while giving little attention to similar gene families in related species. This provides good data on the structures that are common to those genes within that species, but does little to provide information on the conservation of genes with respect to their expression. More emphasis should be put on the analysis of specific types of genes and how they are conserved across the plant kingdom.

The application of molecular biology to agriculture through plant breeding requires that we understand the structures of genes and their functions. The introduction of a gene into the plant is only the first step in the process. That gene will then come under the control of the regulatory mechanisms present in each cell of the plant, and it must be correctly controlled by those mechanisms to be expressed properly. Thus, we must better understand the wide range of regulatory sequence elements and the factors that control them in order to apply this knowledge to the construction of new genes.

REFERENCES

Ananthan, J., Goldberg, A., and Voellmy, R. (1986) *Science* 232, 522–524.

Berry, J., Nikolau, B., Carr, J., and Klessig, D. (1986) *Mol. Cell. Biol.* 6, 2347–2353.

Bilofsky H.S., Burks, C., Fickett, J.W., et al. (1986) *Nucleic Acids Res.* 14, 1–4.

Brown, J. (1986) *Nucleic Acids Res.* 14, 9549–9559.

Chen, Z., Schuler, M., and Beachy, R. (1986) *Proc. Natl. Acad. Sci. USA* 83, 8560–8564.

Das, O., and Messing, J. (manuscript in preparation).

Deveraux, J., Haeberli, P., and Smithies, O. (1984) *Nucleic Acids Res.* 12, 387–395.

Elliston, K., and Messing, J. (1988) in *The Architecture of Eukaryotic Genes* (Kahl, G., ed.), VCH Verlagsgesellshaft: Weinheim, Fed. Rep. Germany.

Fromm, M., Taylor, L., and Walbot, V. (1985) *Proc. Natl. Acad. Sci. USA* 82, 5824–5828.

Hamm, G.H., and Cameron, G.N. (1986) *Nucleic Acids Res.* 14, 5–9.

Heidecker, G., and Messing, J. (1986) *Annu. Rev. Plant Physiol.* 37, 439–466.

Jofuku, K., Okamuro, J., and Goldberg, R. (1987) *Nature* 328, 734–737.

Keith, B., and Chua, N. (1986) *EMBO J.* 5, 2419–2426.

Kirihara, J., and Messing, J. (manuscript in preparation).

Klein, T.M., Fromm, M. Weissinger, A., Tomes, D., Schaaf, S., Sletten, M., and Sanford, J.C. (1988) *Proc. Natl. Acad. Sci.* USA 85, 4305–4309.

Kozak, M. (1984) *Nucleic Acids Res.* 12, 857–873.

Kozak, M (1986) *Cell* 44, 283–292.

Kuhlemeier, C., Green, P., and Chua, N. (1987) *Annu. Rev. Plant Physiol.* 38, 221–257.

Lloyd, C. (1987) *Annu. Rev. Plant Physiol.* 38, 119–139.

Lütcke, H., Chow, K., Mickel, F., et al. (1987) *EMBO J.* 6, 43–48.

McKnight, S., and Tjian, R. (1986) *Cell* 46, 795–805.

McLaughlin, J., Gaffrey, D., Whitton, J., and Clements, J. (1985) *Nucleic Acids Res.* 13, 1347–1368.

Messing, J., Geraghty, D., Heidecker, G., et al. (1983) in *Genetic Engineering of Plants* (Kosuge, T., Meredith, C., and Hollaender, A., eds.), Plenum Press, New York.

Moldave, K. (1985) *Annu. Rev. Biochem.* 54, 1109–1149.

Ohno, M., Sakamato, H., and Shimura, Y. (1987) *Proc. Natl. Acad. Sci. USA* 84, 5187–5191.

Potrykus, I., Paszkowski, J., Saul, M., Petruska, J., and Shillito, R. (1985) *Mol. Gen. Genet.* 199, 169–177.

Proudfoot, N. (1979) *Nature* 279, 376.

Proudfoot, N., and Brownlee, G. (1976) *Nature* 252, 359–362.

Rogers, S., Horsch, R., and Fraley, R. (1986) *Methods Enzymol.* 118, 627–684.

von Hippel, P., Bear, D., Morgan, W., and McSwiggen, J. (1984) *Annu. Rev. Biochem.* 53, 389–446.

Wiebauer, J., Herrero, J., and Filipowicz, W. (1987) in *Abstracts of the 1987 Meeting on RNA Processing.* Cold Spring Harbor Laboratory, Cold Spring Harbor, NY.

Zarkower, D., and Wickens, M. (1987) *EMBO J.* 6, 177–186.

CHAPTER

7

Induction, Commitment, and Progression of Plant Embryogenesis

J.H. Choi
Zinmay R. Sung

The study of plant embryogenesis is fundamental to understanding plant development. During embryogenesis, meristems are formed and the basic plant tissue systems—protoderm, procambium, and ground tissue—are established (Foster and Gifford 1974). The crucial events in embryogenesis are polarity establishment, tissue differentiation, meristem formation, and branching, processes that are fundamental to both embryo and adult development in plants. Embryo development is devoid of the diverse morphological elaborations that distinguish one plant species from another; the morphogenetic events are nearly identical among all dicotyledonous plants. Because of this simplicity and uniformity, plant embryogenesis is a good model system for understanding fundamental strategies of morphogenesis.

The authors thank Dr. Abraham Marcus for his critical reading of the manuscript. The work was supported by grants from the National Science Foundation (DMB 850914) and an unrestricted gift from the Shell Company Foundation to Z.R.S.

that is, embryogenesis and cytodifferentiation. The evidence that 2,4-D plays an important role in this process comes from the study of the abiltiy of single carrot cells to form embryogenic cell clusters (Nomura and Komamine 1985). Single carrot cells exist in spherical, oval, and elongated shapes. While spherical cells were able to divide and form somatic embryos, oval and elongated cells could not. It was possible to rescue the oval-shaped cells by incubating them in 2,4-D for several days. In this way, the single cells divided asymmetrically; the smaller of the two cells divided further into an embryogenic cell cluster, which could undergo embryogenesis upon the removal of 2,4-D. The unequal division resembled the first zygotic division. Polarized distribution of polyadenylated RNA and incorporation of radioactive thymidine were detected in the embryogenic cell clusters as early as the four-cell stage. This polarity was maintained in the globular stage if the cell clusters were transferred into 2,4-D-free medium. But, the polarity disappeared if the clusters were cultured continuously in 2,4-D. Embryogenic potential in the elongated cells in this case could no longer be rescued by 2,4-D.

7.2.2 Developmental Triggers for and Inhibitors of Embryo Morphogenesis

The study discussed above demonstrated that 2,4-D is instrumental in maintaining the embryogenic potential in culture. Moreover, 2,4-D interferes with the establishment of polarity at the globular stage, but not the single cell stage. This finding suggests that the inhibitory effect of 2,4-D is exerted at the transition between globular and heart-stage embryos, when embryogenesis involves polarized growth. One would therefore expect embryogenic cultures to propagate as a mixture of cell clusters and globular-shaped embryos in 2,4-D. This is indeed the case in embryogenic maize, soybean, and other plant cultures (Christianson et al. 1983). Several carrot lines derived from the wild carrot, Queen Ann's lace, can form globular-shaped embryos in 2,4-D, provided the cell clusters are cultured at low density (Sung and Okimoto 1981; Halperin and Wetherell 1975). At high cell density, however, such a cell line can be arrested at the ground state in 2,4-D (Sung and Okimoto 1983).

It has further been found that what triggers the onset of the embryogenic program is the subculture of the cell clusters into fresh medium. This is because, within a few hours, this treatment induces the expression of a set of biochemical properties characteristic of developing embryos: the ability to synthesize the 70-kd protein E_1 (Sung and Okimoto 1981); the ability to inactivate cycloheximide and α-amanitin (Fienberg et al. 1985; Pitto et al. 1985); and the ability to synthesize polyamines (Montague et al. 1978; Fienberg et al. 1984). The presence of 2,4-D in the fresh medium does not prevent the expression of these embryogenic traits until several days later. Thus, if the effect of fresh medium is to remove the suppressor, the suppressor cannot be 2,4-D; rather it could be a factor in the high-density culture medium. However, repeatedly changing the medium of a high-density culture allows the unorganized cell cluster to develop into an organized cell cluster, but not into globular embryos. Thus, we conclude that, rather than removing an inhibitor, fresh medium acts to trigger embryogenesis from the ground state. It does so in a way that is similar to the effect of fertilization on zygotic embryogenesis. In fresh medium, embryogenesis can progress to form the organized cell clusters from unorganized cell clusters. The high-density condition, however, inhibits the transition from the organized cell clusters to globular-shaped embryos.

Morphogenesis is a succession of gain and loss of morphogenetic competence. In this regard, the behavior of the carrot cell line described above can be interpreted as the following: fresh medium provides the "permissive condition" for the onset of embryogenesis; the low-density condition permits the transition from the organized cluster to the globular-stage embryo; and the removal of 2,4-D provides the permissive condition (Christianson 1985) for heart embryo formation.

However, not all carrot cell lines are capable of forming globular-stage embryos in 2,4-D. It seems that embryogenesis in carrot cultures initiated from domesticated carrot varieties is less sensitive to cell density, and more sensitive to inhibition by 2,4-D. In the presence of 2,4-D, these cultures are arrested at stages beyond the ground state, but before the globular stage, that is, the organized cell cluster stage. These organized cell clusters do not respond to fresh medium, because they already express the fresh-medium-induced biochemical traits (Pitto et al. 1985).

Much of the above hypothesis is derived from observation of the morphological characteristics of the culture and the expression of several biochemical phenotypes that are used as landmarks of the developmental pathway. The model serves as a conceptual framework in which to interpret the varied behavior of embryogenic cultures. More landmarks and molecular information are required to substantiate and/or modify the model. Mutants altered in development and in the expression of developmentally regulated genes will be useful in elucidating the developmental program.

7.3 GENE EXPRESSION IN CARROT CULTURE

Because zygotic embryos are inaccessible, carrot somatic embryogenesis has been used as a model system for studying gene expression in embryogenesis. Carrot cultures have the following attributes: (1) Millions of embryos can be produced in liquid culture, and specific morphogenetic stages can be purified. (2) Simple growth conditions, such as the presence or absence of 2,4-D, manipulate growth either to unorganized cell clusters or toward embryogenesis. (3) The cell clusters provide controls for genes and gene products involved in housekeeping functions, cell division, and adaptation to culture conditions.

7.3.1 General Pattern of Protein Synthesis

By focusing on the early period of embryogenesis, investigators studying carrot cultures can examine gene products other than seed storage proteins. In contrast to the protein patterns found in embryos undergoing cotyledon expansion in which the storage proteins dominate, as in cotton embryos (Dure et al. 1980), cotyledons of carrot embryos are small and do not contain highly abundant proteins. This fact makes possible the examination of 1,000 moderately abundant proteins in carrot somatic embryos by two-dimensional gel electrophoresis.

Comparing 15-day-old somatic embryo cultures (consisting of a mixture of somatic embryos up to the torpedo stage) with unorganized cell clusters of the same age, one can resolve 1,000 proteins by two-dimensional gels, and 1–2% of the proteins show quantitative differences between the two morphologically distinct tissues. Extrapolation of these data, based on an estimate of approximately 10,000 polypeptides expressed in a eukaryotic cell, indicates that 100 to 200 proteins are differentially expressed in the two tissues (Choi and Sung 1984).

To identify polypeptide changes that correlate with specific developmental stages, synchronized embryos from each morphogenetic stage were isolated and their total proteins were separated on two-dimensional gels. Again, a few proteins showed quantitative variation. For example, α-tubulin isoforms gradually changed as embryo development advanced, whereas synthesis of the 70-kd heat-shock proteins decreased (Figure 7–1). The vast majority of the proteins, however, remained unchanged through this period. These facts taken together suggest that morphogenesis involves a limited number of changes in moderately abundant gene products. For the identification of temporally regulated genes of known functions, several enzymes have been noted that exhibit higher levels of activity in somatic embryos than in cell clusters (see review by Sung et al. 1984). The cellular concentration of these enzymes is probably too low to render them detectable on two-dimensional gels.

Several biochemical phenotypes show qualitative differences between the two types of tissues. For example, somatic embryos synthesize E_1, which is absent in cell clusters (Sung and Okimoto 1981). Somatic embryos can inactivate cycloheximide (Fienberg et al. 1985) and α-amanitin (Pitto et al. 1985), but not the cell clusters. These traits are called embryogenic traits. In addition, a number of proteins, such as C_1 and C_2, were found to be synthesized only in the cell clusters or callus tissues (Sung and Okimoto 1983). The traits characteristic of embryos were coordinately regulated.

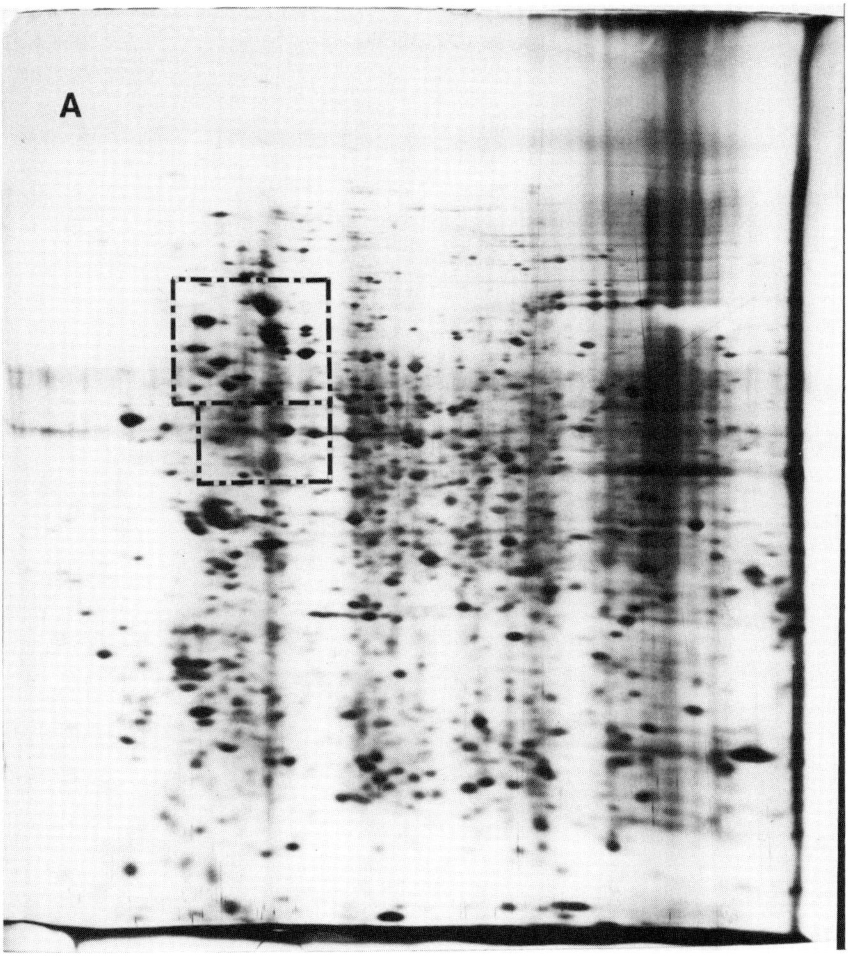

FIGURE 7–1 Two-dimensional gel electrophoresis of proteins from staged embryos. Carrot somatic embryos in suspension were synchronized by the method of Giuliano et al. (1983). Cells were labeled with ^{35}S-methionine and extracts were prepared (Choi and Sung 1984) for isoelectric focusing in the first dimension, followed by 9% polyacrylamide gel electrophoresis in sodium dodecyl sulfate in the second dimension. (A) A full silver-stained gel (Morrissey 1981) of an extract from heart-stage embryos. The dotted lines mark the areas compared in the composites. (B) Arrowheads indicate the 70-kd-protein cluster whose synthesis was elevated upon heat shock at 40°C (data not shown). (C) Arrowheads indicate α-tubulin (a) and β-tubulin (β) subunits. Note the change in position of the α-tubulin subunit as embryogenesis advances. SILVER, visualization by silver-staining; 35S-MET, visualization by autoradiography.

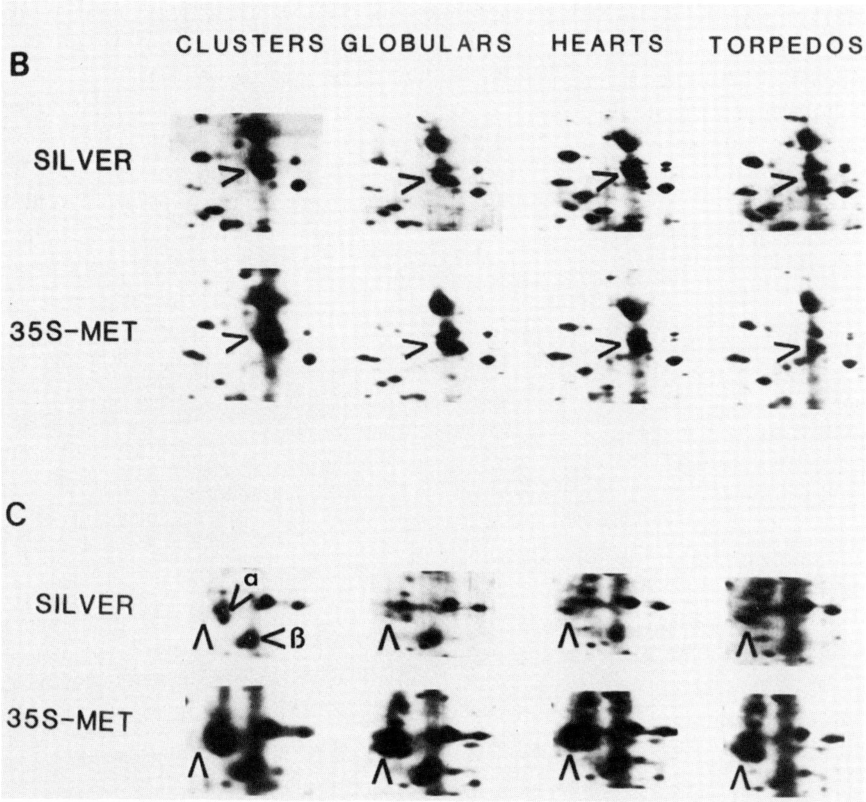

FIGURE 7–1 *Continued*

They appeared to be regulated by the same mechanism that controlled the callus biochemical functions, but in an opposite manner.

7.3.2 Posttranslational Modifications

In carrot cultures, secondary modification of polypeptides occurs to different degrees. We have studied glycosylation in somatic embryos and cell clusters. Tunicamycin preferentially inhibits embryogenesis. Several tunicamycin-induced differences in the synthesis and accumulation of polypeptides were identified by two-dimensional gel electrophoresis. This occurred when embryogenic cultures and cell clusters were compared, but the relationship between these polypeptides and tunicamycin suppression of embryogenesis has not been determined (LoSchiavo et al. 1986).

The pattern of protein phosphorylation can be determined by labeling cells with radioactive phosphate or incubating cell extracts with gamma-labeled ATP. The in vitro and in vivo phosphorylation patterns of the polypeptides are similar, but not identical, to each other. A temperature-sensitive (ts) variant that cannot develop into mature embryos at 32°C was found to show an altered in vivo pattern (Lo Schiavo and Sung 1985). In vitro labeling showed increased phosphorylation in the temperature-sensitive variants. A few specific proteins, including both subunits of tubulin, were heavily phosphorylated in the variant grown at 32°C or in the presence of 2,4-D, when embryogenesis was impaired, but they were only lightly phosphorylated at 24°C in the absence of 2,4-D, when embryogenesis appeared normal (Figure 7-2A). In the parental wild-type cultures, tubulin was never appreciably phosphorylated (Figure 7–2B). It is postulated that this temperature-sensitive variant line may be impaired in some regulatory aspects of protein phosphorylation. Excessive phosphorylation of tubulin could interfere with microtubule formation. Since morphogenesis is correlated with the abundance of cortical microtubules in embryogenic cells, this explanation provides a possible hypothesis for the biochemical basis of the developmental defect.

These results suggested that protein kinases may have regulatory roles in embryogenesis. An antibody against a protein kinase purified from wheat germ (Yan and Tao 1982) cross-reacts with two antigens in somatic carrot embryos (Figure 7–3). No cross-reactivity is found in unorganized cell clusters. Thus, this protein kinase may be responsible for some of the 2,4-D suppressible protein phosphorylation seen in Figure 7–2B. The protein blot in Figure 7–3 also includes extracts from *ts*-59 cells. If the carrot antigens detected here are also protein kinases, they are not impaired in *ts*-59. Their accumulation patterns are identical to those of wild-type under different cultural conditions, and do not correlate with the temperature-sensitive phenotype. The alteration in the variant must therefore involve either a different protein kinase, or perhaps a defect in dephosphorylation.

7.4 IMMUNOLOGICAL APPROACH TO THE IDENTIFICATION OF DEVELOPMENTALLY REGULATED GENES

In an attempt to isolate the few moderate-abundance gene products that are preferentially expressed in somatic embryos, we employed an immunological approach (Choi et al. 1987). The rationale of using immunological rather than cDNA probes was based on the property of variable antigenicity of the proteins. Although a low-abundance protein has less opportunity to elicit antibody production than a high-abundance protein, the low-abundance protein may be more immunogenic than the high-abundance protein. When a rabbit is immunized with crude extracts of somatic embryos, its

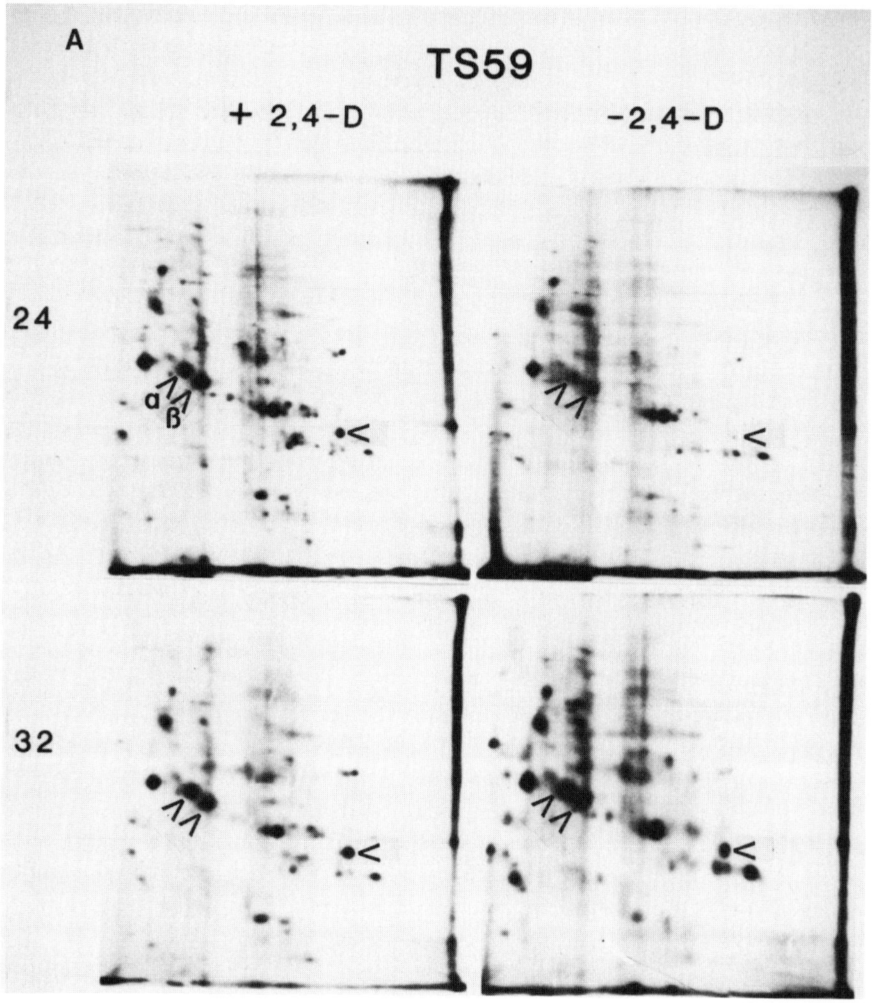

FIGURE 7–2 In vitro protein phosphorylation patterns. Temperature-sensitive mutant *ts-59* cells (A) and parental HA cells (B) were cultured with (+) or without (−) 2,4-D, at 24°C or 32°C as shown for 10 days and then collected on Miracloth by vacuum filtration. Extracts were prepared by powdering and grinding the cells in protein kinase extraction buffer (50 mM Tris-HCl, pH 7.5; 5 mM $MgCl_2$; 50 mM KF; 0.1 mM PMSF; 1 mM DTT) in a mortar and pestle prechilled in dry ice and collecting the supernatant after centrifugation for 10 minutes at 15,000 *g*. In vitro phosphorylation reactions were performed in a volume of 0.1 ml, with 0.12 mg soluble protein in extraction buffer containing 0.5 mM gamma [^{32}P]-ATP (0.5 Ci/mole) for 10 minutes at 32°C. The reaction was terminated by freezing the mixture in dry ice and ethanol. The samples were lyophilized and reconstituted with sample buffer for two-dimensional gel electrophoresis as described in the legend to Figure 7–1. The isotopically labeled phosphate bound to protein was visualized by autoradiography. Arrowheads locate prominent phosphoproteins whose labeling varied with embryogenic capacity and variant phenotype. The α-tubulin (a) and β-tubulin (β) subunits are noted for *ts-59* cells.

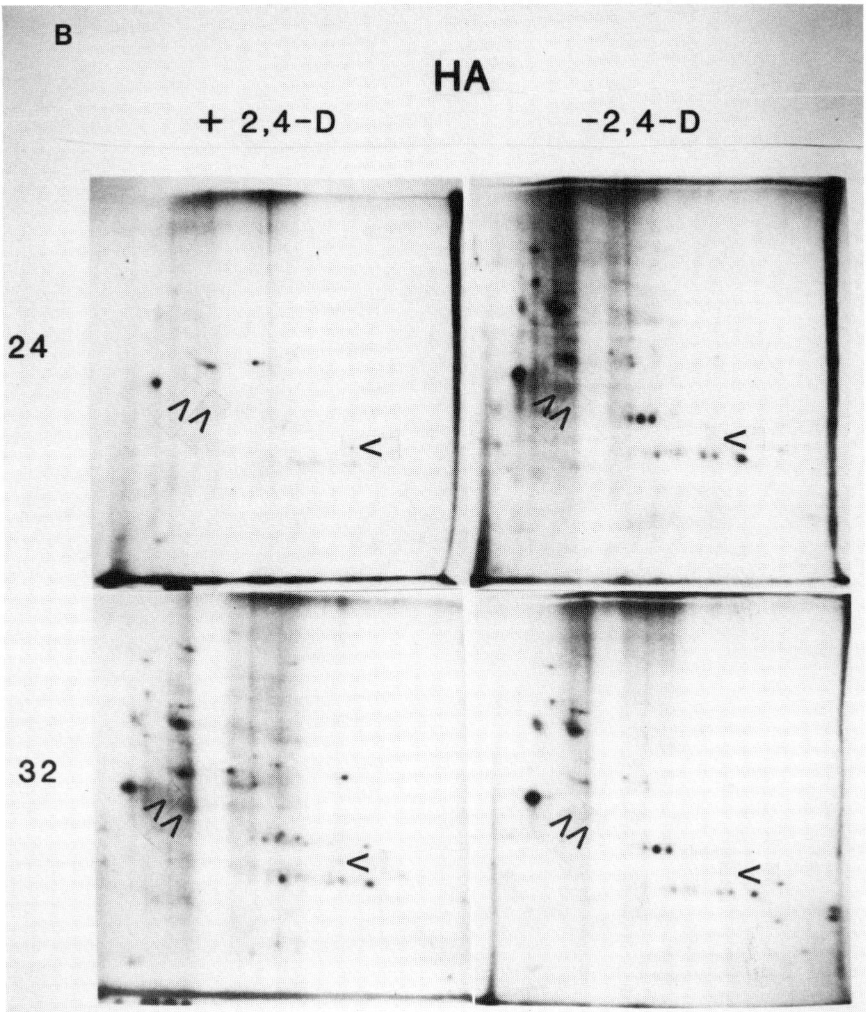

FIGURE 7-2 *Continued*

serum may thus contain antibodies against some highly antigenic but low-abundance proteins. This process can effectively elevate the sensitivity of detecting certain low-abundance molecules in the antigen population. Using immunodiffusion techniques, Khavkin et al. (1979) found similar antigens in maize zygotic embryos and embryogenic callus tissues, which contain a mixture of somatic embryos and undifferentiated cells. The antigens were absent in adult tissues. This finding demonstrated the feasibility of using

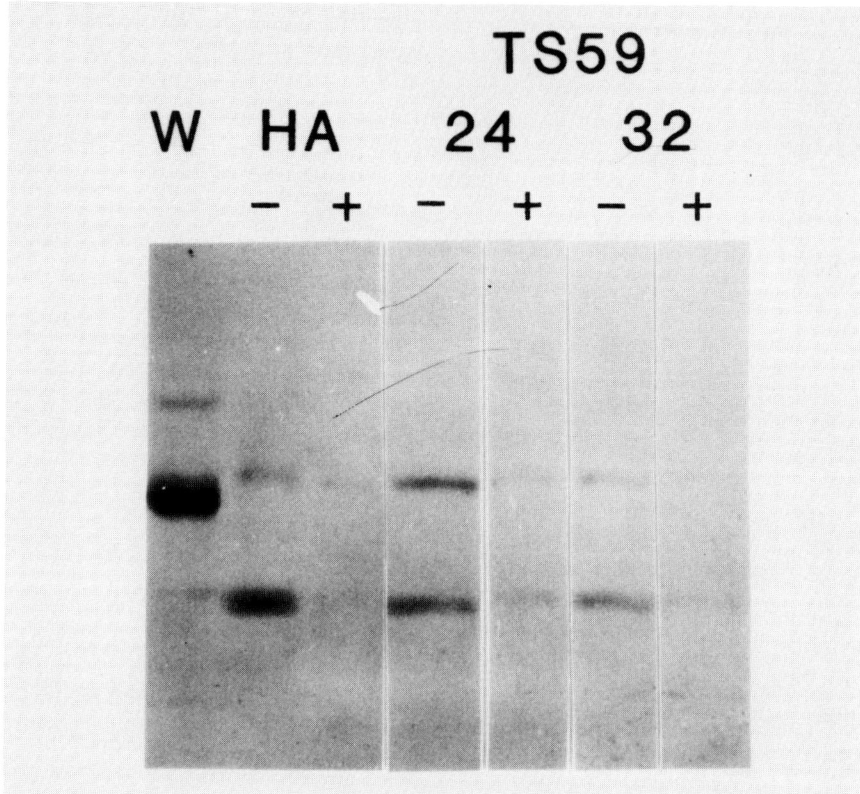

FIGURE 7–3 Protein blot of carrot extracts probed with antibody to wheat germ protein kinase. Protein extracts from wheat germ (W), HA heart-stage somatic embryos (HA), *ts-59* culture grown at 24°C (24), and *ts-59* culture grown at 32°C (32) either with 2,4-D (+) or without 2,4-D (−) were separated by SDS polyacrylamide gel electrophoresis (Laemmli 1970) and blotted onto nitrocellulose paper (Towbin et al. 1979). The blot was probed with rabbit antibody against a protein kinase purified from wheat germ (Yan and Tao 1982), and the bound antibody detected by probing with peroxidase-conjugated secondary antibody.

somatic embryos as the immunogen source to isolate antibodies that would react with the zygotic embryos.

7.4.1 Epitope Selection of cDNA Clones and Monospecific Antibodies

We immunized a rabbit with a crude membrane fraction of somatic embryos at different stages. As expected, the resulting antiserum reacted against a large number of carrot proteins. Figure 7–4 shows a protein blot of two-dimensional gels reacted with the antiserum. The most-reactive spots did not correspond to any spot visible on the silver-stained gel, demonstrating that the antibody population was significantly different from that of the immunogen: The most-abundant antibody species did not represent the high-abundance proteins.

The high-titer serum was adsorbed with cell cluster proteins to remove antibodies against common antigens. The adsorption process effectively reduced the number of antibodies and revealed tissue-specific antibodies. Interestingly, not only embryo-specific, but also cell cluster–specific, antibodies remained in the adsorbed serum (Choi et al. 1985). A possible explanation for the cell cluster–specific antibodies is that the immunogen was contaminated with a small percentage of cell clusters (about 1% in fresh weight). These clusters may have contained a highly antigenic determinant that elicited an abundance of antibodies that could not be completely removed by adsorption.

Such a mixture of polyclonal antibodies enriched for embryogenic determinants was used to isolate cDNA clones from a lambda gt11 library of cDNA from carrot somatic embryos. Following initial screening and subsequent purification, 22 cDNA clones containing epitope-positive hybrid proteins were found (Liu et al. 1985). Northern blot analysis performed with several clones showed that each clone recognized a single polyadenylated RNA species (Choi et al. 1987). These mRNA species were characterized by their temporal pattern of expression. For example, clone 8 was embryo-specific and was expressed in the first one to three days after the onset of embryogenesis. Clone 50 was expressed later in embryogenesis. Clone 66 was expressed in the cell clusters, not in somatic embryos. Clone 18 was expressed both in cell clusters and in somatic embryos (Choi et al. 1987). Based on the hybridization signal, it appeared that these mRNA species belonged to the low- to medium-abundance class messages.

To facilitate the characterization of these clones, monospecific antibodies were affinity purified from the serum using hybrid proteins isolated from the recombinant lambda gt11 clones. Using the antibodies, tissue-specific single antigenic determinants were identified on blots of protein separated by sodium dodecyl sulfate (SDS) gel electrophoresis. For example, clone 8 identified a protein of 66 kd not expressed in the leaf, stem, or flowers of carrots. The tissue specificity suggested that the antigenic determinant was not involved in general cellular metabolism. The fact that proliferating cell clusters did not express this antigen suggested that the antigen was not required for cell division. This narrowed the role of the antigen to the process of embryogenesis. Two of the best-characterized cDNA clones detected genes whose expression was associated with a commitment to embryogenesis and with the acquistion of polarized growth (Borkird et al., in press).

The objectives of a complete analysis of the 22 clones are manifold. We are interested in determining the mRNAs in the carrot culture whose ex-

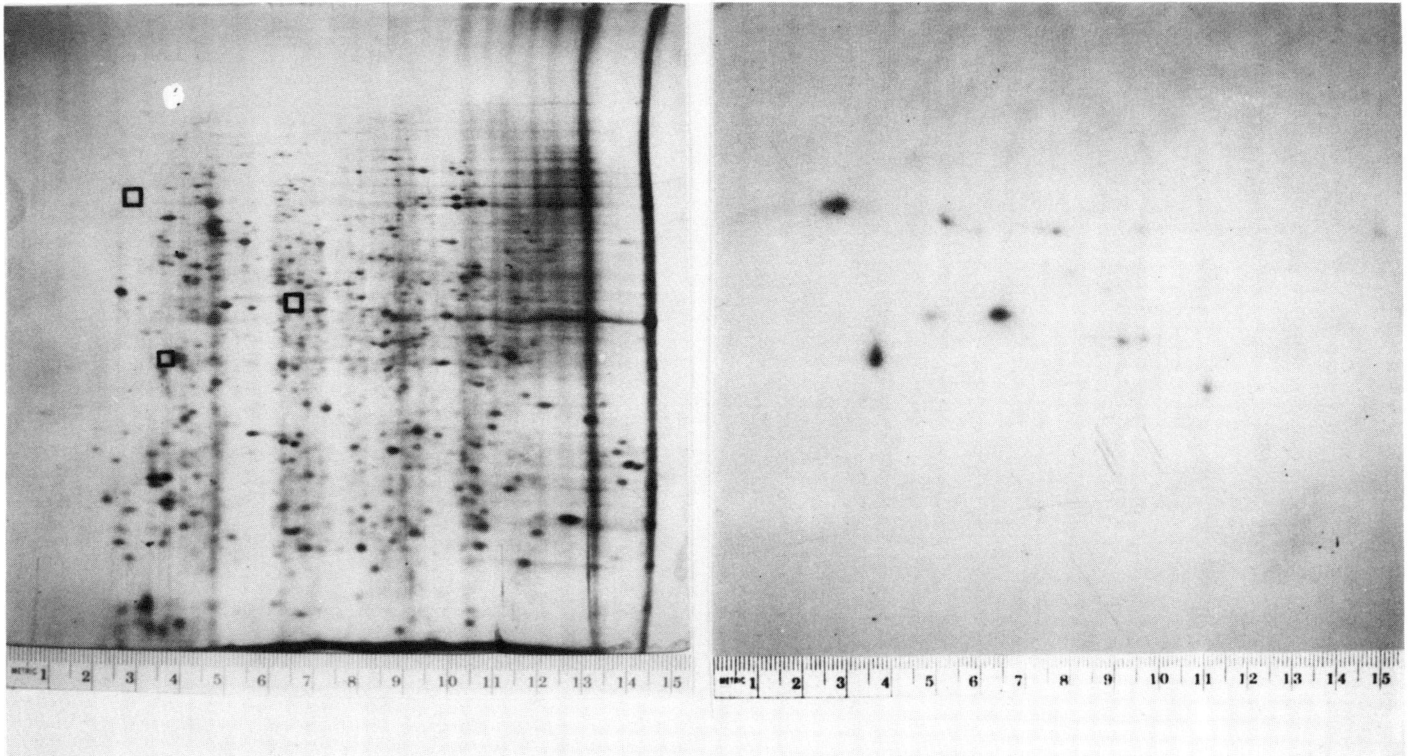

FIGURE 7–4 Two-dimensional gel and protein blot of proteins from globular-stage somatic embryos. A single isoelectric focusing gel was used to make duplicate second-dimension gels on a Biorad® protein gel apparatus with a divider plate. One gel, shown on the left, was silver stained, and the other, blotted. The blot was probed with rabbit antiembryo serum described in the text, and then with iodinated protein A. The autoradiograph of the blot is shown on the right. The open squares on the silver-stained gel note the presumptive location of the three most-prominent autoradiographic spots from the blot.

pression is controlled by 2,4-D and by cell density. We also wish to determine the timing of mRNA expression to provide molecular landmarks in a detailed embryo morphogenetic pathway. By establishing sets of coordinately regulated genes we can also study the mechanism of their regulation. Finally, we can use these probes to study gene expression during zygotic embryogenesis.

7.4.2 Screening of Tissue-Specific Monoclonal Antibodies

A second immunological approach to identifying embryo-specific antigens was designed by screening a large panel of monoclonal antibodies elicited with embryo antigens (Smith et al. 1985). Conceptually, this is similar to the screening of a cDNA library for tissue-specific molecular clones. To look for the clones of low-abundance gene products, a large library is necessary. While the property of variable antigenicity provides an advantage in terms of elevating antibody titer, it is still necessary to screen a large number of hybridoma clones to isolate antibodies against rare antigens.

Balb/c mice were immunized with lyophilized carrot somatic embryos. The antigen source was heterogeneous, consisting of 90% somatic embryos in various stages. Approximately 350 of 2,400 supernatants screened reacted strongly with at least one of the three plant extracts (callus, embryo, chloroplasts). Based on their reaction, the antibodies were divided into seven specificity combinations (Smith et al. 1985).

Approximately 100 clones were further screened by immunoblotting of proteins separated by SDS electrophoresis (Towbin et al. 1979) to confirm the tissue specificity. Many antibodies did not react, suggesting that they did not recognize denaturable epitopes. One-third of the remaining monoclonal antibody reacted with multiple bands on the immunoblots. This suggested that they recognized determinants shared by proteins of several size classes (Smith et al. 1985). Among the multiple bands recognized by a monoclonal antibody, sometimes a few bands showed tissue specificity. The nature of these determinants is still uncertain, but they could be specific carbohydrate moieties that are known to dominate the population of developmentally regulated antigens (Feizi 1985).

The specificity of the monoclonal antibodies permits more-detailed studies of the localization of antigen expression. One antibody, 21D7, reacted with an antigen of 45 kd having an isoelectric focusing point of 6.7. This antigen was present at higher levels in developing somatic embryos than in cell clusters. No antigenic determinants were detected in extracts of carrot leaf, roots, or root tips; nor were any found in dehydrated or mature embryos in the seed. However, the antibody localized the antigen in the nucleus of plant meristems including those of carrots, *Arabidopsis*, and maize (Smith et al., in press). Hence, the antigen recognized by 21D7 is associated with cell division, not with embryogenesis specifically. Several cDNA clones for this antigen have been isolated from the lambda gt11 cDNA library.

7.4.3 Comparison of the Epitope-Selection and Monoclonal Antibody–Screening Procedures

Several comments comparing the epitope-selection and monoclonal antibody–screening approaches are relevant. The monoclonal antibody approach requires constructing a large panel of hybridoma cell lines, laborious screening, and subcloning of a large number of hybridomas to find several clones producing desired antibodies. Another potential difficulty is that most hybridoma lines secrete immunoglobulin (Ig)M, a notoriously sticky molecule prone to nonspecific binding. On the other hand, a good monoclonal antibody is specific for a single epitope, providing fine discrimination between structurally similar molecules, and can be produced in limitless quantities.

The polyclonal antiserum method can be divided into two steps. The first step involves serum adsorption with antigens for the heterologous tissues, resulting in enrichment for tissue-specific antibodies. On a protein blot containing total cellular proteins, the enriched serum can detect tissue-specific antigens otherwise masked by common antigens detectable by the original serum. The second step involves the use of the enriched serum to isolate epitope-specific recombinant DNA clones from a cDNA expression library (Thanh et al. 1985). Monospecific antibodies are subsequently purified from the serum by immunoaffinity techniques that consist of conjugating partially purified β-galactosidase-antigen hybrid protein to an affinity column. This method allows quick production of monospecific antibodies and cDNA probes from complex antigens and semipurified gene products. However the monospecific antibodies thus produced are still polyclonal and can possibly react with more than one epitope on the same or different antigenic molecules. This feature may be advantageous in antigen detection, but reduces resolving power when attempting to discriminate between antigens that are distinctly different, but that share common epitopes.

7.5 CONCLUSION

Using the cDNA and antibody probes, we have identified three classes of embryonic genes. Their expression is associated with (1) the commitment to the cell division process, (2) the commitment to embryogenesis, and (3) polarized growth in the transition from globular- to heart-stage embryos (Borkird et al., in press). As discussed at the beginning of this chapter, cell clusters of carrot cultures are tissues committed to embryogenesis, but prevented from morphogenesis by external factors. Some of the embryonic antigens may represent a class of functions expressed when embryogenic potential is first induced and as long as the tissue remains embryogenic, but the expression declines when morphogenesis commences. Such genes are preferentially expressed in carrot callus tissue (Borkird et al., in press).

Work by other investigators suggests that gametophytes may express a set of genes quite similar to that expressed by the embryogenic tissues. The ease with which microspores alter their programming to embryogenesis implies that the microspores already express many of the genes necessary to reach the committed embryogenic state. In some embryo-lethal mutants of *Arabidopsis* (Meinke 1982), a pleiotropic effect resulting in nonrandom distribution of aborted seeds is observed. The author proposed that this phenomenon is a result of gametophytic expression of the mutant allele. If true, these mutations would provide further evidence that genes essential for embryogenesis are also expressed and functional in the gametophyte. Cytological localization of the embryonic genes may support this contention.

The exact functions of the antigens we have described are unknown. Nevertheless, the specificity and transient nature of the expression pattern of these antigens suggest that they may play important roles in embryogenesis.

The proof for developmental roles for the antigens identified in our study could be obtained by studying cell lines impaired in the expression of these antigens, directly or indirectly. However, obtaining spontaneous or mutagen-induced variants with such specific defects would be difficult since selection schemes are unavailable. Alternatively, the role of the antigen could be assessed either by introducing an antibody or by interrupting its expression with the introduction of an antisense RNA.

The induction of a new developmental fate usually occurs in one cell, or in a group of several cells surrounded by tissues of different developmental potentials. If the induction process does involve new gene expression, it will be difficult to identify the gene products by biochemical and molecular methods that analyze whole tissues. To this end, in situ hybridization experiments involving immunoprobes or cDNA probes can be employed to analyze spatial differentiation of gene products. Furthermore, in situ studies have to be employed to analyze zygotic embryogenesis and gametogenesis in planta. With both molecular and immunological probes, it will be possible to investigate the problem of embryogenic potency, and to ask whether or not somatic and zygotic embryogeneses follow similar rules and "decision-making" processes.

REFERENCES

Borkird, C., and Sung, Z.R. (1987) *Plant Physiol.* 84, 1001–1006.
Borkird, C., Choi, J.H., Jin, F.H., et al. *Proc. Natl. Acad. Sci. USA* (in press).

Choi, J.H., Liu, L.-S., Borkird, C., and Sung, Z.R. (1987) *Proc. Natl. Acad. Sci. USA* 84, 1906–1910.

Choi, J.H., Liu, L., and Sung Z.R. (1985) in *Proceedings of a Workshop on Somatic Embryogenesis* (Terzi, M., Pitto, L., and Sung, Z.R., eds.), pp. 6–13, San Miniato, Italy.

Choi, J.H., and Sung Z.R. (1984) *Plant Mol. Biol. Reptr.* 2, 19–25.

Christianson, M.L. (1985) in *Tissue Culture in Forestry and Agriculture* (Henke, R., Hughes, K., and Hollaender, A., eds.) pp. 83–103, Plenum Press, New York.

Christianson, M.L., Warnick, D.A., and Carlson, P.S. (1983) *Science* 222, 632–634.

Dure, L.S. III, Capdevila, A.M., and Greenway, S.C. (1980) in *Genome Organization and Expression in Higher Plants* (Leaver, C.J., ed.), pp. 127–146. Nato Advanced Study Institute Series, Series A. Life Sciences, vol. 29, Plenum Press, New York.

Feizi, T. (1985) *Nature* 314, 53–57.

Fienberg, A.A., Choi, J.H., Lubich, W.P., and Sung, Z.R. (1984) *Planta* 162, 532–539.

Fienberg, A.A., Shen, P., Borkird, C., et al. (1985) in *Proceedings of a Workshop on Somatic Embryogenesis* (Terzi, M., Pitto, L., and Sung, Z.R., eds.), pp. 22–31, San Miniato, Italy.

Foster, A., and Gifford, E. (1974) *Comparative Morphology of Vascular Plants,* 2nd ed., W.H. Freeman, New York.

Ginliano, G., Rosellini, D., and Terzi, M. (1983) *Plant Cell Reports* 2, 216–218.

Gosh-Wacherle, G., Aviv, L., and Galun, E. (1979) *Z. Pflanzenphysiol.* 91, 267–278.

Green, C.E. (1982) in *Plant Tissue Culture* (Fujiwara, A., ed.), p. 107, Japanese Association of Plant Tissue Culture, Tokyo.

Hagberg, G., and Hagberg, A. (1981) in *Barley Genetics IV. Proceedings of the Fourth International Barley Genetics Symposium,* pp. 686–689, Edinburgh.

Halperin, W. (1970) in *Control Mechanisms in the Expression of Cellular Phenotypes. Symposium of the International Society for Cell Biology,* vol. 9 (Padykula, H.A., ed.), pp. 169–191, Academic Press, Orlando, FL.

Halperin, W., and Jensen, W.A. (1967) *J. Ultrastruct. Res.* 18, 428–443.

Halperin, W., and Wetherell, D.F. (1975) *Science* 147, 756–758.

Henshaw, G.G., O'Hara, J.F., and Webb, K.J. (1982) in *Differentiation In Vitro: British Society for Cell Biology Symposium 4* (Yeoman, M.M., and Truman, D.E.S., eds.), pp. 231–251, Cambridge University Press, New York.

Khavkin, E.E., Misharin, S.I., and Polikarpochkina, R.T. (1979) *Planta* 145, 245–251.

Laemmli, U.K. (1970) *Nature (London)* 227, 680–685.

Liu, L., Borkird, C., and Sung, Z.R. (1985) in *Proceedings of a Workship on Somatic Embryogenesis* (Terzi, M., Pitto, L., and Sung, Z.R., eds.), pp. 70–76, San Miniato, Italy.

LoSchiavo, F., Quesada, L., and Sung Z.R. (1986) *Plant Sci.* 44, 65–71.

LoSchiavo, F., and Sung, Z.R. (1985) in *Proceedings of a Workshop on Somatic Embryogenesis* (Terzi, M., Pitto, L., and Sung, Z.R., eds.), pp. 32–34, San Miniato, Italy.

Meinke, D.W. (1982) *Theor. Appl. Genet.* 63, 381–386.

Montague, M.J., Koppenbrink, J.W., and Jaworski, E.G. (1978) *Plant Physiol.* 62, 430–433.

Morrissey, J.H. (1981) *Anal. Biochem.* 117, 307–310.

Nomura, K., and Komamine, A. (1985) in *Proceedings of a Workshop on Somatic Embryogenesis* (Terzi, M., Pitto, L., and Sung, Z.R., eds.), pp. 1–5, San Miniato, Italy.

Pitto, L., Lo Schiavo, F., and Terzi, M. (1985) *Proc. Natl. Acad. Sci. USA* 82, 2799–2803.

Shafer, E., Groskurt, E., and Neumann, K.H. (1985) in *Proceedings of a Workshop on Somatic Embryogenesis* (Terzi, M., Pitto, L., and Sung, Z.R., eds.) pp. 158–170, San Miniato, Italy.

Smith, J.A., Choi, J.H., Krauss, M.R., Karu, A.E., and Sung, Z.R. (1985) in *Proceedings of a Workshop on Somatic Embryogenesis* (Terzi, M., Pitto, L., and Sung, Z.R., eds.), pp. 86–94, San Miniato, Italy.

Smith, J.A., Krauss, M.R., Borkird, C., and Sung, Z.R. *Planta* (in press).

Steward, F.C. (1958) *Am. J. Bot.* 45, 709–713.

Steward, F.C., Mapes, M.O., Kent, A.E., and Holsten, R.D. (1964) *Science* 143, 20–27.

Sunderland, N. (1974) in *Haploids in Higher Plants: Advances and Potential* (Kasha, K.J., ed.), pp. 91–122, University of Guelph Press, Guelph, Canada.

Sung, Z.R., Fienberg, A., Chorneau, R., et al. (1984) *Plant Mol. Biol. Reptr.* 2, 13–14.

Sung, Z.R. (1985) in *Proceedings of a Workshop on Somatic Embryogenesis* (Terzi, M., Pitto, L., and Sung, Z.R., eds.), pp. 117–121, San Miniato, Italy.

Sung, Z.R., and Okimoto, R. (1981) *Proc. Natl. Acad. Sci. USA* 78, 3683–3687.

Sung, Z.R., and Okimoto, R. (1983) *Proc. Natl. Acad. Sci. USA* 80, 2661–2665.

Thanh, V.H., Young, R.A., and Davis, R.W. (1985) in *DNA Cloning Techniques: A Practical Approach* (Glover, D. ed.), pp. 49–78, IRL Press, Oxford.

Towbin, J., Staehelin, T., and Gordon, J. (1979) *Proc. Natl. Acad. Sci. USA* 76, 4350–4354.

Yan, T.-F.J. and Tao, M. (1982) *J. Biol. Chem.* 257, 7037–7043.

CHAPTER
8

Photoregulation of Gene Expression in Plants

John C. Watson

Several environmental factors influence the growth, development, and reproduction of green plants, but among them light is perhaps the most important. Apart from its role as the energy source for photosynthesis, light provides crucial environmental cues that control virtually every stage of the plant life cycle, from seed germination, through stem and leaf development, to flowering. Perhaps at no time are the effects of light on plant development more apparent then when seedlings are grown in the absence of light. Etiolation, the response to growth in darkness, results in greatly elongated stems and a great reduction or arrest in leaf development. Since the etiolation response has attracted the attention of plant biologists for the last ten decades, we can say with certainty that light is required to control such processes as stem elongation, leaf development, chlorophyll synthesis, and chloroplast development, as well as a plethora of other developmental phenomena. Many of the effects of light on growth and differentiation in etiolated plants can be brought about by quantities of light too small to

I wish to thank the many colleagues who provided reprints and manuscripts prior to publication. I am especially indebted to Drs. L.S. Kaufman, W.R. Briggs, and W.F. Thompson for so many stimulating discussions.

drive photosynthesis and occur even before the photosynthetic machinery is assembled.

Several different photoreceptors allow light to cue plant responses. At the blue end of the spectrum there are at least two photoreceptors. The first absorbs blue and ultraviolet (UV)-A light and is involved in phototropism and stem elongation (for review see Briggs and Iino 1983). This photoreceptor is frequently called cryptochrome. There is almost certainly a separate receptor for UV-B light that controls such events as the induction of flavonoids (for review see Wellmann 1983). At the red end of the spectrum there are also two photoreceptors. One of these is protochlorophyllide (pchlide), a precursor to chlorophyll that is converted to chlorophyll upon absorption of red light, which in part controls chlorophyll synthesis and plastid development (for review see Kasemir 1983). The last photoreceptor is phytochrome, which controls a large number of responses (for review see Smith and Morgan 1983). It is clear that these photoreceptors can interact in the control of a single process, for example, chloroplast biogenesis.

Phytochrome, the most-studied photoreceptor in plants, is a chromoprotein containing a bilitriene chromophore (for review see Rüdiger and Scheer 1983). Phytochrome exists in two photointerconvertible forms. It is synthesized as the red-light-absorbing form, called Pr, and upon illumination with red light is converted to the far-red-light-absorbing form, called Pfr (Figure 8–1). Far-red light drives Pfr back to the Pr form. Based on a large number of physiological experiments, Pfr is generally agreed to be the biologically active form of the pigment. Red light leads to the formation of Pfr, and signal transduction ensues, leading to the observed response (see Figure 8–1). Far-red light given sufficiently soon after the red light cancels the induction events and the response is not observed. Indeed, it is this red/far-red reversibility that gives rise to the idea that Pfr is the active form

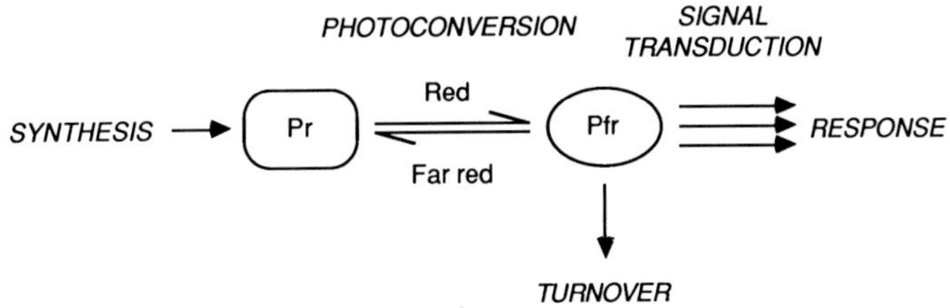

FIGURE 8–1 The phytochrome system. Phytochrome is synthesized in the red-light-absorbing form, Pr. Red light drives Pr to the far red-light-absorbing form, Pfr. Far-red light drives Pfr back to the Pr form. Upon formation of Pfr (the active form of the pigment), signal transduction occurs, unless far-red light is administered sufficiently soon following the formation of Pfr, leading to the response.

and that provides a most-convenient operational definition as to whether a response is controlled by phytochrome.

Two types of phytochrome responses can be defined by the amount (fluence) of red light at which they are initiated and by the ability of far-red light to reverse them (for review see Briggs et al. 1985). The low-fluence (LF) response has a threshold of approximately 1 $\mu mol/m^2$ and is fully reversible with far-red light. In contrast, the very low fluence (VLF) response has a threshold about four orders of magnitude lower than the LF response at approximately 10^{-4} $\mu mol/m^2$. VLF responses are not reversed by far-red light and can in fact be induced by most far-red light sources. This is because VLF responses may require less than 1% of total phytochrome to be in the Pfr form for saturation and far-red light alone can convert approximately 3% of Pr to Pfr. Since most responses that are sensitive to VLF light are also sensitive in the LF range, biphasic fluence response curves are obtained (Briggs et al. 1985). A practical consequence of this exquisite sensitivity to light is that even dim green light, frequently used as safelights in plant research, can induce VLF responses.

In the past, one difficulty in explaining fluence response behavior may have been the idea that phytochrome acts as a monomer. Knowing that phytochrome exists in vitro as a dimer of two identical subunits each containing a chromophore (for review see Quail et al. 1986) has given rise to hypotheses that phytochrome also exists and exerts its effects as a dimer in vivo (VanDerWoude 1985 and 1987; Brockmann et al. 1987). Studies on phytochrome-mediated germination of lettuce seeds (VanDerWoude 1985) and the known physicochemical properties of phytochrome in vivo (Brockmann et al. 1987) are consistent with a dimeric model for phytochrome action. Simulations that assume the interaction of the dimer with the first component of the signal response chain (a specific receptor X) predict that the homodimer complex Pfr/Pfr-X gives rise to the LF response and the heterodimer complex Pr/Pfr-X gives rise to the VLF response (VanDerWoude 1985 and 1987).

It is by now eminently clear that the complex biochemical and morphological events triggered by light are associated with changes in the expression of both nuclear and chloroplast genes (for review see Lamb and Lawton 1983; Tobin and Silverthorne 1985; Thompson et al. 1985; Ellis 1986; Harpster and Apel 1985; Apel et al. 1986; Quail et al. 1986; Kuhlemeier et al. 1987; Silverthorne and Tobin 1987). Up to now most work has centered on genes transcribed into abundant mRNA species that in some cases encode proteins of the photosynthetic apparatus and in other cases encode proteins with yet unidentified functions. In particular, two sets of genes that code for prominent photosynthetic proteins in the chloroplast—the small subunit of ribulose bisphosphate carboxylase (*RbcS*) and the light-harvesting chlorophyll *a*/*b*-binding protein of photosystem II (*Cab*)—have long been known to be encoded by nuclear genes (for review see Kung 1976; Kung et al. 1972) and are the most intensively investigated. This review will focus on the photoregulation of nuclear gene expression in higher plants at the level of transcription and mRNA accumulation. The action and interactions of the major photoreceptor systems will be considered as well as the apparent diversity of gene responses between species and the diverse response patterns of distinct genes within a species. Recent work is beginning to reveal the complex nature of the mechanisms governing gene expression within the nucleus at the level of chromatin organization and transcriptional events as well as the growing awareness of the importance of postranscriptional events. Lastly, by using tranformation to genetically modify the nuclear genome exciting progress has recently been made in defining the DNA sequences responsible for photoregulated gene expression and ways in which these sequences exert their effects.

8.1 EFFECTS OF LIGHT ON rRNA GENE EXPRESSION

8.1.1 Light Effects on rRNA Gene Activity

Phytochrome exerts a three-pronged effect on protein synthesis by increasing ribosome content, increasing ribosome activity, and increasing the amount of mRNA template (for review see Tobin and Silverthorne 1985). In mustard cotyledons it has been shown that the phytochrome-induced increase in rRNA content (Thien and Schopfer 1975) is associated with an increase in rRNA synthesis as shown by pulse-labeling in vivo (Thien and Schopfer 1982). From these observations we might expect to find that light increases the rate of pre-rRNA synthesis as measured by run-on transcription in isolated nuclei, which should reflect the distribution of engaged RNA polymerases in vivo. In this type of measurement, isolated nuclei are pulse labeled with a radioactive RNA precursor; the labeled nuclear RNA is then hybridized to filters containing cloned DNA of the genes of interest. Gallagher and Ellis (1982) were the first to show using the run-on assay that light increases transcription of specific genes. In their comparisons they show that nuclei from light-grown pea seedlings exhibit a twofold higher rate of rRNA gene transcription than nuclei from dark-grown seedlings, although this increase was significantly smaller than the increases found for the *RbcS* and *Cab* genes. Also using the run-on assay, phytochrome-mediated increases in rRNA gene transcription in nuclei have been found in *Lemna* (Silverthorne and Tobin 1984) and barley (Mösinger et al. 1985 and 1988). Interestingly, this is not universally observed: there is no significant effect of light on rRNA gene transcription in soybean (Berry-Lowe and Meagher 1985) or oat (Colbert et al. 1985; Quail et al. 1986) nuclei. The reasons for species differences in this regard are currently unclear. All the studies mentioned above have examined transcription of the nuclear rRNA "operon" that is transcribed by RNA polymerase I in the nucleolus into a 45S rRNA precursor and subsequently processed into the mature 18S, 5.8S, and 25S rRNAs. The 5S RNA species associated with cytoplasmic ribosomes is transcribed by RNA polymerase III from genes that are unlinked to the rRNA operon. In those species in which light does regulate transcription of the 18S and 25S rRNA genes an interesting speculation is that expression of the 5S genes is controlled in a similar fashion, but this has not yet been experimentally demonstrated.

8.1.2 Light Effects on Cytosine Methylation and Chromatin Structure of rRNA Genes

That the plant nuclear rRNA genes are highly reiterated and tandemly arrayed is well established. Although the overall transcriptional organization is similar in plants and animals, the rRNA genes in plants are more highly repeated than their animal counterparts (for review see Long and Dawid 1980). Since the large number of rRNA genes probably exceeds that required to supply rRNA to the cytoplasmic ribosome pool, a mechanism must exist to determine which gene copies are (or will be) active and those that are silent. In those species in which light regulates rRNA gene expression, a recruitment of more genes into the transcriptionally active pool may occur. Although there is not a detailed mechanistic understanding of rRNA gene expression for any plant species, many studies of eukaryotic gene expression suggest that gene regulation results from the concerted regulation of the availability of RNA polymerase (and its attendant factors) as well as structural features of the chromosome template such as chromatin conformation and cytosine methylation.

The methylation of cytosine residues as specific sites in eukaryotic DNA is widely believed to play a role in gene regulation and differentiation (Doerfler 1983; Razin and Szyf 1984). Cytosine methylation status is most frequently studied by gel blot hybridization using restriction enzymes that are differentially sensitive to cytosine methylation within the recognition sequence (Waalwijk and Flavell 1978). Although exceptions exist, the general observation is that cytosine methylation is lower in actively expressed genes. Likewise, chromatin conformation is also believed to contribute significantly to gene control (for review see Reeves 1984). Chromatin containing active (or potentially active) genes exists in an "open" conformation, meaning that the DNA of active genes is more accessible to chemical probes such as DNase I, and hence is more rapidly digested by DNase I. Further, particular sites within active chromatin, called hypersensitive sites, are even more sensitive to digestion by nucleases than bulk active chromatin. Hypersensitive sites in chromatin can be mapped using a method called indirect end labeling (Wu 1980) in many cases to locations in the 5′ and 3′ flanking regions of the gene (Elgin 1981; Reeves 1984). Recent evidence suggests that cytosine methylation status affects active chromatin formation. Keshet et al. (1986) showed that, upon transfection into mouse L cells, unmethylated DNA integrates into the genome in a nuclease-sensitive conformation exhibiting hypersensitive sites, while methylated DNA does not. Their results suggest that cytosine methylation might exert its effect on gene transcription by affecting both specific and nonspecific interactions between nuclear proteins and DNA.

As mentioned above, growth of seedlings in light causes increased transcription of the rRNA genes in isolated pea nuclei (Gallagher and Ellis 1982). Consistent with this is the recent finding that nuclei from buds of dark-grown plants contain on the average of one nucleolus, whereas nuclei from light-treated plants contain an average of three nucleoli (Kaufman, personal communication). Therefore, pea seedlings seem to be a suitable system for exploring the correlation of cytosine methylation and chromatin hypersensitivity to DNase I with rRNA gene expression. Using a well-characterized rDNA genomic clone (Jorgensen et al. 1987) and suitable methods for the isolation of nuclei and DNA for gel blot analysis (Watson and Thompson 1986), genomic rDNA was examined for changes in methylation status using the restriction enzymes HpaII and MspI (Watson et al. 1987), which are differentially sensitive to methylation within the C-C-G-G recognition site, and rDNA chromatin in nuclei was examined for changes in the distribution of DNase I–hypersensitive sites (Kaufman et al. 1987) during light-induced leaf development in peas (Figure 8–2).

In the cytosine methylation studies, the two main length variants present in pea genomic DNA are differentially sensitive to the restriction enzymes employed. For example, the most-prominent HpaII site is cleaved only in the short-length (S) variant but not in the long (L) variant (see Figure 8–2). Further, the decrease in cytosine methylation during light-regulated leaf development occurs predominantly at these sites within the nontranscribed spacer (NTS) region of the S variant (Watson et al. 1987). The magnitude of the change in methylation level of the "promoter" HpaII site is very similar to the change in transcription rate (Gallagher and Ellis 1982). The changes in cytosine methylation (Watson et al. 1987) as well transcriptional and cytological measurements are consistent with the notion that rRNA gene activity is developmentally regulated and inversely related to the levels of cytosine methylation. In the amphibian *Xenopus,* it is the binding of transcription factors to the NTS region that is thought to cause the decrease in cytosine methylation during embryonic development, such that decreased cytosine methylation results from increased transcription by blocking the access of the DNA methyltransferase (see Watson et al. 1987 for references). In this view, the cytosine methylation pattern is in essence a "footprint" resulting from protein binding to the NTS region. A similar situation may exist in peas, but further experimentation is required to clarify the point.

Not only HpaII sensitivity distinguishes the S and L variants; the DNase I hypersensitivity of these variants in chromatin of isolated nuclei is also different (Kaufman et al. 1987). Using isolation conditions that minimize the activity of proteases and nucleases, the chromatin of each major variant is marked by a series of DNase I–hypersensitive sites both within the coding regions and within the NTS region. However, the placement of hypersensitive sites between the subrepeats and the 18S rRNA genes is different for the two variants (see Figure 8–2). Within the NTS region, any differences

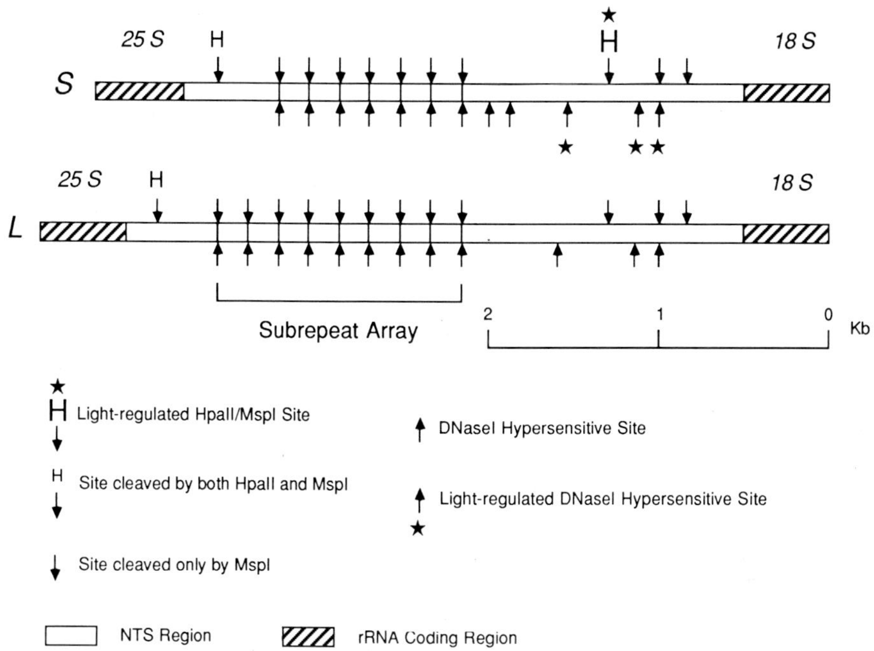

FIGURE 8–2 Comparison of HpaII/MspI sites and DNase I–hypersensitive sites in the pea nontranscribed spacer (NTS) region. The maps of BamHI restriction fragments spanning the NTS region of the short (S) and long (L) variants are compared. The S and L variants differ in length because of differing numbers of 180-bp elements in the subrepeat array and are located on different chromosomes (Jorgensen et al. 1987; Ellis et al. 1984; Polans et al. 1986). HpaII (cleaves C-C-G-G and mC-C-G-G) and MspI (cleaves C-C-G-G and C-mC-G-G) sites (downward-pointing arrows) and DNase I–hypersensitive sites in chromatin (upward-pointing arrows) were mapped using indirect end-labeling procedures (Watson et al. 1987; Kaufman et al. 1987). Comparisons of HpaII and MspI cleavage of DNA extracted from different stages of light-regulated leaf development show that sites to the right of the subrepeat array become less methylated predominantly in the S variant. The HpaII site about 800 bp upstream from the 5′ end of the mature 18S rRNA is cleaved only in the S variant. This site is likely to be in or near the promoter for RNA polymerase I. Virtually all developmental change in HpaII sensitivity is attributable to changes in the methylation status at this site and is thus restricted to the S variant (Watson et al. 1987).

DNase I hypersensitivity was compared in chromatin with nuclei isolated from dark-grown pea seedlings or seedlings greened for three days in white light (Kaufman et al. 1987). The placement of hypersensitive sites between the subrepeats and the 18S rRNA genes is different for the two variants. The S variant possesses three hypersensitive sites within the promoter region that are present in chromatin from the buds of seven-day-old plants exposed to light for three days but absent in chromatin from dark-grown seedlings. The L variant also possesses these three sites, but they are present under both growth conditions. The two sites located farther upstream are present in the S variant but absent from the L variant under both conditions. Further, the hypersensitive sites within the coding region are absent in chromatin from light-treated seedlings (not shown). The hypersensitive sites within the subrepeat array do not map precisely at 180-bp intervals and are aligned as shown only for convenience.

in the distribution of hypersensitive sites in the chromatin from dark-grown or light-treated seedlings occur only within S variant chromatin.

To the extent that HpaII sensitivity is a valid indicator of activity at different rDNA loci within a cell, we might suspect that a higher proportion of S variant genes than of L variant genes is active in pea seedlings (Watson et al. 1987). Since the developmental alterations in methylation status and DNase I hypersensitivity occur predominantly within the S variant, we might further suspect the expression of the S variant to be subject to developmental regulation, whereas the L variant expression may be constitutive (Watson et al. 1987; Kaufman et al. 1987). Further, upon alignment of the maps, there are striking similarities in the pattern of the HpaII/MspI sites and DNase I hypersensitivity within the NTS region (see Figure 8–2). These sites are thought to be in or near the promoter for RNA polymerase I. If in future experiments we are able to demonstrate that the same population of rRNA genes exhibits the decrease in methylation as well as the appearance of DNase I–hypersensitive sites and that the S and L variants are differentially expressed, the belief that these sites are of functional importance for the regulated expression of rRNA genes in pea will be strengthened.

The findings with pea are in contrast to the general situation in wheat, where longer rDNA variants tend to be less methylated and more DNase I sensitive (Flavell et al. 1985 and 1986; Thompson et al. 1987a). In wheat, as in pea, longer variants have more elements in the subrepeat array. An enhancerlike activity for the subrepeat array in plants has been postulated by Flavell et al. (1985). Several remarkable similarities between the wheat subrepeats and the enhancer elements in the NTS of *Xenopus* (Reeder 1984) strengthen this view (Flavell et al. 1985 and 1986). In this model, "dominance" is conferred upon rDNA loci containing repeats with more enhancer elements because longer arrays of enhancers can more effectivlely sequester a limiting transcription factor that binds to the promoterlike sequences in each enhancer (Reeder 1985). Enhancer array length clearly fulfills an important role in determining rRNA gene activity, but other factors, such as the nucleotide sequence of the enchancer elements and the availability of transcriptional machinery, may also participate in this determination. Perhaps by taking advantage of the light-regulated control of rRNA gene activity in pea the role of determinants other than enhancer number can be investigated.

8.2 EFFECTS OF LIGHT ON TRANSCRIPT ABUNDANCE

8.2.1 Regulation by White Light and Phytochrome

For some time now it has been known from solution hybridization studies that different RNA sequences accumulate when dark-grown seedlings are illuminated (Thompson and Cleland 1972) and that a limited number of abundant mRNAs account for the difference in RNA populations (de Vries et al. 1983). The number of abundant mRNAs that are light responsive can be determined by in vitro translation of isolated poly(A)-containing mRNA. For example, in pea shoots 25 mRNAs that are absent in dark-grown seedlings are abundant in white light–grown seedlings, while 9 mRNAs that are present in dark-grown seedlings disappear in the light (de Vries et al. 1982). The combination of antibodies raised against particular proteins to measure specific in vitro translation products and complementary DNA clones to monitor specific mRNAs by hybridization has resulted in a considerable number of studies of the effects of light on the steady-state levels of mRNAs.

It is by now well established that light affects the steady-state level of mRNAs for several different nuclear genes encoding identified proteins, and in several instances regulation by phytochrome has been shown (for review see Tobin and Silverthorne 1985; Thompson et al. 1985). Light-regulated expression has been best characterized in *RbcS, Cab,* and genes for Pchlide reductase and phytochrome. With etiolated seedlings, hybridizable mRNA for some genes, such as *RbcS* and *Cab,* is strongly increased by both white light and red light acting through the phytochrome system, whereas for some genes, such as the Pchlide reductase and phytochrome genes, transcript levels decrease in response to light. A cytoplasmic mRNA encoding a 17-kd chloroplast protein (called ELIP) in peas has been found that increases very rapidly after the onset of white-light illumination and then declines (Meyer and Kloppstech 1984; Scharnhorst et al. 1985). The phytochrome-induction properties of the ELIP mRNA, as well of the RNA encoding the light-harvesting chlorophyll a/b–binding protein associated with photosystem I (Slayton et al. 1988), might provide interesting contrasts to the genes mentioned above.

With a sufficiently high sensitivity of detection, many transcripts that are affected by light can be detected, albeit at low levels, in etiolated seedlings. It is not yet known whether the transcripts are present at very low levels in all responsive cells or exist at somewhat higher levels in only a few rogue cells (Mohr and Schäfer 1983). Perhaps by applying in situ hybridization techniques (Martineau and Taylor 1986) this question can be answered. However, this issue is of importance in elucidating the mechanistic details of phytochrome action. In one case, phytochrome actually turns on a response (induction). Alternatively, phytochrome amplifies processes occurring at very low rates in the dark (modulation). That phytochrome can both induce and modulate physiological responses has been shown in mustard (Oelmüller and Mohr 1984a). Another important point is that sensitivity to phytochrome action can depend on the chronological age of the plant. Thus it is possible to measure the time at which competence—the ability to respond to a stimulus—is acquired (Mohr and Schäfer 1983). The measurements have been made in mustard for *RbcS* and *Cab* RNA accumulation and the data indicate that competence is determined endogenously (Schmidt et al. 1987). Already we can see that factors other than light (e.g., competence) can influence photoregulated gene expression.

Clearly, light induction results in differential gene activity, but in general detailed information exists only for a small number of genes and most reports deal with only one or two of the genes just mentioned. A clear indication of possible coordinate expression and response diversity is obtained when several genes are studied in parallel. Nelson et al. (1984) examined accumulation of three nuclear-encoded transcripts [*RbcS, Cab,* phosphoenolpyruvate (PEP) carboxylase gene] and one chloroplast transcript (gene for the large subunit of ribulose bisphosphate carboxylase—*rbcL*) in maize seedlings grown in either darkness or light and upon illumination of dark-grown seedlings. In the dark, *Cab,* PEP carboxylase gene, and *rbcL* transcripts stay at very low levels, but *RbcS* transcripts accumulate to low levels beginning at day five; all four increase dramatically upon transfer to white light. A clear temporal separation of accumulation of the nuclear-encoded transcripts is found with light-grown seedlings, with *RbcS* RNA beginning to accumulate earliest at day three, then PEP carboxylase at day five, then *Cab* at day seven (Nelson et al. 1984). Furthermore, these transcripts accumulated to different extents along the positional gradient of cellular differentiation within maize leaves (Martineau and Taylor 1985). Differential RNA accumulation also occurs during the differentiation of bundle sheath and mesophyll cells in maize leaves (Martineau and Taylor 1986; Sheen and Bogorad 1986a, 1986b, and 1987) and is likely to be important in the specialized photosynthetic properties in these two cell types.

Among monocots one can find examples of diverse response patterns not only for different genes within a given species, but for a particular gene between species as well. As mentioned the *RbcS* and *Cab* RNAs respond to white light in maize (Nelson et al. 1984), but they are also phytochrome responsive, since red/far-red reversible accumulation can be shown (Zhu et al. 1985). A similar situation exists in *Lemna* (Tobin and Silverthorne 1985) and rye (Ernst et al. 1988). However, in barley *RbcS* RNA accumulation is not responsive to light but *Cab* and Pchlide reductase RNAs are (for example, see Batschauer et al. 1986). The reason for a lack of photoresponsiveness of *RbcS* RNA abundance in barley is not yet clear.

A large number of nuclear-encoded specific transcripts have been studied by Thompson and colleagues in peas and mung beans. Complementary DNA clones were isolated for approximately three dozen abundant, light-responsive mRNAs from peas and about one dozen mRNAs from mung beans (Thompson et al. 1983). Among this group are cDNA clones for *RbcS* and *Cab* (Thompson et al. 1983) as well as the ferridoxin I gene (*Fed*) (Dobres et al. 1987). Several of the mRNAs represented by these clones are phytochrome responsive; while most of the specific RNAs increase upon white-light illumination, a few decrease in abundance (Thompson et al. 1983). The effect of light on RNA accumulation in etiolated seedlings of pea and mung bean was clearly different between the two species. The majority of abundant RNAs assayed in mung bean primary leaves were already at relatively high levels in seven-day-old etiolated seedlings. In contrast, the nuclear-coded RNAs assayed in seven-day-old etiolated pea buds were present only at low levels and increased more substantially upon illumination. In general, chloroplast transcript levels were higher in the dark but increased upon illumination in pea but not in mung bean. It seemed reasonable that many of the differences in RNA responses could be explained by the fact that mung bean primary leaves develop more extensively in the dark than do the leaf primordia contained within pea buds. These observations led to the hypothesis that very young mung bean leaves would have very low transcript levels in the dark and that this level would gradually increase during dark development, whereas in pea RNA levels would increase only very slowly during dark growth. Further, we might expect that during growth in the light, transcript levels would reach maximal levels earlier in mung bean than in pea seedlings.

To test these hypotheses, buds or primary leaves of dissected embryos (from dry seed) were examined, as well as seedlings growing either in white light or in darkness and etiolated seedlings transferred to white light (Watson et al., unpublished observations). The levels of several nuclear-encoded transcripts and two chloroplast transcripts (*rbcL* and *psbA*, which encodes the 32-kd quinone binding protein of photosystem II), as well as chloroplast DNA, are compared in Table 8–1. In light-grown seedlings, most of the RNA levels rapidly increase after germination, continue to accumulate until a maximum steady-state level is achieved on days six to eight for peas and day four for mung bean, and then subsequently decline. In both species the *psbA* transcript continues to show a steady increase over the entire growth period with no clear maximum being reached, in contrast to transient patterns other transcripts or chloroplast DNA exhibit. In dark-grown pea seedlings, most of the transcripts are present at very low levels early on following germination (days one to three) and maintain low but detectable levels afterward. Both the *psbA* and *Cab* RNAs show a clear tendency to accumulate during dark growth, but this is much more pronounced for *psbA*. In dark-grown mung bean seedlings, all transcripts but one show a considerable daily increase in steady-state level. Thus by day seven of dark growth, steady-state levels have been reached that are equal to (or even greater than) the levels in light-grown seedlings of the same chronological age. These results tend to confirm the idea that the differences in photoresponsiveness of transcript levels between species are related to the extent of leaf development occurring in the dark.

With light-grown seedlings, additional complexities can be found. Diurnal fluctuations of *Cab* and *RbcS* RNAs (Kloppstech 1985; Spiller et al. 1987) and the ELIP mRNA (Kloppstech 1985) occur in white light–grown peas that persist in continuous white light and so are presumably circadian rhythms. Interestingly, when the seedlings are grown in light/dark cycles,

TABLE 8–1 Transcript and Chloroplast DNA (cpDNA) Accumulation Patterns in Light-Grown and Dark-Grown Peas and Mung Beans[1]

Species	*Transcript or cpDNA*	*Growth in White Light*		*Growth in Darkness*		*Maximum Level at Day 7*
		Time Course	*Peak Days*	*Increase in Darkness*	*Dark-to-Light Shift*	
Pea	*RbcS*	Transient	7–8	No	Increase	Light
	rbcL	Transient	7–8	No	Increase	Light
	pEA25	Transient	7–8	No	Increase	Light
	pEA277	Transient	7–8	No	Increase	Light
	Cab	Transient	7–8	Yes	Increase	Light
	psbA	Linear	12	Yes	Increase	Light
	pEA207	No change	—	Yes	Decrease	Dark
	cpDNA	Transient	5–6	No	Increase	Light
Mung bean	pMB120	Transient	4	Yes	Increase	D ≥ L
	pMB128	Transient	4	Yes	Increase	D ≥ L
	rbcL	Transient	4	Yes	Increase	D ≥ L
	pMB123	Transient	4	No	Increase	Light
	psbA	Linear	11	Yes	Increase	Light
	cpDNA	Transient	5	Yes	Decrease	D ≥ L

Abbreviations: D, dark; L, light.

[1] Seeds of *Pisum sativum* (var. Alaska) and *Vigna radiata* (var. Berken) were imbibed and grown either in continuous darkness or in continuous white fluorescent light as described previously (Thompson et al. 1983). Pea buds were harvested at 24-hour intervals on days 4 through 12 for light-grown plants, or days 1 through 9 for dark-grown plants. With mung bean, primary leaves were harvested at 24-hour intervals on days 3 through 11 for light-grown seedlings or days 1 through 9 for dark-grown plants. For white light induction of greening, 7-day-old etiolated seedlings were transferred to white light. Buds and primary leaves were harvested on days 8 and 9 so that the plants had been exposed to either 24 hours or 48 hours, respectively, of continuous white light. Pea shoot apices and mung bean primary leaves were also dissected from dry seeds. Total cellular RNA and total cellular DNA were separately purified as described (Thompson et al. 1983). DNA or RNA levels were estimated by fixing the nucleic acid to nitrocellulose with a dot blot manifold and hybridizing with plasmid DNA probes labeled in vitro with ^{32}-P-deoxycytidine 5′-triphosphate by nick translation as described (Thompson et al. 1983).

the *Cab* and ELIP mRNAs, but not the *RbcS* RNA, begin to increase during the latter phase of the dark cycle, when Pfr levels are presumably still low (Kloppstech 1985). Further, growth of pea seedlings for 12 days under constant-fluence-rate white-light sources with different contents of far-red light (which changes the Pr : Pfr ratio) has virtually no effect on *RbcS* and *Cab* mRNA levels (Jenkins and Smith 1985). If the pea cultivar used by Jenkins and Smith (1985) exhibits a transient accumulation of *RbcS* and *Cab* RNAs, then one might expect to see more of an effect of Pr : Pfr ratio earlier in development, before the seedlings pass through the time at which they accumulate maximum levels of these transcripts (see above).

A subset of the cDNA clones obtained by Thompson et al. (1983) were selected for more-detailed studies on the induction of transcript accumulation by phytochrome in pea seedlings, and different genes were shown to exhibit quite diverse response patterns (Table 8–2) (for review see Thompson et al. 1985 and 1987b). Like physiological responses to phytochrome, specific RNA accumulation is induced following a single light pulse, but can be differentiated by the fluence of red light at which accumulation begins and by the ability of far-red light to induce or reverse the response (Kaufman et al. 1984 and 1985a). Most of the transcripts (*RbcS* and *Fed,* for example) accumulate after red pulses in the LF range, and the effect of red light is fully reversible with far-red light. Two transcripts (*Cab* and pEA170) are induced to accumulate by much lower fluences of red light, with thresholds in the VLF range; the accumulation is not fully reversed by far-red light and can in fact be induced by far-red light, consistent with a VLF response. The *Cab* response is clearly biphasic, with responses in both the VLF and the LF ranges (Kaufman et al. 1984; Horwitz 1988), while maximum accumulation of pEA170 RNA is achieved with VLF pulses of red light (Kaufman et al. 1985a). Biphasic fluence response curves have been obtained for *Cab* but not Pchlide reductase RNA abundance in barley (Mösinger et al. 1988). Since the *Cab* genes are a small multigene family, an interesting speculation is that some members of the family are induced by VLF red light pulses, while others are induced only in the LF range. If sufficient sequence divergence exists in the two pea *Cab* genes that have already been sequenced (Cashmore 1983; Timko et al. 1985b) and several others that have been cloned and are currently being analyzed (Watson et al., unpublished observations), we should be able to design gene-specific probes to address this question directly. Similar approaches have shown that individuals in the *RbcS* gene family are not equally expressed (Dean et al. 1985 and 1987; Tobin et al. 1985 and 1987; Coruzzi et al. 1984; Fluhr et al. 1986b). In fact, transcripts from different *RbcS* genes in peas accumulate differentially in organs (Coruzzi et al. 1984; Fluhr et al. 1986b) and at different rates during greening (Fluhr et al. 1986b).

Some of the specific transcripts in peas that have similar fluence responses can be differentiated on the basis of the rapidity with which they accumulate after a red light pulse (see Table 8–2) (Kaufman et al. 1986). Some transcripts accumulate in a linear fashion over the 24-hour period following irradiation (e.g., *RbcS* and *Cab*), some reach maximum levels within 2 hours (e.g., *Fed*), while still others do not begin to accumulate until about 16 hours after the pulse. Still another example of response diversity is obtained in far-red light reversal studies when the time between red and far-red light treatments is varied so as to measure the kinetics of loss of photoreversibility (Kaufman et al. 1986). Differences found in the rates of escape from photoreversibility probably reflect differences in the time needed for the inductive events to become irreversible. A comparison of the *Fed* and the pEA215 transcripts is informative. pEA215 RNA does

TABLE 8–2 Summary of Photoregulation of Specific Transcript Levels in Pea Seedlings[1]

cDNA Clone[2]	*Gene Copy Number*[3]	*Time Course*[4]	*Fluence Response*[5]	*Escape Kinetics*[6]	*White Light*[7]	*Continuous Red Light*[8]	*Blue Light*[9]	*Auxin*[10]	*Compartment*[11]
Cab	6–8	Linear	VLF + LF	Slow	Increase	Yes	No	No	C, N
RbcS	6–8	Linear	LF	Slow	Increase	No	Yes	Yes	C, N
pEA25	1–2	Linear	LF	Slow	Increase		Yes	Yes	C, N
pEA277	1	Linear	LF	No	Increase	No	Yes	Yes	ND
Fed	1	Rapid	LF	No	Increase	No	Yes	Yes	C
pEA315	ND	Rapid	LF	No	Increase	Yes	Yes	Yes	ND
pEA170	ND	Delayed	VLF	ND	Increase		No		ND
pEA215	1	Delayed	LF	Fast	Decrease	Yes	Yes	Yes	ND
pEA13	1	Transient	None	None	Increase		No		ND
pEA207	3–6	No change	None	None	Decrease	No	No	No	No change

Abbreviation: ND, not determined.

[1] A summary is presented of several different characteristics of the abundance of specific transcripts and their genes detected by hybridization using the specified cDNA probe.

[2] Data compiled from Thompson et al. 1983, and Dobres et al. 1987.

[3] Gene copy numbers were estimated by probing agarose gel blots of leaf DNA digested separately with several different restriction enzymes (see Watson and Thompson 1986 for methods). Data compiled from Watson, unpublished observations.

[4] Transcript accumulation kinetics were measured in terminal buds over a 24-hour period following a saturating red light pulse of 6-day-old etiolated seedlings. Data compiled from Kaufman et al. 1986.

[5] Fluence response was determined by irradiating 6-day-old etiolated seedlings with different fluences of red light and assaying relative transcript level in buds 24 hours later. Data compiled from Kaufman et al. 1984 and 1985a.

[6] Escape Kinetics were determined by irradiation of 6-day-old etiolated seedlings with a saturating red light pulse followed by a far-red light pulse immediately or 1, 2, or 7 hours after the red pulse. Data compiled from Kaufman et al. 1986.

[7] Data indicate whether the transcript level increases or decreases after exposure of 7-day-old etiolated seedlings to continuous white light for 24 hours. Data compiled from Kaufman et al. 1985a.

[8] Data indicate whether transcript level is higher (Yes) or approximately the same (No) in 7-day-old seedlings grown in continuous dim red light than in 7-day-old etiolated seedlings given a saturating red pulse on day 6. Data compiled from Watson and Kaufman, unpublished observations, and Kaufman et al. 1985b.

[9] Data indicate whether transcript level increases (Yes) or is relatively unchanged (No) when red-light-grown seedlings are given a blue light pulse on day 6 and assayed after 24 hours. Data compiled from Kaufman et al. 1985b.

[10] Data indicate whether transcript level increases (Yes) or is relatively unchanged (No) when red-light-grown seedlings are treated with indole-3-acetic acid on day 6 and assayed after 24 hours. Data compiled from Watson and Kaufman, unpublished observations.

[11] Data indicate whether light treatments cause parallel changes in both the cytoplasmic (C) and nuclear (N) RNA fractions or in the cytoplasmic RNA fraction only. Data compiled from Sagar et al., submitted for publication.

not begin to accumulate until after the effect is no longer far-red light reversible, whereas the accumulation of *Fed* RNA is reversible even after the RNA has reached its maximal level. In the case of the *Fed* transcript, it seems reasonable to suppose that any RNA that has already accumulated must be destroyed following the far-red light treatment, pointing to a potential role of RNA turnover in the control of transcript abundance. Taken together, the extreme diversity of transcript photoresponses in pea seedlings indicates that the manner in which these responses are coupled to the phytochrome signal perception and transduction machinery can be drastically different for different genes.

8.2.2 White Light and Phytochrome Regulation of Nuclear Events

8.2.2.1 Light Effects on Transcription in Isolated Nuclei. Gallagher and Ellis (1982) first demonstrated that nuclei from light-grown plants transcribe specific genes at higher rates than nuclei from dark-grown plants. The light effect was much more pronounced for the *RbcS* and *Cab* genes than for the rRNA genes. During the first 48 hours following transfer of etiolated seedlings to continuous white light, transcription of both the *RbcS* and the *Cab* genes show an initial, transient increase, followed by a slow increase to maximal transcription rates (Gallagher et al. 1985). In plants greened for 48 hours, transcription declines rapidly upon transfer to darkness and increases equally rapidly upon reillumination. Gallagher et al. suggested that the phytochrome-induced increase in transcript levels was due to the transient increase in transcription rate they observed when etiolated seedlings were first illuminated. In contrast, *RbcS* gene transcription declines more slowly in light-grown ten-day-old soybean seedlings shifted to darkness (Berry-Lowe and Meagher 1985). Interestingly, a far-red light treatment immediately prior to darkness dramatically increases the rate of decline of *RbcS* transcription (Berry-Lowe and Meagher 1985). One might speculate that this far-red light treatment may be acting like the "end-of-day" signals that have been characterized physiologically (for review see Smith and Morgan 1983). It seems that light can affect transcriptional activity very rapidly. That phytochrome controls the transcription rates of specific genes under inductive conditions has now been demonstrated using the run-on assay for *Cab* and *RbcS* in *Lemna* (Silverthorne and Tobin 1984), *RbcS* in soybean (Berry-Lowe and Meagher 1985), *Cab* and the Pchlide reductase gene in barley (Mösinger et al. 1985), and *Cab* and the phytochrome gene in oats (Colbert et al. 1985; Quail et al. 1986). For the genes examined so far it seems clear that at least a portion of the phytochrome control of transcript levels occurs at the transcriptional level.

Since *Cab* transcript levels show both VLF and LF responses (Kaufman et al. 1984; Horwitz et al. 1988; Mösinger et al. 1988), an interesting question is whether nuclear transcription of the *Cab* genes is similarly affected by VLF light. An indication that this is the case was obtained by Mösinger et al. (1987) who showed that saturating pulses of either red or far-red light rapidly but transiently changes the ability of barley nuclei to synthesize both the *Cab* and the Pchlide reductase RNAs. *Cab* transcription rates return to dark levels following the transient increase with a far-red light pulse, but with a red light pulse the rate remains higher than that in dark controls. Similarly, Mösinger et al. (1988) found that VLF pulses of red light do cause changes in run-on transcription of these genes if measured sufficiently soon after the red light pulse. Provided that the transcript does not turn over rapidly, it seems likely that a short burst of transcriptional activity induced by the very low amounts of Pfr caused by VLF pulses can account for *Cab* RNA accumulation in this fluence range. As pointed out by Mösinger et al. (1988), it is possible to calculate that as few as 10 to 100 Pfr molecules per cell may be necessary to elicit the changes in transcriptional activity, and so there is presumably an amplification step between phytochrome and transcription. Careful comparisons of the fluence responses and kinetics of *Cab* RNA and chlorophyll accumulation (Horwitz 1988; Briggs et al. 1988; Mösinger et al. 1988) as well as run-on transcription (Mösinger et al. 1988) suggest that while transcription may limit *Cab* mRNA levels, it is not *Cab* mRNA levels that limit chlorophyll accumulation upon illumination with white light.

8.2.2.2 Reconstituted Systems. In most studies of phytochrome regulation, in vivo irradiations have little or no effect on the ability of subsequently isolated nuclei to incorporate radioactive precursors. However, in oats (Mösinger and Schäfer 1984) and barley (Mösinger et al. 1988) there is an effect on total transcription by the nuclei. In fact, in oats, irradiation of homogenized and filtered material yields the effect (Mösinger and Schäfer 1984). Possible explanations are that the signal response chain is very short or that there is an influx of Pfr into nuclei. Ernst and Oesterhelt (1984) found that adding purified phytochrome to nuclei isolated from dark-grown rye stimulates total transcription by about twofold. This effect is specific in that Pfr but not Pr, degraded Pfr, bovine serum albumin, or cytochrome *c* causes the stimulation in activity, but so far little effect on specific transcription of the *RbcS* and *Cab* genes has been found (Ernst, personal communication). Mösinger et al. (1987) have studied this phenomenon by examining the effects of adding oat or rye Pfr to barley nuclei on total incorporation and *Cab* and Pchlide reductase transcription. In contrast to the study by Ernst and Oesterhelt (1984), Mösinger et al. (1987) do not find an effect of adding Pfr to nuclei from dark-grown plants, but do see an effect if the plants are irradiated previously. Interestingly, oat but not rye phytochrome causes transcriptional effects that more closely approximate the effects of in vivo irradiations. Since transcriptional responses were obtained only with nuclei from irradiated plants, Mösinger et al. (1987) reasoned that the addition of purified Pfr to the isolated nuclei is similar to a second pulse of red light just prior to harvest. To this end they administered a red light pulse and then harvested material at intervals over a 12-hour period with or without a second pulse of red light just before harvest. For *Cab* transcription, the first pulse causes a transient increase that reaches a maximum about three hours after the irradiation and then transcription declines; a reciprocal effect is found for the reductase. Second pulses prior to the time the maximum increase or decrease is observed have little effect, but after that time transcription of *Cab* declines much less than with only the first pulse, while the reductase stays near minimum. Adding Pfr to the nuclei from plants receiving only the first pulse gives transcription rates virtually coincident with the kinetics obtained when second red light pulses are given in vivo. These striking results show that adding Pfr to isolated nuclei can indeed mimic the effect of a second irradiation of the plants just before harvest. The close correspondence of the results with in vivo irradiations and the reconstituted system (Mösinger et al. 1987) should allow further insight into the mechanism(s) by which phytochrome regulates nuclear gene transcription.

8.2.2.3 Changes in Chromatin Structure. As discussed previously, chromatin structure is likely to be an important factor in determining gene activity. Possibly the interactions of chromosomal proteins with DNA are altered in light-treated plants and these changes could in turn affect the rates of initiation of RNA chains, chain elongation, or both. Since DNase I sensitivity of chromatin is a useful indicator of chromatin conformation, Steinmüller et al. (1986) examined the DNase I sensitivity of four different sets of genes in barley chromatin from dark-grown and light-grown leaves as well as endosperm. The genes examined were those encoding the *Cab* protein, Pchlide reductase, the seed storage protein hordein, and a 15-kd protein of unknown function whose mRNA is abundant in dark-grown barley. Significantly, three of these genes are substantially more sensitive to DNase I in chromatin isolated from the organs in which their mRNAs accumulate. In contrast, the reductase gene(s) is DNase I sensitive in chromatin from both dark-grown leaves and endosperm but less sensitive in chromatin from light-grown leaves. Further, the single restriction fragment found with the reductase cDNA hybridization probe is digested by DNase I into discrete products, strongly suggesting the existence of DNase I–hypersensitive sites in chromatin containing the gene(s). The work of Steinmüller et al. (1986) shows that the DNase I sensitivity in leaf chromatin correlates extremely well with relative mRNA levels, implying that chromatin structure may very well contribute to the control of transcription by light.

The run-on transcription experiments by Mösinger et al. (1987) may also point to chromatin structure as an important feature in gene regulation. As mentioned above, a red light pulse induces transient change in the run-on transcription of *Cab* and Pchlide reductase gene sequences in barley. Addition of heparin to the run-on reactions obliterates the transient changes in activtity that are found in its absence. In isolated barley nuclei, as in animal systems, heparin stimulates RNA synthesis by engaged RNA polymerase II by removing chromosomal proteins, thus presumably increasing the chain elongation rate (Ananiev and Karagyozov 1984). Provided that initiation (which is inhibited by heparin) is not occurring, one interpretation of the results of Mösinger et al. (1987) is that heparin removes chromosomal proteins that influence the run-on transcription rates. In this view, RNA chain elongation rates may be a major control point in the regulation of transcription by light via changes in chromatin structure.

8.2.3 Posttranscriptional Control by Light

As discussed above, there is ample evidence to show that light can affect relative transcript levels by affecting transcription rates. In theory, the steady-state level of a transcript can be controlled not only by synthesis rates, but by the rates of processing and export from the nucleus and by transcript stability (turnover). A growing body of work indicates that for at least some genes, controls at posttranscriptional levels are also important determinants of relative transcript levels. The first clear indication of this type of control was the autoregulatory control by phytochrome of its own mRNA in oats (Colbert et al. 1985). Single red light pulses rapidly reduce the level of phytochrome mRNA (Colbert et al. 1983 and 1985), but the decline in run-on transcription rates, although significant, is neither rapid enough nor of sufficient magnitude to fully account for the precipitous decline in mRNA levels (Colbert et al. 1985; Quail et al. 1986). Thus it seems extremely likely that phytochrome controls the level of its own mRNA in part by regulating synthesis and in part by affecting mRNA degradation rates. Since phytochrome regulates its own mRNA in dicots as well (Otto et al. 1984; Lissemore et al. 1987), it will be interesting to know whether apparent posttranscriptional control is operative in those systems also.

Other examples of apparent posttranscriptional control are appearing. Two reports deal with *RbcS* expression and one with Pchlide reductase. In soybean, Shirley and Meagher (unpublished observations) compared relative transcript levels and transcription rates for the SRS1 (Berry-Lowe et al. 1982) and SRS4 (Grandbastien et al. 1986) *RbcS* genes. Transcription of both genes is induced by light, but SRS4 transcription is about twice as active as that of SRS1; transcription rates increase more upon light treatments than do steady-state levels of the RNAs. However, in total RNA, SRS1-specific transcripts are about fourfold more prevalent than SRS4-specific transcripts, but in the poly(A)-containing fraction of RNA, SRS1-specific transcripts are considerably underrepresented. Upon consideration of these facts, Shirley and Meagher (personal communication) conclude that the *RbcS* transcripts may turn over more quickly in the light than in the dark, that the SRS1 mRNA may be more stable than the SRS4 mRNA, and that the two RNAs differ in polyadenylation status, perhaps as a consequence of differential stability. Possible posttranscriptional control of *RbcS* RNA levels was found in rye by Ernst et al. (1988). They show that both *RbcS* and *Cab* relative transcript levels were controlled by phytochrome and white light. In run-on transcription measurements, clear photoinduction of synthesis rates occurred for *Cab* but not for *RbcS* (Ernst et al. 1988). Perhaps in rye the increase of *RbcS* transcript levels in response to light is caused by RNA stabilization. In barley, Mösinger et al. (1988) show that the majority of the decrease in run-on transcription of Pchlide reductase sequences occurs with VLF red light pulses, whereas relative transcript abundance is virtually unaffected by VLF pulses. Once again a posttranscriptional control must be postulated to account for the differential effect of VLF light on reductase transcription and mRNA level.

The idea that light may affect posttranscriptional events also comes from studies of the intracellular distribution of specific transcripts in peas by Sagar et al. (submitted for publication). The relative transcript levels in nuclear and cytoplasmic RNA were compared after different light treatments using *Cab, RbcS, Fed,* and three other cDNA clones as hybridization probes. Of the five light-responsive RNAs tested, three cytoplasmic and nuclear RNA levels change in parallel under the different light regimens, whereas two of the RNAs change dramatically only in the cytoplasmic fraction with very little change in nuclear RNA levels (summarized in Table 8–2). The *Fed* transcript distribution is of the latter type, while *RbcS* and *Cab* transcript distributions are of the former type. One might speculate that in those cases in which RNA does not accumulate in the nuclear fraction (*Fed,* for example), light may act to stabilize the transcripts in the cytoplasm. Recall that a possible role for stability of the *Fed* RNA is indicated by its escape from photoreversibility (see above) (Kaufman et al. 1986). Alternatively, export from the nucleus to the cytoplasm may be different in these cases. Whatever the ultimate cause, the results of Sagar et al. (submitted for publication) indicate that posttranscriptional events can strongly influence the response patterns of particular genes. Taken together with evidence for translational control of polypeptide accumulation (Slovin and Tobin 1982; Berry et al. 1986), the reports discussed above indicate that light regulates expression at both transcriptional and posttranscriptional levels, but much remains to be learned as to the mechanisms by which such control is exerted.

8.2.4 Regulation by Blue Light

The effects of blue light on mRNA levels have been less intensively studied than the responses mediated by phytochrome. One system in which blue light has been shown to regulate transcript abundance is suspension cultures of tobacco cells. Several studies have shown that upon transfer to light dark-grown cultures undergo extensive chloroplast development; chlorophyll accumulates, as do *rbcL* and *psbA* transcripts (for review see Richter et al. 1984; Richter 1984). More recently, Richter and Wessel (1985) found that red light cannot substitute for blue light in eliciting the response, indicating that cryptochrome is the active photoreceptor. Further, blue light causes both *RbcS* (Richter and Wessel 1985) and *Cab* (Richter et al. 1986) transcripts to accumulate with a high degree of coordination.

In the cell culture system, blue light responses are measured over the course of several days. It is also of interest to know whether transcript responses can be found with inductive pulses of blue light. To be sure the responses to blue light are acting through cryptochrome and not phytochrome, seedlings can be grown in continuous dim red light (that saturates the phytochrome system) supplemented with inductive blue light pulses (see Briggs and Iino 1983). Several transcripts in peas that were already known to be phytochrome regulated were tested (Kaufman et al. 1985b) in such an irradiation scheme. Some of the transcript levels are dramatically increased by growth in continuous red light over the levels induced by a single red light pulse given to etiolated seedlings (*Cab* RNA, for example; see Table 8–2). *Cab* RNA level is relatively constant over a wide range of fluence rates of red light for growth (Horwitz et al. 1988). With one exception, those transcripts that do not accumulate to higher levels in continuous red light increase substantially when a blue light pulse is administered to the seedlings (see Table 8–2) (Kaufman et al. 1985b). For example, the blue light pulse increases *RbcS* RNA sevenfold, whereas *Cab* RNA increases only by 10–15%, so the blue light exerts quite differential effects. Experiments like those mentioned above must be approached carefully because recent results suggest that the red light growth conditions can affect responsiveness. In more recent work, blue light pulses cause the expected *RbcS* RNA changes, but *Cab* RNA increases about twofold, while pEA207 RNA decreases, and appropriate changes in run-on transcription rates occur (Kaufman, personal communication). One might speculate that the differences between more-recent and previous findings might arise from slight differences in the spectral output of the light sources used. Since in these experiments the phytochrome system has already acted to increase transcripts to a level higher than that present in etiolated seedlings, and blue pulses given to etiolated seedlings are no more effective than red light pulses (Kaufman and Watson, unpublished observations), some degree of phytochrome action may be required to observe the blue light effect. Red light might render the plants competent to respond to blue light, in a way analogous to the way in which mustard seedlings become competent to accumulate transcripts in response to phytochrome (Schmidt et al. 1987). Alternatively there may be an interdependence of blue light and light acting through phytochrome such that blue light causes an intensification of the Pfr-mediated transcript accumulation, similar to the coaction of the two photoreceptors in glyceraldehyde-3-phosphate dehydrogenase accumulation in milo (Oemüller and Mohr 1984) and anthocyanin synthesis in mustard (Oelmüller and Mohr 1985).

Fluhr and Chua (1986) have shown that photoregulated expression of two *RbcS* genes in peas is quite different in mature green leaves than in etiolated organs. In mature leaves, four days of dark adaptation reduces *RbcS* 3A and *RbcS* 3C RNA levels considerably. Although a single red light pulse has no effect, 24 hours of white light greatly increases the RNA level. To test involvement of phytochrome and cryptochrome, dark-adapted plants were given eight hours of either red or blue light; to test far-red light reversibility, the long-term blue and red light treatments were supplemented with eight minutes of far-red light every twenty minutes. With both genes, they show that blue light is more effective than red light in increasing accumulation of transcripts. With both blue and red light, the increase observed was reversed by supplementing with far-red light. Fluhr and Chua (1986) conclude that light responses depend on the developmental stage of the leaf. With etiolated leaves, phytochrome control can be shown with a single red light pulse, but this is not the case with mature leaves. Even though a long red-light treatment exerts a modest effect of *RbcS* transcript levels, blue light is more effective. Since both the blue light and the red light effects are far-red light reversible, full induction of *RbcS* expression probably requires both cryptochrome and phtyochrome. So although blue light is necessary, it in itself is not sufficient for maximal expression. This situation is reminiscent of the photoregulation of anthocyanin production in milo, which Oelmüller and Mohr (1985) have shown to require the coaction of both the red and the blue light photoreceptors.

Similar blue light effects on mature leaves are found with other members of the *RbcS* gene family (Fluhr et al. 1986a) or when the *RbcS* 3A and 3C genes are expressed in transgenic petunia plants. A different approach has been taken by Shirley et al. (1987) in their studies of petunia. Instead of measuring steady-state levels of gene products, they examined specific run-on transcription in isolated nuclei for several genes (a petunia actin gene, a petunia extensinlike gene, a glycine-rich protein gene, the rRNA genes, the petunia *RbcS* genes). High transcription rates were found for all these genes in the fully expanded leaves of light-grown plants. No specific transcription of any of these genes was found in the mature leaves of plants dark-adapted for four days. Nuclei from plants undergoing these treatments are equally active in incorporating radioactive precursors into RNA but are drastically different in transcription of the genes tested (Shirley et al. 1987). It seems reasonable to assume that if neither light-regulated nor constitutive genes are transcribed after dark adaptation (Shirley et al. 1988), then upon re-illumination transcription of both classes of genes should increase. If reillumination reactivates specific transcription of many genes, then the findings of Fluhr and Chua (1986), showing that both cryptochrome and phytochrome are required for the steady-state levels of *RbcS* transcripts to increase when mature dark-adapted leaves are exposed to light, may mean that action of both photoreceptors is necessary for this hypothetical reactivation. Alernatively, both phytochrome and cryptochrome may be required for only certain genes (like *RbcS*), whereas others may respond only to one or the other photoreceptor.

8.2.5 Regulation by UV Light

The extensive characterization of the induction of flavonoids by UV light in cultured parsley cells by Halhbrock and colleagues makes this system one of the best-understood examples of the effects of light on gene expression in plant cells. One of the reasons this system is so interesting is that dark-grown cultures treated with UV light accumulate a class of phenolic compounds known as flavonoids whereas they respond to "fungal elicitor" by producing other phenolics called furanocoumarins. Induction of this latter class of phenolic compounds represents a major defense response in higher plants. Work has centered on three key enzymes of the biosynthetic pathway and their mRNAs: phenylalanine ammonia lyase (PAL), 4-coumarate : CoA ligase (4 CL), and chalcone synthase (CHS). PAL and 4 CL participate in the part of the general phenylpropanoid pathway that is common to both flavonoids and furanocoumarins and are induced by both elicitor and UV, as are their mRNAs (Kreuzaler et al. 1983). Hence it is possible to study PAL and 4 CL induction by entirely different stimuli. On the other hand CHS is specific to the flavonoid glycoside pathway, and the enzyme and its mRNA are induced only by UV (Kuhn et al. 1984). Further, run-on transcription measurements of these three show the same specificity (Chappell and Hahlbrock 1984). Importantly, Chappell and Hahlbrock (1984) show that run-on transcription of PAL and CHS sequences in nuclei is very similar to synthesis rates measured in vivo by pulse labeling. Thus in parsley cells, UV light leads to increases in transcription that are followed in sequence by increases in mRNA levels, increases in enzyme levels, and finally flavonoid accumulation. Since rates of synthesis and degradation are known for the different steps, it is clear that this orderly and highly coordinated process is almost certainly limited solely by the kinetics of gene transcription (Chappell and Hahlbrock 1984 and references therein). The fluence dependence of UV induction of CHS mRNA as well the effects of blue and far-red light have shown that three photoreceptors (a UV-B receptor, a blue light receptor, and phytochrome) all participate in regulating CHS mRNA accumulation (Bruns et al. 1986). Recently the two 4 CL genes from the parsley genome have been cloned and sequenced (Douglas et al. 1987).

Interestingly, run-on transcription measurements show that both genes are induced equally by either elicitor or UV light. We can look forward to future work to establish whether there are distinct or common regulatory sequences in these genes for the induction of transcription by two very dissimilar stimuli.

8.3 EFFECTS OF PLANT HORMONES

Light and plant growth hormones have been observed frequently to influence the same physiological processes, and this is often taken as circumstantial evidence that hormones play a role in photomorphogenesis. In a few cases, direct links have indeed been established between light and hormone effects on physiological responses (for review see O'Brien et al. 1985; De Greef and Frédéricq 1983). For example, a series of elegant experiments has recently shown that the hormonal status of coleoptile sections from oats and maize has a profound influence on red light–induced growth (Shinkle and Briggs 1984 and 1985; Shinkle 1985). Coleoptile sections grow only in response to LF red light when incubated on buffered sucrose; inclusion of 6 μM indole-3-acetic acid (IAA) shifts the fluence response entirely into the VLF range, whereas in 0.6 μM IAA both VLF and LF responses are found just as with intact seedlings (Shinkle and Briggs 1984). For several reasons the IAA sensitization of growth to red light is likely to be the consequence of an altered transduction of the Pfr signal such that some process between Pfr and the growth response is susceptible to amplification by auxin or by a physiological response to IAA (Shinkle and Briggs 1984). We do not yet know if such simple changes in the physiological status of the plant can so drastically alter the transduction of a Pfr signal leading to gene responses, but evidence is beginning to appear concerning the effects of hormones on light-regulated gene expression.

In experiments with red light–grown pea seedlings, treatment with exogenous IAA can alter levels of some of the same transcripts that are known to be light regulated (see Table 8–2) (Watson et al. unpublished observations). Quite differential effects are found: A single spraying with IAA 24 hours prior to harvest increases *RbcS* transcripts about fourteenfold, with only a 20–30% increase in *Cab* RNA. Interestingly, those transcripts that exhibit a blue light response (Kaufman et al. 1985b) also show a response to exogenous auxin (see Table 8–2). We do not yet know if IAA increases the sensitivity to red light as in oat coleoptiles (see above) or even if IAA causes ethylene production, and ethylene actually causes the changes in transcript abundance. Even though a full understanding awaits the results of experiments that are in progress, these data raise the possibility that plant hormones may in some way affect (or interact with) the processes by which light controls gene expression.

Better understood is the effect of cytokinins on gene expression in *Lemna* (Flores and Tobin 1986 and 1987). Cytokinin treatment alone causes increases in *Cab* and *RbcS* transcripts in dark-grown *Lemna* and can magnify the effect of a red light pulse in increasing transcript levels. The effect is specific in that other transcripts are unaffected. The amplification by cytokinin of the red-light–induced increases occurs when total RNA, but not when nuclear RNA, is analyzed (Flores and Tobin 1986). Further, cytokinin treatment of the plants has little effect on the rate of run-on transcription in isolated nuclei (Flores and Tobin 1987). In contrast to phytochrome-induced increases, in which transcripton does increase (Silverthorne and Tobin 1984), cytokinin appears to exert its effects at a posttranscriptional level. Possible connections with tissue-specific accumulation or chloroplast biogenesis are interesting speculations (Flores and Tobin 1986). Whether posttranscriptional control is involved in the cytokinin-dependent light/dark differences in *RbcS* and *Cab* RNA levels in petunia tissue cultures (Funckes-Shippy and Levine 1985) or in the cytokinin effect on *RbcS* and *rbcL* RNA accumulation in cucumber cotyledons (Lerbs et al. 1985) is not known.

8.4 CHLOROPLAST TRANSCRIPT ACCUMULATION

Reports of light effects on chloroplast transcript levels are numerous (Tobin and Silverthorne 1985). Many early reports did not take into account that both white light and light acting through the phytochrome system change chloroplast DNA levels (Thompson et al. 1983; Sasaki et al. 1984 and 1986), so it was not clear to what extent changes in chloroplast transcripts were related to gene dosage. When specific chloroplast transcripts and chloroplast DNA are measured in parallel, gene dosage in part accounts for the *rbcL* RNA level (see Table 8–1)(Sasaki et al. 1984 and 1987), whereas other transcripts, such as the *psbA* RNA, accumulate quite differently than chloroplast DNA (see Table 8–1). Although dissimilar patterns of abundance are sometimes used as indirect evidence for transcriptional control (Sasaki et al. 1987), two recent reports indicate that posttranscriptional events are of importance in controlling chloroplast transcript levels. In both reports (Mullet and Klein 1987; Deng and Gruissem 1987) transcription of specific genes is measured in a run-on assay using lysed plastids in which elongating RNA chains are labeled with radioactive precursors and compared to changes in steady-state levels of the corresponding transcripts. These comparisons show that in both barley (Mullet and Klein 1987) and spinach (Deng and Gruissem 1987), posttranscriptional control, presumably mRNA stability, plays a large role in determining transcript abundance. Together with evidence for translational control of particular proteins (Fromm et al. 1985;

Inamine et al. 1985; Berry et al. 1986; Klein and Mullet 1986), we see that gene product levels are regulated at a variety of steps, as is the case with nuclear genes.

8.5 PLASTID DEVELOPMENT AND NUCLEAR GENE EXPRESSION

Evidence has accumulated in recent years that early stages of choloroplast development can influence the expression of certain nuclear genes encoding chloroplast proteins. This conclusion emerges primarily from studies using pigment-deficient seedlings. Mayfield and Taylor (1984) first reported that blocking carotenoid accumulation either by mutation or by treatment with the herbicide norflurazon results in no accumulation of *Cab* protein or *Cab* mRNA in light-grown maize seedlings. Carotenoid deficiency in the light results in the photo-oxidation of chlorophyll and the production of rudimentary plastids. In contrast, chlorophyll-deficient mutants, in which chloroplast development is arrested at a later stage, accumulate normal levels of *Cab* mRNA. Similar results were obtained by Batschauer et al. (1986) with carotenoid- and chlorophyll-deficient mutants and norflurazon-treated seedlings of barley. Both mustard (Oelmüller and Mohr 1986) and maize (Mayfield and Taylor 1987) carotenoid-deficient seedlings also fail to accumulate *RbcS* mRNA, while PEP carboxylase mRNA (a cytoplasmic protein) is unaffected in maize. However, in barley *RbcS* mRNA and Pchlide reductase mRNA levels are unaffected by carotenoid deficiency (Batschauer et al. 1986). The difference in the effect on *RbcS* mRNA may reflect the lack of a light effect on *RbcS* mRNA in barley (Batschauer et al. 1986), whereas this mRNA is both white light and phytochrome responsive in maize (Nelson et al. 1984; Zhu et al. 1985) and mustard (Oelmüller and Mohr 1986; Oelmüller et al. 1986b). Importantly, Batschauer et al. (1986) have shown that *Cab* mRNA fails to accumulate in barley seedlings lacking carotenoids because the rate of run-on transcription in isolated nuclei does not increase upon illumination.

Carotenoid deficiency itself does not cause the failure of the mRNAs to accumulate, since in norflurazon-treated barley (Batschauer et al. 1986) and mustard (Oelmüller and Mohr 1986) or mutant maize (Mayfield and Taylor 1987) the mRNAs accumulate to normal levels with light conditions under which photo-oxidation of chlorophyll does not occur (dim white light or continuous far-red light). Interestingly, carotenoid-deficient barley mutants do not accumulate *Cab* mRNA even in dim light (Batschauer et al. 1986). Thus it seems unlikely that photo-oxidation per se causes the failure of certain mRNAs to accumulate. A more likely cause is that photo-oxidation prevents some important transition in the light-dependent conversion of proplastids or etioplasts to chloroplasts. If so, it is equally likely that perturbing this transition blocks the production of a signal or factor of plastid origin that affects the expression of certain nuclear genes at the transcriptional level. Consistent with this notion are other findings that indicate that expression of certain light-regulated nuclear genes is correlated with the presence of chloroplasts in the expressing organ or tissue (Herrara-Estrella et al. 1984; Eckes et al. 1985; Simpson et al. 1986b and 1988). At present it seems reasonable to suppose that the induction of some nuclear genes requires not only phytochrome, but one or more "plastid-derived factors" as well.

At present, the molecular nature of the plastid-derived factor is unknown, but the physiological characterization of the factor by Oelmüller et al. (1986b) allows further insight into its action. In experiments with norflurazon, they find that even brief periods of photo-oxidative damage cause *RbcS* and *Cab* mRNAs to decline rapidly, and that if damage occurs no recovery is found upon transfer of the mustard seedlings to non-photo-oxidizing light conditions. They conclude that the signal from the plastid is short lived and required continually and that even brief periods of damage prevent production of the signal. Interestingly, Oelmüller et al. (1986a and 1986b) show that inhibiting plastid protein synthesis with chloramphenicol also inhibits *RbcS* and *Cab* mRNA accumulation, but only when present early on—the drug is without effect once a certain stage of plastid development is reached. They conclude that a particular stage of plastid development must be reached before the signal is released. Further, they find that translatable *RbcS* and *Cab* mRNAs are detected only during that period when plastid development is occurring in the mesophyll cells of the cotyledon, suggesting that the signal is released during that time.

Oelmüller et al. (1986a) point out that the chloramphenicol experiments make it seem unlikely that the signal is a protein. In this regard, it has been suggested that plastid-encoded RNA may regulate the cytoplasmic synthesis of chloroplast proteins (Bradbeer et al. 1979). Although RNA transport across plastid membranes has never been directly shown, a set of transfer RNAs (tRNAs) associated with cytoplasmic polysomes in *Euglena* were found to hybridize with chloroplast DNA (McCrea and Hershberger 1978), implying that the tRNAs were exported from the plastid. A recent report dealing with mammalian mitochondria also suggests that nucleic acids might indeed cross organelle membranes. Chang and Clayton (1987) have shown that the RNA moiety of a mitochondrial endoribonuclease involved with primer RNA processing is not encoded in the mitochondrial genome, implying that the RNA is nucleus encoded and imported into the mitochondrion. Although the plastid-derived factor is presumably exported from the organelle, in contrast to the mitochondrial situation, the work of Chang and Clayton (1987) at least raises the possibility that RNA can cross the membranes of semi-autonomous organelles.

8.6 GENE TRANSFER EXPERIMENTS

8.6.1 Expression in Tranformed Calli

The ability to introduce genes into plants using *Agrobacterium* Ti plasmid vectors has stimulated many studies on the transfer of light-regulated genes using this technology. If the introduced genes are properly expressed in their new host cells, then it should be possible to define the DNA sequence elements (cis-acting elements) responsible for the regulation of gene expression by light. The first generation of these experiments involved the use of oncogenic Ti plasmids as vectors to study photoregulated expression of the two different *RbcS* genes from peas in petunia and tobacco (Broglie et al. 1984; Herrara-Estrella et al. 1984) and were shortly followed by reports of the expression of the soybean *RbcS* SRS1 gene (Berry-Lowe et al. 1982) in soybean (Facciotti et al. 1985) and kalanchöe (Shaw et al. 1986) tumors.

Broglie et al. (1984) transferred the entire *RbcS* E9 gene (Coruzzi et al. 1984), containing 1,059 bp of upstream sequence, into petunia protoplasts by cocultivation. Herrara-Estrella et al. (1984) took a different approach by fusing the −973 to −4 region of the pea *RbcS* ss3.6 gene (Cashmore 1983) to the CAT reporter gene and using these constructs to induce tumors on decapitated tobacco seedlings. In both cases gene products are more abundant in light-grown calli than in dark-adapted calli and increase after shifting dark-adapted calli to white light. This light effect is specific for chloroplast-containing tissue: White calli (grown initially without cytokinin) grown either in the light or in the dark contain no detectable CAT activity (Herrara-Estrella et al. 1984). With both the entire gene (Broglie et al. 1984) and similar *RbcS* fusions containing the signal sequence (Van den Broeck et al. 1985; Schreier et al. 1985), polypeptides that are processed and incorporated into chloroplasts are produced. Following up on these observations, attempts were made to localize the upstream sequence elements responsible for light regulation of transcript level using deletions in the 5′ flanking region (Morelli et al. 1985; Timko et al. 1985a). In general, as progressively more of the 5′ flanking region is removed, light-dark ratios decline. Constructs containing only the promoter region still show light-dark differences but the ratios are considerably lower than those for full-length 5′ flanking regions.

Timko et al. (1985a) also tested whether the ss3.6 upstream region could confer light regulation independently of orientation and thereby resemble the enhancers that have been found in animal systems (for review see Maniatis et al. 1987). The −973 to −90 region placed in either orientation upstream of the ss3.6 promoter (−92 to −4) has light-dark ratios virtually identical to those for the wild-type. Placing the −973 to −90 region in either orientation downstream of the ss3.6 promoter–CAT fusion gives light-dark ratios of about one fifth that of the wild type. This same upstream region also conferred light regulation on a heterologous promoter (a *NOS* promoter–CAT fusion) in either orientation with light-dark ratios of about one half of that of the wild type. The exciting conclusion from these studies is that sequences upstream of the ss3.6 TATA box contain the information required for photoregulation and do indeed resemble in many respects the enhancer sequences of animal origin.

8.6.2 Expression in Transgenic Plants

Even though many important conclusions emerge from studies of transformed calli, these systems have certain limitations, as pointed out by Fluhr et al. (1986a) and Nagy et al. (1986a). For example, light induction in calli is slow compared with that in etiolated leaves, and calli do not respond to single flashes of red light as do etiolated tissues, so that it is difficult to assess phytochrome involvement. Further, high endogenous phytohormone levels probably result from transformation with oncogenic Ti plasmids, and this may have an impact on light regulation. Transformed calli do not elaborate normal organs, resulting in difficulty in evaluating the connection between light and organ and tissue specificity.

To overcome these problems, disarmed Ti plasmid vectors have been used that allow the regeneration of transgenic, but otherwise normal, plants. Because some species have thus far proved refractory to regeneration from tissue culture or *Agrobacterium* infection, the range of suitable recipient species is limited. This means that in most (but not all) cases the transferred genes are studied in heterologous backgrounds. So far, most workers have chosen tobacco or petunia as the recipient for gene constructs. These species have relatively small seeds that upon dark germination present etiolated organs quite different from the leaf primordia or primary leaves found in legumes, thus making studies on etiolated seedlings more difficult. For example, studies of transgenic, etiolated seedlings have thus far been limited to tobacco seeds germinated on agar plates supplemented with nutrient salts and sucrose. It is not yet clear whether the exogenous carbon source affects light regulation. More frequently, mature plants are grown in light/dark cycles and then placed in the dark (dark adapted) for several days, and gene products are measured at the end of this time or after reillumination. This highly unnatural situation is necessary to reduce the steady-state levels of the gene products sufficiently so that light/dark differences can be measured. With insertion of a single copy of the foreign gene into the recipient (which is generally desirable), after selfing, the introduced gene will segregate in the progeny in the 3 : 1 ratio expected for a normal Mendelian locus. Therefore, the S_1 seedlings necessarily represent a heterogeneous population with respect to the gene of interest. Although chromosome doubling of transgenic haploids would generate homogeneous S_1 progeny, this would certainly make an already laborious process even more difficult.

The results of many recent studies of *Cab* and *RbcS* gene expression in transgenic plants are summarized in Table 8–3. Some general conclusions emerge from these experiments. First, the introduced *RbcS* or *Cab* genes are faithfully expressed in a light-dependent manner in the recipient plant.

Second, the introduced genes are expressed in an organ-specific manner, that is, highest expression is found in leaves, much less expression in stems, and no detectable expression in roots. As is natural, more detailed studies have been conducted on the donor species than on the transgenic plants, but it would seem that the photoregulation and organ distribution of the gene in the donor is faithfully recapitulated in the recipients. Lastly, because some of the studies have used 5′ flanking sequences fused to reporter genes, sometimes coupled to heterologous promoters, it is clear that the upstream sequences from *RbcS* and *Cab* genes contain sufficient information not only to confer light-responsive expression but to direct the expression in the appropriate organ as well. Similar analyses are under way with a UV-light-regulated gene encoding chalcone synthase (Kaulen et al. 1986).

8.6.2.1 Variation in Expression Levels. Most of the work exploring variation in expression levels to date has used either an intact *RbcS* or *Cab* gene or the upstream sequence (containing the homologous promoter) fused to a reporter gene. In the former case, the levels of expression of the gene in the donor and in the recipient can be compared. The results indicate that the transferred genes are not always expressed in transgenic plants at as high a level as in the donor (Lamppa et al. 1985a; Nagy et al. 1985; Fluhr and Chua 1986; Fluhr et al. 1986a). One factor determining the level of expression is the amount of upstream sequence present on the construct. Simpson et al. (1985) found that chimeric genes containing 2.5 kb of 5′ sequence from pea *Cab* gene AB80 (Cashmore 1984) result in higher expression than constructs with 400 bp of upstream sequence. Quantitative differences in expression possibly result from transferring the gene into a heterologous background, even though expression is qualitatively similar. Since the wheat *Cab*-1 gene (Lamppa et al. 1985b) expression is photoregulated and organ specific (Lamppa et al. 1985a; Nagy et al. 1986b), it can be argued that the mechanisms of expression are conserved during evolution. However, the wheat *RbcS* WS4.3 gene (Broglie et al. 1983) upon introduction into tobacco is not expressed at the RNA level when driven by its own promoter (Keith and Chua 1986).

Expression level can differ not only between donor and recipient, but also between individual transgenic plants harboring the same construct. This clonal variation has been observed with pea and *Nicotiana plumbaginifolia RbcS* (Nagy et al. 1985; Poulsen et al. 1986) and petunia *Cab* (Jones et al. 1985) gene constructs as well as the potato leaf/stem specific gene ST-LS1 (Eckes et al. 1986). As with expression level, clonal variation in organ specificity occurs (Poulsen et al. 1986; Simpson et al. 1986b). Clonal variation is often ascribed to "position effects," since it is generally assumed that the site of integration in individual transformants is different. Although the cause of clonal variation is unknown, inclusion of prokaryotic vector sequences, insertion into heterochromatin or near enhancerlike elements, and DNA methylation have been mentioned as possible explanations of this phenomenon (Fluhr et al. 1986a; Simpson et al. 1986b). Alternatively, the observed qualitative changes in expression may be caused by structural changes in the upstream region of the chimeric gene. Until it is possible to direct the insertion to a specific chromosomal location, obviously either some caution must be exercised in the selection of individual transgenic plants for further study or a large number of individuals must be examined.

8.6.2.2 The Role of Chloroplasts. Simpson et al. (1986b) examined expression of *RbcS*-NPTII and *Cab*-NPTII fusions in several different organs. With the *RbcS* (leaves > sepals ≈ stigmas >> stems > roots, petals) and *Cab* (leaves > stems >> sepals > stigmas, roots, petals) fusions the highest NPTII activity is in leaves but are otherwise differentially expressed. Neither construct is expressed in anthers or seeds. Since the *RbcS* and *Cab* chimeric genes are expressed predominantly in organs containing chloroplasts, it is of interest to determine whether expression of these genes is correlated with the presence of chloroplasts at the tissue level within the leaf. They compared NPTII activity in leaf tissues of transgenic plants containing *NOS*-NPTII fusion with activity in those plants containing *RbcS* and *Cab* fusions. With the *NOS*-NPTII construct, high activity is found in both the upper and the lower epidermis and midrib, with very low activity in the mesophyll. In sharp contrast, the *Cab* and *RbcS* fusions exhibit high activity only in the mesophyll and very low levels in the lower epidermis and midrib. Thus, expression of the *RbcS* and *Cab* chimeric genes is correlated with the presence of chloroplasts at both the organ and the leaf tissue levels. Further, expression of the *RbcS* and *Cab* fusions in transgenic plants is affected by norflurazon (Simpson et al. 1986b and 1988) in much the same way as expression in normal seedlings (see Section 8.5). The results with transgenic plants (Simpson et al. 1986b and 1988) are consistent with the notion that a plastid-derived factor is needed for expression of the *RbcS* and *Cab* genes. Moreover, they indicate that DNA sequences necessary for the effect of the plastid-derived factor must reside within the 5′ flanking regions present in the chimeric genes.

8.6.2.3 Light-Regulated Expression. Only in a few instances has gene induction mediated by phytochrome been demonstrated in transgenic plants. Nagy et al. (1986b) show that expression of the wheat *Cab*-1 gene or *Cab*-CAT gene fusions in etiolated transgenic tobacco seedlings could be induced strongly by red light; the induction could be largely reversed with far-red light, and far-red light alone results in expression levels slightly higher than those for dark controls. Similar findings were obtained for total *Cab* RNAs in both etiolated wheat and tobacco seedlings and *Cab*-1-specific RNA in wheat. The increase in RNA level found with far-red light alone might indicate a VLF component in the response, but fluence response curves similar to those for pea (Kaufman et al. 1984 and 1985a; Horwitz et al. 1988) and barley (Mösinger et al. 1988) are required to clarify this point. Another study with etiolated transgenic tobacco seedlings is that by Simpson et al. (1986b), who found phytochrome control of *Cab* AB80–NPTII fusions but not *RbcS* ss3.6–NPTII fusions. Fluhr et al. (1986b) state that the *RbcS* ss3.6 gene is phytochrome regulated in pea (although the response is exceedingly small) and suggest that the NPTII assay may not be sufficiently sensitive to detect the weak induction by red light. Fluhr and Chua (1986) found red-light–induced and red/far-red-light reversible expression of the *RbcS* 3A and *RbcS* 3C genes in transgenic petunia grown as tip cultures in darkness or intermittent red light. Taken together, these three studies indicate that at least the major features of phytochrome regulation of the *RbcS* and *Cab* genes are preserved in transgenic plants.

Most studies, by far, of *RbcS* and *Cab* gene expression in transgenic plants have dealt with white light effects. Several studies have shown the steady-state levels of gene products to be higher in light-grown plants than in dark-adapted plants. A smaller number of reports show that the gene products increase after reillumination of dark-adapted plants (see Table 8–3). As discussed previously, Fluhr and Chua (1986) have shown that the spectral sensitivity of *RbcS* expression is quite different in mature green leaves and in etiolated organs. Further, the light effects in petunia are very similar to the responses in peas, in that with etiolated leaves, phytochrome control can be shown with a single red light pulse, but this is not the case with dark-adapted mature leaves. Even though continuous red light exerts a modest effect on *RbcS* transcript levels, blue light is more effective in mature leaves (Fluhr and Chua 1986).

In their studies of expression of SRS1-NPTII fusions in transgenic petunia, Shirley et al. (1987) measured run-on transcription in isolated nuclei of the chimeric gene and a cotransferred *Nos* gene, as well as several other genes, instead of measuring steady-state levels of gene products. For young expanding leaves, dark-adaptation for four days, compared with growth in light, reduces transcription of the SRS1-NPTII fusion and the endogenous petunia *RbcS* genes to near zero but has no effect on rRNA gene transcription. Shifting light-grown plants to darkness for two hours has little effect on transcription. As in soybean (Berry-Lowe and Meagher 1985), when the shift to darkness is immediately preceded by a far-red light treatment, there is no effect on rRNA gene transcription, but a rapid reduction of transcription of both the endogenous *RbcS* genes and the SRS1-NPTII fusion occurs. The selective effect of far-red light on the decline in transcription from *RbcS* promoters is almost certainly mediated by phytochrome, although it has not been shown directly that red light can reverse the effect. Surprisingly different results were obtained by Shirley et al. (1987) with fully expanded leaves, where specific transcription of several additional endogenous genes were tested (a petunia actin gene, a petunia extensinlike gene, and a glycine-rich protein gene, in addition to the rRNA genes, SRS1-NPTII fusion, *Nos* gene, and the petunia *RbcS* genes). As with young leaves, high transcription rates are found for all these genes in light-grown plants, but no specific transcription of any of these genes is found in dark-adapted plants. Far-red light treatment of light-grown plants shifted to darkness for 15 minutes has little effect on rRNA gene transcription, but transcription of all the other genes is reduced, with the largest effect being on the *RbcS* gene and the SRS1-NPTII fusion. After two hours in the dark there is a one-hundred-fold reduction in the transcription of all genes tested. Since nuclei from these plants subjected to different treatments are equally active in incorporating radioactive precursors into RNA, the differences cannot be attributed to overall transcriptional activity. In sharp contrast to seedlings and young leaves, transcription of genes normally considered constitutive with respect to light somehow becomes sensitive to phytochrome in mature leaves (Shirley et al. 1987). The developmental or metabolic differences between young and mature leaves that cause these transcriptional alterations are currently unknown.

These findings (Shirley et al. 1987) lead to many interesting speculations concerning light responses in the mature leaves of dark-adapted plants where steady-state levels of gene products are measured. First, many of the studies summarized in Table 8–3 cotransfer a constitutive gene (such as the *NOS* gene) with the light-regulated gene construct so that data can be normalized. This is useful, since in general the steady-state levels of these constitutive genes do not markedly decline during dark adaptation, as do products derived from *Cab* or *RbcS* constructs. If active transcription from both the constitutive genes and their light-regulated counterparts has ceased in mature, dark-adapted leaves (Shirley et al. 1987), then the large difference in steady-state levels of gene products between these two types of genes may reflect differences in their turnover rates in darkness. Perhaps there is a balancing of synthesis rate and turnover rate such that under any light regimen the steady-state level of products remains relatively constant. As discussed previously, if reillumination reactivates specific transcription of many genes in mature dark-adapted leaves, then perhaps both cryptochrome and phytochrome are necessary for this hypothetical reactivation, as is the case for *RbcS* transcripts (Fluhr and Chua 1986).

8.6.2.4 *Cab* and *RbcS* Upstream Sequences Fused to Heterologous Promoters. At the time of this writing, there are two reports of fusing upstream sequences from *Cab* (Simpson et al. 1986a) or *RbcS* (Fluhr et al. 1986a) genes to heterologous promoters to examine light-regulated expression of chimeric genes in transgenic plants. The approach is quite similar to that used by Timko et al. (1985a) with transformed calli. Simpson et al (1986a) linked the upstream element from a pea *Cab* gene to the *Nos* promoter to

TABLE 8–3 Upstream Sequences of *Cab* and *RbcsS* Genes Conferring Light Responsiveness in Transgenic Plants

						Response To		
Gene	*Donor Species*	*Recipient Species*	*Promoter*	*Reporter Gene*	*Upstream Sequences*	*Pfr*	*White Light*	*Reference*
Cab-1	Wheat	Tobacco[1]	*Cab*	*Cab*	−4,400	NT	Yes[2]	Lamppa et al. 1985a
Cab-1	Wheat	Tobacco	*Cab*	*Cab*	−4,400	Yes[3]	Yes[3]	Nagy et al. 1986b
Cab-1	Wheat	Tobacco[1]	*Cab*	CAT	−1,816 to +31	Yes[3]	Yes[3]	Nagy et al. 1986b
AB80	Pea	Tobacco[1]	*Cab*	NPTII	−2,500 to +27	NT	Yes[4]	Simpson et al. 1985
AB80	Pea	Tobacco	*Cab*	NPTII	−400 to +27	NT	Yes[2]	Simpson et al. 1985
AB80	Pea	Tobacco[5,6]	*Cab*	NPTII	−2,500 to +27	Yes[3]	Yes[3]	Simpson et al. 1986b
AB80	Pea	Tobacco[1]	*Cab*	NPTII	−400 to +24	NT	NT	Simpson et al. 1986b
AB80	Pea	Tobacco[1]	*NOS*	NPTII	−317 to −100[7]	NT	Yes[4]	Simpson et al. 1986a
ss3.6	Pea	Tobacco	*RbcS*	NPTII	−973 to +341	NT	Yes[4]	Schreier et al. 1985
ss3.6	Pea	Tobacco[5,6]	*RbcS*	NPTII	−850 to +80	No[3]	Yes[3]	Simpson et al. 1986b
RbcS E9	Pea	Tobacco	*RbcS*	*RbcS*	−1,052	NT	Yes[2]	Nagy et al. 1985
RbcS E9	Pea	Petunia[1]	*RbcS*	*RbcS*	−1,052	NT	Yes[2]	Nagy et al. 1985
RbcS E9	Pea	Petunia[1]	*RbcS*	*RbcS*	−352	NT	Yes[4]	Nagy et al. 1985
RbcS 3A	Pea	Petunia	*RbcS*	*RbcS*	−410	Yes[8]	Yes[9,10]	Fluhr and Chua 1986
RbcS 3C	Pea	Petunia	*RbcS*	*RbcS*	−2,000	Yes[8]	Yes[9,10]	Fluhr and Chua 1986
RbcS 3A	Pea	Tobacco[1]	*RbcS*	*RbcS*	−410	NT	Yes[4]	Fluhr et al. 1986a
RbcS 3A	Pea	Tobacco[1]	*RbcS*	CAT	−410 to +15	NT	NT	Fluhr et al. 1986a
RbcS 3A	Pea	Tobacco[1]	CaMV 35S	CAT	−327 to −48[7]	NT	Yes[4]	Fluhr et al. 1986a
RbcS 3A	Pea	Tobacco	CaMV 35S	CAT	−327 to −48[11]	Yes[3]	Yes[3]	Fluhr et al. 1986a
RbcS E9	Pea	Petunia	*NOS*	CAT	−317 to −82[7]	NT	Yes[4]	Fluhr et al. 1986a
RbcS 8B	*N. plumbaginifolia*	*N. plumbaginifolia*[1]	*RbcS*	CAT	−1,038 to +32	NT	Yes[9]	Poulsen et al. 1986
RbcS 8B	*N. plumbaginifolia*	Tobacco[1]	*RbcS*	CAT	−1,038 to +32	NT	Yes[9]	Poulsen et al. 1986
RbcS 8B	*N. plumbaginifolia*	*Petunia*[5]	*RbcS*	CAT	−1,038 to +32	NT	Yes[9]	Poulsen et al. 1986
SRS1	Soybean	Petunia	*RbcS*	NPTII	−740 to +14	Yes[12]	Yes[4]	Shirley et al. 1987

Abbreviation: NT, not tested.

[1] Organ-specific expression in regenerated plants.

[2] Expression is greater in light-grown than in dark-adapted regenerated plants and increases upon reillumination of dark-adapted plants.

[3] Assays performed with etiolated S_1 seedlings.

[4] Expression is greater in light-grown than in dark-adapted regenerated plants (reillumination not tested).

[5] Organ-specific expression in regenerated plants, but some clonal variation was observed.

[6] Tissue-specific expression in regenerated plants.

[7] Upstream element functions in both orientations.

[8] Phytochrome-regulated expression in dark-grown tip cultures derived from regenerated plants.

[9] Expression increased by reillumination of dark-adapted regenerated plants (light grown versus dark-adapted not tested).

[10] Blue-light-responsive expression in mature green leaves of dark-adapted regenerated plants.

[11] Upstream element tested in only one orientation.

[12] Far-red light treatment of light-grown regenerated plants just before shifting to darkness greatly decreases run-on transcription compared with nontreated controls.

drive expression of the NPTII reporter gene. Since the light-dark ratios of NPTII activity are approximately equal regardless of orientation of the upstream element, the *Cab* AB80 −347 to −100 region clearly has enhancerlike properties. Further, this construct is not expressed in roots, whereas a cotransferred *NOS*-NPTII fusion is expressed equally well in roots and in leaves, suggesting that the *Cab* upstream element not only enhances expression in leaves but silences expression in roots. To account for the silencing effect of this enhancerlike element, Simpson et al. (1986a) propose that in the absence of developed plastids, some factor acting on the upstream element prevents transcription. This negatively acting factor becomes inactivated during chloroplast development, and a second factor (that probably interacts with phytochrome) acts on the upstream element to enhance transcription upon illumination. As these authors point out (Simpson et al. 1986a), the situation is likely to be somewhat more complicated than this, since regions upstream farther than −400 increase expression levels (Simpson et al. 1985).

Fluhr et al. (1986a) determined that upstream elements from pea *RbcS* genes also have enhancerlike properties. For example, the upstream element from *RbcS* 3A linked to a CaMV 35S promoter–CAT fusion confers organ-specific CAT activity in either orientation as well as higher levels of CAT RNA in light-grown than in dark-adapted leaves. Importantly, the *RbcS* 3A enhancerlike element conferred phytochrome control on the CaMV 35S promoter–CAT fusion in etiolated tobacco seedlings (Fluhr et al. 1986a). These authors stressed that gene transfer experiments show that two upstream regions are important for light-regulated expression. First, a short conserved region surrounding the TATA box (e.g., the promoter) exerts effects, although small in magnitude, in transformed calli (Timko et al. 1985a; Morelli et al. 1985). Second, the enhancerlike elements upstream of the promoter contain sufficient information to confer both light-responsive and organ-specific expression. By analogy with animal systems, Fluhr et al. (1986a) proposed that within the enhancerlike sequences there are one or more light-responsive elements (LREs) so that light-regulated expression may result from the concerted action of the promoter and LREs and nuclear proteins that interact with them. Further exploitation of gene transfer technology should allow a more-complete definition of LRE sequences. In this regard, Chua and colleagues have noted that higher-plant *RbcS* sequences contain a conserved "GT sequence motif" in the −140 region (Fluhr et al. 1986a; Nagy et al. 1986a; Kuhlemeier et al. 1987). If the enhancerlike elements in the *RbcS* and *Cab* genes resemble the inducible enhancers of animal systems, then one wonders if genes that are negatively regulated by light (e.g., the Pchlide reductase gene) contain "repressible" enhancer elements. Once the structure of such genes is sufficiently characterized, using expression in transgenic plants it should be possible to determine if repressible enhancers exist.

8.7 TRANS-ACTING FACTORS AND TRANSCRIPTION IN VITRO

By analogy with animal genes it seems reasonable to expect that multiple nuclear protein factors (trans-acting factors) will be found to bind in a sequence-specific manner both within and upstream from the promoter (for review see Dynan and Tjian 1985; Maniatis et al. 1987). In some cases, sequence-specific DNA binding proteins have been shown to be transcription factors that affect various functions of RNA polymerase. In all likelihood, trans-acting factors will be shown to play important roles in gene regulation in plants just as they have in animals. A protein from pea nuclei that binds to the upstream region of a pea *RbcS* gene has already been detected (Green et al. 1987) using gel electrophoresis of DNA-protein complexes (Hendrickson 1985) and by DNase I footprinting (Galas and Schmitz 1978). By careful comparisons of binding activities in nuclei from plants grown under different light regimens, and by correlating the sequences to which the proteins bind with sequences shown to be important in expression in transgenic plants, much useful insight into the role of the DNA binding proteins in gene regulation can be inferred. Eventually, it may even be possible to use protection from methylation by dimethylsulfate in vivo (Ephruzzi et al. 1985) or DNase I footprinting at single nucleotide resolution in nuclei (Zinn and Maniatis 1986) to explore light-regulated changes in DNA binding proteins. Trans-acting factors can also be identified in protein gel blots by their ability to bind specific sequences (Miskimins et al. 1985; White et al. 1985).

At some point it will be useful to determine whether certain sequence-specific DNA binding proteins are transcription factors. To accomplish this, whole cell (Manley et al. 1980) or nuclear extracts (Dignam et al. 1983) that can accurately initiate transcription on an exogenous DNA template are required. Soluble extracts from chloroplasts have been obtained (Orozco et al. 1986; Gruissem et al. 1986). Generally, a useful starting point for isolating active extracts is isolated nuclei that are themselves able to initiate RNA chains in vitro (Manley et al. 1978). A good indication that initiation by RNA polymerase II is occurring is the incorporation of β-^{32}P-ribonucleoside triphosphates (rNTPs) into RNA and cap structures (Manley et al. 1978; Bunick et al. 1982). Cap structures can be isolated from plant poly(A)-containing RNA (Haugland and Cline 1978), so presumably the synthesis of radiolabeled cap structures in nuclei would be an indication of initiation in plant nuclei. Since RNA polymerase II transcripts do not retain the γ-phosphate from the initiating nucleotide after capping (Bunick et al. 1982), incorporation of γ-^{32}P-rNTP into RNA is more likely to reflect initiation by RNA polymerase I or III or both. Incorporation of γ-^{32}P-NTP into RNA by tobacco nuclei has been reported (Mennes et al. 1978). More importantly, Guilfoyle et al. (1986) have shown that accurate initiation of transcription of 5S rRNA and tRNA genes occurs in isolated wheat nuclei. The ability of the nuclei to initiate synthesis is particularly sensitive to the presence of the detergent Triton X-100 (Guilfoyle et al. 1986)—a substance frequently used to remove other membranous organelles from nuclear preparations. A combination of rapid cell proliferation and low nuclease and protease contents facilitates the production of active transcription extracts from certain animal cells. The lack of these characteristics in plant cells may certainly hinder the preparation of such extracts from plant nuclei.

Underlying reasons for studying initiation of transcription in vitro are that (1) we know little about the biochemistry of initiation in plants, (2) transcription in vitro allows for rapid identification of the DNA sequences required for proper initiation, and (3) with such systems at least some of the control elements involved in developmental gene regulation may be elucidated by reconstitution experiments. Recent developments in animal systems where promoters are accurately transcribed and regulated in a cell type–specific manner in soluble extracts (Bazett-Jones et al. 1985; Mizushima-Sugano and Roeder 1986; Suzuki et al. 1986) are most encouraging. At present no definitive statements can be made concerning how closely DNA-protein interactions are linked to photoperception, but since several models for the regulation of transcription factors by light come readily to mind that are similar to the models of inducible enhancer activation proposed by Maniatis et al. (1987), this area of research seems especially promising.

8.8 CONCLUSION

The efforts of several laboratories have greatly increased our knowledge of photoregulated gene expression over the last few years. We now have a much clearer idea of the complexity and diversity inherent in the regulation of gene expression by light. For example, a particular gene in different species may be regulated in quite distinct ways. Perhaps these differences reflect the divergent patterns that plants have evolved to cope with etiolation. Likewise, light requirements and gene product accumulation patterns for different genes within a species can be quite diverse and presumably reflect an underlying diversity in signal transduction mechanisms. Possible relationships between transcriptional regulation and alterations in chromatin structure and cytosine methylation have yet to be fully elucidated. In some cases, it seems likely that different signal transduction chains exist and that these may diverge rather early on, perhaps even at the photoreceptor itself. That both transcriptional and posttranscriptional events are controlled by light may be suggestive of different signal transduction mechanisms. Clearly, much remains to be learned about signal transduction. Further, we are beginning to understand that hormonal status and the interaction of distinct photoreceptors have an impact upon gene regulation.

Efforts are underway to define further the DNA sequences and proteins concerned with transcriptional processes. Future efforts might provide similar data on the factors and sequences necessary for the control of RNA processing and turnover. Although these types of studies will not necessarily provide an understanding of signal transduction per se or how different responses are coordinated during development, it is clear that a much more detailed knowledge of regulatory sequences and proteins, and the biochemistry of transcription, is a prerequisite for future studies. Work such as this, together with studies on the photoreceptors themselves and their interactions with cellular components, should ultimately lead to a detailed understanding of signal transduction and photoregulated gene expression in what promises to be a bright future for this field of endeavor.

REFERENCES

Ananiev, E., and Karagyozov, L. (1984) *Physiol. Veg.* 22, 555–563.

Apel, K., Bauschauer, A., Dehesh, K., Kruez, K., and Steinmüller, K. (1986) *Philos. Trans. R. Soc. Lond.* B 314, 481–492.

Batschauer, A., Mösinger, E., Kreuz, K., Dörr, I., and Apel, K. (1986) *Eur. J. Biochem.* 154, 625–634.

Bazett-Jones, D.P., Yechel, M., and Gottesfeld, J.M. (1985) *Nature* 317, 824–828.

Berry, J.O., Nikolau, B.J., Carr, J.P., and Klessig, D.F. (1986) *Mol. Cell. Biol.* 6, 2347–2353.

Berry-Lowe, S.L., McKnight, T.D., Shah, D.M., and Meagher, R.B. (1982) *J. Appl. Mol. Genet.* 1, 483–498.

Berry-Lowe, S.L., and Meagher, R.B. (1985) *Mol. Cell. Biol.* 5, 1910–1917.

Bradbeer, J.W., Atkinson, Y.E., Börner, T., and Hagemann, R. (1979) *Nature* 279, 816–817.

Briggs, W.R., and Iino, M. (1983) *Philos. Trans. R. Soc. Lond.* B 303, 347–359.

Briggs, W.R., Mandoli, D.F., Shinkle, J.R., et al. (1985) in *Sensory Perception in Aneural Organisms* (Colombetti, G., and Song, P.-S., eds.), pp. 265–280, Plenum Press, New York.

Briggs, W.R., Mösinger E., and Schäfer, E. (1988) *Plant Physiol.* 86, 435–440.

Brockmann, J., Rieble, S., Kazarinova-Fukshansky, N., Seyfried, M., and Schäfer, E. (1987) *Plant Cell Environ.* 10, 105–111.

Broglie, R., Coruzzi, G., Fraley, R.T., et al. (1984) *Science* 224, 838–843.

Broglie, R., Coruzzi, G., Lamppa, G., Keith, B., and Chua, N.-H. (1983) *Biotechnol.* 1, 55–61.

Bruns, B., Hahlbrock, K., and Schäfer, E. (1986) *Planta* 169, 393–398.

Bunick, D., Zandomeni, R., Ackerman, S., and Weinmann, R. (1982) *Cell* 29, 877–886.

Cashmore, A.R. (1983) in *Genetic Engineering of Plants* (Kosuge, T., Meridith, C.P., and Hollaender, A., eds.), pp. 29–38, Plenum Press, New York.

Cashmore, A.R. (1984) *Proc. Natl. Acad. Sci. USA* 81, 2960–2964.

Chang, D.D., and Clayton, D.A. (1987) *Science* 235, 1178–1184.

Chappell, J., and Hahlbrock, K. (1984) *Nature* 311, 76–78.

Colbert, J.T., Hershey, H.P., and Quail, P.H. (1983) *Proc. Natl. Acad. Sci. USA* 80, 2248–2252.

Colbert, J.T., Hershey, H.P., and Quail, P.H. (1985) *Plant Mol. Biol.* 5, 91–101.

Coruzzi, G., Broglie, R., Edwards, C., and Chua, N.-H. (1984) *EMBO J.* 3, 1671–1679.

Dean, C., van den Elzen, P., Tamaki, S., Dunsmuir, P., and Bedbrook, J. (1985) *EMBO J.* 4, 3055–3061.

Dean, C., van den Elzen, P., Tamaki, S., et al. (1987) *Nucleic Acids Res.* 206, 465–474.

De Greef, J.A., and Frédéricq, H. (1983) in *Encyclopedia of Plant Physiology New Series,* vol. 16A (Shropshire, W. Jr., and Mohr, H., eds.), pp. 401–427, Springer-Verlag, New York.

Deng, X.-W., and Gruissem, W. (1987) *Cell* 49, 378–387.

de Vries, S.C., Springer, J., and Wessels, J.G.H. (1982) *Planta* 156, 129–135.

de Vries, S.C., Springer, J., and Wessels, J.G.H. (1983) *Planta* 158, 42–50.

Dignam, J.D., Lebovitz, R.M., and Roeder, R.G. (1983) *Nucleic Acids Res.* 11, 1475–1489.

Dobres, M.S., Elliot, R.C., Watson, J.C., and Thompson, W.F. (1987) *Plant Mol. Biol.* 8, 53–59.

Doerfler, W. (1983) *Annu. Rev. Biochem.* 52, 93–124.

Douglas, C., Hoffman, H., Schulz, W., and Hahlbrock, K. (1987) *EMBO J.* 6, 1189–1195.

Dynan, W.S., and Tjian, R. (1985) *Nature* 316, 774–778.

Eckes, P., Rosahl, S., Schell, J., and Willmitzer, L. (1986) *Mol. Gen. Genet.* 205, 14–22.

Eckes, P., Schell, J., and Willmitzer, L., (1985) *Mol. Gen. Genet.* 199, 216–224.

Elgin, S.R.C. (1981) *Cell* 27, 413–415.

Ellis, R.J. (1986) *Biosci. Rep.* 6, 127–136.

Ellis, T.H.N., Davies, D.R., Castleton, J.A., and Belford, I.D. (1984) *Chromosoma* 91, 74–81.

Ephruzzi, A., Church, G.M., Tonegawa, S., and Gilbert, W. (1985) *Science* 227, 134–140.

Ernst, D., and Oesterhelt, D. (1984) *EMBO J.* 3, 3075–3078.

Ernst, D., Pfeiffer, F., Vojacek, R., Schefbeck, K., and Oesterhelt, D. (1988). *Plant Mol. Biol.* 10, 21–33.

Facciotti, D., O'Neal, J.K., Lee, S., and Shewmaker, C.K. (1985) *Biotechnol.* 3, 241–246.

Flavell, R.B., O'Dell, M., Smith, D.B., and Thompson, W.F. (1985) in *Molecular Form and Function of the Plant Genome* (van Vloten-Doting, L., Groot, G.S.P., and Hall, T.C., eds.), pp. 1–14, Plenum Press, New York.

Flavell, R.B., O'Dell, M., Thompson, W.F., et al. (1986) *Philos. Trans. R. Soc. Lond.* B 314, 385–397.

Flores, S., and Tobin, E.M. (1986) *Planta* 168, 340–349.

Flores, S., and Tobin, E.M. (1987) in *Molecular Biology of Plant Growth Control* (Fox, J.E., and Jacobs, M., eds.), pp. 123–132, Alan R. Liss, New York.

Fluhr, R., and Chua, N.-H. (1986) *Proc. Natl. Acad. Sci. USA* 83, 2358–2362.

Fluhr, R., Kuhlemeier, C., Nagy, F., and Chua, N.-H. (1986a) *Science* 232, 1106–1112.

Fluhr, R., Moses, P., Morelli, G., Coruzzi, G., and Chua, N.-H. (1986b) *EMBO J.* 5, 2063–2071.

Fromm, H., Devic, M., Fluhr, R., and Edelman, M. (1985) *EMBO J.* 4, 291–295.

Funckes-Shippy, C.L., and Levine, A.D. (1985) in *Molecular Biology of the Photosynthetic Apparatus* (Steinback, K.E., Bonitz, S., Arntzen, C.J., and Bogorad, L., eds.), pp. 407–412, Cold Spring Harbor Laboratory, Cold Spring Harbor, NY.

Galas, D., and Schmitz, A. (1978) *Nucleic Acids Res.* 5, 3157–3170.

Gallagher, T.F., and Ellis, R.J. (1982) *EMBO J.* 1, 1493–1498.

Gallagher, T.F., Jenkins, G.I., and Ellis, R.J. (1985) *FEBS Lett.* 186, 241–245.

Grandbastien, M.A., Berry-Lowe, S.L., Shirley, B.W., and Meagher, R.B. (1986) *Plant Mol. Biol.* 7, 451–465.

Green, P.J., Kay, S.A., and Chua, N.-H. (1987) *EMBO J.* 6, 2543–2549.

Gruissem, W., Greenberg, B.M., Zurawski, G., and Hallick, R.B. (1986) *Methods Enzymol.* 118, 253–270.

Guilfoyle, T.J., Suzich, J., and Lindberg, M. (1986) *Plant Mol. Biol.* 7, 95–104.

Harpster, M., and Apel, K. (1985) *Physiol. Plant.* 64, 147–152.

Haughland, R.A., and Cline, M.G. (1978) *Plant Physiol.* 62, 838–840.

Hendrickson, W. (1985) *Biotechniques* 3, 198–207.

Herrara-Estrella, L., Van den Broeck, G., Maenhaut, R., et al. (1984) *Nature* 310, 115–120.

Horwitz, B.A., Thompson, W.F., and Briggs, W.R. (1988) *Plant Physiol.* 86, 299–305.

Inamine, G., Nash, B., Weissbach, H., and Brot, N. (1985) *Proc. Natl. Acad. Sci. USA* 82, 5690–5694.

Jenkins, G.I., and Smith, H. (1985) *Photochem. Photobiol.* 42, 679–684.

Jones, J.D.K., Dunsmuir, P., and Bedbrook, J. (1985) *EMBO J.* 4, 2411–2418.

Jorgensen, R.A., Cuellar, R.A., Thompson, W.F., and Cavanaugh, T.A. (1987) *Plant Mol. Biol.* 8, 3–12.

Kasemir, H. (1983) in *Encyclopedia of Plant Physiology New Series,* vol. 16B (Shropshire, W. Jr., and Mohr, H., eds.), pp. 662–686, Springer-Verlag, New York.

Kaufman, L.S., Briggs, W.R., and Thompson, W.F. (1985a) *Plant Physiol.* 78, 388–393.

Kaufman, L.S., Roberts, L.L., Briggs, W.R., and Thompson, W.F. (1986) *Plant Physiol.* 81, 1033–1038.

Kaufman, L.S., Thompson, W.F., and Briggs, W.R. (1984) *Science* 226, 1447–1449.

Kaufman, L.S., Watson, J.C., Briggs, W.R., and Thompson, W.F. (1985b) in *Molecular Biology of the Photosynthetic Apparatus* (Steinback, K.E., Bonitz, S., Arntzen, C.J., and Bogorad, L., eds.), pp. 367–372, Cold Spring Harbor Laboratory, Cold Spring Harbor, NY.

Kaufman, L.S., Watson, J.C., and Thompson, W.F. (1987) *Proc. Natl. Acad. Sci. USA* 84, 1550–1554.

Kaulen, H., Schell, J., and Kreuzaler, F. (1986) *EMBO J.* 5, 1–8.

Keith, B., and Chua, N.-H. (1986) *EMBO J.* 5, 2419–2425.

Keshet, I., Lieman-Hurwitz, J., and Cedar, H. (1986) *Cell* 44, 535–543.

Klein, R.R., and Mullet, J.E. (1986) *J. Biol. Chem.* 261, 11138–11145.

Kloppstech, K. (1985) *Planta* 165, 502–506.

Kreuzaler, F., Ragg, H., Fautz, E., Kuhn, D.N., and Hahlbrock, K. (1983) *Proc. Natl. Acad. Sci. USA* 80, 2591–2593.

Kuhlemeier, C., Green, P.J., and Chua, N.-H. (1987) *Annu. Rev. Plant Physiol.* 38, 221–257.

Kuhn, D.N., Chappell, J., Boudet, A., and Hahlbrock, K. (1984) *Proc. Natl. Acad. Sci. USA* 81, 1102–1106.

Kung, S.D. (1976) *Science* 191, 429–434.

Kung, S.D., Thornber, J.P., and Wildman, S.G. (1972) *FEBS Lett.* 24, 185–188.

Lamb, C.J., and Lawton, M.A. (1983) in *Encyclopedia of Plant Physiology New Series,* vol. 16A (Shropshire, W. Jr., and Mohr, H., eds.), pp. 213–311, Springer-Verlag, New York.

Lamppa, G., Nagy, F., and Chua, N.-H. (1985a) *Nature* 316, 750–752.

Lamppa, G.K., Morelli, G., and Chua, N.-H. (1985b) *Mol. Cell. Biol.* 5, 1370–1378.

Lerbs, S., Lerbs, W., Wollgiehn, R., and Parthier, B. (1985) in *Molecular Form and Function of the Plant Genome* (van Vloten-Doting, L., Groot, G.S.P., and Hall, T.C., eds.), pp. 267–276, Plenum Press, New York.

Lissemore, J.L., Cobert, J.T., and Quail, P.H. (1987) *Plant Mol. Biol.* 8, 485–496.

Long, E.O., and Dawid, I.B. (1980) *Annu. Rev. Biochem.* 74, 727–764.

McCrea, J.M., and Hershberger, C.L. (1978) *Nature* 274, 717–719.

Maniatis, T., Goodbourn, S., and Fischer, J.A. (1987) *Science* 236, 1237–1245.

Manley, J.L., Fire, A., Cano, A., Sharp, P.A., and Gefter, M.A. (1980) *Proc. Natl. Acad. Sci. USA* 77, 3855–3859.

Manley, J.L., Sharp, P.A., and Gefter, M.L. (1978) *Proc. Natl. Acad. Sci. USA* 76, 160–164.

Martineau, B., and Taylor, W.C. (1985) *Plant Physiol.* 78, 399–404.

Martineau, B., and Taylor, W.C. (1986) *Plant Physiol.* 82, 613–618.

Mayfield, S.P., and Taylor, W.C. (1984) *Eur. J. Biochem.* 144, 79–84.

Mayfield, S.P., and Taylor, W.C. (1987) *Mol. Gen. Genet.* 208, 309–314.

Mennes, A.M., Bouman, H., Van Den Burg, M.P.M., and Libbenga, K.R. (1978) *Plant Sci. Lett.* 13, 329–339.

Meyer, G., and Kloppstech, K. (1984) *Eur. J. Biochem.* 138, 201–207.

Miskimins, W.K., Roberts, M.P., McClelland, A., and Ruddle, F.H. (1985) *Proc. Natl. Acad. Sci. USA* 82, 6741–6744.

Mizushima-Sugano, J., and Roeder, R.G. (1986) *Proc. Natl. Acad. Sci. USA* 83, 8511–8515.

Mohr, H., and Schäfer, E. (1983) *Philos. Trans. R. Soc. Lond.* B 303, 489–501.

Morelli, G., Nagy, F., Fraley, R.T., Rogers, S.G., and Chua, N.-H. (1985) *Nature* 315, 200–204.

Mösinger, E., Batschauer, A., Apel, K., Schäfer, E., and Briggs, W.R. (1988) *Plant Physiol.* 86, 706–710.

Mösinger, E., Batschauer, A., Schäfer, E., and Apel, K. (1985) *Eur. J. Biochem.* 147, 137–142.

Mösinger, E., Batschauer, A., Vierstra, R., Apel, K., and Schäfer, E. (1987) *Planta* 170, 505–514.

Mösinger, E., and Schäfer, E. (1984) *Planta* 161, 444–450.

Mullet, J.E., and Klein, R.R. (1987) *EMBO J.* 6, 1571–1579.

Nagy, F., Fluhr, R., Kuhlemeier, C., et al. (1986a) *Philos. Trans. R. Soc. Lond.* B *314, 493–500.*

Nagy, F., Kay, S.A., Boutry, M., Hsu, M.-Y., and Chua, N.-H (1986b) EMBO J. 5, 1119–1124.

Nagy, F., Morelli, G., Fraley, R.T., Rogers, S.G., and Chua, N.-H. (1985) *EMBO J.* 4, 3063–3068.

Nelson, T., Harpster, M.H., Mayfield, S.P., and Taylor, W.C. (1984) *J. Cell Biol.* 98, 558–564.

O'Brien, T., Beall, F.D., and Smith, H. (1985) in *Encyclopedia of Plant Physiology New Series,* vol. 11 (Pharis, R.P., and Reid, D.M., eds.), pp. 282–307, Springer-Verlag, New York.

Oelmüller, R., Dietrich, G., Link, G., and Mohr, H. (1986a) *Planta* 169, 260–266.

Oelmüller, R., Levitan, I., Bergfeld, R., Rajasekar, V.K., and Mohr, H. (1986b) *Planta* 168, 482–492.

Oelmüller, R., and Mohr, H. (1984a) *Planta* 161, 165–171.

Oelmüller, R., and Mohr, H. (1984b) *Plant Cell Environ.* 7, 29–37.

Oelmüller, R., and Mohr, H., (1985) *Proc. Natl. Acad. Sci. USA* 82, 6124–6128.

Oelmüller, R., and Mohr, H. (1986) *Planta* 167, 106–113.

Orozco, E.M. Jr., Mullet, J.E., Hanley-Bowdoin, L., and Chua, N.-H. (1986) *Methods Enzymol.* 118, 232–253.

Otto, V., Schäfer, E., Nagatani, A., Yamamoto, K.T., and Furuya, M. (1984) *Plant Cell Physiol.* 25, 1579–1584.

Polans, N.O., Weeden, N.F., and Thompson, W.F. (1986) *Theor. Appl. Genet.* 72, 289–295.

Poulsen, C., Fluhr, R., Kauffman, J.M., Boutry, M., and Chua, N.-H. (1986) *Mol. Gen. Genet.* 205, 193–200.

Quail, P.H., Cobert, J.T., Peters, N.K., et al. (1986) *Philos. Trans. R. Soc. Lond.* B 314, 469–480.

Razin, A., and Szyf, M. (1984) *Biochim. Biophys. Acta* 782, 331–342.

Reeder, R.H. (1984) *Cell* 38, 349–351.

Reeder, R.H. (1985) *J. Cell Biol.* 101, 2013–2016.

Reeves, R. (1984) *Biochim. Biophys. Acta* 782, 343–393.

Richter, G. (1984) in *Blue Light Effects in Biological Systems* (Senger, H., ed.), pp. 253–263, Springer-Verlag, New York.

Richter, G., Einspanier, R., Hüsemann, W., Dudel, A., and Wessel, K. (1986) in *Regulation of Chloroplast Differentiation* (Akoyunoglou, G., and Senger, H., eds.), pp. 549–558, Alan R. Liss, New York.

Richter, G., Hundriesen, J., Gross, M., et al. (1984) in *Blue Light Effects in Biological Systems* (Senger, H., ed.), pp. 387–396, Springer-Verlag, New York.

Richter, G., and Wessel, K., (1985) *Plant Mol. Biol.* 5, 175–182.

Rüdiger, W., and Scheer, H. (1983) in *Encyclopedia of Plant Physiology New Series,* vol. 16A (Shropshire, W. Jr., and Mohr, H., eds.), pp. 119–151, Springer-Verlag, New York.

Sasaki, Y., Nakamura, Y., and Matsuno, R. (1986) *FEBS Lett.* 196, 171–174.

Sasaki, Y., Nakamura, Y., and Matsuno, R. (1987) *Plant Mol. Biol.* 8, 375–382.

Sasaki, Y., Tomoda, Y., and Kamikubo, T. (1984) *FEBS Lett.* 173, 31–35.

Scharnhorst, C., Heinze, H., Meyer, G., et al. (1985) *Plant Mol. Biol.* 4, 242–245.

Schmidt, S., Drumm-Herrel, H., Oelmüller, R., and Mohr, H. (1987) *Planta* 170, 400–407.

Schreier, P.H., Seftor, E.A., Schell, J., and Bohnert, H.J. (1985) *EMBO J.* 4, 25–32.

Shaw, C.H., Sanders, D.M., Bates, M.R., and Shaw, C.H. (1986) *Nucleic Acids Res.* 14, 6603–6612.

Sheen, J-Y., and Bogorad, L. (1986a) *EMBO J.* 5, 3417–3422.

Sheen, J-Y., and Bogorad, L. (1986b) *Proc. Natl. Acad. Sci. USA* 83, 7811–7815.

Sheen, J-Y., and Bogorad, L. (1987) *Plant Mol. Biol.* 8, 227–238.

Shinkle, J.R. (1986) *Plant Physiol.* 81, 533–537.

Shinkle, J.R., and Briggs, W.R. (1984) *Proc. Natl. Acad. Sci. USA* 81, 3742–3746.

Shinkle, J.R., and Briggs, W.R. (1985) *Plant Physiol.* 79, 349–356.

Shirley, B.W., Berry-Lowe, S.L., Rogers, S.G., et al. (1987) *Nucleic Acids Res.* 15, 6501–6514.

Silverthorne, J., and Tobin, E.M. (1984) *Proc. Natl. Acad. Sci. USA* 81, 1112–1116.

Silverthorne, J., and Tobin, E.M. (1987) *BioEssays* 7, 18–24.

Simpson, J., Schell, J., Van Montagu, M., and Herrara-Estrella L. (1986a) *Nature* 323, 551–554.

Simpson, J., Timko, M.P., Cashmore, A.R., et al. (1985) *EMBO J.* 4, 2723–2729.

Simpson, J., Van Montagu, M., and Herrara-Estrella, L. (1986b) *Science* 233, 34–38.

Simpson, J., Van Montagu, M., and Herrara-Estrella, L. (1988) *Plant Cell Rep.* (in press).

Slayton, M.M., Brosio, P., and Dunsmuir, P. (1988) *Plant Mol. Biol.* 10, 127–138.

Slovin, J.P., and Tobin, E.M. (1982) *Planta* 154, 465–472.

Smith, H., and Morgan, D.C. (1983) in *Encyclopedia of Plant Physiology New Series,* vol. 16B (Shropshire, W. Jr., and Mohr, H., eds.), pp. 491–517, Springer-Verlag, New York.

Spiller, S.C., Kaufman, L.S., Thompson, W.F., and Briggs, W.R. (1987) *Plant Physiol.* 84, 409–414.

Steinmüller, K., Batschauer, A., and Apel, K. (1986) *Eur. J. Biochem.* 158, 519–525.

Suzuki, Y., Tsuda, M., Takiya, S., et al. (1986) *Proc. Natl. Acad. Sci. USA* 83, 9522–9526.

Thien, W., and Schopfer, P. (1975) *Plant Physiol.* 56, 660–664.

Thien, W., and Schopfer, P. (1982) *Plant Physiol.* 69, 1156–1160.

Thompson, W.F., and Cleland, R.E. (1972) *Plant Physiol.* 50, 289–292.

Thompson, W.F., Everett, M., Polans, N.O., Jorgensen, R.A., and Palmer, J.D. (1983) *Planta* 158, 487–500.

Thompson, W.F., Flavell, R.B., Watson, J.C., and Kaufman, L.S. (1987a) in *The Architecture of Eukaryotic Genes* (Kahl, G., ed.), pp. 385–396, VCH, New York.

Thompson, W.F., Kaufman, L.S., Horwitz, B.A., et al. (1987b) in *Biomechanisms Regulating Growth and Development: Keys to Progress* (Steffens, G.L., and Rumsey, T.S., eds.), pp. 269–284, Kluwer, Boston.

Thompson, W.F., Kaufman, L.S., and Watson, J.C. (1985) *BioEssays* 3, 153–159.

Timko, M.P., Kausch, A.P., Castresana, C., et al. (1985a) *Nature* 318, 579–582.

Timko, M.P., Kausch, A.P., Hand, J.M., et al. (1985b) in *Molecular Biology of the Photosynthetic Apparatus* (Steinback, K.E., Bonitz, S., Arntzen, C.J., and Bogorad, L., eds.), pp. 381–397, Cold Spring Harbor Laboratory, Cold Spring Harbor, NY.

Tobin, E.M., and Silverthorne, J. (1985) *Annu. Rev. Plant Physiol.* 36, 569–593.

Tobin, E.M., Silverthorne, J., Flores, S., Leutwiler, L.S., and Karlin-Neumann, G.A. (1987) in *Molecular Biology of Plant Growth Control* (Fox, J.E., and Jacobs, M., eds.), pp. 401–411, Alan R. Liss, New York.

Tobin, E.M., Wimpee, C.F., Karlin-Neumann, G.A., Silverthorne, J., and Kohorn, B.D. (1985) in *Molecular Biology of the Photosynthetic Apparatus* (Steinback, K.E., Bonitz, S., Arntzen, C.J., and Bogorad, L., eds.), pp. 373–380, Cold Spring Harbor Laboratory, Cold Spring Harbor, NY.

Van den Broeck, G., Timko, M.P., Kausch, A.P., et al. (1985) *Nature* 313, 358–363.

VanDerWoude, W.J. (1985) *Photochem. Photobiol.* 42, 655–661.

VanDerWoude, W.J. (1987) in *Phytochrome and Photoregulation in Plants* (Furuya, M., ed.), pp. 249–258, Academic Press, Orlando, FL.

Waalwijk, C., and Flavell, R.A. (1978) *Nucleic Acids Res.* 5, 3231–3236.

Watson, J.C., Kaufman, L.S., and Thompson, W.F. (1987) *J. Mol. Biol.* 193, 15–26.

Watson, J.C., and Thompson, W.F. (1986) *Methods Enzymol.* 118, 57–75.

Wellmann, E. (1983) in *Encyclopedia of Plant Physiology New Series,* vol. 16B (Shropshire, W. Jr., and Mohr, H., eds.), pp. 745–756, Springer-Verlag, New York.

White, B.A., Preston, G.M., Lufkin, T.C., and Bancroft, C. (1985) *Mol. Cell. Biol.* 5, 2967–2974.

Wu, C. (1980) *Nature* 286, 854–860.

Zhu, Y.S., Kung, S.D., and Bogorad, L. (1985) *Plant Physiol.* 79, 371–376.

Zinn, K., and Maniatis, T. (1986) *Cell* 45, 611–620.

CHAPTER

9

Hormonal and Stress Regulation of Gene Expression in Cereal Aleurone Layers

Tuan-hua David Ho

9.1 THE CEREAL ALEURONE LAYERS

The aleurone layers of barley and wheat seeds play a crucial role in the mobilization of endosperm nutrients to support the postgermination seedling growth. This tissue is usually one to three cell layers-thick and lies directly underneath the seed coat and forms the outermost layers of endosperm. Like the rest of endosperm, aleurone cells are triploid, with one set of chromosomes from the male parent and two sets from the female parent. After the onset of germination the aleurone layers respond to the hormone gibberellins (GA) from the embryo by synthesizing and secreting several hydrolytic enzymes—including α-amylases (Filner and Varner 1967), proteases (Jacobsen and Varner 1967), 1,3;1,4-β-glucanase (Stuart et al. 1986), xylanase (Dashek and Chrispeels 1977), and nuclease (Brown and Ho 1986)—to the endosperm, where these enzymes hydrolyze the stored starch,

The research in the author's lab has been supported by grants from the National Science Foundation, the United States Department of Agriculture, and Anheuser Busch Company.

proteins, cell wall polysaccharides, and remnant nucleic acids. Another hormone, abscisic acid (ABA), which induces seed dormancy, prevents all the known GA effects in this tissue. The aleurone layers have been considered a convenient system in which to study the mode of action of GA and ABA because of the following advantages: First, in the one-cell-thick wheat aleurone layers, there are apparent homogeneous cell populations that all respond to the two hormones. In barley, the three layers of cells may respond to the hormones with slightly different timing. Second, at least for GA, the source (embryo) and the target tissue (aleurone layers) of the hormone can be physically separated by dissecting the dry seeds, thus the target tissue can be treated with known concentrations of exogenously applied hormones. Third, many enzymes and proteins that can serve as convenient biochemical markers for the studies of hormone actions are available (Table 9–1). Fourth, protoplasts of aleurone cells that still respond to GA can be prepared (Jacobsen et al. 1985). The protoplasts are ideal for studying the effect of hormones on secretion. Besides, organelles such as nuclei can be isolated from protoplasts easier than from intact cells (Jacobsen and Beach 1985). Fifth, the antagonism between GA and ABA provides a new dimension in which to investigate the role of hormone interactions on plant development. Last, genetic mutants with altered sensitivities to these hormones are available (Ho et al. 1980). The study of these mutants

TABLE 9–1 Hormonal Regulation of Gene Expression in Barley Aleurone Layers

	Hormone Treatment			
Gene	*None*	*GA*	*ABA*	*GA + ABA*
α-Amylase, high pI	–[2]	+++++	–	–
α-Amylase, low pI	+	+++++	+	+
Thioprotease	–	+++	–	–
Nuclease (RNase + DNase +3′-nucleotidase)	+	+++	–	ND
1,3; 1,4-β-Glucanase	–	+++	–	ND
Actin	++	++	++	++
Nondifferential[1]	++	++	++	++
Alcohol dehydrogenase	+	–	+	+
GA suppressed[1]	+++	–	+++	+
ABA-induced p27	+	–	+++	+
α-Amylase inhibitor	+	–	+++	+

Abbreviations: GA, gibberellic acid; ABA, abscisic acid; ND, not determined.
[1] Unidentified cDNA clones.
[2] See text for details.

should be a valuable complement to the molecular studies of hormone action.

Because of the homogeneous cell population in aleurone layers, this tissue has also become a convenient system in which to study other processes. In the last few years, several papers have been published concerning the use of aleurone layers to investigate plants' responses to environmental stresses (Belanger et al. 1986; Hanson and Jacobsen 1984). Interestingly, information obtained from stress research has also revealed insights into the regulation of hormone-induced genes. Since more work has been performed with barley than with wheat or oat aleurone layers, most of the information discussed in this chapter deals with barley unless stated otherwise.

9.2 EFFECT OF GIBBERELLINS ON GENE EXPRESSION

Both GA and ABA alter the expression of several sets of genes in barley aleurone layers. As summarized in Table 9–1, gibberellic acid (GA_3, one of the most active gibberellins in barley aleurone layers) not only induces α-amylase isozymes, two thioproteases, a nuclease, and a few other enzymes, but also suppresses the expression of several other genes. In contrast, ABA suppresses the expression of the aforementioned hydrolytic enzymes, yet it induces another set of new proteins (see Section 9.3.2).

9.2.1 Complexity of α-Amylases and Their Genes

Because of the abundance of α-amylase, the effect of hormones on its synthesis has been most extensively studied. α-Amylase is composed of two sets of isozymes that are similar in size (44 kd) yet differ in net charges (Jacobsen and Higgins 1982; Callis and Ho 1983). Since several other classifications have been used in the literature, it is crucial to point out that the high-pI isozymes are equivalent to isozyme group A, or 1, and the low-pI isozymes are equivalent to isozyme group B, or 2. Using chromosome addition lines Brown and Jacobsen (1982) and Muthukrishnan et al. (1983b) have demonstrated that the high-pI and low-pI isozymes are encoded by a "set" of structural genes on chromosomes 6 and 1, respectively. Whittier et al. (1987) and Knox et al. (1987) have isolated and studied several α-amylase genomic clones by restriction mapping and sequence analysis. Four of five genomic clones contain high-pI isozyme sequences and one contains the low-pI isozyme sequence. The DNA sequences of the two types of α-amylase genes are divergent, with the high-pI gene containing two introns and the low-pI gene containing three introns. Although the putative promoter regions of the two types of α-amylase genes show little sequence homology, both contain pairs of inverted repeats. It is not yet known whether these repeats are involved in the regulation of expression of these genes.

9.2.2 Expression of α-Amylase Genes

The expression of high-pI α-amylase is not detectable before the addition of GA_3 to aleurone layers, yet the low-pI α-amylase is expressed at a low level in the same tissue. After the onset of hormone treatment the expression of both groups of α-amylase is enhanced within a couple of hours (Figures 9–1 and 9–2). The GA_3-enhanced expression of high-pI α-amylase reaches a maximum around 20 hours and then declines. Very little of this isozyme is synthesized beyond 30 hours of GA_3 treatment. In contrast, the synthesis of low-pI α-amylase continues to about 40 hours of GA_3 treatment. This differential expression of α-amylase isozymes in GA_3-treated barley aleurone layers can be observed at the protein level by analyzing newly synthesized proteins with native gel electrophoresis (see Figure 9–1). Similar results have been obtained at the RNA level by dot blot analysis using cDNA specific for the two groups of isozymes (see Figure 9–2). The fact that the results of protein analysis match well with those of RNA analysis indicates that the regulation of GA_3 induction of α-amylase isozymes is mainly at the transcriptional level. Results similar to those described here have also been reported by several other laboratories (Huang et al. 1984; Chandler et al. 1984; Rogers 1985). More recently, Jacobsen and Beach (1985) have performed in vitro run-on transcription experiments with nuclei isolated from GA_3-treated barley aleurone protoplasts and shown that GA_3 enhances the rate of transcription of α-amylase genes by about tenfold, and ABA treatment of aleurone can reverse this GA_3 effect. A similar conclusion has been reached by Zwar and Hooley (1986) using nuclei isolated from oat aleurone protoplasts.

Regulation of α-amylase synthesis at levels other than transcription also exists in barley aleurone layers. For example, α-amylase mRNAs appear to be quite stable. It has been shown that transcription inhibitors such as cordycepin (3′-deoxyadenosine) are effective in inhibiting the GA_3-induced α-amylase synthesis (Ho and Varner 1974). However, cordycepin fails to prevent the continued synthesis of α-amylases when it is added 12 hours or later after GA_3 administration, indicating that the α-amylase mRNAs are synthesized during the first 12 hours of hormone treatment and the turnover rate of this mRNA is slow (Ho and Varner 1974). The half-life of these mRNAs has been estimated at longer than 100 hours (Ho et al. 1987). However, the stability of α-amylase mRNA does not appear to be an intrinsic property of this message.

Investigating the effect of heat shock in barley aleurone layers recently, Belanger et al. (1986) revealed a potential mechanism underlying mRNA stability. A heat-shock treatment (40°C) of barley aleurone layers not only induces the heat-shock proteins but also effectively diminshes the synthesis

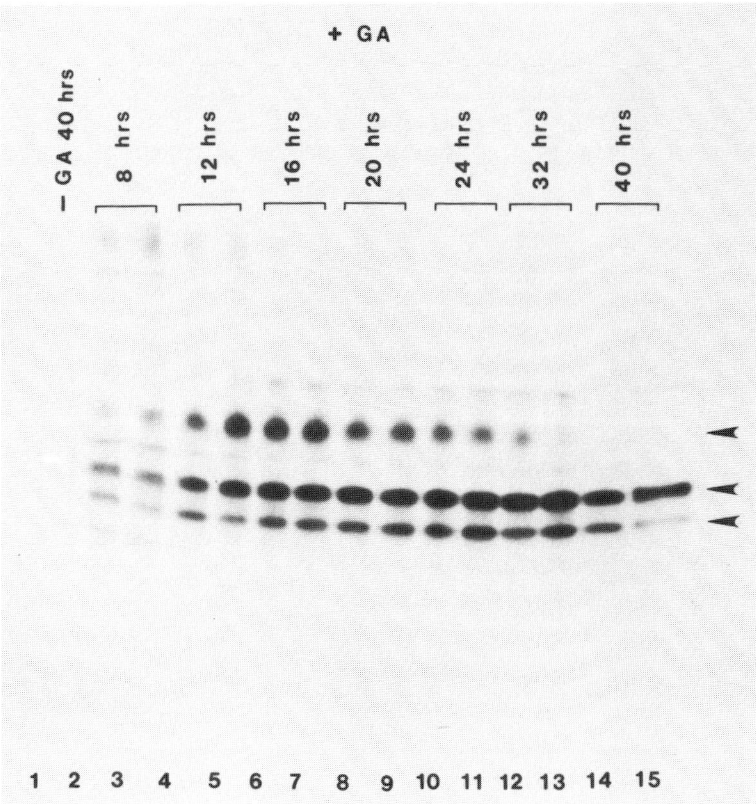

FIGURE 9–1 Time course of the GA_3 induction of α-amylase isozymes in barley aleurone layers. Isolated aleurone layers were incubated with 10^{-6} M GA_3 for the lengths of time indicated on top of the gel. The tissues were labeled with ^{35}S-methionine during the last hour of incubation. Duplicate samples of newly synthesized proteins were analyzed by nondenaturing gel electrophoresis and the fluorogram of the gel is shown. The top arrowhead on the right indicates the position of the high-pI isozymes (the two major high-pI isozymes run to the sample position on this gel) and the two lower arrowheads indicate the low-pI isozymes. From R.C. Nolan and T.H.D. Ho (1988) *Planta* (in press). With permission.

of GA_3-induced α-amylases. Analyzing the levels of α-amylase mRNA by Northern blots probed with specific cDNA, it is observed that the normally stable α-amylase mRNAs in heat-shocked barley aleurone layers are actively degraded (Figure 9–3). This is contrary to what happens in heat-shocked *Drosophila* cells, in which normal mRNAs are sequestered and preserved, and only heat-shock protein mRNAs are translated (Storti et al. 1980).

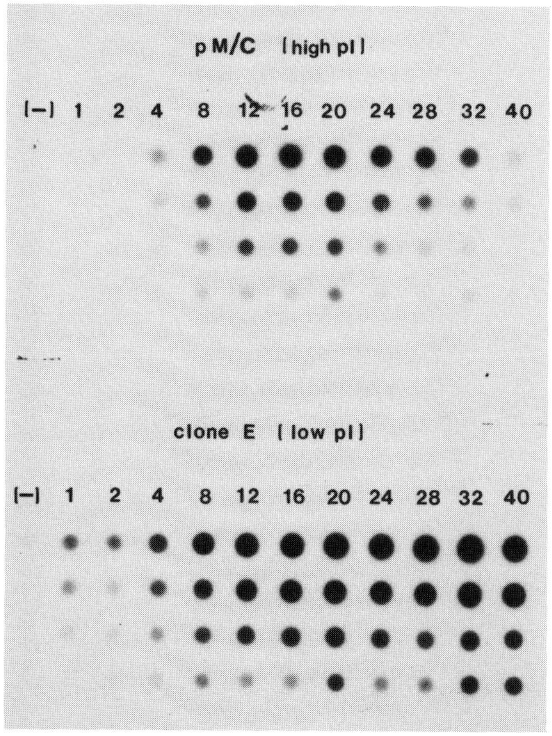

FIGURE 9–2 Dot blot analysis of the levels of mRNA sequences for α-amylase isozymes in GA_3-treated barley aleurone layers. Isolated aleurone layers were incubated with 10^{-6} M GA_3 for the number of hours shown on top of the blots. Total RNA was isolated and applied to Gene Screen Plus membrane (New England Nuclear, Boston, MA) after a series of twofold dilutions (the highest concentration was on the very top). The membrane was then probed with two nick-translated cDNA probes, pM/C for the high-pI isozyme mRNA sequences and clone E for the low-pI isozyme sequences. The autoradiograph of the blots is shown. –, aleurone layers incubated without GA_3 for 40 hours. From R.C. Nolan and T.H.D. Ho, unpublished observations.

The timing of the heat shock–induced α-amylase mRNA destruction is closely correlated with another heat stress–induced phenomenon: a fast delamellation of endoplasmic reticulum (ER) (Figure 9–4). Both of these events start within 30 minutes of heat shock and reach the maximum within three to four hours. When the tissue is recovered from heat-shock treatment, the ER is reformed and the synthesis of α-amylases is resumed (see Figures 9–3A and 9–4). The resumption of α-amylase synthesis, is, of course, dependent on the new synthesis of its mRNAs (Belanger et al. 1986). It was

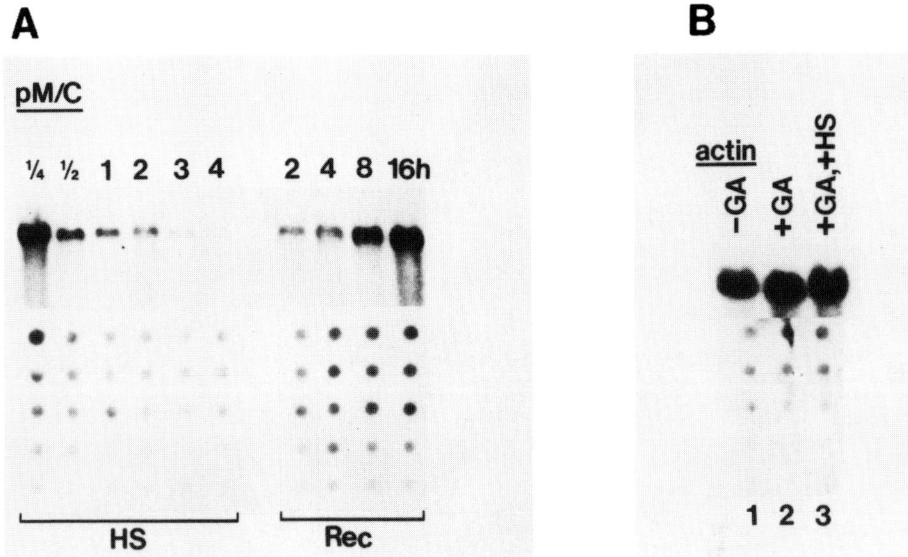

FIGURE 9–3 Effect of heat shock (HS) on the levels of mRNA is barley aleurone layers. The aleurone layers were incubated with 10^{-6} M GA_3 for 16 hours before they were heat shocked at 40°C for the number of hours indicated on top of the blots. Some of the heat-shocked layers were recovered at 25°C (Rec) for up to 16 hours, as indicated. Total RNA was isolated and analyzed by dot blot (bottom) and Northern gel (top) hybridizations. The probes used were pM/C, a cDNA clone for the high-pI α-amylase isozymes, and a soybean actin cDNA clone that cross-hybridized with barley RNA. From M.R. Brodl and T.H.D. Ho, unpublished observations.

found that not only the synthesis of α-amylase but that of all the other secretory protein is suppressed in heat-shocked barley aleurone layers (Brodl 1987). However, the mRNA encoding actin, a protein synthesized on free polyribosomes, is unaffected by the heat-shock treatment (see Figure 9–3B). Since ER is the site for the synthesis and processing of secretory proteins, the association of mRNA encoding secretory proteins with this organelle may lead to the stabilization of these mRNAs. Once ER is destroyed by heat shock, mRNAs normally associated with ER become unstable and are degraded quickly. It is well documented that GA_3 also induces the formation of ER in barley aleurone layers (Jones 1969; Evins and Varner 1971), and ER appears to be essential for synthesis and processing of α-amylase (Jones and Jacobsen 1982). Thus, besides inducing the transcription of α-amylase genes, GA_3 treatment may also stabilize α-amylase mRNAs.

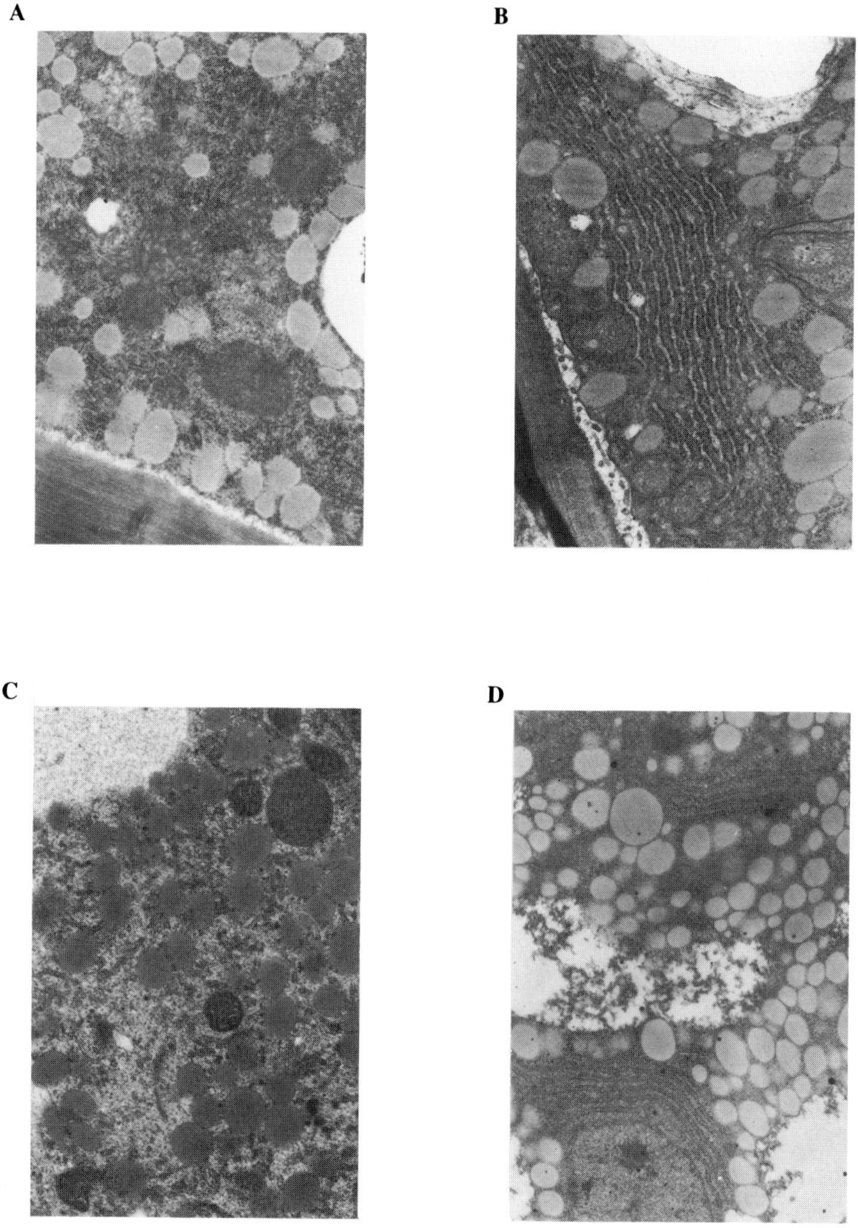
A
B
C
D

9.2.3 Regulation of Expression of Other Genes

As mentioned above, GA_3 is known to induce several hydrolytic enzymes besides α-amylase isozymes. Among them endoxylanase (Benjavongkulchai and Spencer 1986), 1,3:1,4-β-glucanase (Woodward and Fincher 1982), nuclease (Brown and Ho 1986), and a thioendoprotease (Hammerton and Ho 1986; Koehler and Ho 1988) have been purified to homogeneity, and antibodies against them are available. In addition, cDNA clones for 1,3:1,4-β-glucanase (Fincher et al. 1986) and thioprotease (Rogers et al. 1985) are available. Thus, the molecular mechanisms underlying the expression of these genes could be investigated, and the information could be compared with that on α-amylase induction. The induction of the thioprotease aleurain is also regulated at the level of its mRNA, and the timing of its induction by GA_3 is similar to that of the low-pI α-amylase.

Besides induction of hydrolytic enzymes, GA_3 suppresses quite a few other genes. One of the GA_3-suppressed genes has been identified as alcohol dehydrogenase (Nolan and Ho 1988). The expression of a third type of genes, such as actin genes, remains unaltered by the hormone treatment (Brodl and Ho, unpublished observations). Thus, the GA_3 suppression of genes is not the consequence of induction of hydrolytic enzymes due to some sort of competition for transcriptional machinery. However, the exact mechanism of GA_3 suppression of genes remains unknown.

9.2.4 Early Events Induced by Gibberellins

Since it takes about two to four hours of GA_3 treatment to detect the induction of α-amylase genes, it is crucial to investigate the early GA_3-induced events that lead to the later expression of α-amylase. It has been shown that the induction of α-amylase mRNAs can be blocked if aleurone layers are treated with the protein synthesis inhibitor cycloheximide or the amino acid analogue aminoethylcysteine during the first two hours of hormone treatment (Muthukrishnan et al. 1983a). This observation indicates that the expression of a gene or a group of genes during the early hours of

FIGURE 9–4 Effect of heat shock on the ultrastructure of barley aleurone layers. The aleurone layers were incubated with 10^{-6} M GA_3 for 16 hours before they were heat shocked at 40°C for 3 hours. The tissues were fixed and examined under an electron microscope. Panel A shows tissue not treated with GA_3 and not heat shocked. Panel B shows tissue treated with GA_3 but not heat shocked. Note the massive proliferation of ER that was induced by GA_3. Panel C shows GA_3-treated tissue that had been heat shocked. Note that the GA_3-induced ER was destroyed. Panel D shows the same tissue as that shown in Panel C, except the tissue was recovered from heat shock at 25°C for 8 hours. Note that some ER reappeared. From M.R. Brodl and T.H.D. Ho, unpublished observations.

GA_3 treatment is necessary for the later expresion of α-amylase. To date, this type of "early gene" has not been further characterized.

Another approach to studying the early events induced by GA_3 is to remove the hormone after a few hours. Chrispeels and Varner (1967) concluded that GA_3 is continuously required for the synthesis of α-amylase. However, it has recently been observed that GA_3 is required only for a short period of time in order to stimulate the continuous synthesis of α-amylase (Lu and Ho, unpublished observations). As shown in Figure 9–5, when GA_3 is removed after four hours (or less) of treatment, aleurone layers fail to synthesize as much α-amylase as the tissue that has been continuously treated with GA_3, indicating that early removal of the hormone stops the sequence of events that is triggered by the hormone. On the other hand, removal of hormone after eight hours has no effect on the continuous synthesis of α-amylase. The system has somehow reached the point of no return at this stage, and the synthesis of α-amylase is no longer dependent on the presence of hormone molecules. It is not known whether a second messenger is involved in this process. More research is certainly needed to characterize the early changes induced by GA_3.

9.3 EFFECT OF ABA ON GENE EXPRESSION

9.3.1 Suppression of Gibberellin-induced Genes

ABA is effective in inhibiting the synthesis of both high-pI and low-pI α-amylase isozymes when it is added at the same time as GA_3 (Figure 9–6). However, it has a much more noticeable effect on the synthesis of the high-pI isozymes than on the synthesis of the low-pI isozymes when it is added 20 hours longer after GA_3 (see Figure 9–6) (Nolan et al. 1987). This differential effect of ABA on α-amylase synthesis can be easily observed by nondenaturing gel electrophoresis, yet not by SDS gel electrophoresis, because the isozymes are not resolved by the latter technique. The physiological significance of this differential inhibition of the synthesis of α-amylase isozymes is still unclear. Probing mRNA sequences on Northern gels, Nolan et al. (1987) have observed that the ABA inhibition of α-amylase synthesis correlates with the decrease in mRNA sequences for the individual groups of isozymes. Thus, ABA appears to inhibit α-amylase synthesis at the level of mRNA synthesis rather than at the translational level, as suggested before (Ho and Varner 1976; Mozer 1980). The data of Jacobsen and Beach (1985) also suggest that ABA inhibits the transcription rate of α-amylase genes. Therefore, a major action of ABA in this system is to counter the action of GA_3 at the transcriptional level. These two hormones probably do not compete with each other for a common site of action; it has been shown that high concentrations of GA_3 cannot totally alleviate the inhibition of ABA (Chrispeels and Varner 1966).

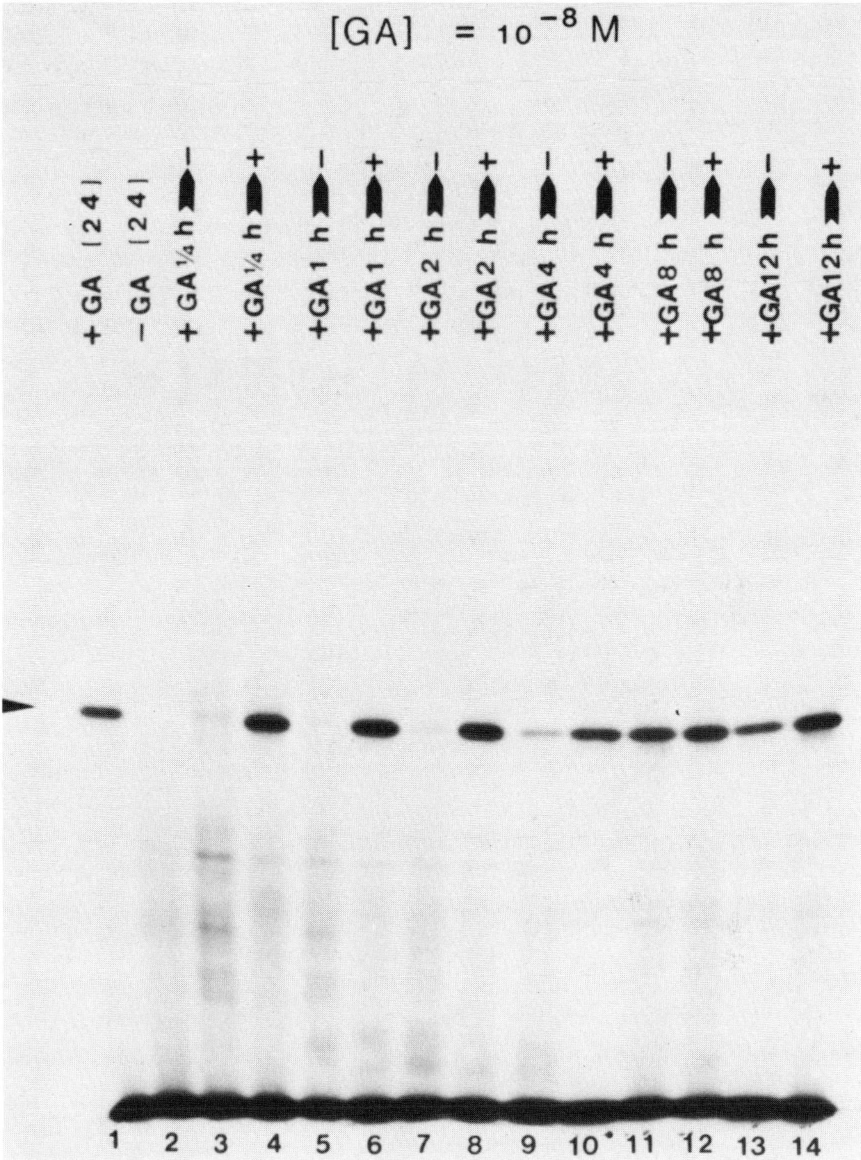

FIGURE 9–5 Effect of mid-course removal of GA_3 on the continuous synthesis of α-amylase in barley aleurone layers. Barley aleurone layers were treated with 10^{-8} M GA_3 for the lengths of times indicated on top of the gel. The tissues were then washed with (+) or without (−) GA_3 for 2 hours and continued to incubate in the presence (+) or absence (−) of GA_3 until 24 hours after the initial addition of GA_3. The tissues were pulse-labeled with ^{35}S-methionine during the last hour of incubation, and the extracted proteins were analyzed by SDS polyacrylamide gel electrophoresis. The fluorogram of the gel is shown. The arrowhead on the left indicates the position for α-amylase. Lanes 1 and 2 are samples continuously incubated with (+) or without (−) GA_3 for 24 hours. From H.Y. Lu and T.H.D. Ho, unpublished observations.

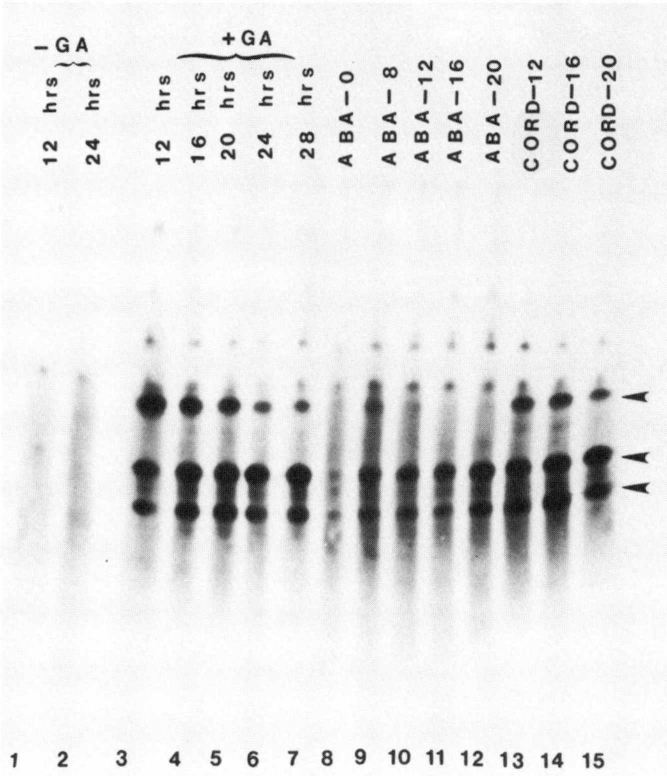

FIGURE 9–6 Effect of ABA on the synthesis of α-amylase isozymes in barley aleurone layers. Aleurone layers were incubated with various hormones and inhibitors as indicated, and labeled with ^{35}S-methionine during the last hour of incubation. The newly synthesized proteins were analyzed by nondenaturing gel electrophoresis, and the fluorogram of the gel is shown. GA_3-treated samples were incubated for the number of hours shown. ABA (2×10^{-5} M) or cordycepin (CORD) (10^{-4} M) was added to various samples at various times (numbers, in hours) after GA_3 addition. GA_3 (10^{-6} M) was present continuously for these sample, and the total incubation time was 25 hours. The top arrowhead on the right indicates the position of high-pI α-amylase isozymes and the two lower arrowheads indicate the low-pI isozymes. From R.C. Nolan et al. (1987). *Plant Mol. Biol.* 8, 13–22. With permission.

Besides acting at the transcription level, ABA also appears to affect the stability of α-amylase mRNA. As mentioned previously, α-amylase mRNAs appear to be quite stable after more than 12 hours of GA_3 treatment and the mid-course addition of ABA alters the levels of α-amylase mRNA. As shown in Figure 9–7, treatment with ABA apparently has a much stronger effect on the stability of high-pI mRNA than on the stability of low-pI

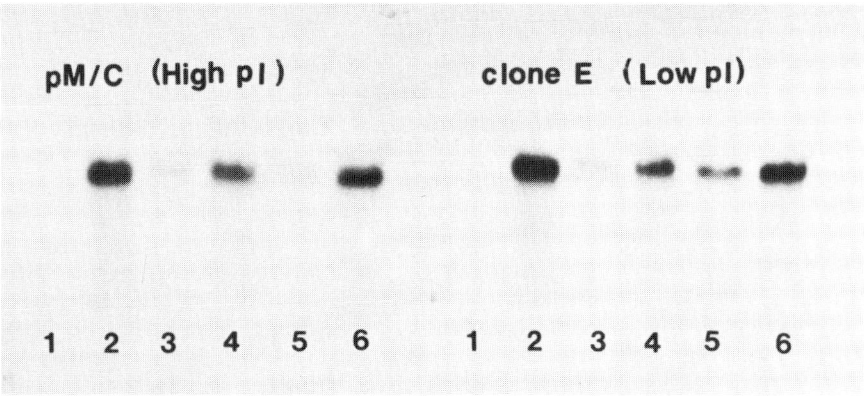

FIGURE 9–7 Effect of ABA on the levels of mRNA for α-amylase isozymes. Aleurone layers were incubated for 25 hours in the absence of GA_3 (lane 1); 25 hours in the presence of 10^{-6} M GA_3 (lane 2); 25 hours in the presence of GA_3 and 2×10^{-5} M ABA with a second addition of ABA at 12 hours (lane 3); 25 hours with ABA added 10 hours after GA_3 (lane 4); 25 hours with ABA added 20 hours after GA_3 (lane 5); or 25 hours with ABA and 10^{-4} M cordycepin added 20 hours after GA_3 (lane 6). Total RNA from these samples was analyzed by Northern gels probed with pM/C for high-pI isozymes and clone E for low-pI isozymes. Note that cordycepin prevented the effect of ABA (compare lanes 5 and 6). From R.C. Nolan and T.H.D. Ho, unpublished observations.

mRNA. Within four hours of ABA treatment, the level of high-pI mRNA is reduced to less than 10% of the control.

9.3.2 Induction of New Proteins

In the course of investigating the action of ABA on α-amylase mRNA stability, Nolan et al. (1987) observed an interesting phenomenon: the action of ABA itself is dependent on the continuous synthesis of some RNA and proteins. As shown in Figure 9–7, inhibition of RNA or protein synthesis by cordycepin or cycloheximide blocks the effects of ABA on the decrease of α-amylase mRNA sequences. One cannot argue that the effects of these metabolic inhibitors are the consequence of induced cellular toxicity, because the α-amylase mRNA sequences are even more abundant in the presence of these compounds. Thus, the action of ABA probably relies on the expression of one or more other genes.

In order to study the gene or genes whose expression is crucial to the action of ABA in the inhibition of α-amylase synthesis, newly synthesized proteins in ABA-treated aleurone layers have been analyzed. There are nine ABA-induced protein bands on a one-dimensional gel, and the number increases to sixteen when the samples are analyzed on two-dimensional gels

(Lin and Ho 1986). Treatment of aleurone layers also increases the levels of mRNA encoding the ABA-induced proteins, as determined by in vitro protein synthesis. ABA concentrations as low as 10^{-8} M are able to induce some of these proteins. The identities of these ABA-induced proteins have not been totally resolved. However, some interesting features of them have been revealed. An ABA-induced 36-kd protein can be precipitated with antiserum against a barley lectin specific for glucosamine, galactosamine, and mannosamine (Figure 9–8) (Partridge et al. 1976; Lin and Ho 1986). A

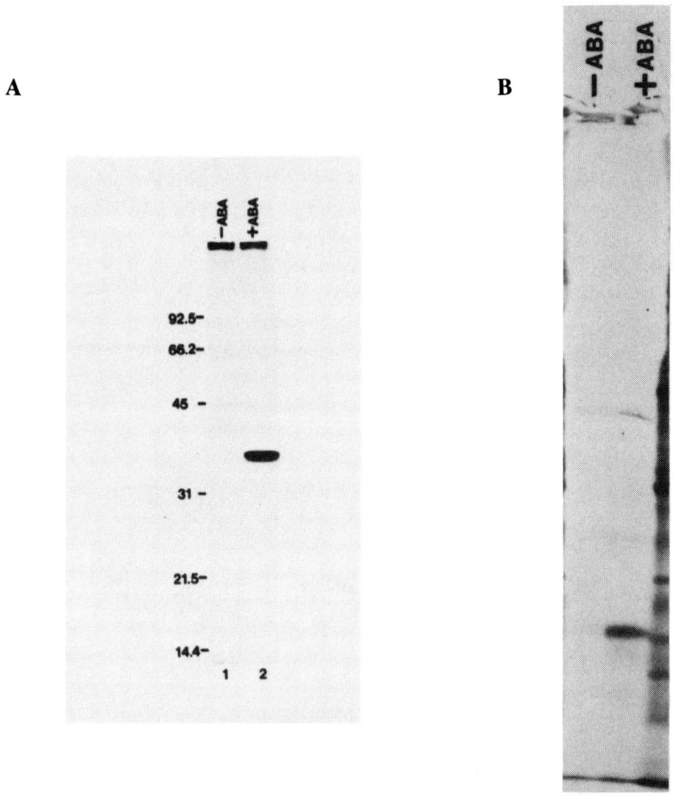

FIGURE 9–8 Immunoprecipitation of barley aleurone proteins with antiserum against a barley lectin specific for glucosamine, galactosamine, and mannosamine (A), and with antiserum against α-amylase inhibitor (B). Aleurone layers were incubated with (+) or without (−) 2×10^{-5} M ABA for 24 hours and labeled with ^{35}S-methionine for the last hour. Proteins were extracted and reacted with antiserum against the lectin and precipitated with *Staphylococcus aureus* Cowan strain 1. The pellets were washed, resuspended, and analyzed by SDS gel electrophoresis. The fluorogram of the gel is shown. Molecular weight markers are indicated by bars at left in A.

21-kd ABA-induced protein appears to be a specific inhibitor of high-pI amylase (Mundy 1984; Lin and Ho, unpublished observations). The most-abundant among all of the ABA-induced proteins is 29 kd in size and is soluble in 0.1 M HCl (pH 1) (L.S. Lin, personal communication). The cDNA of a 27-kd ABA-induced protein has recently been cloned and analyzed (Hong and Ho, unpublished observations). The amino acid sequence from the DNA sequence indicates that this protein is rich in lysine and alanine and contains nine 11-amino-acid repeats. Two-dimensional gel electrophoresis indicates that the pI of this protein is higher than 8.5. The expression of this protein can be induced by ABA at a concentration as low as 10^{-9} M. The level of mRNA of this 27-kd protein increases within 30 minutes of ABA treatment, and reaches a maximum around 8 to 12 hours (Figure 9–9). Although the exact function of this 27-kd protein remains unknown, its amino acid composition resembles that of some DNA binding proteins such as histones and the high-mobility-group proteins. A few of the ABA-induced proteins can be also induced by ABA in developing barley seeds. Although none of the ABA-induced proteins resemble the major barley storage proteins, hordeins, it has not been ruled out that some ABA-induced proteins are less-abundant seed storage proteins.

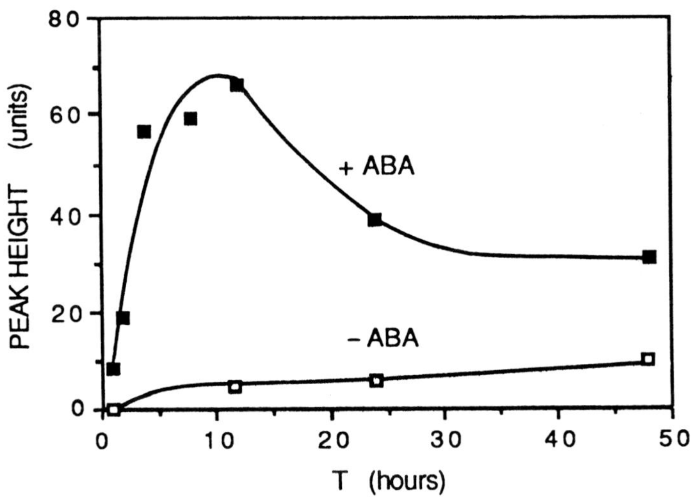

FIGURE 9–9 Time course of the static levels of gene transcripts encoding ABA-induced p27. Aleurone layers were treated with (+) or without (−) ABA (20 μM) for the lengths of time indicated before total RNA was extracted. The level of p27 mRNA was analyzed by Northern blot probed with cDNA clone pHVA-1. Hybrid-select translation studies with pHVA-1 indicated that it encodes a basic 27-kd protein with pI higher than 8.5. The fluorograms of the Northern blot were scanned in a densitometer and the relative density for each sample is presented. From B. Hong, S.J. Uknes, and T.H.D. Ho, unpublished observations.

9.3.3 Role of the ABA Metabolite Phaseic Acid

It has been known for some time that ABA can be metabolized quickly in barley aleurone layers. Besides glucose conjugates of ABA, two metabolites have been identified (Dashek et al. 1979). The first stable metabolite of ABA is phaseic acid (PA), which is effective in inhibiting α-amylase synthesis, yet has little effect on the induction ABA-inducible proteins (Figure 9–10) (Dashek et al. 1979; Ho et al. 1985). Like ABA, when PA is added 12 to 16 hours after GA_3, the synthesis of α-amylase is inhibited. Since the most significant effect of ABA at this stage is the destabilization of α-amylase mRNA, the effect of PA is probably also on mRNA stability. It has been shown by Uknes and Ho (1984) that the in vivo conversion of ABA to PA in barley aleurone layers is enhanced by pretreating the tissue with ABA (Figure 9–11). Thus, it seems that ABA is capable of enhancing its own metabolism to PA, which is similar to the substrate induction of nitrate reductase by nitrate, that has been well documented in many plant tissues. On the other hand, the conversion of PA to the next metabolite, dihydrophaseic acid, is affected neither by ABA nor by PA. Isolated dihydrophaseic acid has little or no effect on the GA_3-induced synthesis of α-amylase. The ABA enhancement of its own metabolism is unlikely to be a scavenging mechanism to remove excssive amounts of ABA, because PA is also biologically effective. An effective scavenging mechanism would have to enhance the metabolism of ABA to dihydrophaseic acid or later metabolites to diminish the bioligical effect due to ABA.

The conversion of ABA to PA is dependent on the RNA and protein synthesis, and the enzyme involved in this process is thought to be a cytochrome P450–linked mono-oxygenase (Walton 1980). Two mono-oxygenase inhibitors, paclobutrazol and tetcylacis, which block the conversion of ABA to PA, also alleviate the inhibitory effect of ABA on α-amylase synthesis (Lin and Ho, unpublished observations). These results strongly suggest that ABA has to be converted to PA in order to exert its regulatory effect on α-amylase mRNA. Based on these observations, PA may be the active component in ABA's action on the stability of α-amylase mRNA. On the other hand, the effect of ABA on transcription, either to inhibit α-amylase gene transcription or to induce new proteins, may not be mediated by PA.

9.3.4 Mode of Action of ABA

As shown in the diagram in Figure 9–12, ABA can thoroughly nullify the effect of GA_3 on the induction of α-amylase. At the transcriptional level, ABA slows down the transcription rate of α-amylase genes. The 27-kd ABA-induced protein has properties similar to DNA binding proteins; thus, it could be involved in the suppression of α-amylase genes. Since both α-amylase and its mRNA appear to be quite stable, simply preventing the transcription of α-amylase genes cannot totally reverse the effect of GA_3.

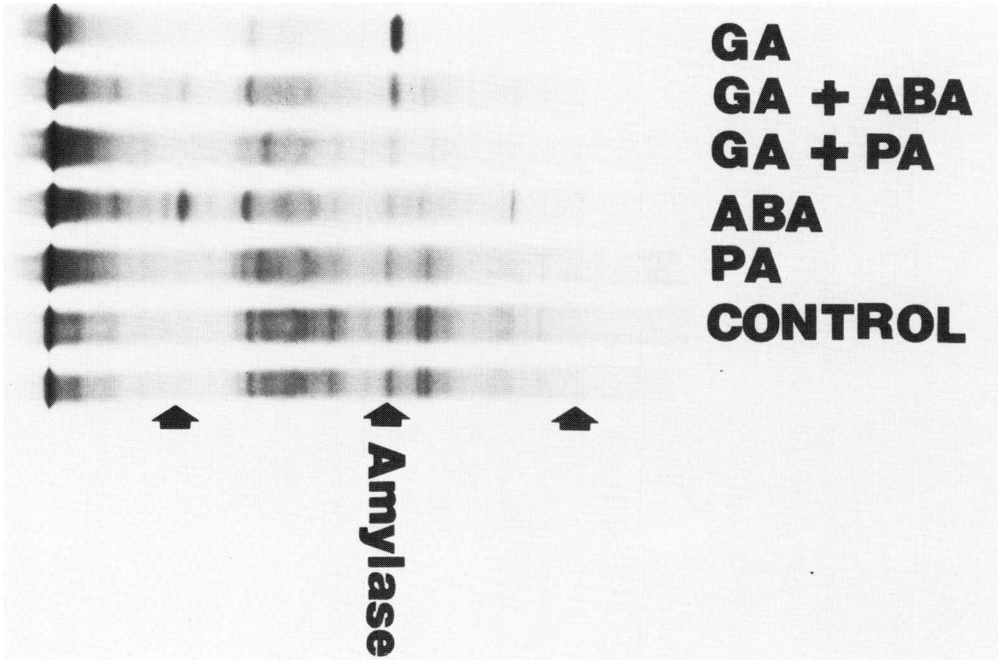

FIGURE 9–10 SDS polyacrylamide gel electrophoresis analysis of newly synthesized proteins from barley aleurone layers The tissues were treated with various combinations of hormones as indicated for 24 hours and labeled with ^{35}S-methionine for the last 2 hours. The fluorogram of the gel is shown. The two arrows above and below α-amylase indicate the position of two of the ABA-inducible proteins. Note that PA was effective in inhibiting α-amylase synthesis, yet it had little effect on the induction of ABA-inducible proteins.

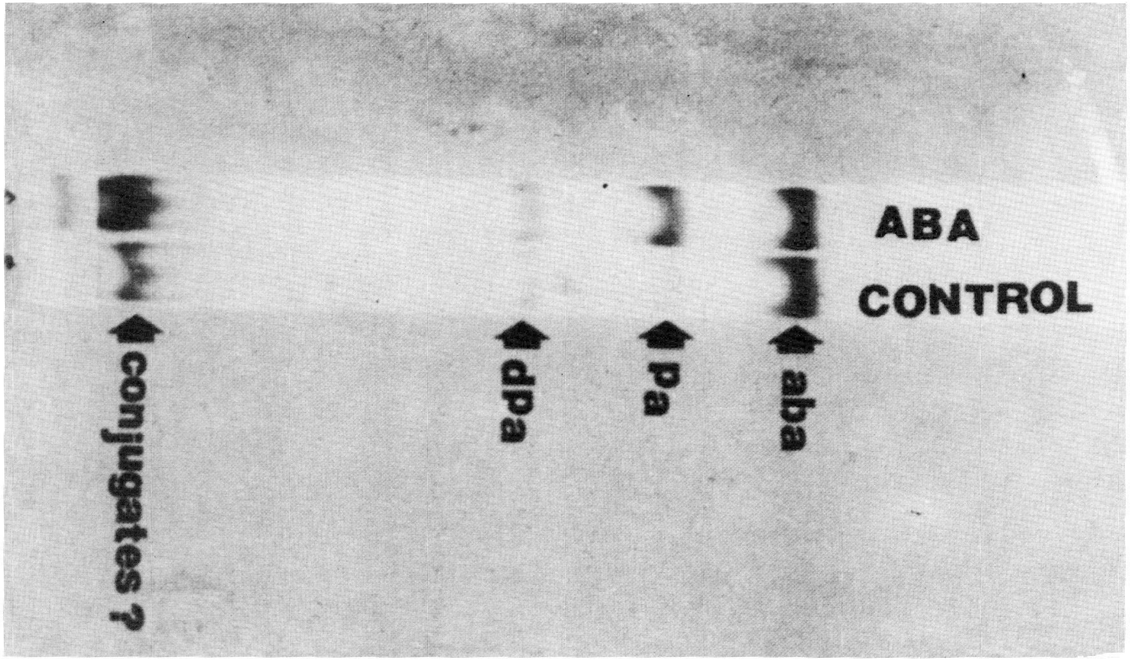

FIGURE 9–11 Analysis of the metabolism of ^{3}H-ABA in barley aleurone layers after pretreatment with ABA. Barley aleurone layers were incubated with 10 μM ABA for 24 hours. The tissue was then labeled in fresh medium containing ^{3}H-ABA for 4 hours. ABA and its metabolites were extracted with 90% ethanol. The extracts were combined and analyzed on a high-resolution thin-layer chromatography (TLC) system (Uknes and Ho 1984). The TLC plate was sprayed with Enhance (New England Nuclear, Boston, MA) and exposed to Xar-5 x-ray film. The fluorogram is shown. From Ho et al. (1985). *Curr. Topics Plant Biochem. Physiol.* 4, 118–125. With permission.

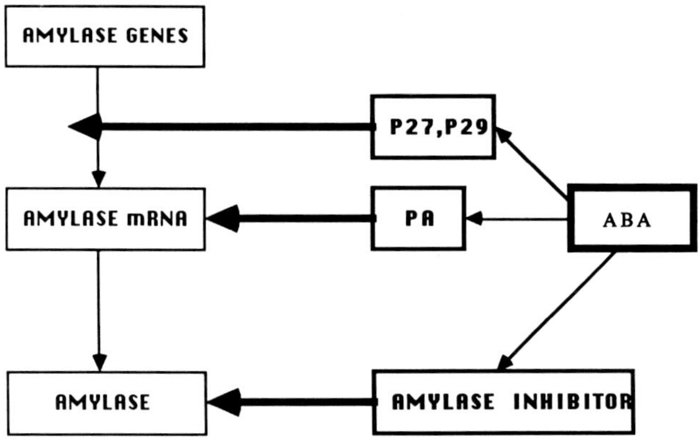

FIGURE 9–12 Diagram illustrating the proposed mode of action of ABA in barley aleurone layers. See text for details.

Preexisting α-amylase and its mRNA, which were formed before ABA was added (or synthesized; see Section 9.3.5), would continue to function. Therefore, ABA has to exert its regulation at other levels. The effect of ABA on the stability of α-amylase mRNA is probably mediated by its metabolite, PA, whose formation is enhanced by ABA. It is conceivable that ABA induces the mono-oxygenase responsible for the conversion of ABA to PA. However, evidence supporting this notion is still lacking. To remove the activity of preexisting α-amylase, ABA induces the 21-kd inhibitor, which is specific for barley α-amylase. Although cereal grains contain many amylase and protease inhibitors, almost all of them act against animal enzymes as a means of self-defense. The ABA-induced 21-kd inhibitor is the only one that acts against the endogenous α-amylase, indicating a role for it in regulating the physiology of developing or germinating seeds or both. By working at all three levels—inhibition of transcription of α-amylase genes, destabilization of α-amylase mRNAs, and inhibition of α-amylase activity—ABA can effectively perform its physiological role to stop the action of GA.

9.3.5 Stress-induced Synthesis of ABA

ABA appears to play an essential role in regulating seed development and preventing premature seed germination. The level of ABA has to decrease after seed maturation to allow successful seed germination. It is crucial to investigate whether the levels of ABA ever increse again in germinating seeds.

Since it has been reported that water stress can enhance the synthesis of ABA in leaf and root tissues, we have investigated whether aleurone layers under stress also have elevated levels of ABA. Water or salt stress induces the same group of proteins in barley aleurone layers as ABA does (L.S. Lin and T.H.D. Ho, unpublished observation). The stress induction of these proteins can be effectively blocked by the ABA biosynthesis inhibitor, fluridone (L.S. Lin and T.H.D. Ho, unpublished observation). This observation suggests that the stress induction of new proteins is most likely to occur via the elevated synthesis of ABA in the stressed tissue. Furthermore, we have found that water stress (treatment with 0.6 M sorbitol) indeed causes a fivefold increase in the level of ABA (from 2.6 to 12 nM), and this increase in ABA level can be effectively blocked by fluridone. Therefore, it is conceivable that after the onset of seed germination, the levels of ABA can increase when a germinating seed encounters stressful conditions. The stress-induced ABA will then in turn slow down the postgermination growth of the seedling by downregulating the production and activity of α-amylase.

9.4 SUMMARY AND PERSPECTIVE

Plants, like all other organisms, are constantly monitoring alterations in their external and internal environments and making necessary adjustments in response to them. Hormones are probably the most important signals integrating the functions of various parts of plants. The physiological role of hormonal regulation in the aleurone layers in cereal grains is well established. Many of the hormone-regulated enzymes (and proteins) have been purified and characterized, and they have been used as biochemical markers in the investigation of hormonal action. In recent years, molecular biological techniques have been successfully applied to the isolation and characterization of hormone-responsive genes. Through deletion and transformation studies, it is hoped that the putative GA_3- and ABA-responsive promoter regions will be defined in the future. Besides the cis-acting sequences, trans-acting elements, such as regulatory proteins bound to the promoter sequences, can also be studied by in vitro DNA bindings. With this type of approach, the goal is eventually to investigate the early events that are induced by hormone treatments. The primary action of hormones in aleurone layers is probably the interaction between the hormone molecules and some kind of receptor molecules. Unfortunately, the search for hormone receptors in this system has not been successful in the past. Recently, new techniques, such as hormone affinity chromatography (Lobler and Klamdt 1986), photoaffinity labeling (Hornberg and Weiler 1984), and immunochemistry with monoclonal antibodies (Jacobs and Gilbert 1983), have been applied in other plant systems to study the hormone receptors. The same type of techniques should be applied to study GA_3 and ABA receptors in aleurone layers.

It is generally accepted that hormone treatments trigger a sequence of events leading to the production of hydrolytic enzymes in aleurone layers. From the biochemical and molecular data, it is possible to postulate this sequence of events. However, the causality among the individual steps cannot be firmly established unless genetic analysis is employed. Mutants with altered sensitivity to GA and ABA have been isolated and characterized. For example, a viviparous maize fails to respond to ABA, and several GA insensitive mutants have been reported (Robichaud et al. 1980; Ho et al. 1980). A more-interesting mutant is the slender barley that always functions as if the hormone concentrations are oversaturating. The aleurone layers of the slender mutant constitutively synthesize α-amylase (even in the absence of GA) (Lanahan and Ho 1988). The study of these mutants should be used to supplement molecular and biochemical studies to further our knowledge of the action of hormones.

The external signals a plant receives from its environment are usually in the form of stress conditions. The stressed cells reprioritize their biosynthetic activities so that nonessential processes are temporarily curtailed and the synthesis of metabolites or macromolecules that are crucial in coping with the stress is enhanced. In heat-stressed aleurone layers, ER is destroyed, leading to cessation of synthesis and secretion of hydrolytic enzymes. On the other hand, the synthesis of heat-shock proteins is enhanced in this tissue, a phenomenon that is probably related to the establishment of thermotolerance. The synthesis of ABA is enhanced in water-stressed aleurone layers, and the elevated ABA levels counteract the action of GA to slow down the postgermination growth. Once the stress is over, ABA is no longer synthesized, and the synthesis of hydrolytic enzymes is resumed. In this case, the action of a hormone, ABA, is integrated with a plant's response to its external environment. Information concerning the biosynthesis and action of ABA would also contribute to our knowledge on how plants interact with their environment.

REFERENCES

Belanger, F.C., Brodl, M.R., Ho, T.H.D. (1986) *Proc. Natl. Acad. Sci. USA* 83, 1354–1358.

Benjavongkulchai, E., and Spencer, M.S. (1986) *Planta* 169, 219–277.

Brodl, M.R. (1987) Ph.D. thesis, Washington University, St. Louis, MO.

Brown, A.H.D., and Jacobsen, J.V. (1982) *Genet. Res.* 40, 315–324.

Brown, P.H., and Ho, T-H.D. (1986) *Plant Physiol.* 82, 801–806.

Callis, J., and Ho, T.H.D. (1983) *Arch. Biochem. Biophys.* 224, 224–234.

Chandler, P.M., Zwar, J.A., Jacobsen, J.V., Higgins, T.J.V., and Inglis, A.S. (1984) *Plant Mol. Biol.* 3, 407–418.

Chrispeels, M.J., and Varner, J.E. (1966) *Nature* 212, 1066–1067.

Chrispeels, M.J., and Varner, J.E. (1967) *Plant Physiol.* 42, 398–406.

Dashek, W.V., and Chrispeels, M.J. (1977) *Planta* 134, 251–256.

Dashek, W.V., Singh, B.N., and Walton, D.C. (1979) *Plant Physiol.* 42, 43–48.

Evins, W.H., and Varner, J.H. (1971) *Proc. Natl. Acad. Sci. USA* 68, 1631–1633.

Filner, P., and Varner, J.E. (1967) *Proc. Natl. Acad. Sci. USA* 58, 1520–1526.

Fincher, G.B., Lock, P.A., Morgan, M.M., et al. (1986) *Proc. Natl. Acad. Sci. USA* 83, 2081–2085.

Hammerton, R.W., and Ho, T-H.D. (1986) *Plant Physiol.* 80, 692–697.

Hanson, A.D., and Jacobsen, J.V. (1984) *Plant Physiol.* 75, 566–572.

Ho, D.T-H., and Varner, J.E. (1974) *Proc. Natl. Acad. Sci. USA* 71, 4783–4786.

Ho, D.T-H., and Varner, J.E. (1976) *Plant Physiol.* 57, 175–178.

Ho, T.H.D., Nolan, R.C., Lin, L.-S., Brodl, M.R., and Brown, P.H. (1987) in *Molecular Biology of Plant Growtn Control* (Jacobs, M., and Fox, E., eds.), pp. 35–49, Alan R. Liss, New York.

Ho, T.H.D., Nolan, R.C., and Uknes, S.J. (1985) *Curr. Topics Plant Biochem. Physiol.* 4, 118–125.

Ho, T.H.D., Shih, S-C., and Kleinhofs, A. (1980) *Plant Physiol.* 66, 153–157.

Hornberg, C., and Weiler, E.W. (1984) *Nature* 310, 321–324.

Huang, J-K., Swegle, M., Dandekar, A.M., and Muthukrishnan, S. (1984) *J. Mol. Appl. Genet.* 2, 579–588.

Jacobs, M., and Gilbert, S.F. (1983) *Science* 220, 1297–1300.

Jacobsen, J.V., and Beach L.R. (1985) *Nature* 316, 275–277.

Jacobsen, J.V., and Higgins, T.J.V. (1982) *Plant Physiol.* 70, 1647–1653.

Jacobsen, J.V., and Varner, J.E. (1967) *Plant Physiol.* 42, 1596–1600.

Jacobsen, J.V., Zwar, J.A., and Chandler, P.M. (1985) *Planta* 163, 430–438.

Jones, R.L. (1969) *Planta* 87, 119–133.

Jones, R.L., and Jacobsen, J.V. (1982) *Planta* 156, 421–432.

Knox, C.A., Sonthayanon, B., Chandra, R.C., and Muthukrishnan, S. (1987) *Plant Mol. Biol.* 9, 3–17.

Koehler, S., and Ho, T.H.D. (1988) *Plant Physiol.* 87, 95–103.

Lanahan, M.B., and Ho, T.H.D. (1988) *Planta* 175, 107–114.

Lin, L-S., and Ho, T-H.D. (1986) *Plant Physiol.* 82, 289–297.

Lobler, M., and Klamdt, D. (1986) *J. Biol. Chem.* 260, 9848–9853.

Mozer, T.J. (1980) *Cell* 20, 479–485.

Mundy, J. (1984) *Carlsberg Res. Commun.* 49, 439–444.

Muthukrishnan, S., Chandra, G.R., and Albaugh, G.P. (1983a) *Plant Mol. Biol.* 2, 249–258.

Muthukrishnan, S., Gill, B.S., Swegle, M., and Chandra, G.R. (1983b) *J. Biol. Chem.* 259, 13637–13639.

Nolan, R.C., and Ho, T.H.D. (1988) *Planta* 174, 551–560.

Nolan, R.C., Lin, L-S., and Ho, T.H.D. (1987) *Plant Mol. Biol.* 8, 13–22.

Patridge, J., Shannon, L., and Grumpf (1976) *Biochim. Biophys. Acta* 451, 470–483.

Robichaud, C.S., Wong, J., and Sussex, I.M. (1980) *Dev. Genet.* 1, 325–330.

Rogers, J.C. (1985) *J. Biol. Chem.* 260, 3731–3738.

Rogers, J.C., Dean, D., and Heck, G.R. (1985) *Proc. Natl. Acad. Sci. USA* 82, 6512–6516.

Storti, R.V., Scott, M.P., Rich, A., and Pardue, M.L. (1980) *Cell* 22, 825–834.

Stuart, I.M., Loi, M.L., and Fincher, G.B. (1986) *Plant Physiol.* 80, 310–314.

Uknes, S.J., and Ho, T.H.D. (1984) *Plant Physiol.* 75, 1126–1132.

Walton, D.C. (1980) *Annu. Rev. Plant Physiol.* 31, 453–489.

Whittier, R.F., Dean, D.A., and Rogers, J.C. (1987) *Nucleic Acids Res.* 15, 2515–2535.

Woodward, J.R., and Fincher, G.B. (1982) *Eur. J. Biochem.* 121, 663–669.

Zwar, J.Z., and Hooley, R. (1986) *Plant Physiol.* 80, 459–463.

CHAPTER 10

Auxin-Regulated Gene Expression in Plants

Athanasios Theologis

There are five main types of plant hormones: auxins, cytokinins, gibberellins, abscisic acid, and ethylene. The fundamental questions in plant hormonology are the following: What is the primary mechanism of action of the plant hormones and how do they mediate their various responses? Auxins were the first plant hormones to be discovered and are the most widely studied. The classic effect of auxins, typified by indole acetic acid (IAA), is the regulation of plant cell growth, a process that requires extension of the cell wall (Cleland 1971; Ray 1967). Two main theories have been proposed to explain auxin-induced cell elongation. First, the "gene activation hypothesis" suggested in the early 1960s that auxins initiate the growth process by regulating the expression of specific mRNAs coding for polypeptides necessary for the growth process (Key 1969). Second, the "acid growth theory" proposed that cell enlargement is regulated by auxin-induced proton secretion (Rayle and Cleland 1977). Both theories are unable to explain the physiological and biochemical observations associated with the growth process of plant cells.

The support of the National Institutes of Health (GM 35447) and the National Science Foundation (DCB 84-21157) is gratefully acknowledged.

With the development of in vitro translation systems for mRNAs and two-dimensional polyacrylamide gel electrophoresis (PAGE) of proteins, it became possible to reexamine the gene activation hypothesis and to search for specific mRNA changes brought about by auxin treatment (Theologis and Ray 1982b; Zurfluh and Guilfoyle 1982). Furthermore the development of gene cloning techniques allowed the isolation of DNAs complementary to some auxin-regulated mRNAs from pea and soybean tissues (Theologis 1986).

Herein, the experimental work that led to the isolation of DNAs complementary to mRNAs rapidly induced by IAA in the pea epicotyl tissue will be presented and discussed. The results strongly indicate that IAA acts at the transcriptional level. However, the mechanism of this effect, as well as the role of the proteins coded by the induced mRNAs, is unknown at the present time.

10.1 ENHANCEMENT OF SPECIFIC TRANSLATIONAL PRODUCTS BY IAA IN PEA TISSUE

Autoradiographs of two-dimensional separations of ^{35}S-methionine-labeled polypeptides obtained by in vitro translation of pea stem RNApoly(A) in the wheat germ system are shown in Figure 10–1. Autoradiographs of translation products from third internode segments incubated with IAA for six hours show that IAA causes a substantial increase (Figure 10–1B), compared with the untreated control (Figure 10–1A), in translational activity of mRNAs for polypeptides 1 through 6 (see Figure 10–1). Time course experiments (Theologis and Ray 1982a) have shown that translation products 1 and 2 increase after two hours of exposure to IAA ("late IAA–regulated mRNAs"). On the other hand, translation product 6 (doublet) increases after 30 minutes, and products 3, 4, and 5 increase within 20 minutes of exposure to IAA ("early IAA–regulated mRNAs").

These results clearly indicate that auxin potentiates specific mRNAs rapidly, almost concomitant with the initiation of cell elongation. This potentiation can be brought about by an activation of transcription or by a posttranscriptional event. The possibility that the hormone alters the translatability of preexisting mRNA also exists (Theologis and Ray 1982b). To find out what is actually happening, it is necessary to isolate DNA sequences complementary to some of these mRNAs.

10.2 ISOLATION OF DNA SEQUENCES COMPLEMENTARY TO SOME IAA-REGULATED mRNAS IN PEA

In order to facilitate the isolation of DNA clones complementary to some IAA-inducible mRNAs, the RNA was enriched for these sequences by fractionating total RNApoly(A). Equal amounts of RNApoly(A) from con-

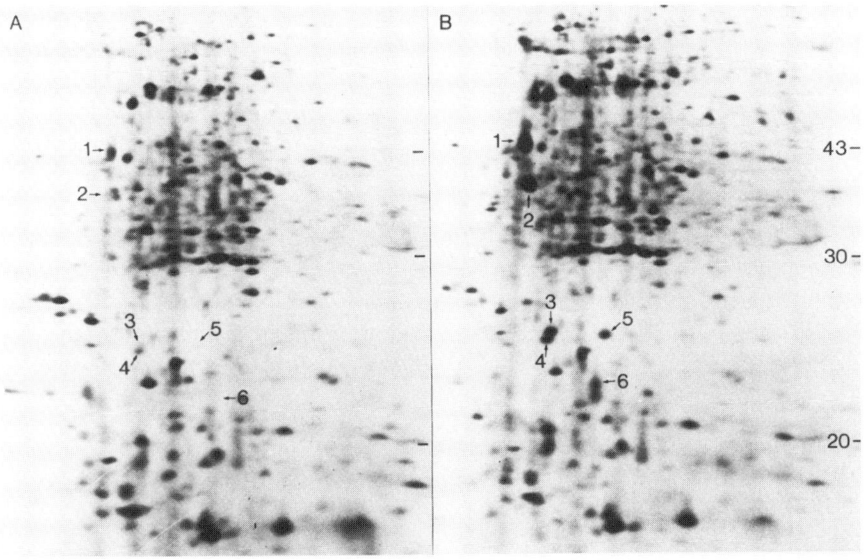

FIGURE 10–1 Enhancement of specific translational products by IAA in etiolated pea tissue segments. Portions of autoradiographs of electrophoretically separated ^{35}S-methionine-labeled in vitro translation products specified by RNApoly(A) from pea stem segments. First (horizontal) direction, nonequilibrium pH gradient; second (vertical) direction, SDS-PAGE. Translation products' molecular weight ranges from 14,000 to 50,000 and their pH ranges from 4.0 (left) to 8.0 (right). RNApoly(A) was from segments kept without auxin for two hours after cutting and subsequently incubated for an additional six hours without (A) or with (B) 20 μM IAA. Numbers refer to specific proteins; see text for discussion. (Reproduced with permission of the *Annual Reviews of Plant Physiology.)*

trol pea tissue and pea tissue treated for six hours with IAA were fractionated on two almost identical 5–20% (w/w) sucrose gradients (Theologis et al. 1985).

Figure 10–2 compares the Ab_{260} absorption profile (solid line) of the control gradient (Figure 10–2A) with that of the IAA gradient (Figure 10–2B). The profiles were almost identical, except in the 10–13S region, where an upward shift of the Ab_{260} was noticeable in the IAA gradient. IAA-regulated mRNAs were localized in that region by in vitro translation of the RNAs from each fraction and separation of the translational products by two-dimensional PAGE. Comparison of the translational products of fraction 11 from both the control and the IAA gradients by two-dimensional PAGE revealed that the early sequences 6 and 4 and a small amount of 5

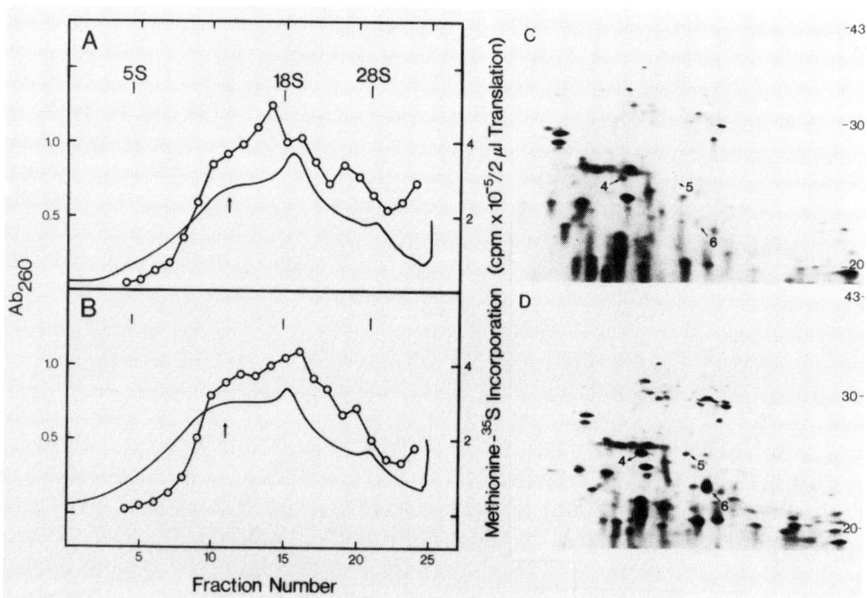

FIGURE 10–2 Subfractionation of mRNA. Sucrose gradient centrifugation of RNA-poly(A) from six hour control (A) and IAA-treated (B) pea tissue. IAA concentration was 20 μM. RNApoly(A) (2 mg) was applied to an exponential sucrose density gradient (5–20% w/w) and centrifuged as previously described (Theologis et al. 1985). The Ab_{260} profile across the gradient is shown by the solid line. The mRNAs of each fraction were translated in a wheat germ extract and their translational activity is shown by open circles. Translational products derived from fraction 11 (arrows) of the control and IAA-treated gradients were analyzed by two-dimensional PAGE. Portions of autoradiographs of the electrophoretically separated ^{35}S-methionine-labeled products are shown in C (control) and D (IAA-treated). Numerals refer to specific proteins; see text for discussion. Electrophoretic conditions for the first and second dimension were as described in the legend to Figure 10–1.

were localized in this fraction [compare Figure 10–2C (control) with Figure 10–2D (IAA)]. Fractions 13 and 14 contained primarily the late sequences 1 and 2 and the early sequences 3 and 5.

Fractions 11 and 14 of the IAA gradient were the mRNA source for synthesis of double-stranded cDNA, which was subsequently cloned into λgt10. Differential plaque filter hybridization, as shown in Figure 10–3, allowed the isolation of recombinant phages to three early IAA-regulated mRNAs.

Two cDNAs clones, pIAA4/5 and pIAA6, were isolated. The RNA-poly(A) that hybridized to clone pIAA4/5 was enhanced after two hours of incubation with 20 μM IAA [Figure 10–4A, compare lane 1 (IAA) with lane 2 (control)]. Similarly, clone pIAA6 hybridized to an IAA-inducible mRNA after two hours of hormone treatment [Figure 10–4B, compare lane 1 (IAA) with lane 2 (control)].

The question immediately arises as to whether these hormonally inducible mRNAs are the ones that have previously been identified by in vitro translation and two-dimensional PAGE (see Figure 10–1). Hybrid-selected translations and subsequent two-dimensional PAGE revealed that clone

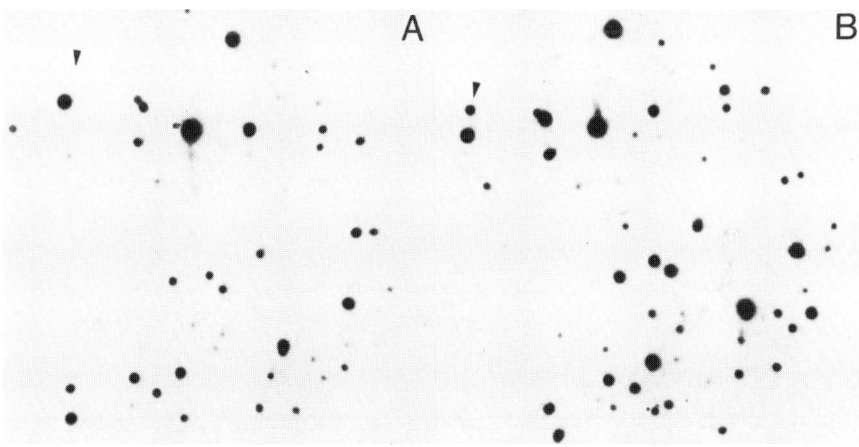

FIGURE 10–3 Differential plaque filter hybridization. Duplicate plaque filter replicas were prepared from a plate containing plaques of λ-pea cDNA hybrids clones in the vector λgt10. The λ-pea cDNA hybrids were constructed as described by Theologis et al. (1985) using RNApoly(A) of fraction 11 from the IAA-treated gradient. The filters were hybridized to ^{32}P-labeled cDNA probes prepared from the RNApoly(A) of fraction 11 of the control (A) or the IAA-treated gradient (B). A comparison of the two autoradiographs shows one clear example (arrowhead) of positive differential hybridization to the IAA-treated probe.

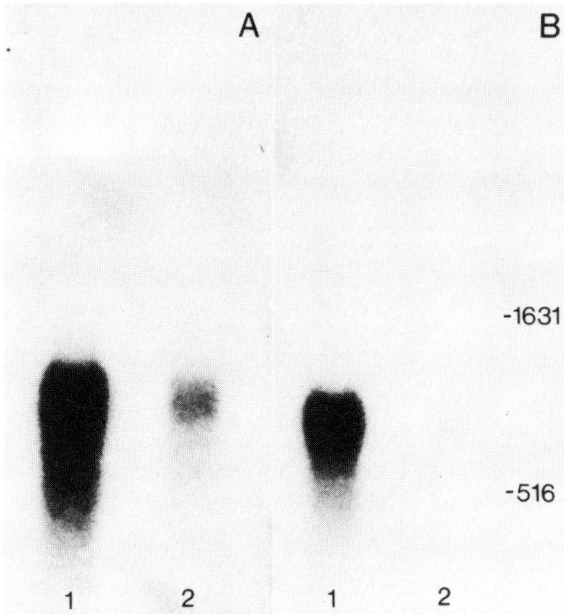

FIGURE 10–4 Size of the IAA-inducible RNAs. Autoradiographs of RNA filter paper hybridized successively with ^{32}P-labeled pIAA4/5 (A) and pIAA6 (B) plasmid DNAs. RNApoly(A) was isolated from pea segments and incubated for six hours with (lane 1) or without (lane 2) 20 μM IAA.

pIAA6 selected an mRNA coding for the previously identified polypeptide 6 (compare Figure 10–5B with Figure 10–5A). On the other hand, clone pIAA4/5 selected two mRNAs coding for the previously identified polypeptides 4 and 5 (compare Figure 10–5C with Figure 10–5A). Surprisingly, clone pIAA4/5 selected two RNAs, one for polypeptide 4 and the other for polypeptide 5. This suggests that there is sequence homology between the two mRNAs. This mRNA heterogeneity will eventually be resolved by structural characterization of the gene.

10.3 CHARACTERIZATION OF THE HORMONAL RESPONSE

The isolation of cDNA clones allowed a detailed characterization of the hormonal response in pea. In studying the initial steps in hormone action, three criteria have to be fulfilled for the response under investigation to be considered the result of a primary action of the hormone (Ringold 1979).

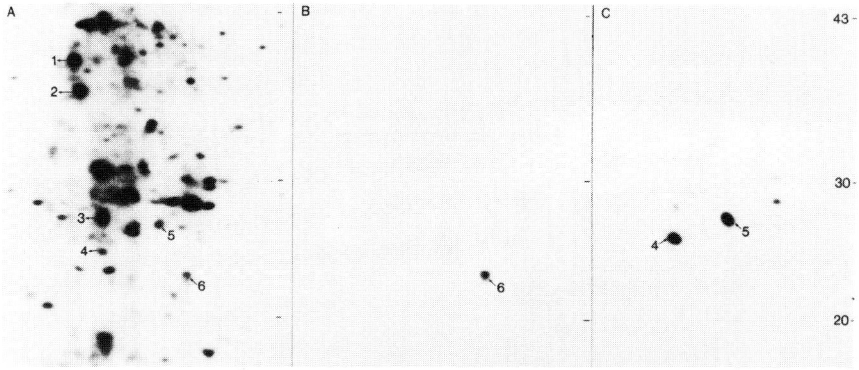

FIGURE 10–5 Hybrid-selected translation. Portions of autoradiographs of electrophoretically separated ^{35}S-methionine-labeled in vitro translation products specified by RNApoly(A) purified by hybridization selection to recombinant plasmid DNAs. Polypeptides were fractionated by two-dimensional PAGE and detected by fluorography (Theologis et al. 1985). First and second dimensions were as described in the legend to Figure 10–1. (A) In vitro translation products derived by total RNApoly(A) from pea segments treated for six hours with 20 μM IAA. (B) In vitro translation products derived by RNApoly(A) selected from total RNApoly(A) of six hour IAA-treated segments by the plasmid DNA pIAA6. (C) In vitro translation products derived by RNApoly(A) selected from total RNApoly(A) of six hour IAA-treated segments by the plasmid DNA pIAA4/5. IAA concentration was 20 μM. Numbers refer to specific proteins; see text for discussion. (Reproduced with permission of the *Journal of Molecular Biology.*)

1. The induction of the mRNA should occur rapidly. Figure 10–6 shows that an increase in the pIAA4/5 mRNA was detected after a lag period of ten minutes and continued to accumulate over a period of two hours. On the other hand, the pIAA6 mRNA started to accumulate after 20 minutes of hormone administration. A nondifferential clone, pWI1, gave the same level of hybridization in all lanes with or without auxin (see Figure 10–6). This response was considerably faster than the induction of α-amylase mRNA by gibberellic acid in aleurone cells (Higgins et al. 1976). The response was comparable to the most rapid mRNA inductions by mammalian and insect hormones (Asburner et al. 1973; Ringold et al. 1984). The increase in translational activity of pIAA4/5 and pIAA6 mRNAs clearly was due to an increase in the amount of the corresponding RNA. This enhancement could not be attributed to polyadenylation of preexisting RNAs because hybridization analysis with total nucleic acids revealed a great enhancement in the amount of the IAA4/5 and IAA6 mRNAs after two hours

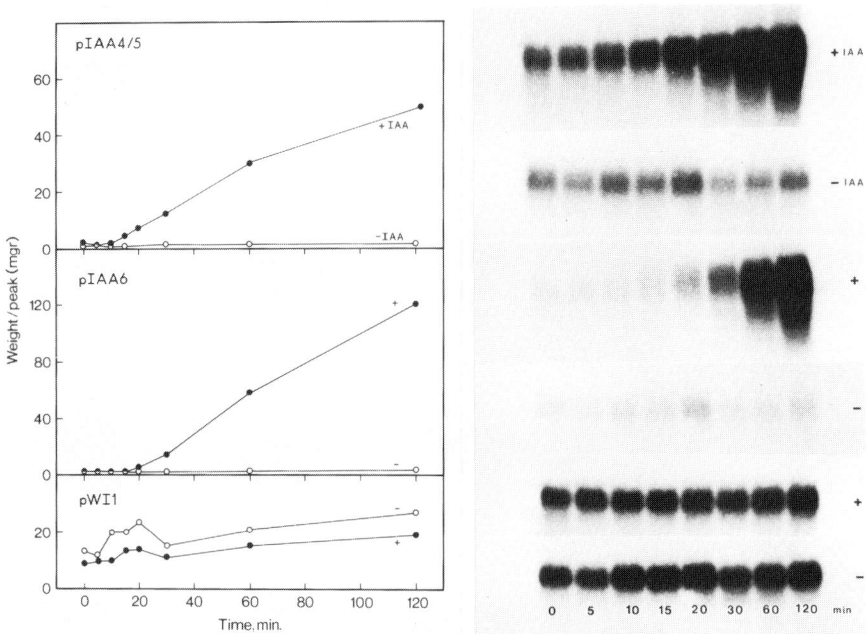

FIGURE 10–6 Induction kinetics of the IAA-regulated RNAs. Endogenous IAA-depleted pea segments were treated with (+) or without (−) 20 μM IAA. RNA-poly(A) was isolated at various time intervals from 2 g fresh weight (100 segments) of control or IAA-treated tissue. Twenty micrograms of RNApoly(A) from 0, 5, 10, 20, 30, 60, and 120 minute incubations were electrophoretically separated and transferred to aminophenylthioether paper. The papers were successively hybridized to ^{32}P-labeled pIAA4/5 (top pair, right), pIAA6 (middle pair, right), and pWI1 (bottom pair, right) plasmid DNAs. pWI1 is a nondifferential clone. (Reproduced by permission of the *Journal of Molecular Biology*.)

of treatment with IAA compared with the control. (Compare lane 3 with lane 4 in Figure 10–7A and 10–7B). However this mRNA accumulation may have been due to activation of transcription, activation of posttranscriptional processing, or selective stabilization of mRNA by the hormone. α-Amanitin, an inhibitor of RNA polymerase II, completely abolished the mRNA induction (Figure 10–8), indicating that the hormonally mediated mRNA accumulation depended on transcription. Recent experimental evidence indicates that the hormone does not affect the stability of the RNA, suggesting that the hormone acts at the transcription initiation level (Koshiba and Theologis 1987).

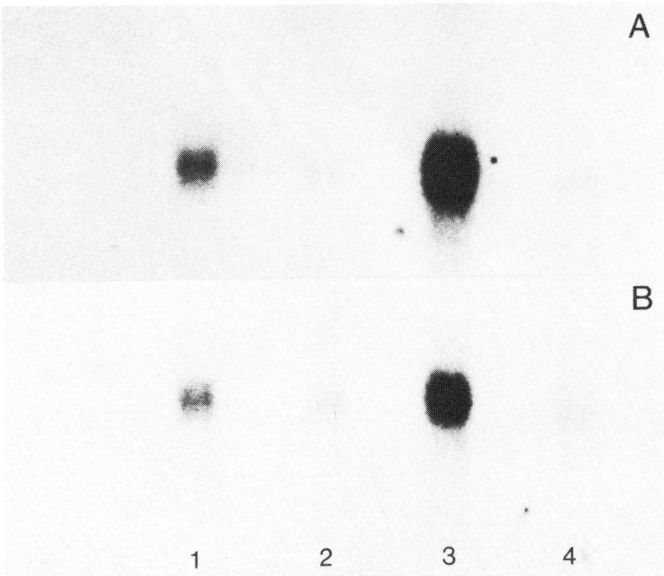

FIGURE 10–7 RNA hybridization of total nucleic acids isolated from pea segments. (A) pIAA4/5 mRNA. (B) pIAA6 mRNA. Lane 1 held total intact third internode tissue. Lane 2 held segments depleted for two hours. Lane 3 held tissue treated with 20 μM IAA for 2 hours. Lane 4 was the two hour control. (Reproduced by permission of the *Journal of Molecular Biology*.)

2. The induction of the hormonally regulated mRNA should be specific. The specificity of the response is shown in Figure 10–9. The induction was specific to auxins [IAA, 2,4D; naphthalene-1-acetic acid (NAA)], whereas nonauxin analogues were ineffective at inducing the genes (Figure 10–9A). The accumulation of the pIAA4/5 and pIAA6 mRNAs was also unaffected by other plant hormones: kinetin, gibberellic acid, abscisic acid, or ethylene (Figure 10–9B). Finally, stress conditions, such as heat and cold and anaerobiosis, were also unable to induce the mRNAs (Figure 10–9C).

3. The induction of mRNAs should require no concurrent protein synthesis. The effect of inhibitors of protein synthesis on the induction of the pIAA4/5 and pIAA6 mRNAs is shown in Figure 10–10. Three different protein synthesis inhibitors (Pestka 1971)—cycloheximide, emetine, and anisomycin—mimicked the IAA-induced mRNA accumulation. Puromycin and the lysine analogue 2-S-aminoethylcysteine were without effect (see Figure 10–10). None of the inhibitors interfered with the IAA induction of the mRNAs. In all cases in which the inhibitors were ineffective at inducing

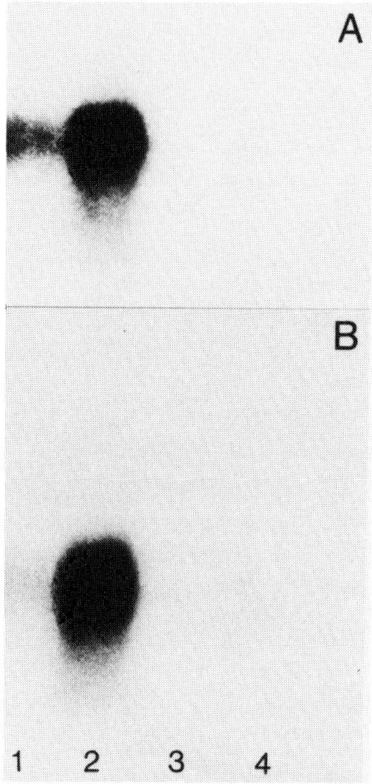

FIGURE 10–8 Effect of α-amanitin on the expression of the IAA-inducible mRNAs in pea tissue segments. (A) pIAA4/5 mRNA. (B) pIAA6 mRNA. Lane 1, control; lane 2, 20 μM IAA; lane 3, 5 μM α-amanitin, and lane 4, IAA + α-amanitin. Time of incubation was two hours.

the RNAs, protein synthesis inhibition was higher than 80% (Theologis et al. 1985). Protein synthesis inhibitors have been previously been shown to enhance the level of expression of various hormonally regulated mammalian genes (Makino et al. 1984; Ringold et al. 1984). Although the mechanism by which cycloheximide induces the IAA genes is unknown, the effect is specific because neither was the level of the nondifferential mRNA pWI1 increased (see Figure 10–10) nor was the translational activity of RNAs from cycloheximide-treated tissue altered significantly (Theologis, unpublished observations).

10.4 DOSE RESPONSE CURVE

It has been axiomatic that the dose responses of plant tissues to hormones are linear functions of the logarithm of the applied hormone concentration (Trewavas 1981). However, the pIAA4/5 and pIAA6 mRNAs in pea tissue

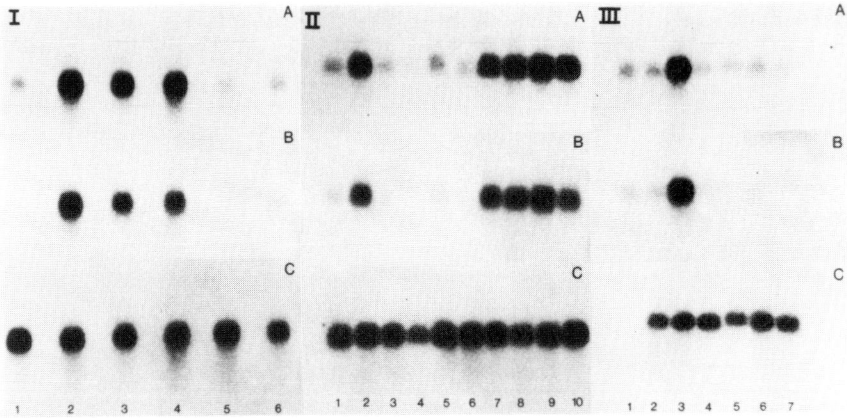

FIGURE 10–9 Specificity of the hormonal response. (A) Effect of IAA analogues on the expression of IAA-inducible mRNAs. Autoradiographs of RNA filter paper hybridized successively with ^{32}P-labeled pIAA4/5 (**a**), pIAA6 (**b**), and pWI1 (**c**) plasmid DNAs. Total nucleic acids were prepared from segments (100 per sample) incubated for two hours without IAA (lane 1), and with 20 μM IAA (lane 2), 20 μM 2,4D (lane 3), 20 μM NAA (lane 4), 20 μM phenylacetic acid (lane 5), and 20 μM p-chlorophenoxyisobutyric acid (lane 6). (B) Effect of various plant hormones on the induction of the IAA-inducible mRNAs. Autoradiographs of RNA filter paper hybridized successively with ^{32}P-labeled pIAA4/5 (**a**), pIAA6 (**b**), and pWI1 (**c**) plasmid DNAs. Total nucleic acids were prepared from segments (100 per sample) treated for two hours without IAA (lane 1) and with 20 μM IAA (lane 2), 20 μM gibberellic acid (lane 3), 20 μM abscisic acid (lane 4), 10 μl/l C_2H_4 (lane 5), 0.4 M mannitol (lane 6), IAA + gibberellic acid (lane 7), IAA + abscisic acid (lane 8), IAA + C_2H_4 (lane 9), and IAA + mannitol (lane 10). (C) Effect of various stress conditions on the inducibility of the IAA-regulated genes. Autoradiographs of RNA filter paper hybridized successively with ^{32}P-labeled pIAA4/5 (**a**), pIAA6 (**b**), and pWI1 (**c**) plasmid DNAs. Total nucleic acids were isolated from intact third internode tissue (lane 1); two-hour depleted segments (lane 2); segments incubated for two hours after depletion with 20 μM IAA (lane 3), without IAA (lane 4), or with N_2 (lane 5); and segments incubated at 40°C (lane 6), or for two hours at 4°C (lane 7). (Reproduced by permission of the *Journal of Molecular Biology*.)

are exceptions and show a dose-dependent increase in their induction up to 10^{-4} M IAA, as illustrated in Figure 10–11. The response curve shown in Figure 10–11 is reminiscent of that obtained for the induction of prolactin mRNA in response to estradiol (Lieberman et al. 1978), except that prolactin mRNA induction requires four times less hormone. This could be due to

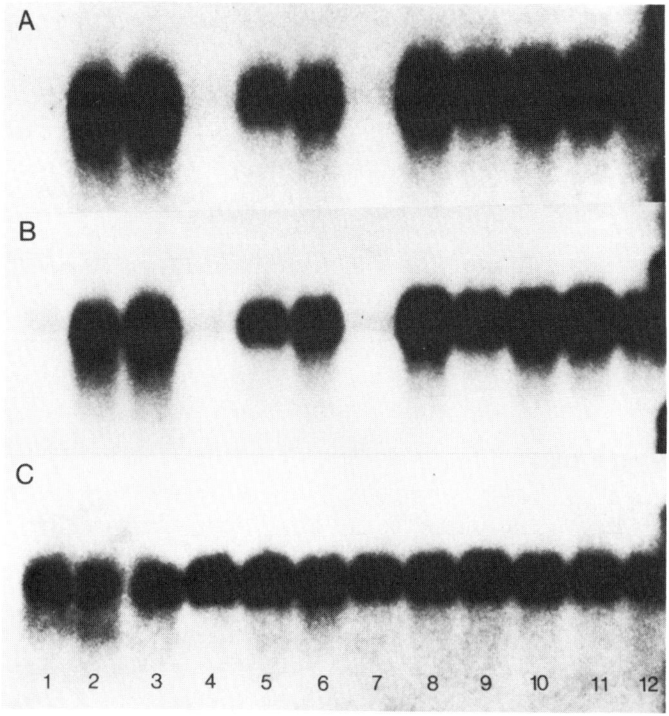

FIGURE 10–10 Effect of protein synthesis inhibitors on the induction of the IAA-inducible mRNAs. Autoradiographs of RNA filter paper hybridized successively with ^{32}P-labeled pIAA4/5 (A), pIAA6 (B), and pWI1 (C) plasmid DNAs. Total nucleic acids were isolated from abraded segments (100 per sample) treated for two hours without IAA (lane 1) or with 20 μM IAA (lane 2); 20 μM cycloheximide (CH) (lane 3); 300 μM puromycin (PUR) (lane 4); 300 μM emetine (EME) (lane 5); 300 μM anisomycin (ANI) (lane 6); 500 μM 2-S-aminoethylcysteine (S-AEC) (lane 7); IAA + CH (lane 8); IAA + PUR (lane 9); IAA + EME (lane 10); IAA + ANI (lane 11); and IAA + S-AEC (lane 12). (Reproduced by permission of the *Journal of Molecular Biology*.)

differences in effective hormone concentration intracellularly or hormone receptor affinities.

10.5 MODEL FOR REGULATION OF THE AUXIN GENES

The induction of the IAA-regulated genes by protein synthesis inhibitors suggests that the genes are under the control of a rapidly turning over

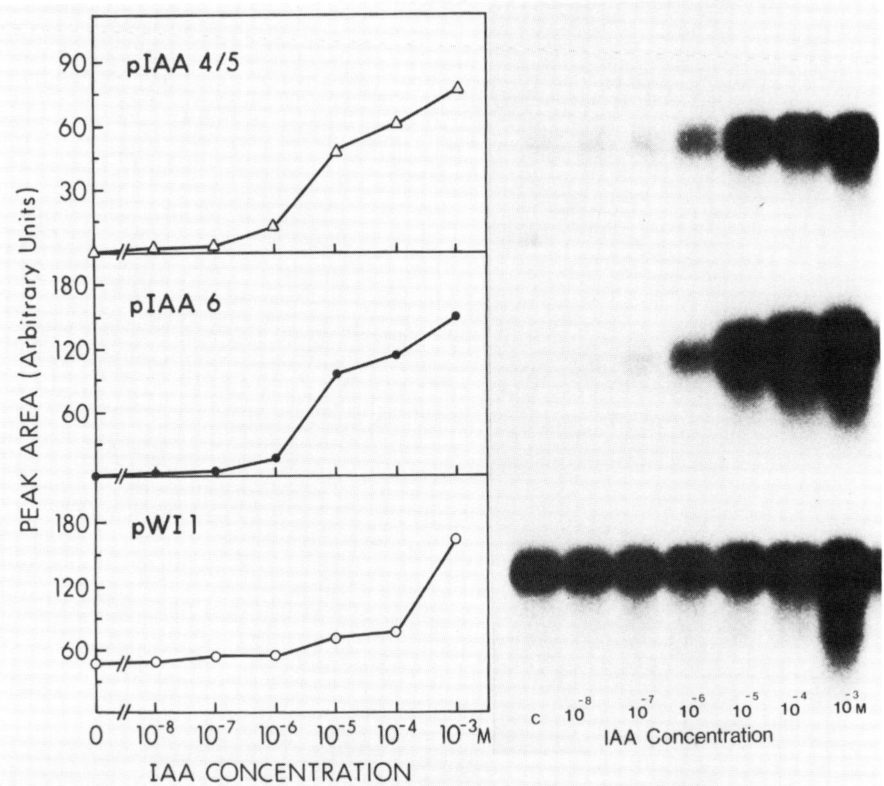

FIGURE 10–11 IAA-dose response. Autoradiographs of RNA filter papers hybridized with ^{32}P-labeled pIAA4/5 (top), pIAA6 (middle), and pWI1 (bottom) plasmid DNAs. pWI1 is a nondifferential clone.

protein. A short-lived protein repressor could specifically interact with the promoter of the auxin genes to limit their transcription. Figure 10–12 shows two possible mechanisms for the transcriptional regulation of the IAA genes. According to the negative control model, a short-lived protein repressor interacts with the promoter region of the auxin-regulated genes and prevents its transcription. Protein synthesis inhibition by cycloheximide results in a rapid decrease in the level of the repressor, releasing its control over gene transcription. Auxin, on the other hand, binds to the repressor and inactivates it, thus altering the functionality of the repressor (whereas cycloheximide alters its concentration). The kinetics of the mRNA accumulation by cycloheximide induction are slower than those by IAA induction (Ko-

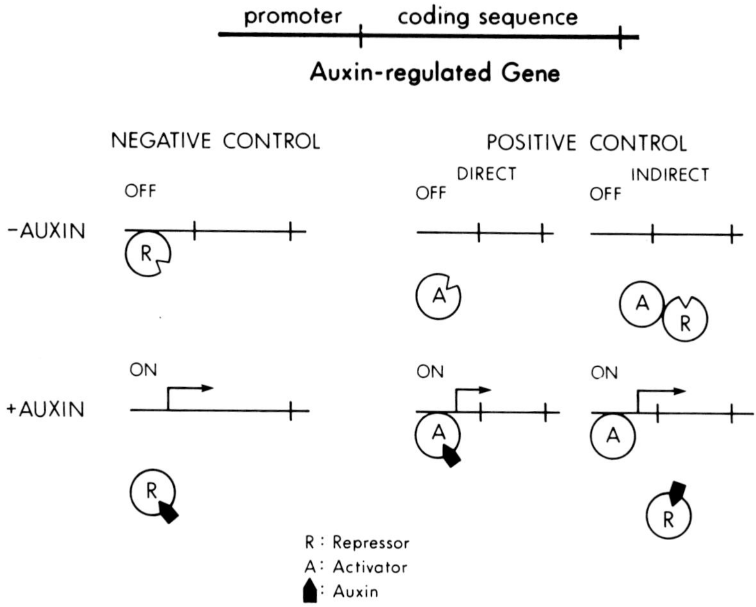

FIGURE 10–12 Model for the regulation of the early auxin-inducible genes in pea tissue. (Reproduced by permission of the *Annual Reviews of Plant Physiology.*)

shiba and Theologis 1987) and support the above view that the mechanisms of mRNA induction by IAA and cycloheximide are different.

The second model presented in Figure 10–12 predicts that the auxin genes are under positive control. Direct positive control requires an activator molecule, a protein with a short half-life that is inactive in the absence of auxin and active in its presence. The data obtained with the protein synthesis inhibitors eliminate the direct positive control mechanism as a possible regulatory mechanism for the auxin genes in pea tissue, because in the presence of cycloheximide the genes are expressed. The indirect positive control mechanism, however, requires an activator molecule that in the absence of hormone interacts with a protein repressor molecule, resulting in the inactivity of the former. Auxin binds to the repressor and releases the activator, which then interacts positively with RNA polymerase II, resulting in transcription of the gene. According to this view cycloheximide lowers the levels of the rapidly turning over repressor molecule. The negative control mechanism presented in Figure 10–12 is a simplification of the regulatory mechanism of the lactose operon of *Escherichia coli* (Miller and Reznikoff 1980). The (indirect) positive control model appears to be a main regulatory mechanism operating in simple eukaryotic organisms such as yeast and neurospora (Metzenberg 1979; Oshima 1982). Recently, an instance of negative control has been demonstrated in yeast (Johnson and Herskowitz 1985).

10.6 CONCLUSIONS AND FUTURE DIRECTIONS

It is clear that auxin induces specific mRNAs in pea tissue and that this induction appears to be close to its primary mechanism of action. The role of the polypeptides coded by the hormonally regulated mRNAs and their involvement in cell growth, if any, remain to be determined. The structural characterization of these rapidly induced genes is of great importance for elucidating the biochemical machinery responsible for the transcriptional activation by auxin.

REFERENCES

Asburner, M., Chibaru, C., Meltzer, P., and Richards, G. (1973) *Cold Spring Harbor Symp. Quant. Biol.* 38, 655.

Cleland, R. (1971) *Annu. Rev. Plant Physiol.* 22, 197–222.

Higgins, T.J.V., Zwar, J.A., and Jacobson, J.V. (1976) *Nature* 260, 166–169.

Johnson, A.D., and Herskowitz, I. (1985) *Cell* 42, 237–247.

Key, J.L. (1969) *Plant Physiol.* 20, 449–474.

Koshiba, T., and Theologis, A (1987) *Plant Physiol.* 83 (suppl.), 153.

Lieberman, M.E., Maurer, R.A., and Gorski, J. (1978) *Proc. Natl. Acad. Sci. USA* 75, 5946–5949.

Makino, R., Hayashi, E., and Sugimura, T. (1984) *Nature* 310, 697–698.

Metzenberg, R.L. (1979) *Microbiol. Rev.* 43, 361–383.

Miller, J.H., and Reznikoff, W.S. (1980) *The Operon,* Cold Spring Harbor Laboratory, Cold Spring Harbor, NY.

Oshima, Y. (1982) in *The Molecular Biology of the Yeast Saccharomyces: Metabolism and Gene Expression* (Strathern, J.N., Jones, E.W., and Broach, J.R., eds.), pp. 159–180, Cold Spring Harbor Laboratory, Cold Spring Harbor, NY.

Pestka, S. (1971) *Annu. Rev. Microbiol.* 25, 487–562.

Ray, P.M. (1967) *J. Cell Biol.* 35, 659–674.

Rayle, D.L., and Cleland, R.E. (1977) *Curr. Top. Dev. Biol.* 11, 187–214.

Ringold, G.M. (1979) *Biochim. Biophys. Acta* 560, 4487–4508.

Ringold, G.M., Dieckmann, B., Vannice, J.L., Trahey, M., and McCormick, F. (1984) *Proc. Natl. Acad. Sci. USA* 81, 3964–3968.

Theologis, A. (1986) *Annu. Rev. Plant Physiol.* 37, 407–438.

Theologis, A., Huynh, T.V., and Davis, R.W. (1985) *J. Mol. Biol.* 183, 53–68.

Theologis, A., and Ray, P.M. (1982a) in *Plant Growth Substances* (Wareing, P.F., ed.), pp. 43–57, Academic Press, Orlando, FL.

Theologis, A., and Ray, P.M. (1982b) *Proc. Natl. Acad. Sci. USA* 79, 418–421.

Trewavas, A.J. (1981) *Plant Cell Environ.* 4, 203–228.

Zurfluh, L.L., and Guilfoyle, T.J. (1982) *Plant Physiol.* 69, 338–340.

CHAPTER
11

Cytokinin-modulated Macromolecular Synthesis and Gene Expression

Chong-maw Chen

Plant growth and development are regulated by various hormones or combinations of hormones. These events are also influenced by environmental factors such as light quality, photoperiod, gravity, temperature, infection, wounding, and anaerobic conditions. While the molecular basis of these responses has not been completely worked out, there are several clear examples of specific gene expression regulated by hormones or the other factors described above or both (for recent reviews see Kuhlemeier et al. 1987; Goldberg 1986; Matters and Scandalios 1986; Tobin and Silverthorne 1985; Sachs and Ho 1986). Recent studies on the mode of cytokinin action suggests that gene expression is a likely site of action of cytokinins (Lerbs et al. 1984; Teyssendier de la Serve et al. 1985; Flores and Tobin 1986; Chen et al. 1987a and 1987b).

Cytokinins, like other plant hormones, elicit a wide range of responses including cell division and differentiation, but the molecular mechanism of cytokinin-modulated gene expression is largely unknown; research into the molecular mechanisms of action of other plant hormones, especially gibberellins and auxin, has been more fruitful. To gain insight into the events in cytokinin-modulated gene expression, a few laboratories have begun to analyze the effects of cytokinin on the levels of specific mRNAs using cDNA probes, and to determine at what level of gene expression cytokinin exerts its effects. Such information would be useful in that it would be a starting point in the dissection of events in plant development regulated by cytokinins and possibly other plant hormones.

11.1 ACTIVE FORMS OF CYTOKININS

A signal question often asked regarding the mode of cytokinin action is, what are the "active forms" of cytokinins?

Structural studies of cytokinins revealed that the expression of cytokinin activity is influenced by structure variations in the cytokinin (Skoog et al. 1967; Hecht 1980). Skoog et al. (1967) tested 69 compounds, mostly purine derivatives and closely related substances, for promotion of growth in the tobacco bioassay to determine relationships between chemical structure and cytokinin activity. In all instances, cytokinin bases were more active than their corresponding nucleosides. However, it was not totally clear whether or not cytokinin metabolites were more active than the added compounds per se, because endogenous as well as exogenous cytokinins are rapidly metabolized to different forms of cytokinins (for recent reviews see Morris 1986; Horgan 1984; Letham and Palni 1983; Chen 1981). Cytokinin structure-activity relationships seem to be related to the metabolism of these hormones by tissues. Mok et al. (1982) demonstrated that differences in cytokinin structure-activity relationships correlate with differences in the metabolism of various cytokinins in different strains of *Phaseolus* callus tissues. Therefore, one cannot easily determine the active form or forms of cytokinins in a specific tissue until one can clearly identify cytokinin "receptor molecules" that directly transduce the physiological activity of cytokinins.

Cytokinin bases and their corresponding nucleosides and nucleotides are interconverted in plant tissues, and these conversions are regulated by various enzymes (Chen and Eckert 1977; Chen and Petschow 1978; Chen and Melitz 1979; Chen and Christopeit 1981a and 1981b; Chen et al. 1982; Chen and Leisner 1984; Laloue and Fox 1985; Chatfield and Armstrong 1987; Morris et al. 1985). Possible cytokinin interconversion pathways regulated by these enzymes are illustrated in Figure 11–1. Cytokinin metabolism is much more complicated than these oversimplified pathways suggest. Side-chain modification such as hydroxylation and hydrogenation, side-chain cleavage (oxidase action), glucosidation, xylosylation (Mok and Mok 1987), and amino acid conjugation are examples of extensive modifications of cytokinin molecules that are mediated by enzymes in plant cells (for reviews see Letham and Palni 1983; Horgan 1984). There has been considerable speculation as to the physiological significance of conjugates of cy-

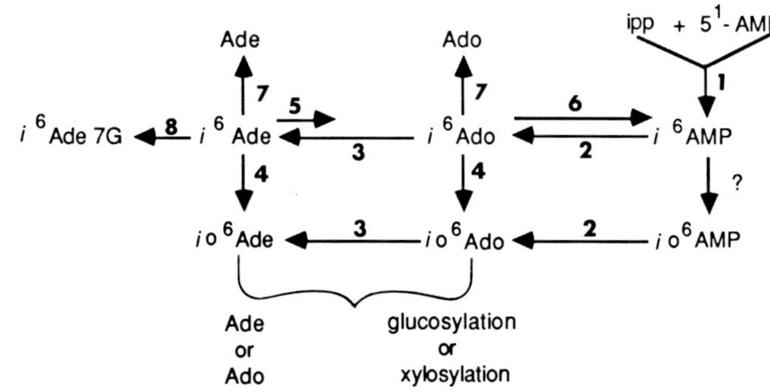

FIGURE 11–1 A scheme for enzymatic regulation of cytokinin biosynthesis and interconversion. Some metabolites are not shown. Ade, adenine; Ado, adenosine; AMP, adenosine monophosphate; ipp, isopentenylpyrophosphate; i^6 Ade, N^6-(Δ^2-isopentenyl)adenine; i^6 Ado, N^6-(Δ^2-isopentenyl)adenosine: io^6 Ade, zeatin; io^6 Ado, zeatin riboside. i^6 Ade 7G, i^6Ade-7-glucoside. Enzymes involved are as follows: **1**, Δ^2-isopentenylpyrophosphate: 5′-AMP-Δ^2isopentenyltransferase; **2**, 5′-nucleotidases; **3**, adenosine nucleosidase; **4**, microsomal hydroxylases; **5**, adenosine phosphorylase; **6**, adenosine kinase; **7**, cytokinin oxidases; **8**, cytokinin glycosyltransferase.

tokinins and sugars and amino acids. It is thought that conjugates of cytokinins are storage or transported forms of the active compounds. However, much more evidence is needed before functional roles can be assigned to them. Hence, we await an answer to the important question, what are the active forms of cytokinins?

11.2 CYTOKININ-REGULATED SYNTHESIS OF MACROMOLECULES

Specific macromolecules whose syntheses are under the regulation of cytokinin can serve as marker molecules; their identification is necessary for further elucidation of cytokinin action at molecular levels.

Cytokinin enhances the synthesis of various enzymes. For example, hydroxypyruvate reductase, which catalyzes the conversion of hydroxypyruvate to glycerate in the glycolate pathway, is significantly stimulated by benzyladenine (BA) (Longo et al. 1981; Chen and Leisner 1985). Investigating cytokinin-modulated gene expression in excised pumpkin cotyledons, Chen and Leisner (1985) demonstrated that hydroxypyruvate reductase prepared from these cotyledons consists of five isozymes. Isozyme activity in cotyledons treated with 10^{-4} M BA was about fourfold higher than activity in the 10^{-9} M BA–treated ones, and about fivefold higher than activity in the water-treated control. Furthermore, all of the isozymes were uniformly enhanced by the hormone. Addition of BA (10^{-8}–10^{-4} M) to a cell-free enzyme reaction mixture containing hydroxypyruvate reductase did not affect the enzyme activity; thus, hydroxypyruvate reductase activity is not directly influenced by BA in a cell-free system. In a detached radish cotyledon system, Howard and Witham (1983) demonstrated that kinetin treatment of the cotyledons results in increased levels of two invertase forms, type I and type II. Type I was soluble and type II was insoluble in 10 mM Tris-HCl buffer, pH 7. The differences in invertase levels between kinetin-treated and untreated cotyledons did not appear until after ten hours of incubation, and the authors suggested that the increased activity was due to de novo synthesis of the enzyme.

Nitrate reductase, a key enzyme in the pathway of nitrogen assimilation, has been shown to be regulated by cytokinin by Borris (1967), Kulaeva (1980), and Kende et al. (1971). These investigators suggested that cytokinins and nitrate induce the synthesis of nitrate reductase independently. However, a recent report by Schmerder and Borris (1986) indicated that in *Agrostemma* embryos, nitrate reductase activity is also induced by ethylene, and that, contrary to nitrate reductase induction by nitrate, nitrate reductase induction by cytokinins requires the presence of ethylene.

Cytokinins also enhance the activity of endopeptidase, acid phosphatase, ribulosediphosphate carboxylase, alkaline pyrophosphatase, and chromatin-bound RNA polymerase (Kulaeva 1980). However, it is unknown whether these enzyme activities are enhanced directly or indirectly by the hormone. Ananiev et al. (1987) demonstrated that cytokinins cause a marked enhancement of nuclear rRNA synthesis accompanied by an increase in the amount of RNA polymerase I in excised pumpkin cotyledons. Analysis of the uridine monophosphate–uridine ratio of the alkali-digested nascent pre-rRNA chains showed that the stimulation of RNA polymerase I activity was the consequence of an increase in the polyribonucleotide elongation rate. No significant change in the number of transcribing enzyme molecules was detected after the hormonal treatment. Similar treatment of soybean hypocotyls with auxin, another plant hormone, also resulted in a prominent enhancement of rRNA elongation rate (Olszewski and Guilfoyle 1980). In cytokinin-deprived cotyledons, cycloheximide caused a rapid inhibition of endogenous RNA polymerase I activity. These results suggest that the control of rRNA elongation rate by cytokinins may involve continuous synthesis of some proteins. Synthesis of rRNA is also similarly regulated by phenylurea-type cytokinins, N_1-(1-chloro-4-pyridyl)-N_2-phenylurea. This indicates possibly similar modes of action for purine- and phenylurea-type cytokinins in plant cells.

Turnover of phosphatidylinositol and phosphatidic acid was stimulated by kinetin in cytokinin-starved cell culture (Connett and Hanke 1987) and the stimulation in phospholipid synthesis could be monitored by an increment of the incorporation of ^{32}Pi into these phospholipids within 15 minutes.

Contrarily, Falkenau et al. (1987) demonstrated that incorporation of ^{32}P into phosphorylated phosphatidylinositol and phosphatidic acid was reduced to 50% of the controls in the presence of 1 μM kinetin in suspension cultured *Catharanthus roseus* cells. The use of different cultured cells may account for these contradictory results.

11.3 THE COMPLEX NATURE OF CYTOKININ-REGULATED GENE EXPRESSION

A single gene may be modulated by more than one environmental factor or by combinations of hormonal and environmental factors. Furthermore, cytokinin-modulated plant genes often respond to the same hormone in different ways. This was illustrated by two-dimensional PAGE analysis of proteins isolated from BA-treated and untreated pumpkin cotyledons grown in the dark (Chen and Leisner 1985). Cytokinin enhanced, reduced, induced, or inhibited synthesis of specific proteins. Analysis of radioactive polypeptides translated from poly(A)RNA prepared from BA-treated and untreated cotyledons by two-dimensional PAGE also indicated that specific sequences of mRNAs are enhanced, reduced, induced, or suppressed by the hormone (Chen et al. 1987b). Differential expression of plant genes regulated by other plant hormones was also observed (Zurflush and Guilfoyle 1982; Theologis and Ray 1982; Hong et al. 1987; Lin and Ho 1986; Galau et al. 1986; Christoffersen and Laties 1982). Whether a unifying mode of action for all plant hormones underlies the diversity of activity remains an open question. Hypotheses regarding plant hormone action have generally failed to present a comprehensive description of how the hormones induce their many effects.

Although there are several reports that show that gene expression is modulated by cytokinin, maximum responses are not detected until hours after cytokinin application (24 hours or more). Thus, it is not known whether a particular macromolecular (often protein) response is primary or secondary to cytokinin action. To gain insight into the earliest events in cytokinin-regulated gene expression, Chen et al. (1987a and 1987b) and Chang et al. (1987 and unpublished observations) investigated the early effects of cytokinin application on the levels of specific mRNAs and proteins. These investigators demonstrated the temporal relation between gene expression and the rate of uptake of BA. In suspension culture, the earliest observable mRNA changes occurred within 60 minutes of the application of BA to excised pumpkin cotyledons; in filter paper culture, 7.5 hours of BA incubation was required to induce detectable change in the mRNA population (Chen et al. 1987b).

A cDNA clone, pCEC 1, which contains an insert of about 900 bp and hybridizes strongly with an mRNA of 2.25 kb has recently been isolated. The pCEC 1–corresponding mRNA displayed a rapid increase in concentration after cytokinin application to excised pumpkin cotyledons (Chen et al. 1987a; Chang et al., unpublished observations). In these experiments dot blot analysis was used to determine the effect of BA on mRNA levels; suspension culture technique was employed for investigating how quickly the cytokinin-responsive mRNA responded to the hormone. Dot blot analysis indicated that the mRNA level for pCEC 1 increased (about threefold) within 60 minutes of BA application. After 12 hours of BA treatment, this mRNA sequence was enhanced about twelvefold by the hormone. This cytokinin-responsive mRNA was also present in every organ of 20-day-old pumpkin plants, with root having the highest concentration. These results suggested that some cytokinin-mediated changes in gene expression might be common to tissues and organs that ultimately respond quite differently to the hormone. Furthermore, these data also indicated that the increased levels of the specific mRNA corresponded to increased transcriptional activity of its gene. Additional experiments, such as nuclear run-off experiments, are needed to determine whether cytokinin regulation of the pCEC 1 gene is at the transcriptional or the posttranscriptional level or at both levels.

To understand the complex nature of cytokinin-modulated gene expression, sequencing of cytokinin-responsive genes, preparation of genomic clones, and determination of the enhancer elements modulated by the hormone are needed. There is some evidence suggesting that cytokinin may also regulate gene expression posttranscriptionally. While the precise mechanism for hormonal regulation of posttranscriptional events remains to be elucidated, one possibility is the stabilization of specific mRNAs.

11.4 REGULATION OF GENE EXPRESSION BY COMBINATIONS OF HORMONES

In many instances developmentally regulated genes respond to a combination of hormones. For example, regeneration of shoots in vitro requires the presence of a high cytokinin-auxin ratio (Skoog and Miller 1957). Employing normal, habituated, and crown gall–transformed tissues of *Nicotiana tabacum* L. cv. Havana 425, Eichholz et al. (1983) provided evidence that the synthesis of a specific abundant polypeptide with an apparent molecular weight of approximately 33 kd is regulated by cytokinin and auxin. The 33-kd polypeptide is drastically inhibited by kinetin, in some cell lines at concentrations as low as 1.3×10^{-8} M. The polypeptide is a major component of the protein fraction obtained from nonhabituated clones, cytokinin-habituated clones, and revertant subclones of crown gall–transformed clones, but cannot be detected in clones habituated for both auxin and cytokinin or in crown gall–transformed clones. These authors postulated that cytokinin in the presence of auxin regulates the production of a specific polypeptide.

The reports of Mohnen et al. (1985) and Shinshi et al. (1987) provided additional examples of regulation of plant gene expression by a combination of auxin and cytokinin. The gene for β-1,3-glucanase was induced by the absence of hormones or by the separate addition of either auxin or cytokinins. When both hormones were present, expression was reduced to a basal level (Mohnen et al. 1985). Regulation of another defense-related enzyme, chitinase, was also shown to be coordinated by combinations of cytokinin and auxin. Shinshi et al. (1987) demonstrated that chitinase accumulation could be blocked by auxin and cytokinin and that this effect correlated with the content of chitinase mRNA in tissues, as judged by RNA blot analysis. It will be of interest to know how the regulatory regions of these genes are modulated by these hormones, and how the cis-acting elements of these genes interact coordinately to provide a harmonious response to both hormones.

11.5 ENHANCEMENT OF LIGHT-REGULATED GENE EXPRESSION BY CYTOKININ

Plant growth and differentiation requires light. Light-regulated enzyme activities have been reviewed (Lamb and Lawton 1983; Tobin and Silverthorne 1985), and gene expression regulated by light has also been widely reported (for reviews see Quail 1983; Tobin and Silverthorne 1985; Thompson et al. 1983; Kuhlemeier et al. 1987). Photoregulated plant gene expression is a complex phenomenon. Coordinated action of cytokinin and light on gene expression increases the complexity even more. The light responses of photogregulated genes differ at many levels, and in nearly all cases phytochrome is assumed to play a pivotal role. Realizing that identification of the DNA sequences mediating light-regulated transcription is an essential step in understanding the molecular biology of light induction, Chua et al. (for review see Kuhlemeier et al. 1987) probed and characterized the cis-acting and trans-acting elements for light-regulated transcription. However, the mechanism of transcriptional and posttranscriptional regulation of gene expression by light remains to be elucidated.

There is ample evidence that both cytokinin and light promote plastid or chloroplast development (for reviews see Parthier et al. 1981; Longo et al. 1981; Mohr 1984). However, the interaction of light and cytokinin on gene expression has only recently been reported. The effects of BA and light on gene expression related to plastogenesis have been studied in detached pumpkin cotyledons (Lerbs et al. 1984). Using ribulose-1,5-bisphosphate carboxylase (rbc) as a marker molecule, these workers provided evidence that the formation of the large and small subunits of rbc is stimulated by BA and light, probably at the level of mRNA transcription. Their results also indicated that both light and cytokinin coact in an additive rather than a multiplicative manner in the gene expression of the enzyme.

However, whether cytokinin and light promote chloroplast development via a common or different mode of action is still unknown. Many tobacco cell suspension cultures respond to the application of cytokinin by differentiating chloroplasts. Axelas et al. (1984) observed that cytokinin stimulates the appearance of some plastid proteins. One of these cytokinin-responsive proteins is the light-harvesting chlorophyll a/b–binding protein (LHCP). Using a pea LHCP cDNA as a probe, Teyssendier de la Serve et al. (1985) reported a tenfold increase of RNApoly(A) obtained from kinetin-treated tobacco cells cultured under continuous white light for nine days. They concluded that the regulation of LHCP gene expression by cytokinin in the presence of light occurred at the transcriptional level, and not at the translational or posttranslational level.

Evidence presented by Flores and Tobin (1986) suggested that cytokinin regulation of transcriptional events may be primarily at the posttranscriptional level in *Lemna gibba* L. (duckweed plants). They examined the effects of red light and BA on the levels of mRNA and demonstrated that the amounts of hybridizable RNA coding for both the major LHCP and for the small subunit of rbc decrease to a low level when white-light-grown duckweed plants are placed in the dark. Upon addition of BA to the dark-grown plants, the levels of these two mRNAs increased. Furthermore, BA treatment magnified the extent of the response to red light treatment. Therefore, combinations of cytokinin and light enhanced gene expression. Slot blot analysis of nuclear RNAs isolated from plants treated with red light and cytokinin revealed that cytokinin did not enhance LHCP RNA transcription. Thus, these authors suggested that cytokinin regulated expression of LHCP RNA at a posttranscriptional level, possibly by affecting the stability of the RNA. It will be of interest to investigate this further by performing a nuclear run-off experiment in which nuclei are isolated from appropriately treated tissues and then incubated in vitro in the presence of radioactive nucleotide triphosphates. It is thought that these isolated nuclei do not initiate new RNA in vitro, but rather continue elongation of RNA initiated prior to isolation. Such a procedure has been used to show that hormones can act in vitro to increase transcription.

11.6 CYTOKININ-BINDING MOLECULES

Another interesting aspect of cytokinin-regulated gene expression is the substantial information that has been gathered over the years concerning cytokinin-binding proteins. A widespread view regarding the mechanism of action of plant hormones is that the initial event of hormone action involves binding of a hormone receptor complex to a hormone-responsive element of the enhancer region. Consequently, it is speculated that the cytokinin-binding factors may be components of the signal transduction chain leading to cytokinin-induced gene expression. The most intensively studied cytokinin-binding protein is the cytokinin-binding fraction 1 (CBF-1) protein associated with ribosomes (Fox and Erion 1975). CBF-1 was shown to have a relatively high affinity for 6-substituted purines. It has a dissociation constant, Kd, of 6.5×10^{-7} for BA. Subsequently, the subunit structure of CBF-1 has been shown to be a trimer composed of 54-kd polypeptides (Brinegar and Fox 1985). It is unlikely that CBF-1 is a receptor molecule, because the abundance of CBF-1 seems inconsistent with it being a receptor protein. The authors postulated that CBF-1 may function as a cytokinin-sequestering protein and an embryo storage protein.

There have been several other reports of the isolation of relatively high–affinity cytokinin-binding proteins (Polya and Davies 1978; Moore 1979; Sussman and Kende 1978; Chen et al. 1980). However there is no direct evidence supporting a physiological role for any of the cytokinin-binding proteins mentioned above.

11.7 CONCLUDING REMARKS

In this article, an attempt has been made to integrate biological and molecular aspects of cytokinin-regulatory gene expression. Clearly, any progress in the knowledge of the mechanism of cytokinin action must depend heavily on the remarkable advances in the molecular analysis of gene structure, function, and regulation. The true impact of these analyses will be felt only if the molecular aspect can be integrated into phenomena and mechanisms of development exhibited on a whole-organismal level.

Regulation of gene expression by plant hormones during development includes a cascade of events with alternating sequences of structural and regulatory gene expression. Cytokinins and other plant hormones are likely to control regulatory genes both positively and negatively. We do not know with what plant hormones react in the cell, nor have any plant hormone receptors been purified. Nevertheless, it is now possible to analyze cis-acting DNA sequences that mediate gene expression (for example see Knox et al. 1987).

There are several obvious directions for research that will aid elucidation of plant gene regulation by cytokinins and possibly other plant hormones: (1) characterization of the genes and their flanking sequences that might be involved in making the genes responsive to the action of cytokinins; (2) investigation of the intermediate regulatory factors operating between cytokinin application and transcriptional response, including trans-acting factors and chromatin structure changes induced by the hormone; (3) study of the coordination between a hormone, or combined hormones, and other regulatory factors. Genetic, biochemical, and molecular approaches, when combined, may shed light on hormonal regulation of plant growth.

The wisdom gained from past experiences tempers one's enthusiasm for a generalized scheme for plant hormone action. Nonetheless, the heuristic value of such a scheme allows the formulation shown in Figure 11–2. The regulatory gene might be located on the same or on a different chromosome, and the action of the regulatory protein may not be targeted at the cytokinin-responsive sequences directly but may cause a conformational change in the chromatin domain, which in turn may affect cytokinin-responsive sequences. A signal of transcription may require the binding of a specific cytokinin receptor to an enhancer element.

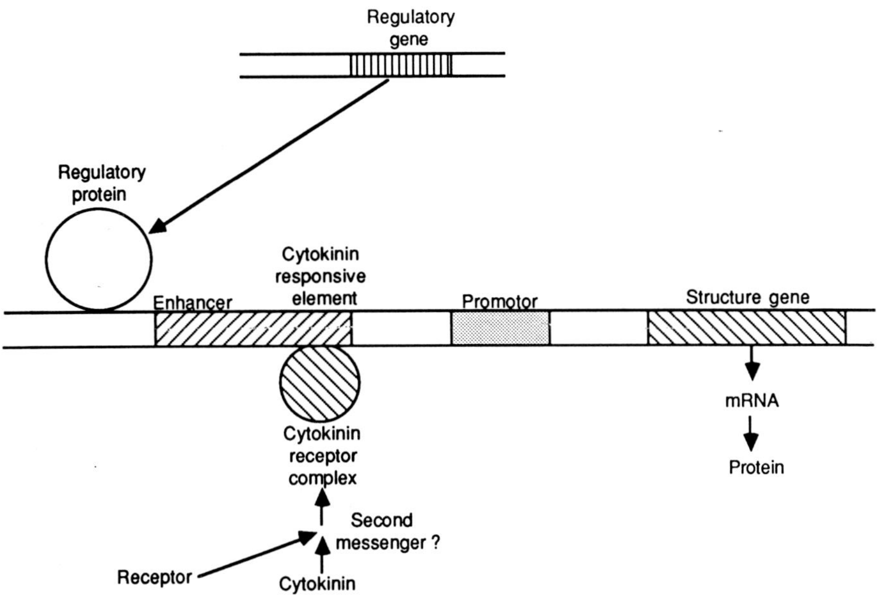

FIGURE 11–2 A hypothetical scheme of cytokinin regulation of plant gene expression. A possible posttranscriptional regulation by cytokinin and a role of a second messenger (Hepler and Wayne 1985) that interacts with the hormone are not shown.

REFERENCES

Ananiev, E.D., Karazyozov, L.K., and Karnov, E.N. (1987) *Planta* 170, 370–378.
Axelas, M., Barbet, J., and Peaud-Lenoel, C. (1984) *Plant Sci. Lett.* 33, 202–212.
Borris, H. (1967) *Wiss. Z. Univ. Rostock., Math. Naturwiss. Reihe* 16, 629–639.
Brinegar, A.C., and Fox, J.E. (1985) *Biol. Plant* 27, 100–104.
Chang, C.C., Ertl, J., and Chen, C.M. (1987) *Plant Physiol.* 83 (suppl.), 96.

Chatfield, J.M., and Armstrong, D.J. (1987) *Plant Physiol.* 84, 726–731.
Chen, C.M. (1981) in *Metabolism and Molecular Activities of Cytokinins* (Guern, J., and Peaud-Lenoel, C., eds.), pp. 34–43, Springer, New York.
Chen, C.M., and Christopeit, S.M. (1981a) *Plant Physiol.* 67, 494–498.
Chen, C.M., and Christopeit, S.M. (1981b) *Plant Physiol.* 68, 1020–1023.
Chen, C.M., and Eckert, R.L. (1977) *Plant Physiol.* 59, 443–447.
Chen, C.M., Ertl, J., and Chang, C.C. (1987a) XIV International Botany, Congress, p. 118, Berlin, July 24–August 1.
Chen, C.M., Ertl, J., Yang, M.S., and Chang, C.C. (1987b) *Plant Sci.* 52, 169–174.
Chen, C.M., and Leisner, S.M. (1984) *Plant Physiol.* 74, 442–446.
Chen, C.M., and Leisner, S.M. (1985) *Plant Physiol.* 77, 99–103.
Chen, C.M., and Melitz, D.K. (1979) *FEBS Lett.* 107, 15–20.
Chen, C.M., Melitz, D.K., and Clough, F.W. (1982) *Arch. Biochem. Biophys.* 214, 634–641.
Chen, C.M., Melitz, D.K., Petschow, B., and Eckert, R.L. (1980) *Eur. J. Biochem.* 108, 379–387.
Chen, C.M., and Petschow, B. (1978) *Plant Physiol.* 62, 871–874.
Christofferson, R.E., and Laties, G.G. (1982) *Proc. Natl. Acad. Sci. USA* 79, 4060–4063.
Connett, R.J.A., and Hanke, D.E. (1987) *Planta* 170, 161–167.
Eichholz, R., Harper, J., Felix, G., and Meins, F. (1983) *Planta* 158, 410–415.
Falkenau, C., Heim, S., and Wagner, K.G. (1987) *Plant Sci.* 50, 173–178.
Flores, S., and Tobin, E.M. (1986) *Planta* 168, 340–349.
Fox, J.E., and Erion, J.E. (1975) *Biochem. Biophys. Res. Commun.* 64, 694–700.
Galau, G.A., Hughes, D.W., and Dure L. III (1986) *Plant Mol. Biol.* 7, 155–170.
Goldberg, R.B. (1986) *Philos. Trans. R. Soc. Lond.* [*Biol.*] 314, 343–353.
Hecht, S.M. (1980) in *Plant Growth Substances 1979* (Skoog, F., ed.), pp. 144–158, Springer-Verlag, New York.
Hepler, P.K., and Wayne, R.O. (1985) *Annu. Rev. Plant Physiol.* 36, 397–439.
Hong, J.C., Nagao, R.J., and Key, J.L. (1987) *J. Biol. Chem.* 262, 8367–8376.
Horgan, R. (1984) in *Advanced Plant Physiology* (Wilkins, M.B., ed.), pp. 53–75, Pitman Publishing, London.
Howard, H.F., and Witham, F.H. (1983) *Plant Physiol.* 73, 304–308.
Kende, H., Hahn, H., and Kays, E. (1971) *Plant Physiol.* 48, 702–706.
Knox, C.A.P., Sonthayanon, B., Ram Chandra, G., and Muthukrishnan, S. (1987) *Plant Mol. Biol.* 9, 3–18.
Kuhlemeier, C., Green, P.J., and Chua, N-H. (1987) *Annu. Rev. Plant Physiol.* 38, 221–257.
Kulaeva, O.N. (1980) in *Plant Growth Substances 1979* (Skoog, F., ed.), pp. 119–128, Springer-Verlag, New York.
Laloue, M., and Fox, J.E. (1985) 12th International Conference on Plant Growth Substances, p. 23, Heidelberg.
Lamb, C.J., and Lawton, M.A. (1983) in *Encyclopedia of Plant Physiology: Photomorphogenesis,* vol. 16A (Shropshire, W., and Mohr, H., eds.), pp. 213–243, Springer-Verlag, New York.
Lerbs, S., Klyachko, N.L., Romanko, E.G., et al. (1984) *Planta* 162, 289–298.
Letham, D.S., and Palni, L.M.S. (1983) *Annu. Rev. Plant Physiol.* 34, 163–197.
Lin, L-S., and Ho, T-H.D. (1986) *Plant Physiol.* 82, 289–297.

Longo, C.P., Longo, G.P., Lampugnani, M.G., Rossi, G., and Servettaz, O. (1981) in *Metabolism and Molecular Activities of Cytokinins* (Guern, J., and Peaud-Lenoel, C., eds.), pp. 261–266, Springer, New York.

Matters, G.L., and Scandalios, J.G. (1986) *Dev. Genet.* 7, 167–175.

Mohr, H. (1984) in *Chloroplast Biogenesis* (Baker N.R., and Barber, J., eds.), pp. 305–347, Elsevier, Amsterdam.

Mohnen, D., Shinshi, H., Felix, G., and Neins, F. (1985) *EMBO J.* 4, 1631–1635.

Mok, M.C., and Mok, D.W.S. (1987) *Plant Physiol.* 83 (suppl.), 29.

Mok, M.C., Mok, D.W.S, Dixon, S.C., Armstrong, D.J., and Shaw, G. (1982) *Plant Physiol.* 70, 173–178.

Moore, F.H. (1979) *Plant Physiol.* 64, 594–599.

Morris, R.O. (1986) *Annu. Rev. Plant Physiol.* 37, 509–538.

Morris, R.O., Powell, G.K., Beaty, J.S., et al. (1985) in *Plant Growth Substances 1985* (Mopp, M., ed.), pp. 185–196, Springer-Verlag, New York.

Olszewski, N., Guilfoyle, T.J. (1980) *Biochem. Biophys. Res. Commun.* 94, 553–559.

Polya, G.M., and Davies, J.R. (1978) *Planta* 139, 139–147.

Quail, P.H. (1983) in *Encyclopedia of Plant Physiology: Photomorphogenesis* (Shropshire, W., and Mohr, H., eds.), pp. 178–212, Springer-Verlag, Berlin.

Sachs, M.M., and Ho, T-H.D. (1986) *Annu. Rev. Plant Physiol.* 37, 363–376.

Schmerder, B., and Borris, H. (1986) *Planta* 169, 589–593.

Shinshi, H., Mohnen, D., and Meins, F. (1987) *Proc. Natl. Acad. Sci. USA* 84, 89–93.

Skoog, F., Hamzi, H.Q., Szweykowska, A.M., et al. (1967) *Phytochem.* 6, 1169–1192.

Skoog, F., and Miller, C.O. (1957) *Symp. Soc. Exp. Biol.* 15, 118–131.

Sussman, M.R., and Kende, H. (1978) *Planta* 140, 251–259.

Teyssendier de la Serve, B., Axelas, M., and Peaud-Lenoel, C. (1985) *Plant Mol. Biol.* 5, 155–163.

Theologis, A., and Ray, P. (1982) *Proc. Natl. Acad. Sci. USA* 79, 418–421.

Thompson, W.F., Everett, M., Polans, N.O., Jorgensen, R.A., and Palmer J.D. (1983) *Planta* 158, 487–500.

Tobin, E.M., and Silverthorne, J. (1985) *Annu. Rev. Plant Physiol.* 36, 569–593.

Zurflush, L.L., and Guilfoyle, T.J. (1982) *Plant Physiol.* 69, 332–337.

CHAPTER 12

Organization and Expression of Genes for Photosynthetic Pigments–Protein Complexes in Photosynthetic Bacteria

Yu Sheng Zhu
John E. Hearst

Photosynthesis, the conversion by living organisms of light into chemical energy, is a fundamental biological process on earth. One of the major goals of plant biotechnology is to alter genetically the photosynthetic apparatus of crop plants to increase the efficiency of photosynthesis and ultimately to design an artificial photosynthetic system for harvesting solar energy. This strategy depends on a full understanding of the structure, organization, and function of the photosynthetic apparatus. Over the last few years a great deal of knowledge about the basic physics, chemistry, and molecular biology of photosynthesis has accumulated (Govindjee 1982; Sybesma 1984; Staehelin and Arntzen 1986; Steinback et al. 1985). However, the molecular

We are grateful to Dr. S. Kaplan for the use of his unpublished data. We thank D. Cook and G. Armstrong for their comments in preparation of this review. Part of this work was supported by the U.S. Department of Energy and by a grant from the National Institutes of Health.

events involved in biosynthesis, assembly, and regulation of the photosynthetic apparatus are not well understood.

Photosynthetic bacteria, genus *Rhodobacter,* including *Rhodobacter capsulatus* and *Rhodobacter sphaeroides* (formerly *Rhodopseudomonas capsulata* and *R. sphaeroides,* respectively) afford an unusual opportunity to study the molecular events of photosynthesis. These purple nonsulfur bacteria have a photosynthetic system that is less-complex than that of green plants and algae. Unlike higher photosynthetic organisms, they possess only one photosystem and lack the ability to oxidize water to oxygen. Their photochemical reaction center (RC) is similar to, but simpler than, that of photosystem II, and can thereby provide fundamental information about the light reactions in more complex organisms (Kaplan and Arntzen 1982). For example, a new model of photochemical function of photosystem II components D_1 and D_2 has resulted from a study of DNA sequence homology between genes coding for RC subunits L and M in *R. capsulatus* and genes coding for the 32-kd thylakoid membrane polypeptides in spinach and tobacco (Hearst and Sauer 1984a and 1984b; Hearst 1986). Study of *Rhodobacter* has been greatly advanced by several well-developed genetic techniques, particularly for *R. capsulatus,* including the use of the gene-transfer agent (Marrs 1974; Solioz et al. 1975), marker rescue (Taylor et al. 1983), transposon mutagenesis (Youvan et al. 1982; Zsebo and Hearst 1984), and interposon mutagenesis (Scholnik and Haselkorn 1984). Photosynthetic mutants are viable because of the remarkable metabolic versatility of purple nonsulfur bacteria. *Rhodobacter* can grow either chemoheterotrophically or photoheterotrophically, depending on oxygen tension and light intensity in the environment (Kaplan 1978; Drews 1978; Drews and Oelze 1981). This dual lifestyle makes the organism an excellent laboratory tool with which to study the regulation of photosynthetic genes by light and oxygen as well as the formation and development of the photosynthetic membrane. Thus, in comparison with algae and higher plants, both the biophysical chemistry and the genetics of the photosynthetic apparatus are more advanced in the genus *Rhodobacter.*

The initial experiments that led to the discovery of the photosynthetic gene cluster of *R. capsulatus* were conducted by Marrs et al. They utilized the gene transfer agent, a particle resembling a generalized transducing phage, to map a gene cluster for bacteriochlorophyll (Bchl) and carotenoid (Crt) biosynthesis (Yen and Marrs 1976; Solioz et al. 1975; Yen et al. 1979). Using the R factor mobilization technique, they isolated an R-prime plasmid, pRPS404, bearing 46 kb of chromosomal DNA that contained most of the photosynthetic genes of *R. capsulatus* (Marrs 1981). This key step, achieved by genetic methods, is the basis for studying the molecular biology of photosynthesis in *R. capsulatus.* A subsequent breakthrough in the study of the photosynthetic genes of *R. capsulatus* was achieved by Hearst et al. when they reported the complete DNA sequence of the structural genes encoding the subunuits RC-L, RC-M, and RC-H (designated *puf*L, *puf*M, and *puh*A, respectively) as well as the light-harvesting (LH) I polypeptides

LH-I β and LH-I α (*puf*B and *puf*A, respectively) (Youvan et al. 1984a and 1984b). Subsequently, the DNA sequence of two of the three LH-II polypeptides, LH-II β and LH-II α (*pucB* and *puc*A, respectively), was determined by Youvan and Ismail (1985). Additionally, 17 genes for Bchl and Crt biosynthesis have been aligned with a detailed restriction map (Taylor et al. 1983; Zsebo and Hearst 1984; Zsebo et al. 1984). Paralleling this achievement, the *R. sphaeroides* genes for RC-L, RC-M, RC-H (Williams et al. 1983 and 1984; Donohue et al. 1986a), LH-I and LH-II (Kiley et al. 1987; Kiley and Kaplan 1987) have also been cloned and sequenced. Recently, the DNA sequences of the genes for cytochrome c_2 (*cyc*A)(Daldal et al. 1986; Donohue et al. 1986b) and the iron sulfur ubiquinol bc_1, oxidoreductase complex (*fbc*) have been determined in both *R. capsulatus* and *R. sphaeroides* (Gabellini and Sebald 1986; Davidson et al. 1987). Even more recently the structural genes for the Form I ribulose bisphosphate carboxylase, phosphoribulose kinase, and cytochrome *c′* (*cyd*A) have also been identified (Kaplan, personal communication). Furthermore, the sequences of the genes for the B880 holochrome from *Rhodospirillum rubrum* (Bérard et al. 1986) and the genes for H, L, and M subunits of the RC (Michel et al. 1985 and 1986) from *Rhodopseudomonas viridis* have been determined. In summary, thus far the structural genes for LH complexes (except the γ subunit of LH-II), RCs, and secondary electron carriers, as well as a few other genes involved in carbon dioxide fixation have been isolated and subjected to detailed structural analyses in both *R. capsulatus* and *R. sphaeroides*.

At the protein level, Deisenhofer et al. (1985a) have recently presented a detailed crystal structure of the RC from *R. viridis* at 3-Å resolution. A three-dimensional structure of the *R. sphaeroides* RC (Allen and Feher 1984) has also been reported. This approach will lead to significant progress in our understanding of how protein structure facilitates primary photochemistry in the RC.

In addition to gene and protein structure, studies on transcriptional and posttranscriptional regulation of the photosynthetic genes, namely, the synthesis and processing of mRNA, are becoming an important branch in the study of the molecular biological properties of photosynthetic bacteria. Although early studies indicated the importance of transcriptional regulation (Biel and Marrs 1983 and 1985; Clark et al. 1984; Klug et al. 1984), the isolation and characterization by Northern blots and S-1 nuclease analysis of discrete mRNAs for LH-I, RC-L, and RC-M were shown only by Belasco et al. (1985) for *R. capsulatus,* and by Zhu and Kaplan (1985) and Zhu et al. (1986b) for *R. sphaeroides*. Recently, we have described a more-detailed study of the regulation by oxygen and light of the genes coding for the structural proteins and pigment biosynthetic enzymes in *R. capsulatus* (Zhu and Hearst 1986; Zhu et al. 1986a).

In this chapter we have reviewed the recent progress in research aimed at the organization and regulation of photosynthetic genes primarily in *R. capsulatus* and *R. sphaeroides*. For other reviews regarding the biosynthesis and assembly of the photosynthetic apparatus in purple photosynthetic bacteria, see the following articles: Drew 1978 and 1985; Kaplan 1978; Drews and Oelze 1981; Kaplan and Arntzen 1982; Youvan and Marrs 1984 and 1985; Drews et al. 1986.

12.1 METABOLIC VERSATILITY

Purple nonsulfur bacteria normally inhabit muddy lake bottoms and sewage lagoons. They have a remarkable ability to adapt metabolically in response to changes in oxygen tension, light intensity, and available nutrients in their environment. Such changes are likely to occur in the mud of ponds and ditches; in the littoral zones of lakes, rivers, and seas; and in all kinds of sewage lagoons (Imhoff et al. 1984).

The multiple modes of growth seem to parallel the evolution of the earth's atmosphere from an initial reducing environment of hydrogen, ammonia, and methane through the synthesis of organic acids and depletion of ferrous iron in the oceans, to the development of an oxidizing environment through the action of the cyanobacteria. Typical purple nonsulfur bacteria, *R. capsulatus* and *R. sphaeroides,* are capable of six distinct modes of growth; anaerobic photoautotrophy growing on hydrogen and carbon dioxide under light; anaerobic photoheterotrophy in the light using organic compounds and carbon dioxide; anaerobic chemoheterotrophy in the dark, utilizing a variety of organic carbon sources plus a terminal electron acceptor such as DMSO; dark aerobic chemoautotrophy, using hydrogen as the sole source of energy and reducing power; aerobic chemoheterotrophy growing on a diverse group of carbon compounds and generating ATP through oxidative phosphorylation; and anaerobic dark fermentation in the absence of oxygen, utilizing substrate-level phosphorylation. Among the various environmental variables, oxygen tension and light intensity, which change frequently in the vertical ecological zone, are the most important in regulating these metabolic shifts (Drews 1978; Kaplan 1978; Drews and Oelze 1981; Kaplan and Arntzen 1982; Chory et al. 1984; Drews et al. 1986; Youvan and Marrs 1984 and 1985).

When growing chemoheterotrophically, purple nonsulfur bacteria contain a typical Gram-negative outer membrane and a cytoplasmic membrane. Under these conditions cell growth is supported by an aerobic respiratory chain with components structurally and functionally similar to those found in mitochondria (Zannoni and Baccarini-Melandri 1980). Upon the removal of oxygen, the cell develops an extensive intracytoplasmic membrane (ICM) that comprises the photosynthetic apparatus. The amount of ICM per cell and the composition of the ICM are functions of the light intensity (Drews 1978 and 1986; Kaplan 1978). Increase in the light intensity depresses the formation of the photosynthetic apparatus. The quick response of facultative phototrophic bacteria to environmental changes provides an attractive system for studying the regulation of the expression of photosynthetic genes by oxygen and light.

12.2 PHOTOSYNTHETIC APPARATUS

In most photosynthetic bacteria the photosynthetic apparatus is localized in the ICM in structures known as chromatophores (Schachman et al. 1952). These double-membraned vesicles or tubules are clustered to form grainlike stacks and bundles. For *R. capsulatus* and *R. sphaeroides* the ICM is the result of invaginations of the cytoplasmic membrane in response to a lowering of the oxygen tension below 10 mm Hg. The ICM contains three major pigment protein complexes that function in the capture of light energy and electrochemical gradient formation: LH-II (B800-850), LH-I (B870), and RC (Figure 12- 1).

The LH-II antenna complex consists of three polypeptides—α (12 kd), β (10 kd), and γ (14 kd)—three molecules of Bchl, and one molecule of Crt. Two molecules of Bchl, associated with the α subunit, and one molecule of Bchl, associated with the β subunit, are believed to be responsible for the absorption maxima at 850 and 800 nm, respectively (Cogdell and Crofts 1978; Sauer and Austin 1978). The Crt is probably associated with the β subunit (Webster et al. 1980). Cross-linking experiments support the idea that the LH-II complex in the membrane has an oligomeric (possibly a tetrameric) structure (Drews et al. 1986). An α and a β polypeptide dimerize with α-helical structures crossing through the membrane. This dimer is considered to be the basic structural unit of the LH complexes. The oligomerization of the α-β dimers is proposed to optimize the efficiency of exciton migration by correctly orienting the chromophores. The LH-II γ subunit is less hydrophobic and exposed on the cytoplasmic surface of the membrane. It does not bind Bchl (Drews et al. 1986). Although the function of the LH-II γ subunit is unknown, it may be necessary for the assembly or stability of the complexes in *R. capsulatus*. However, the LH-II γ subunit is absent in *R. sphaeroides* and other photosynthetic bacteria. The LH-II complex from *R. capsulatus* has been crystallized, but its fine structure has not yet been resolved (Welte et al. 1985).

The LH-I antenna complex (B870) from *R. capsulatus* consists of two polypeptides—α (12 kd) and β (7 kd)—complexed with two molecules of Bchl and one molecule of Crt (Drews 1985). They show very low homology to one another (7–13%), but are very homologous (76–78%) to the corresponding polypeptides in *R. sphaeroides*. As with the LH-II complex, a pair of α and β subunits span the membrane via a central, hydrophobic, α-helical domain. Conserved histidine residues in the hydrophobic regions are thought to bind to a pair of Bchl molecules. The polar C-terminal and N-terminal domains of each subunit are distributed toward the periplasm and

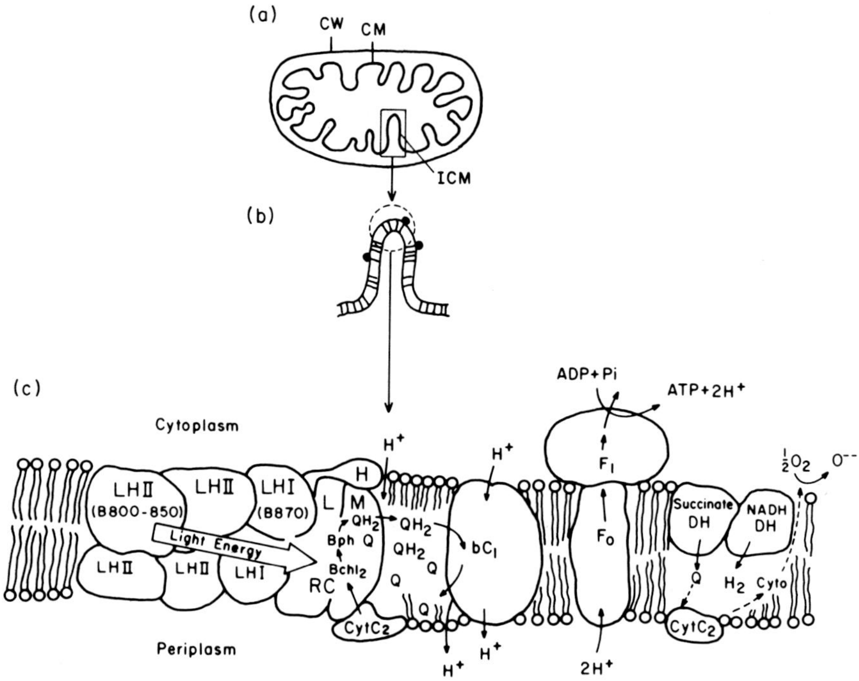

FIGURE 12–1 Model of the topology of the photosynthetic membrane of purple bacteria. (A) A cross-section of a representative cell grown photosynthetically. Cell wall (CW), and intracytoplasmic membrane (ICM) are indicated. (B) A magnified ICM portion. (C) A model of the spatial arrangement of functional units and the photosynthetic electron transport chain in ICM. Light energy absorbed by the antenna LH-II–Bchl is transferred through the LH-I–Bchl complex to the RC, which is composed of H, L, and M subunits. As a result, a Bchl dimer in the RC is excited and gives up an electron to bacteriopheophytin (Bph). Bph in turn reduces a bound quinone molecule (Q) entering a Q cycle within the adjacent membrane-bound cytochrome bc_1 oxidoreductase that pumps protons across the membrane. The electrons finally return to the RC through cytochrome c_2 located on the periplasm of the ICM. The proton gradient across the membrane drives ATP synthesis via an ATPase (F_1-F_0) in the membrane. The respiratory electron transport chain from succinate dehydrogenase (DH) and NADH DH through terminal oxidase cyt$_o$ to oxygen is also indicated, although it is not active in the photosynthetically grown cells.

cytosol, respectively. This basic structural unit is stabilized by forming oligomers, providing the specific binding and arrangement of the Bchl in the polypeptide-lipid environment for efficient energy transfer (Drews et al. 1986; Zuber 1986).

The RC is an integral membrane protein complex where charge separation occurs. It consists of three polypeptides designated RC-H, RC-M, and RC-L; four molecules of Bchl; two bacteriopheophytins (Bphs); two ubioquinones; one non-heme Fe^{2+}; and one molecule of the carotenoid spheroidene (Feher and Okamura 1978; Cogdell et al. 1976). The exact molecular weights of RC-H, RC-M, and RC-L from *R. capsulatus* based on DNA and deduced amino acid sequence data are 28,534; 34,440; and 31,565 daltons, respectively. The RC-L and RC-M subunits have some similar sequences. Each polypeptide chain consists of five hydrophobic regions that are believed to span the membranes as α-helices. RC-L and RC-M bind Bchl and are involved in the primary photochemistry. The RC-H subunit has only one hydrophobic section near the amino terminal (Youvan et al. 1984a and 1984b) and has its carboxyl terminal domain exposed on the cytoplasmic surface of the membrane. At least in *R. sphaeroides* and *Rhodospillum rubrum* the H subunit can be removed without loss of photosynthetic electron transport (Feher and Okamura 1978; Wiemken and Bachofen 1984). The function of the H subunit is unknown. It has been suggested that it may play a role in initiating the assembly of the L-M complex (Chory et al. 1984) or it may coordinate the interaction between RC and LH-I (Takemoto et al. 1982; Peters et al. 1983). It has been also reported that loss of the H subunit changes proper electron transport rates through Q_A and Q_B (Debus et al. 1985).

Recently, a model of the arrangement of the protein backbone and the prosthetic groups of the RC of *Rps. viridis* has been reported (Deisenhofer et al. 1984, 1985a, and 1985b; Michel and Deisenhofer 1986). The L and M polypeptides, 274 and 320 amino acid residues, respectively, span the membrane with five helical segments similar to those hypothesized in *R. capsulatus* and *R. sphaeroides*. The N-terminals of both the L and M subunits are on the cytoplasmic side of the membrane, while the C-terminals of both proteins are on the periplasmic side of the membrane. The prosthetic groups are embedded in a central cylinder formed by the L and M proteins. The folding of L and M are similar, and they are related to one another by an axis of dyad symmetry oriented normal to the membrane and passing through the Bchl special pair and the Fe^{2+}. The transmembrane helices are connected by segments that are partly helical, β-sheet, or irregular in conformation. These segments presumably lie on the membrane surface and mediate the contact to the H subunit. The H subunit (258 amino acid residues) is the cytoplasmic cover of the L-M complex. It is folded in a globular C-terminal part and a loosely folded N-terminal part containing a prominent transmembrane helix. Most of the H subunit is in contact with M and L polypeptides. One difference between the RCs of *R. viridis* and *R. capsulatus* is the presence of a tightly bound *c*-type cytochrome on the periplasmic surface of the membrane of *R. viridis*. The N-terminal transmembrane helix of the H subunit interacts with a segment of the cytochrome and acts as a transmembrane clamp.

In addition to the RC and LH complexes, the photosynthetic membrane contains a ubiquinol-cytochrome c_2 oxidoreductase complex. The functional oxidoreductase that is part of the cyclic photosynthetic and respiratory electron chain of *R. capsulatus* and *R. sphaeroides* comprises three main subunits—cytochrome *b*, cytochrome c_1, and a high-potential FeS protein—carrying a cluster of four redox centers (Gabellini et al. 1982). Furthermore, ATPase, another membrane component, couples proton flow across the membrane to the synthesis of ATP. This high-molecular-weight, multisubunit complex consists of two primary domains: F_o and F_1. The F_1 portion, consisting of five subunits, is hydrophilic and sits on the cytoplasmic surface of the membrane. The F_o part consists of three polypeptides—denoted a, b, and c—and is buried in the membrane. The c subunit from *R. rubrum* has been isolated and its sequence has been determined (Bachofen and Wiemken 1986).

The proper assembly and packaging of photochemical pigments (Bchl, Crt), cofactors (quinone), and polypeptides in the LH and RC complexes are a prerequisite for efficient energy transfer from molecule to molecule. The detailed description of assembly and topology of the photosynthetic membranes of purple bacteria has recently been published (Drews et al. 1983 and 1986; Bachofen and Wiemken 1986; Donohue and Kaplan 1986; Sprague and Varga 1986; Zuber et al. 1986; Loach et al. 1986). Studies on the topography of RC polypeptides using proteolysis, site-specific labeling, and x-ray and neutron diffraction have shown that the RC is asymmetrically arranged (Drews 1985). The RC-H subunit is exposed on the cytoplasmic surface of the membrane; the RC-M and RC-L subunits are buried in the membrane (Deisenhofer et al. 1984; Youvan et al. 1984b). All three subunits of the RC interact. The LH-I complexes are localized around the RC and are structurally and functionally interconnected to the RC. The LH-II complexes are arranged peripherally, interconnecting many RC–LH-I units (Monger and Parson 1977).

The pathway of the photochemical reaction is depicted diagrammatically in Figure 12–1: Light energy absorbed by the antenna LH-II–Bchl is transformed into mobile electronic singlet states, which migrate as excitons by a random walk through the LH-I complex (Drews et al. 1986). These two LH complexes function to increase the cross-sectional area for absorption of light. The excitation energy is finally captured with high efficiency by RC-Bchl molecules, where the energy is transduced into a charge-separated state. A photon of light normally causes the oxidation of a Bchl dimer (the special pair) in the RC with the resultant transfer of an electron to a tightly bound quinone molecule (Q_A). Q_A is then oxidized by a secondary quinone (Q_B) to form a stable semiquinone (Q_B^-). A second photoreduction results in a second electron transfer to Q_B, resulting in production of the fully reduced quinol, which is released into the membrane. The reduced quinol is subsequently reoxidized by an adjacent membrane-bound cytochhrome bc_1 oxidoreductase that pumps protons across the membrane from the cytoplasmic side to the periplasmic side. The electrons finally return to the RC through cytochrome c_2, which is located on the periplasm of the ICM. The proton gradient across the membrane drives ATP synthesis via an ATPase in the membrane.

12.3 ORGANIZATION OF GENES CODING FOR LH, RC, AND PIGMENT BIOSYNTHETIC ENZYMES AND CYTOCHROMES

The genetic control of photosynthesis and photosynthetic membrane formation has been an important issue in the last few years. Recombinant DNA technology has made possible the isolation of most of the genes associated with the photosynthetic apparatus of the photosynthetic bacteria. The isolated genes have been subjected to a detailed analysis of nucleotide sequence and structure. Therefore, a great deal of information about the structure and function of the photosynthetic apparatus both in *R. capsulatus* and in *R. sphaeroides* has been obtained.

The study of photosynthetic genes in *R. capsulatus* is aided by an indigenous gene-transfer agent (Yen and Marrs 1976) and isolation of a conjugative R-prime plasmid, pRPS404, containing *R. capsulatus* 46-kb DNA, which complements most mutations affecting the differentiation of the photosynthetic apparatus (Marrs 1981). A procedure for the transposon mutagenesis of pRPS404 in *Escherichia coli* that allows for mapping of the positions of the transposon insertions was developed in our laboratory (Youvan et al. 1982). In order to improve the stability of the mutants generated in this way, a new transposon was created that had the transposition characteristics of Tn5 and the antibiotic resistant characteristics of Tn7 (Zsebo et al. 1984). By screening both the earlier mutants generated by Marrs's laboratory and the transposon mutants generated in this laboratory for enhanced near-infrared fluorescence indicating a defect in the utilization of absorbed photons, two restriction fragments from the R-prime plasmid pRPS404 were found to complement all enhanced-fluorescence mutants defective in LH-I and RC genes. The complete nucleotide sequence (8,867 bp) and a deduced polypeptide sequence for these two restriction fragments showed that they encoded 11 proteins from the photosynthetic gene cluster of *R. capsulatus* (Youvan et al. 1984b). Four structural genes for LH-I β (*puf*B), LH-I α (*puf*A), RC-L (*puf*L), RC-M (*puf*M), and two putative genes, C2397 and C2814, were found in the *Bam*HI-C-*Eco*RI-B restriction fragment, whereas four putative genes—F108, F460, F1025, and F1696—and one structural gene for RC-H (*puh*A) were found in the *Bam*HI-F fragment (Figure 12–2). The *puf*B, *pufA*, *puf*L, *puf*M, and C2397 genes are arranged contiguously and comprise a single operon: Anoxygenically induced mRNA hybridizes to this region (Clark et al. 1984; Belasco et al. 1985) and a specific oligonucleotide (17 mer) complementary to the C2397 gene hybridizes to a 2.6-kb long transcript from the *puf* operon (Zhu and Hearst 1986). From the deduced amino acid sequences of the RC subunits it was concluded that both the L and the M subunits are very hydrophobic proteins with 282 and 307 amino acids, respectively. As mentioned earlier, hydropathy plots suggest that the L and M subunits are transmembrane proteins that may cross the membrane five times. On the other hand, the H subunit is a hydrophilic polypeptide (254 amino acids long) with a very hydrophobic amino terminal

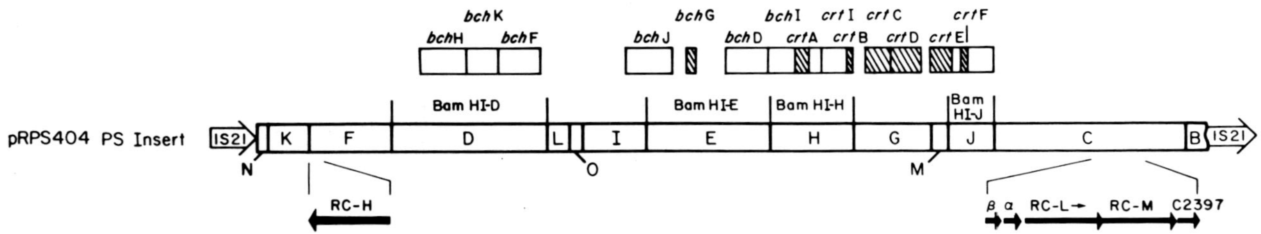

FIGURE 12–2 Photosynthetic gene cluster map. *Bam*HI restriction sites are shown within a 46-kb photosynthetic gene cluster from *R. capsulatus* carried by the R′ plasmid pRPS404. The genes coding for LH-I (β and α) and RC-L and RC-M subunits and an open reading frame C2397 are located in the *Bam*HI-C fragment, whereas the gene coding for the RC-H subunit resides in the *Bam*HI-F fragment. The arrows indicate the direction of transcription. The *Bam*HI fragments between these two genetic loci contain a number of *bch* genes (in *Bam*HI-D and -E) and *crt* genes (in *Bam*HI-H, -G, -M, and -J). The shaded areas indicate the genes determined by cluster point mutations conferring the same phenotype. The blank areas represent genes determined by transposon mutagenesis and complementation. From Y.S. Zhu, D.N. Cook. F. Leach, et al. (1986) *J. Bacteriol.* 168, 1180–1188. With permission.

of 30 amino acids. All five structural and six putative genes possess a Shine-Dalgarno sequence (A/C)GGAG(A/G) N_{3-10} ATG that is complementary to the 3′ terminal of *R. capsulatus* 16S rRNA. However, no *E. coli*-like consensus promoters were found over the entire 9 kb of sequenced DNA. Three potential hairpins have been found between the genes for LH-1 α and RC-L, after the gene for C2397, and between the genes for RC-H and F3981.

The *puf* and the *puh* operons are separated by 35 kb and are transcribed in the opposite direction. A large number of genes (at least 17) associated with Bchl and Crt biosynthetic enzymes are located between the *puf* and the *puh* operons (see Figure 12–2). Studies of the structure and the regulation of genes for pigment biosynthesis are in progress in our laboratory.

The genes coding for another LH complex and the β (*puc*B) and α (*puc*A) subunits of LH-II were isolated using a deoxyoligonucleotide probe and sequenced (Youvan and Ismail 1985). This completed the sequencing of all seven structural genes coding for the LH and RC polypeptides that bind the pigments and cofactors participating in the primary light reaction of photosynthesis. The *puc*A and the *puc*B genes are at least 11 kb away from *puf*L and *puf*M, or at least 7 kb away from *puh*A. Since the organization of the β and α structural genes for both LH-I and LH-II is identical, and very high homologies exist between the two β-polypeptides and between the two α-polypeptides, it has been suggested that both complexes arose by gene duplication from a single ancestral LH gene. A highly conserved sequence, Ala-x-x-x-His, is believed to bind Bchl (Youvan and Ismail 1985).

Hearst and Sauer (1984a) first reported at the VIth International Congress on Photosynthesis in 1983 that the RC subunits L and M of *R. capsulatus* share amino acid sequence homology with the Q_B protein of spinach. Particularly, a strong homology has been found between the L subunit and the Q_B protein (residues 196–221). It was proposed that this sequence of amino acids is involved in quinone binding and function (Youvan et al. 1984a and 1984b; Hearst and Sauer 1984a and 1984b).

The *puf*M (Williams et al. 1983), *puf*L (Williams et al. 1984), *puh*A (Donohue et al. 1986a), *puf*B, *puf*A (Kiley et al. 1987a), and *puc*B and *puc*A (Kiley et al. 1987b) in *R. sphaeroides* have also been isolated and sequenced. The same arrangement of genes coding for LH-I (β, α) and RC (L, M) proteins and a striking homology of deduced amino acid sequences of the LH and RC polypeptides were found between *R. sphaeroides* and *R. capsulatus*. The considerable sequence homology between the L and M subunits indicates that their genes arose by duplication of a single gene. The RC-L and RC-M polypeptides of *R. sphaeroides* also share homology with the 32-kd thylakoid membrane protein in spinach (Williams et al. 1983 and 1984) and are proposed to span the membrane five times, like the RC-L and RC-M polypeptides in *R. capsulatus*. One of the histidine residues (His-219) in a hydrophobic segment of the RC-M subunit is a candidate for a

Bchl or a quinone binding site (Williams et al. 1983). The genes *puf*B and *puf*A are contiguous with *puf*L and *puf*M. A region about 170 bp upstream from the 5′ end of the structural gene is sufficient for the in vitro expression of LH-I β and α polypeptides (Kiley et al. 1987a). In addition, three regions of dyad symmetry upstream from *puf*B have been found. These inverted repeats may serve as recognition sites for DNA binding proteins in regulating expression of this operon.

Unlike most of the other photosynthetic bacteria, *R. rubrum* contains a single holochrome antenna complex, designated B880, that also consists of β and α subunits. Genes for B880 holochrome have recently been isolated and sequenced (Bérard et al. 1986). The genes for the β and α polypeptides are contiguous and show high homologies to *puf*B and *puf*A in *R. capsulatus*. They are presumed to be transcribed as a single operon. An additional C-terminal tail of 10 and 13 amino acid residues on the α and β subunits, respectively, may serve as a means of orienting these polypeptides in the membrane.

The *puf*L and *puf*M (Michel et al. 1986) and the *puh*A (Michel et al. 1985) genes in *R. viridis* have been cloned using oligonucleotide hybridization and expression vectors, respectively, and sequenced. The arrangement of *puf*L and *puf*M is the same as in *R. capsulatus* and *R. sphaeroides*. However, due to a deletion of five bases at the end of *puf*L in *R. viridis*, *puf*L and *puf*M do not overlap. Therefore, overlap of these two genes is not necessary for 1:1 stoichiometry of the L and M polypeptides (Williams et al. 1984). Another structural feature different from *R. capsulatus* and *R. sphaeroides* is that the gene encoding the RC cytochrome c_2, which is tightly bound to the RC and unique in *R. viridis*, is located at the end of *puf*M and overlaps *puf*M by one base pair. In *R. capsulatus puf*M is followed by an unknown gene C2397 (Youvan et al. 1984b). Furthermore, the M subunit from *R. viridis* has 17 more amino acids than *R. capsulatus* and 18 more amino acids than *R. sphaeroides* at its carboxy-terminal. This extension binds the cytochrome subunit (Deisenhofer et al. 1985a and 1985b).

Cytochrome c_2 is an electron carrier that is common to the photosynthetic and respiratory machinery of *R. capsulatus* (Bartsch 1978). The structural gene for cytochrome c_2 (*cyc*A) of *R. capsulatus* has been cloned and sequenced (Daldal et al. 1986). Comparison of the known amino acid sequence of the purified cytochrome c_2 and the nucleotide sequence corresponding to the N-terminal part of the *cyc*A product indicates the presence of a putative 21-amino-acid signal sequence. Cytochrome c_2 may be synthesized as a precursor that is processed during its secretion to the periplasm. Studies of the insertion and insertion-deletion mutations have led Daldal et al. to conclude that cytochrome c_2 is not essential for photosynthetic growth. Donohue et al. (1986b) have analyzed *cyc*A in *R. sphaeroides* and have also found a 21-amino-acid signal sequence. Small (740 nucleotide) and large (920 nucleotide) *cyc*A-specific mRNA species are found in various growth conditions. In contrast to the result of *R. capsulatus* (Daldal et al. 1986), using cartridge insertion coupled with partial deletions of the *cyc*A gene, they have found that cytochrome c_2 obligatory for photosynthetic growth of *R. sphaeroides*.

Recently the genes encoding the three main subunits of the cytochrome bc_1 complex in a strain of *R. sphaeroides* have been cloned (Gabellini et al. 1985). This strain was identified later by Davidson and Daldal (1987) as most likely a strain of *R. capsulatus*. The genes are arranged in the order (5′)*fbc*F, *fbc*B, and *fbc*C, encoding the FeS protein, cytochrome *b*, and cytochrome c_1, respectively. The *fbc* genes are coordinately transcribed in one polycistronic mRNA and constitutively expressed under both photosynthetic and respiratory growth conditions. The nucleotide sequence of 3,874 bp of cloned chromosomal DNA, including the three structural genes *fbc*F, *fbc*B, and *fbc*C, has been determined (Gabellini and Sebald 1986). The *fbc*F gene codes for 191 amino acids of the FeS protein, starting from a GTG codon that is preceded by an adenine and guanine–rich sequence complementary to the 3′ end of the 16S ribosomal subunit of *R. sphaeroides*. The *fbc*F gene is followed by the *fbc*B gene with a 12 nucleotide space that includes a Shine-Dalgarno–like sequence. The *fbc*B gene codes for 437 residues of the cytochrome *b* subunit. The *fbc*C gene starts 20 bp after the *fbc*B gene and is preceded by a Shine-Dalgarno–like sequence. The *fbc*C gene codes for a 280-residue preapocytochrome c_1 that has a transient leader sequence (21 amino acids). S-1 nuclease protection analysis indicates that the transcription of the *fbc* operon starts approximately 240 bp upstream from the start codon of the *fbc*F gene and terminates 120 bp downstream from the stop codon of the *fbc*C gene. The deduced amino acid sequences, homologies with other similar proteins, and hydropathy plots indicate the following structural features: (1) FeS protein, carrying a high-potential 2Fe-2S cluster, has a large hydrophilic domain and is exposed to the aqueous phase; (2) cytochrome *b* has nine or ten possible membrane-spanning regions, and four conserved histidine residues are postulated as ligands of the two heme groups; (3) cytochrome c_1 is synthesized as a precursor form including a transient leader sequence of 21 amino acids. The typical sequence that binds the heme covalently is located near the N-terminal corresponding to Cys-55, Cys-88, and His-50.

Similar results of a detailed analysis of the *fbc* operon in *R. capsulatus* were recently reported by Davidson et al. (1987). They designated the *fbc* operon as the photosynthetic electron transport (*pet*) operon. A possible promoter region was found to be located upstream from the first *pet*A structural gene coding for Rieske FeS protein. It is concluded that unlike cytochrome c_2 (Daldal et al. 1986), a functional bc_1 complex is essential for photosynthetic growth.

rRNA from both *R. sphaeroides* (Marrs and Kaplan 1970) and *R. capsulatus* (Yu et al. 1982) consists of 23S, 16S, and 5S species. The 23S rRNA, however, is not stable during isolation and is cleaved into 16S and 14S rRNAs probably by a specific endonuclease (Yu et al. 1982; Klug et al. 1984; Zhu and Kaplan 1985). The rRNA genes of *R. capsulatus* have been cloned (Yu et al. 1982). At least seven rRNA operons are present in the *R. capsulatus* chromosome. The arrangement of 16S, 23S, and 5S rRNAs is the same as that in *E. coli*. However, the spacer (0.91 kb) between the 16S and the 23S rRNA genes is longer than that (0.44 kb) in *E. coli*.

Although ribulose-1,5-bisphosphate carboxylase (RuBPCase) is not a membrane protein, a great deal of attention has been paid to it because it is the most abundant protein in the cytoplasm and it plays a key role in carbon dioxide fixation in the Calvin cycle in photosynthetic bacteria, algae, and plants. It is of particular interest that two forms of RuBPCase are present in *R. sphaeroides* and *R. capsulatus* (Gibson and Tabita 1977a and 1977b). The form I RuBPCase resembles that found in plants and most other bacteria. It is a large oligomeric protein composed of eight large (52 kd) and eight small (11 kd) subunits (Gibson and Tabita 1977b). The form II enzyme is a hexamer of only large subunits (52 kd) and does not have a small subunit. Recently, the gene for the form II RuBPCase (*rbcL*) was isolated from *R. rubrum* (Somerville and Somerville 1984) and *R. sphaeroides* (Fornani and Kaplan 1983; Quivey and Tabita 1984; Muller et al. 1985). The nucleotide sequence of the *rbcL* gene from *R. rubrum* has been reported (Nargang et al. 1984). No consensus prokaryotic promoter sequence was found in the *rbcL* gene from *R. rubrum*. It is likely that the promoters for highly regulated genes such as those involved in carbon dioxide fixation and the photosynthetic apparatus are quite different from typical bacterial promoters (Quivey and Tabita 1984). A region of dyad symmetry 5′ to the coding region of the *R. rubrum* RuBPCase gene similar to the proposed attenuator regions was found, indicating that additional proteins may be involved in transcription (Nargang et al. 1984).

One of the consequences of the knowledge of structure and organization of nucleotide sequences of the photosynthetic genes in purple photosynthetic bacteria is the ability to mutate the genes that are of interest by site-directed mutagenesis and to examine the effect of the resulting mutation on the function of the photosynthetic apparatus (Youvan and Ismail 1985). Another consequence that will be described here in more detail is to facilitate the elucidation of structure and function of the photosynthetic apparatus in plants and algae as well as in purple bacteria. Proteins can be studied based on sequence homologies that may shed some light on functionally important structural features (Youvan et al. 1984a and 1984b; Hearst and Sauer 1984a and 1984b; Hearst 1986). Amino acid sequence comparison between photosynthetic membrane proteins of purple bacteria and those of blue-green algae and higher oxygen evolvers remains an extraordinarily valuable and insightful exercise. A simple generalization is that there is strong sequence homology between proteins of identical function in all blue-green algae and higher plants. The purple bacteria are very different but have small regions of homology with plant proteins. Such homologous regions provide special insight regarding the essential functional sites in the corresponding proteins.

A weak, but significant, sequence homology exists between the L and M subunits of the RC of *R. capsulatus* and the 32-kd thylakoid membrane protein (also known as D1, Q_B, herbicide-binding protein) of the RC of photosystem II in higher plants and algae (Zurawski et al. 1982; Youvan et al. 1984b; Hearst and Sauer 1984b; Hearst 1986). Hearst and Sauer (1984b) proposed a model suggesting that a highly conserved sequence of amino acids between the L and M subunits of the RC of *R. capsulatus* and 32-kd thylakoid membrane protein of spinach and tobacco is involved in quinone binding and function. Figure 12–3A reveals that a sequence PFHMLG – – – –F– – – – – –AMHG–LV–S is common to both the L subunit of *R. capsulatus* and the 32-kd (D1) proteins of spinach, *Nicotiana, Chlamydomonas,* and *Euglena* and starts at the one hundred seventy-second amino acid from the N-terminal of the L subunit (282 amino acids) and at the one hundred ninety-sixth amino acid from the N-terminal of the 32-kd (D1) protein (353 amino acids). Later another photosystem II RC protein, the D2 protein, was also found to possess sequence homologies to the D1 protein and the L and M subunits (Rochaix et al. 1984; Alt et al. 1984; Holschuhk et al. 1984). A weaker homology between the *R. capsulatus* M subunit (from the one hundred ninety-eighth amino acid of a 306-amino-acid molecule) and the pea and *Chlamydomonas* D2 protein also reveals a conserved sequence PFHM– – – –G–L–A–LLC– – –GATV (Figure 12–3B). The data strongly indicate that the L and M subunits and the 32-kd protein arose from a common precursor. The fact that selection pressure preserved such a precise amino acid sequence over three billion years (Olson 1981) suggests that this region is functionally very important. Subsequently, the crystal structure of the RC from *R. viridis* implicated the conserved histidine residues in forming the site for the non-heme Fe^{2+} (Deisenhofer et al. 1985a and 1985b). The discovery of sequence homologies between the L and D1 proteins and between the M and D2 proteins suggests that the D1 and D2 proteins are involved in photochemical activity, a hypothesis in conflict with the old concept that two other larger polypeptides of 47–51 kd and 43–45 kd are the core of the photosystem II.

This hypothesis is further supported by the finding that both the L subunits and D1 proteins bind the photo affinity–labeled herbicide azido atrazine (Pfister et al. 1981; DeVitry and Diner 1984). Through the use of a tritium-labeled quinone alkylating reagent, Worland et al. (1987) have demonstrated the affinity labeling of a 38-kd protein that they suggest may be D2 in spinach photosystem II complexes. In addition, the x-ray structure analysis has shown that both L and M subunits are needed to establish the primary electron donor, which is made up of two nearly parallel Bchl moleucles (the special pair) and the electron-accepting quinone-iron complex. All the evidence supports the proposal of Hearst (1986) and Michel

```
          10        20        30        40        50        60        70        80        90
MT.AILERRESESLWGRFCNWI.TSTENRLYIGWFGVLMIPTLLTATSVF.IIAFIAAPPVDIDGIREPVSGSLLYGNNIISGAIIPTSAAIG   SPINACH, NICOTIANA
MT.AILERRENSSLWARFCEWI.TSTENRLYIGWFGVIMIPCLLTATSVF.IIAFIAAPPVDIDGIREPVSGSLLYGNNIITGAVIPTSNAIG   CHLAMYDOMONAS
MISPVLKKYARPSLWYRFCAWV.ASKKNRLYVGWFGVLMIPTLLTAATVF.IIAFIAAPPVDIDGIREPVSGSLFYGNNIITGAVVPTSNAIG   EUGLENA
MALLSFERKYRVPGGTLIGGSLFDFWVGPFYVGFFGVTTI...FFATLGFLLILWGAAMQGTWNP.......................QLI   R. CAPSULATUS
          10        20        30        40        50        60

         100       110       120       130       140       150       160       170       180
LHFYPIWEAASVDEWLYNGGPYELIVLHFLLGVACYMGREWELSFRLGMRPWIAVAYSAPVAAATAVFLIYPIGQGSFSDGMPLGISGTF   SPINACH, NICOTIANA
LHFYPIWEAASLDEWLYNGGPYQLIVCHFLLGVYCYMGREWELSFRLGMRPWIAVAYSAPVAAASAVFLVYPIGQGSFSDGMPLGISGTF   CHLAMYDOMONAS
LHFYPIWEATSLDEWLYNGQPNQLIVCHFFIGICCYMGREWELSFRLGMRPWIAVAYSAPVAAASAVFIVYPLGQGSFSDGMPLGISGTF   EUGLENA
SIFPPPVENGLNVAALDKGGLWQVITVCATGAFCSWALREVEICRKLGIGFHIPVAFSMAIFAYLTLVVIRPMMMGSWGYAFPYGIWTHL   R. CAPSULATUS
   70        80        90       100       110       120       130       140       150

         190       200       210       220       230       240       250       260       270
NFMIVFQAEH.NILMHPFHMLGVAGVFGGSLFSAMHGSLVTSSLIRETTENESANEGYRFGQEEETYNIVAAHGYFGRLIFQYASFNNSRS   SPINACH, NICOTIANA
NFMIVFQAEH.NILMHPFHMLGVAGVFGGSLFSAMHGSLVTSSLIRETTENESANEGYRFGQEEETYNIVAAHGYFGRLIFQYASFNNSRS   CHLAMYDOMONAS
NFMIVFQAEH.NILMHPFHMLGVAGVFGGSLFSAMHGSLVTSSLLRETTENESINVGYKFGQEEETYNIIAAHAYFGRLIFQYASFNNSRS   EUGLENA
DWVSNTGYTYGNFHYNPFHMLGISLFFTTAWALAMHGALVLS.....AANPVKGKTMRTPDHEDT........YFRDLMGYSVGTLGIHR   R. CAPSULATUS
   160       170       180       190            200       210                220       230

         280       290       300       310       320       330       340       350
LHFFLAAWPVVGIWFTALGISTMAFNLNGFNFNQSVVDSQGRVINTWADIINRANLGMEVMHERNAHNFPLDLAAIEAPSTNG   SPINACH, NICOTIANA
LHFFLAAWPVIGIWFTALGLSTMAFNLNGFNFNQSVVDSQDRVLNTWADIINRANLGMEVMHERNAHNFPLDLASTNSSSNN.   CHLAMYDOMONAS
LHFFLAVWPVVGIWFTALGVSTMAFNLNGFNFNQSVIDSQGRVINTWADIINRANLGMEVMHERNAHNFPLDLA.........   EUGLENA
LGLLLALNAVFWSACCMLVSGTIYFDLWSDWWYWWV......NMPFWAD.........................MAGGING   R. CAPSULATUS
   240       250       260            270                                280
```

SPINACH, NICOTIANA	(353)	: R. CAPSULATUS	76 (21.5%)	: SPINACH	353 (100%)
CHLAMYDOMONAS, psbA	(352)	: R. CAPSULATUS	74 (21.0%)	: SPINACH	325 (92.3%)
EUGLENA	(345)	: R. CAPSULATUS	62 (18.0%)	: SPINACH	305 (86.4%)

```
MTIALGKFTKDQNDLFDIMDDWLRRDRFVFVGWSGLLLFPCAYFAVGGWFTGTTFVTSWYTHGLASSYLEGCNFLTAAVSTP      PEA
MTIAIGTYQEKRT.WFDDADDWLRQDRFVFVGWSGLLLFPCAYFALGGWLTGTTFVTSWYTHGLATSYLEGCNFLTAAVSTP      CHLAMYDOMONAS
MAEYQNFFNQVQVAGAPEMGLKEDVDTFERTPAGMFNILGWMGNAQIGPIYLGIAGTVSLAFGAAWFFTIGVWYWYQAGFDP     R. CAPSULATUS

 ANSLAHSLLLLWGPEAQGDLTRWCQLGGLWTFVALHGAFGLIGFMLRQFELARSVQLRPYNAIAFSGPIAVFVSVFLIYPLGQSGWFFAPSF      PEA
 ANSMAHSLLFVWGPEAQGDFTRWCQLGGLWAFVALHGAFGLIGFMLRQFEIARSVNLRPYNAIAFSAPIAVFVSVFLIYPLGQSGWFFAPSF      CHLAMYDOMONAS
FIFMRDLFFFSLEPPPAEYGLAIAPLKQGGVWQIASLFMAISVIAWWVRVYTRADQLGMGKHMAWAFLSAIWLWSVLGFWRPILMGSWSVAPPY    R. CAPSULATUS

GVAAIFRFILFFQGFH.NWTLNPFHMMGVAGVLGAALLCAIHGATVENTLFEDGDGANTFRAFNPTQAE.ETYSMVTANRFWSQIFG      PEA
GVAAIFRFILFFQGFH.NWTLNPFHMMGVAGVLGAALLCAIHGATVENTLFEDGDGANTFRAFNPTQAE.ETYSMVTANRFWSQIFG      CHLAMYDOMONAS
GIFSHLDWTNQFSLDHGNLFYNPFHGLSIAALYGSALLFAMHGATILAVTRFGGERELEQIVDRGTASERAALFWRWTMGFNATMEG     R. CAPSULATUS

VAFSNKRWLHFFMLFVPVTGLWMSALGVVGLALNLRAYDFVSQEIRAAEDPEFETFYTKNILLNEGIRAWMATQDQPHENLIFPEEVLPRGNAL      PEA
VAFSNKRWLHFFMLLVPVTGLWMSAIGVVGLALNLRAYDFVSQEIRAAEDPEFETFYTKNILLNEGIRAWMAAQDQPHERLVFPEEVLPRGNAL      CHLAMYDOMONAS
....IHRWAIWMAVMVTLT.......GGIGILLSGTVVDNWYVWAQVHGYAPVTP                                          R. CAPSULATUS

D2 PEA      (353)     MATCHES    M:D2 PEA 64   (20.8%)     M:D2 60   (19.5%)     D2:D2 PEA 287   (84.7%)
D2   psbD   (339)
M SUBUNIT   (307)
```

FIGURE 12–3 (A) An amino acid sequence alignment of the RC-L subunit of *R. capsulatus* and the D1 proteins of spinach, *Nicotiana, Chlamydomonas,* and *Euglena* that are encoded by the *psb*A gene of the chloroplast genome. The italic bold letters that appear in all of the four lines represent the homologous sequences among these organisms. The brackets indicate the potential regions involved in oxygen evolution. There are 76 matches and 21.5% homology between RC-L and the spinach, *Nicotiana* D1; 74 matches and 21% homology between RC-L and the *Chlamydomonas* D1; and 62 matches and 18% homology between RC-L and the *Euglena* D1. (B) An amino acid sequence alignment of the RC-M subunit of *R. capsulatus* and the pea and *Chlamydomonas* D2 proteins that are encoded by the *psb*D gene of their respective chloroplast genome. The italic bold letters that appear in all of the three lines illustrate the homologous sequences. There are 64 matches and 20.8% homology between RC-M and the pea D2, and 60 matches and 19.5% homology between RC-M and the *Chlamydomonas* D2.

and Deisenhofer (1986) that the D1 and D2 proteins form the core of the photosystem II RC and the L and M subunits form the RC core of purple photosynthetic bacteria in similar ways. The L subunit is assumed to be the equivalent of D1, and M to be the equivalent of D2.

Figure 12–3A provides a sequence alignment that clearly accentuates the additional sequences in the higher plants that are not found in the corresponding protein in the purple bacteria. The region of particular interest to us at the present time is the carboxy terminal of the D1 or herbicide-binding protein. This region is highly conserved among the oxygen evolvers and might be associated with the oxygen-evolving manganese center. Evidence for such a hypothesis comes from several sources. First, Metz et al. (1987) have identified a D1 mutant that is incapable of oxygen evolution. Second, D1 in higher plants is rapidly turned over because of photolability, while the L subunit in purple bacteria is quite stable (Kyle et al. 1984; Ohad et al. 1984; Bishop 1987). This photolability might well be the consequence of the presence of the oxygen-evolving site. Finally, the amino acid composition of this carboxy terminal region is unusual, containing three acidic residues, four asparagines, two arginines, two methionines, one phenylalanine, and one proline in the twenty-one residues from position 322 to position 342 on the alignment in Figure 12–3A, as indicated by the bracket. This suggests a function for this region that must operate on the periplasmic side of the photosynthetic membrane, where oxygen evolution is known to occur, if the analogy with the purple bacteria RC structure of Michel and Deisenhofer (1986) is valid.

Other regions of this protein that are believed to be in the periplasm and that may also be involved with the site of oxygen evolution include amino acids 56 to 76 and 92 to 108, the aspartic acid at position 170, and the amino acids from 308 to 321, as marked with brackets in Figure 12–3A.

12.4 REGULATION OF EXPRESSION OF THE GENES CODING FOR LH, RC, BCHL, AND CRT BIOSYNTHESIS

The biosynthesis of the photosynthetic apparatus requires a coordinated expression of the genes for LH and RC pigment complexes in response to environmental stimuli, that is, oxygen tension and light intensity. In addition, these components must be correctly assembled into the membrane. In this section we focus on the aspects of internal and external regulation of the photosynthetic genes.

12.4.1 Coordination and Interaction of the Genes for LH, RC, Bchl, and Crt Biosynthesis

The expression of the photosynthetic genes for LH and RC proteins and the accumulation of pigments are tightly coupled and interdependent. Mutation within these genes causes a pleiotropic effect. It has been shown that a mutation in any of the genes for LH-I, RC-H, RC-L, Bchl, or Crt biosynthetic enzymes results in different degrees of reduction in the amount of mRNAs from the rest of the genes (O'Brien et al., unpublished observations; Zhu and Hearst 1987). O'Brien et al. have used a series of transposon mutants created in this laboratory (Zsebo and Hearst 1984) that carry their defect within or near the structural genes for the RC to study the expression of these genes during photosynthetic development. Insertions within ORF F1696, which is located upstream from the gene for RC-H, caused severe defects in the photosynthetic apparatus. These defects included reduced levels of mRNA for LH-I, RC-L, RC-M, and RC-H, along with decreased levels of Bchl mRNA and diminished absorption by LH-II. Since we have been unable to detect transcription from the putative F1696 gene (Zhu and Hearst 1986), it seems logical to assume that these insertions blocked the regulatory sequences of the gene for RC-H. Inactivation of the gene for RC-H by an internal mutation resulted in a total loss of LH-II absorption and a dramatic decrease in the amount of the RC-L and RC-M proteins as determined by Western blots. Mutations within the gene for the RC-H protein seemed to have a more dramatic effect than those within the gene for the RC-L protein. Such a difference may have resulted from more stringent requirements for the RC-H protein than for the RC-L and RC-M proteins in photosynthetic development (O'Brien et al., unpublished observations). On the other hand, mutations in *bch* (*bch*C) or *crt* (*crt*I) genes substantially reduced levels of mRNA for LH-II, LH-I, RC-L, RC-M, and RC-H. A close link was found to exist between the genes for LH-II proteins and Crt biosynthesis (Zhu and Hearst 1987). We have also performed experiments to determine whether the decreased levels of mRNA for LH and RC proteins in these mutants are due to reduced transcription or to increased degradation of mRNA. A comparison of the decay rate of mRNAs for LH-I, LH-II, RC-L, RC-M, and RC-H after treatment with proflavin as an inhibitor of transcription reveals that the mRNAs in the mutants are at least as stable as those in wild-type, although they are present at very low levels. Therefore, the reduction in mRNA levels in these mutants is due to transcriptional control. We suggest that proper and accurate assembly of the photosynthetic components in the ICM is necessary for transcription. Absence of one of these components (either proteins or pigments) may turn off the transcription of the corresponding genes.

12.4.2 Environmental Stimuli

12.4.2.1 Oxygen. In purple nonsulfur bacteria a decrease in oxygen tension is the signal for a switch of metabolism from aerobic mode to photosynthetic mode. Upon the removal of oxygen, the cells develop an extensive ICM, mainly consisting of LH-I, LH-II, and RC proteins. This process is accompanied by an immediate biosynthesis of Bchl (Drews and Oelze 1981; Chory et al. 1984). The signal chain, however, between variation of oxygen tension in the medium and regulation of gene expression is unknown.

Recent studies (Clark et al. 1984) using Southern hybridization have shown that the transcripts from the genes for LH-I (β, α), RC-L and RC-M in *R. capsulatus* are greatly enhanced (fortyfold) in cells grown at low oxygen tension, whereas the transcripts from the genes coding for RC-H and Bchl biosynthetic enzymes respond to a lesser extent, and transcripts encoding several carotenoid biosynthetic enzymes seem unresponsive to changes in oxygen concentration. Lac fusion experiments support the observation that the rate of Bchl synthesis is regulated transcriptionally to a small degree (Biel and Marrs 1983) and that the *crt* genes are not directly regulated by oxygen tension (Biel and Marrs 1985). The steady-state concentration of mRNA specific for LH-II and the level of rRNA also increase after lowering oxygen tension (Klug et al. 1984). Klug et al. (1985) have demonstrated that a maximum in the concentration of mRNA for LH-I and RC proteins occurs 30 minutes after induction under low oxygen concentration, while the maximal expression of mRNA for LH-II proteins appears to lag the LH-I/RC mRNA maximum by about 25 minutes. This differential expression of the genes for LH-I and LH-II agrees with the sequential accumulation of LH-I and LH-II complexes in the membrane of *R. capsulatus* under decreasing oxygen tension (Schumacher and Drews 1978). These observations support the transcriptional control of the genes for LH and RC proteins. None of these studies, however, showed discrete mRNA species until Northern hybridization and S-1 nuclease analysis, using highly defined probes, were applied to *R. capsulatus* (Belasco et al. 1985) and *R. sphaeroides* (Zhu and Kaplan 1985).

Belasco et al. (1985) have observed two anaerobically induced transcripts of 0.5 kb and 2.7 kb from the *puf* operon of *R. capsulatus*. the 2.7-kb transcript coding for LH-I (β, α), RC-L, and RC-M, is less stable than the 0.5-kb transcript coding for LH-I (β, α) only. Similar observations were made in *R. sphaeroides* (Zhu and Kaplan 1985). In addition, we have found that these two mRNAs (0.5 kb and 2.6 kb) respond differently to oxygen concentration and light intensity. The 2.6-kb mRNA is more sensitive to oxygen, whereas the 0.5-kb mRNA is more responsive to light intensity. The response of the 0.5-kb transcript to light intensity can be functionally rationalized, since it acts as a template for LH-I antenna proteins. None of these mRNAs are significantly affected by the redox potential of the organic acids used in the medium. In contrast to the membrane proteins, the cellular level of mRNA specific for the large subunit of RuBPCase (form II), a cytoplasmic protein and a key enzyme in carbon dioxide fixation, is markedly affected by both oxygen concentration and the redox potential of the substrate. We have found that the expression of the *rbcL* gene in *R. sphaeroides* is completely turned off by high oxygen tension under conditions where the mRNAs for LH and RC are still observable.

The regulation by oxygen of the mRNAs for LH-II (β, α), LH-I (β, α), and RC (-L, -M, -H) and for the Bchl and Crt biosynthetic enzymes in *R.*

capsulatus has recently been investigated in our laboratory using Northern and dot hybridization (Zhu and Hearst 1986 and 1987a; Zhu et al. 1986b) (Figure 12–4). We have demonstrated that while the small transcript (0.5 kb) codes only for LH-I (β, α), the large transcript (2.6 kb) codes for LH-I (β, α), RC-L, and RC-M, as well as an ORF C2397. These five genes thus comprise a single operon. The C2397 gene is contranscribed and coregulated with *puf*L and *puf*M in response to oxygen and light. It is 249 bp long and contains the information for a small hydrophobic protein. The function of the C2397 gene product is as yet unknown. However, it may be functionally associated with the LH-I and RC genes, based on their tight cotranscription and coregulation as one operon. We have found that the gene for RC-H, *puh*A, has two transcripts of 1.2 kb and 1.4 kb, that are initiated within the ORF F1696 and whose ratio is light intensity dependent. The *puh*A gene is coordinately regulated with *puf*L and *puf*M in response to oxygen (Zhu and Hearst 1986), an observation different from that of Clark et al. (1984). It is important to note that the proteins RC-L, RC-M, and RC-H are present in a virtual stoichiometry of 1 : 1 : 1 in the membrane. Recently, Donohue et al. (1986) made the similar observation in *R. sphaeroides* that the *puh*A gene produced 1.13-kb and 1.4-kb transcripts. The mRNA specifying the LH-II polypeptides is 0.5 kb long and is more abundant and more sensitive to oxygen change, indicating that the genes for the LH-II and LH-I/RC are independently regulated (Zhu and Hearst 1986). Furthermore, we have found that oxygen accelerates the degradation of the mRNA for LH-II (Zhu et al. 1986a) (Figure 12–5). In *R. sphaeroides* high oxygen tension (25%) also accelerates the degradation of the mRNAs for LH-I, RC-L, and RC-M (Zhu et al. 1986b). We hypothesize that oxygen represses the transcription of the photosynthetic genes and stimulates the degradation of their mRNAs. Such a combination of transcriptional and posttranscriptional regulations provides the cells with a mechanism to shift to a new metabolic mode quickly in response to oxygen. Recently, a similar result has been reported for the *nif* genes of *Klebsiella pneumoniae* (Collins et al. 1986).

One discrepancy between our observation and that of Clark et al. is that we have found only a threefold to fourfold or a sixfold increase in the mRNA levels for LH-I and RC, depending on light intensity, in contrast to the fortyfold increase they reported when comparing anaerobically and aerobically grown cells. A similar threefold to fourfold increase in mRNA levels has been observed in *R. sphaeroides* (Zhu and Kaplan 1985). One argument for such discrepancy is the method utilized to normalize the results to the total RNA. If the rRNA undergoes a dramatic change under different conditions, it would indeed affect the results. Klug et al. (1984) reported that the total rRNA per cell of *R. capsulatus* increased sevenfold in the transition period of 140 minutes from high to low oxygen tensions. However, it is difficult to understand how they could have gotten sevenfold difference in rRNA amounts when they applied equal amounts of total RNA of which the majority was rRNA for hybridization. As a matter of fact, an early study by Gray (1967) demonstrated that the total RNA, DNA, and

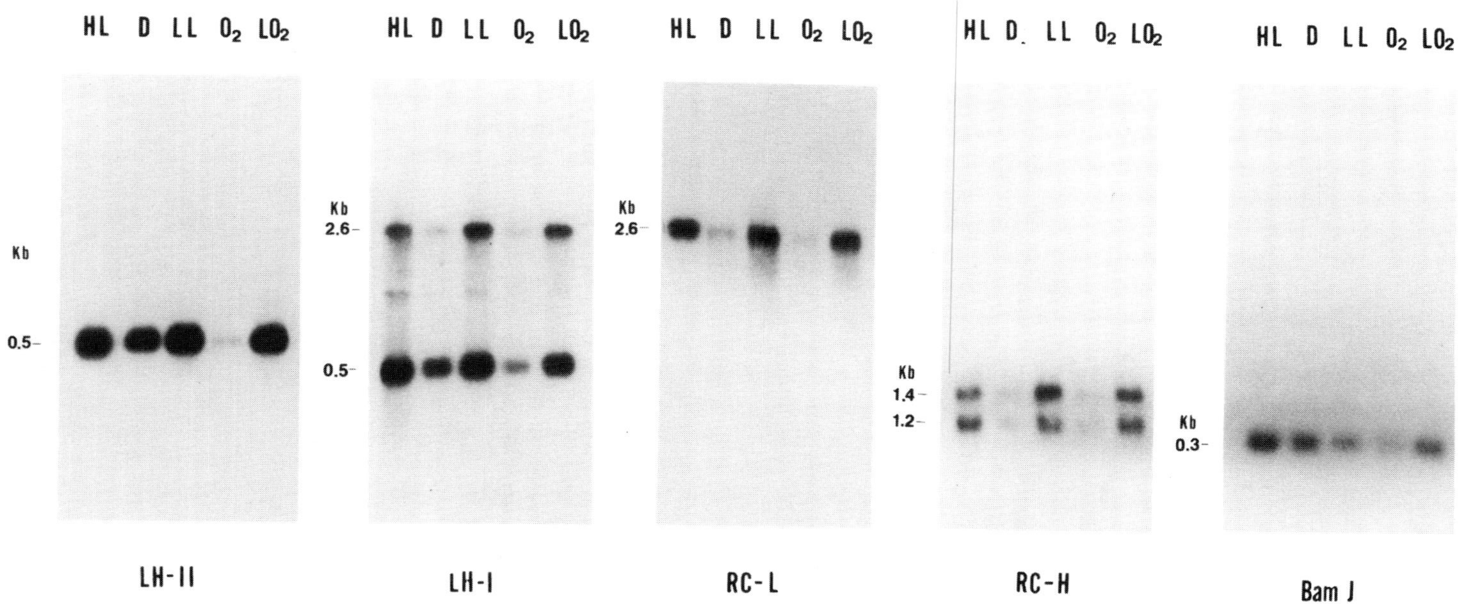

FIGURE 12–4 Northern hybridization of *R. capsulatus* RNA with cloned probes for LH-II, LH-I, RC-L, RC-H, and the *Bam*HI-J fragment of pRPS404 containing *crt*F, *crt*E, and an unidentified gene. Cells were grown under different conditions. HL, high light (30 W/m^2); D, a shift of HL-grown cells to the dark for one hour; LL, low light (5 W/m^2); O_2, aerobic conditions; LO_2, semiaerobic conditions. Adapted from Y.S. Zhu and J.E. Hearst (1987). With permission.

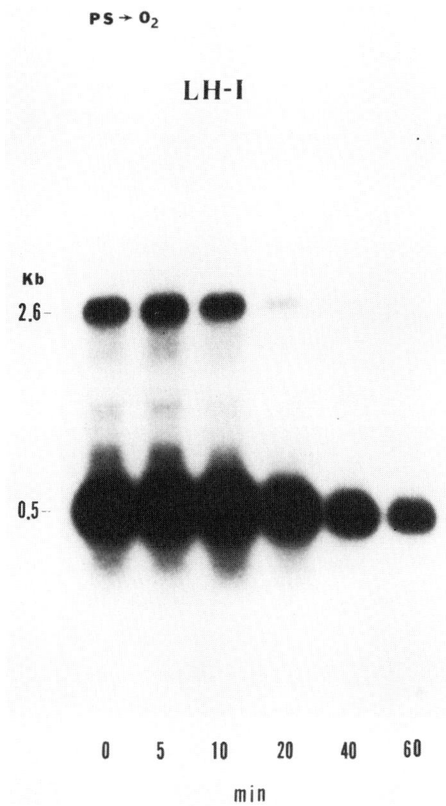

FIGURE 12–5 Northern hybridization of *R. capsulatus* RNA and DNA probes specific for the LH-I polypeptides. The mRNAs are degraded when the photosynthetic cells are exposed to oxygen at various times.

protein per cell of *R. sphaeroides* remain constant during the transition from high to low oxygen tension. Our investigation in *R. capsulatus* confirms Gray's result. The total RNA amounts for the photosynthetic cells and for the aerobic cells are $2.0 \times 10^{-2} \pm 0.3 \times 10^{-2}$ and $2.1 \times 10^{-2} \pm 0.4 \times 10^{-2}$ pg per cell, respectively, averaged over four experiments. These values are in reasonable agreement with the measurements by Gray (1967). Since the growth rate is only slightly different under our anaerobic and aerobic conditions, it would also be surprising to see a dramatic change in the rRNA amounts. The discrepancy of the results among different groups may be due to the experimental differences in methods, including the growth conditions, the accuracy of measurements, and the specificity of the probes used for hybridization. In our experiments we have used specific probes for LH and RC proteins. as well as Northern hybridization to show the discrete mRNA species, which provides direct information about specific mRNA size, quality, etc. Clark et al. used Southern hybridization and a large probe for the entire region, including DNA sequences other than the *puf* operon. We expect that the difference in methodology has caused many discrepancies between the results. We will not attempt to discuss all of them in this review.

The genes for pigment biosynthesis appear to be regulated by oxygen in somewhat different ways. While the levels of *bch* mRNA from the *bch* J, G, D, H, K, and F genes are lower under steady-state aerobic growth conditions than under anaerobic photosynthetic growth conditions, the extent of difference is less than that for the LH and RC mRNAs under those conditions. However, the levels of mRNA from *bch* genes decrease even more under steady-state low oxygen (semiaerobic) conditions than under aerobic conditions. When the *R. capsulatus* cells are shifted from anaerobic to high-oxygen conditions, the mRNAs from *bch* genes decrease rapidly during the first 20 minutes (Zhu et al. 1986a). These results were obtained using several *Bam*HI fragments containing more than one *bch* gene for studies, as were earlier results (Biel and Marrs 1983; Clark et al. 1984). The same results, however, were observed recently when we used a series of probes specific for *bch*C and *bch*A for hybridization (Zhu and Hearst, unpublished observations).

In contrast to the genes for LH, RC, and Bchl biosynthesis, *crt* genes respond to oxygen in an opposite fashion. We have found that the amount of mRNA from *crt* genes, relative to the amounts of mRNA for LH and RC proteins, actually increases during the shift from anaerobic to aerobic conditions (Zhu et al. 1986a) (Table 12–1). Although the *Bam*HI-H and the *Bam*HI-J fragments contain several *crt* genes, the *Bam*HI-M fragment is an internal region specific for *crt*E gene (Zhu and Hearst 1987). On the whole the results are consistent. While most photosynthetic genes are turned off by oxygen, the *crt*A gene responsible for the oxidation of spheroidene to spheroidenone, is activated by oxygen, because a transposon mutation of *crt*A gene abolished the twofold increase in mRNA level by oxygen. Furthermore, the activation of *crt*A gene by oxygen and light was also observed when a subfragment of the *Bam*HI-H fragment specific for *crt*A gene was used as a hybridization probe (Zhu and Hearst, unpublished observations). The reaction catalyzed by an oxygenase encoded by *crt*A may scavenge toxic singlet oxygen, which is formed in the photosynthetic apparatus. These results appear to disagree with Biel and Marrs' argument (1985) that the *crt* genes are not directly regulated by oxygen.

12.4.2.2 Light Intensity. Light intensity is another external signal that regulates the development of the photosynthetic apparatus. The photosyn-

TABLE 12–1 Effects of Light and Oxygen on the mRNA Levels of Genes for LH, RC, Bchl, and Crt Biosynthesis in *R. capsulatum*

	mRNA (%)[1]			
Genes	*HL*	*D*	*LL*	O_2
*puc*B, A	55	19	100	2
*puf*B, A	59	30	100	17
*puf*L	77	19	100	21
*puf*M	72	20	100	23
*puh*A	57	16	100	17
*bch*J, G, D (*Bam*E)	81	20	100	62
*bch*H, K, F (*Bam*D)	84	30	100	118
*crt*A, I, B, *bch*I (*Bam*H)	125	39	100	131
*crt*C, D (*Bam*G)	132	67	100	72
*crt*E (*Bam*M)	132	91	100	95

[1] Cells were grown under various conditions (HL, D, and LL were anaerobic conditions) and mRNA levels were measured by dot hybridization using various specific probes. The relative amounts of mRNA are expressed as percentages of the mRNA levels for low-light conditions (set as 100%). HL, high light; D, shift from HL to dark for one hour; LL, low light; O_2, aerobic growth.

thetic bacteria have developed mechanisms to adapt to different light intensities in order to maintain a more constant rate of photosynthetic activity. Cells grow competently in an environment of dim light, and high intensity light represses the formation of the photosynthetic apparatus. Earlier studies indicated that the cellular Bchl content of *R. sphaeroides* varied in inverse proportion to the incident light intensity (Cohen-Bazire and Kunisawa 1963). The molar ratio of LH-I to RC Bchl is fixed at about 20 to 25 (Agaard and Sistrom 1972). A shift of *R. sphaeroides* cells from high light (30 W/m^2) to low light (3 W/m^2) causes an immediate cessation of cell growth, Bchl accumulation, and protein synthesis. After a short period of adaptation, the cells resume their growth, Bchl synthesis commences at a new and higher rate, and two LH proteins and two additional high-molecular-weight photosynthetic membrane proteins increase threefold to fivefold (Chory and Kaplan 1983). The increased synthesis of LH antenna proteins as a result of reduction in light intensity results in an increase in the number and size of photosynthetic units per ICM and the number of ICMs per cell (Drews and Oelze 1981; Drews 1986). Chory and Kaplan (1982) have developed an in vitro transcription-translation system in which the levels of translatable mRNA are correlated with the increased levels of ICM proteins.

The expression of the genes coding for pigment-binding proteins in *R. sphaeroides* is also inversely correlated to the light intensity (Zhu and Kaplan 1985). When *R. sphaeroides* cells were shifted from the light to the dark, the levels of mRNA for LH-I, RC-L, and RC-M increased in the first five minutes after shifting because of derepression of transcription, although they eventually declined, presumably as a result of shortage of energy generation. Recently we have found that decreased light intensity results in increased levels of mRNAs for LH-I, LH-II, and RC-L, -M, and -H, as well as Bchl biosynthetic enzymes in *R. capsulatus* (Zhu and Hearst 1986). An interesting result is that the *crt* genes (A, I, B, C, D, E) in *R. capsulatus* again respond to light intensity in a way opposite to the other photosynthetic genes: increased light intensity leads to increased mRNA levels from these *crt* genes (Zhu and Hearst 1986) (see Table 12–1). Since photo-oxidative killing occurs only in high light and oxygen (Drews and Oelze 1981), and since it is well documented that Crt functions both to harvest light as an auxiliary pigment and to protect the photosynthetic apparatus in all photosynthetic organisms (Cogdell 1978; Halliwell 1984), we proposed that the transcriptional response of *crt* genes to high light is essential to the function of protecting cells from photo-oxidative damage. It is interesting to note that a very abundant transcript (0.4 kb) from *Bam*HI-J, whose location does not correspond to the genetically mapped *crt*F or *crt*E genes in this fragment as shown by Armstrong and Hearst (unpublished observations), also increases (threefold) in response to high light (Zhu and Hearst 1986).

In contrast to the pigment-binding protein genes, the *rbcL* gene is positively regulated by light intensity. The increase in light intensity results in an enhancement of both the levels of mRNA and enzymatic activity of RuBPCase in *R. sphaeroides* (Zhu and Kaplan 1985). Obviously, the gene for RuBPCase represents another type of regulation in *R. sphaeroides*. The mechanism by which cells adapt to light intensity is as yet unknown. From the results described here we suggest that the adaptation may take place at the transcriptional level. Another speculation is that variations of light intensity regulate the biosynthesis of the pigment complexes via the redox state of enzymes or regulatory proteins (Drews and Oelze 1981). Recently a sequence region of at least 300–700 bp 5′ of promoter function of the *puf* operon was found to be essential for light regulation (Kaplan, personal communication).

12.4.3 Regulation of the *puf* Operon

The LH-I complexes are found in the photosynthetic membrane in a fixed molar excess (approximately twelvefold) over the RC complex (Chory and Kaplan 1983). However, each of the two polypeptides within an LH-I complex (β and α) or within the RC (L and M) are present in a 1 : 1 stoichiometry. As these four genes plus C2397 are closely linked in a single operon both in *R. sphaeroides* and *R. capsulatus*, it would be of particular interest to elucidate how this stoichiometry and differential expression are achieved. Belasco et al. (1985) have reported that a 2.7-kb mRNA containing information for the LH-I(β, α) and RC (-L, -M) proteins and a small 0.5-kb mRNA containing information for LH-I (β, α) in *R. capsulatus* are derived from the same transcriptional initiation site. They came to the conclusion that the ratio of the small to large transcripts encoding the LH-I and RC proteins in *R. capsulatus* results predominantly from the selective degradation of a specific segment of the initial large transcript at the 3′ end of the molecule, resulting in a stabilized small transcript. The final ratio of small to large transcripts (10–20 : 1) would then reflect the relative 3′ processing activity.

In *R. sphaeroides* Zhu and Kaplan (1985) have also observed the presence of two transcripts in a similar 10–20 : 1 ratio. Additionally, it has been demonstrated that the half life of the large transcript is 9 minutes and that of the small transcript is 20 minutes (Zhu et al. 1986b). The difference in the stability of the two mRNAs indicates that some type of posttranscriptional control is involved in the differential expression of *puf*B and A and *puf*L and M, but it cannot be the major event responsible for the 10–20 : 1 ratio of small to large transcripts. Further, we have demonstrated that the two transcripts have different 5′ terminals as measured by both S-1 nuclease protection analysis and the primer extension technique (Zhu et al. 1986b). The small transcript has a 5′ end 104 bp upstream from the first ATG codon of *puf*B, whereas the large transcript has a 5′ end 75 bp upstream from the same start codon. A model is presented as follows: When the RNA polymerase binds to the sequence 104 bp upstream from the ATG start codon of *puf*B, the transcription terminates between *puf*A and *puf*L; whereas when the RNA polymerase binds to the sequence 75 bp upstream from the start codon of *puf*B, it reads through the first termination site between *puf*A and *puf*L, and then stops at the second termination site at the end of the *puf* operon. The mRNA secondary structure of the intercistronic region between the *puf*B and A and *puf*L and M is compatible with a rho factor–dependent termination signal (Zhu et al. 1986b). Based on the model described it is suggested that a combination of differential transcriptional initiation coupled with transcription termination led to the 10–20 : 1 ratio of small to large transcripts. These results seem to conflict with those of Belasco et al. (1985) who found that both the large and the small messengers initiate from the same initiation site about 35 bp downstream of the *Eco*RI site. This conclusion is based mainly on the S-1 mapping in which only one strong band was observed under low-oxygen conditions, although another band corresponding to an initiation site 70 bp downstream of the *Eco*RI site was noticed. Since this minor band accounted for only 1–2% of the *puf* mRNA 5′ ends, it was neglected (Belasco et al. 1985). By carefully maintaining the growth under strictly anaerobic photosynthetic conditions instead of low-oxygen conditions, which may cause the degradation of the labile large mRNA, we have been able to obtain similar results in *R. capsulatus* and in *R. spharoides* using the S-1 mapping technique. Two bands were detected with a correct ratio of 10–20 : 1 and an approximately 35-bp space between their 5′ ends (Zhu and Hearst, unpublished observations). We suggest that the transcripts of the *puf* operon in *R. capsulatus* also have two 5′ ends.

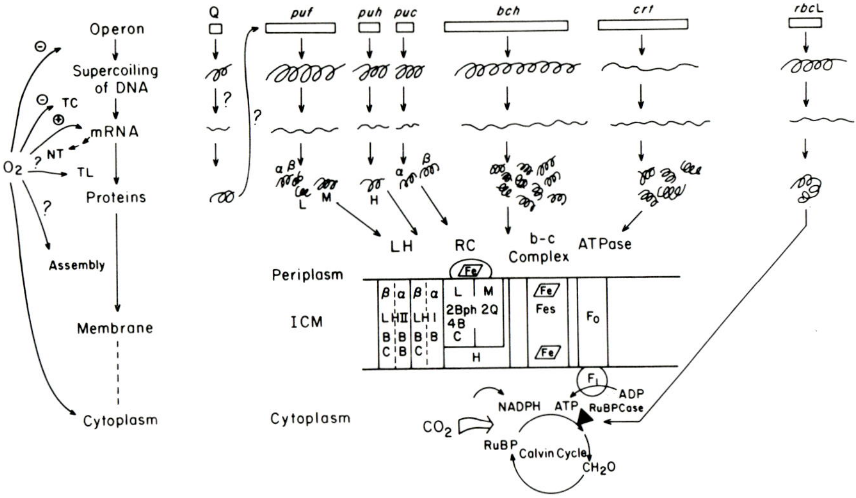

FIGURE 12–6 Schematic diagrams illustrating transcriptional (TC), posttranscriptional, and posttranslational (TL) regulation of photosynthetic genes in *R. capsulatus*. In development of the photosynthetic apparatus photosynthetic genes are expressed coordinately and differentially in response to environmental signals such as oxygen and light intensity. Current information indicates three levels of regulation: transcriptional regulation of mRNA synthesis, posttranscriptional processing of mRNA, and posttranslational assembly and insertion of pigment-protein complexes. The photosynthetic apparatus in ICM is comprised of LH complexes (LH-I, LH-II), RC polypeptides (L, M, H), cyctochrome bc_1 complexes, and ATPase. The polypeptides LH-I (β, α); RC subunits L, M, and H, and LH-II (β, α) are encoded by the *puf, puh* and *puc* operons, respectively. Bchl (B) binding to LH and RC is the pigment participating in light harvesting (in LH) and photochemical reaction (in RC). A large number of enzymes involved in Bchl biosynthesis are encoded by various *bch* genes. Pigments Crt (C) in LH and RC play a role both in light harvesting and in protection of the photosynthetic apparatus from photooxidation. The enzymes catalyzing Crt biosynthesis are encoded by various *crt* genes arranged as a cluster (see Figure 12–2). *rbcL* is the gene coding for the large subunit of form II RuBPCase, which is a cytoplasmic enzyme catalyzing carbon dioxide fixation in the Calvin cycle driven by reducing power NADPH and ATP derived from the photochemical reaction and coupled with ATPase in the ICM (see Figure 12–1). Oxygen has a major effect on the transcription of photosynthetic genes. Under anaerobic conditions the photosynthetic genes are activated. The exact mechanism for this activation has not been determined. Some *crt* genes, such as *crt*A, are activated by oxygen (Zhu et al. 1986a; Zhu and Hearst 1987). Oxygen also affects the stabilty of mRNA, accelerating the degradation of some mRNA (i.e., LH-II) to produce nucleotides (NT). It is not clear if oxygen acts on the translational process or assembly and stability of the proteins in the membrane. There is some evidence supporting the theory that oxygen can directly affect the activity of enzymes such as those catalyzing tetrapyrrole synthesis (not shown in this diagram) (Lascelles 1975). The transcriptional, posttranscriptional, and posttranslational regulations may also be influenced by light.

regulatory sequences and transcriptional machinery, to defining the molecular mechanism of transcriptional and posttranscriptional regulation, and to investiguating the interactions of regulatory proteins, DNA, and mRNA. Such information will eventually facilitate the development of a biotechnology for increasing the efficiency of photosynthesis and food production.

REFERENCES

Agaard, J., and Sistrom, W.R. (1972) *Photochem. Photobiol.* 15, 209–255.

Allen, N.P., and Feher, G. (1984) *Proc. Natl. Acad. Sci.USA* 81, 4795–4799.

Alt, J., Morris, J., Westhoff, P., and Herrmann, R.G. (1984) *Curr. Genet.* 8, 597–606.

Bachofen, R., and Wiemken, V. (1986) in *Encyclopedia of Plant Physiology: Photosynthesis III,* vol. 19 (Staehelin, L.A., and Arntzen, C.J., eds.), pp. 620–631, Springer-Verlag, New York.

Bartsch, R.G. (1978) in *The Photosynthetic Bacteria* (Clayton, R.K., and Sistrom, W.R., eds.), pp. 249–279,, Plenum Press, New York.

Bauer, C.E., Eleuterio, M., Young, D.A., and Marrs, B.L. (1987) in *Proceedings of the VII International Congress of Photosynthesis,* vol. 4 (Biggins, J., ed.), pp. 699–705, Martinus Nijhoff, Boston.

Bauer, C.E., Young, D.A., and Marrs, B.L. (1988) *J. Biol. Chem.* 263, 4820–4827.

Beatty, J.T., Adams, C.W., and Cohen, S.N. (1985) in *V International Symposium on Photosynthetic Prokaryotes,* Grindelwald, Switzerland, September 22–28, 1985, Abstract, p. 100.

Beatty, J.T., Adams, C.W., and Cohen, S.N. (1986) in *Microbial Energy Transduction* (Youvan, D.C., and Daldal, F., eds.), pp. 27–29, Cold Spring Harbor Laboratory, Cold Spring Harbor, NY.

Belasco, J.G., Beatty, J.T., Adams, C.W., Gabain, A.V., and Cohen S.N. (1985) *Cell* 40, 171–181.

Bérard, J., Bélanger, G., Corriveau, P., and Gingras, G. (1986) *J. Biol. Chem.* 261, 82–87.

Biel, A.J., and Marrs, B.L. (1983) *J. Bacteriol.* 156, 686–694.

Biel, A.J., and Marrs, B.L. (1985) *J. Bacteriol.* 162, 1320–1321.

Bishop, N.I. (1987) in *Proceedings of the VII International Congress of Photosynthesis,* vol. 3 (Biggins, J., ed.), pp. 795–798, Martinus Nijhoff, Boston.

Chory, J., Donohue, T.J., Varga, A.R., Staehelin, L.A., and Kaplan, S. (1984) *J. Bacteriol.* 159, 540–554.

Chory, J., and Kaplan, S. (1982) *J. Biol. Chem.* 257, 15110–15121.

Chory, J., and Kaplan, S. (1983) *J. Bacteriol.* 153, 465–474.

Clark, W.G., Davidson, E., and Marrs, B.L. (1984) *J. Bacteriol.* 157, 945–948.

Cogdell, R.J. (1978) *Philos. Trans. R. Soc. Lond.* B 248, 569–579.

Cogdell, R., and Crofts, A. (1978) *Biochim. Biophys. Acta* 502, 409–416.

Cogdell, R., Parson, W.W., and Kerr, M.A. (1976) *Biochim. Biophys. Acta* 430, 83–93.

Cohen-Bazire, G., and Kunisawa, R. (1963) *J. Cell Biol.* 16, 401–419.

Collins, J.J., Roberts, G.P., and Brill, W.J. (1986) *J. Bacteriol.* 168, 173–178.

Daldal, F., Cheng, S., Applebaum, J., Davidson, E., and Prince, R. (1986) *Proc. Natl. Acad. Sci. USA* 83, 2012–2016.

Davidson, E., and Daldal, F. (1987) *J. Mol. Biol.* 195, 25–29.

Davidson, E., Rook, S., and Daldal, F. (1987) in *Proceedings of the VII International Congress of Photosynthesis,* vol. 3 (Biggins, J., ed.), pp. 713–716, Martinus Nijhoff, Boston.

Debus, R.J., Feher, G., and Okamura, M.Y. (1985) *J. Biochem.* 24, 2488–2500.

Deisenhofer, J., Epp, O., Miki, K., Huber, R., and Michel, H. (1984) *J. Mol. Biol.* 180, 385–398.

Deisenhofer, J., Epp, O., Miki, K., Huber, R., and Michel, H. (1985a) *Nature* 318, 618–624.

Deisenhofer, J.R., Michel, H., and Huber, R. (1985b) *TIBS* 3, 243–248.

DeVitry, C., and Diner, B. (1984) *FEBS Lett.* 167, 327–331.

Donohue, T.J., Hoger, J.M., and Kaplan, S. (1986a) *J. Bacteriol.* 168, 953–961.

Donohue, T.J., and Kaplan, S. (1986) in *Encyclopedia of Plant Physiology: Photosynthesis III,* vol. 19 (Staehelin, L.A., and Arntzen, C.J., eds.), pp. 632–639, Springer-Verlag, New York.

Donohue, T.J., McEwan, A.G., and Kaplan, S. (1986b) *J. Bacteriol.* 168, 962–972.

Drews, G. (1978) *Curr. Top. Bioenerg.* 8, 161–199.

Drews, G. (1985) *Microbiol. Rev.* 49, 59–70.

Drews, G. (1986) *TIBS* 11, 255–257.

Drews, G., Kaufmann, N., and Klug, G. (1986) in *Molecular Biology of the Photosynthetic Apparatus* (Steinback, K.E., Bonitz, S., Arntzen, C.J., and Bogorad, L., eds.), pp. 211–222, Cold Spring Harbor Laboratory, Cold Spring Harbor, NY.

Drews, G., and Oelze, J. (1981) in *Advances in Microbial Physiology,* vol. 22, pp. 1–93, Academic Press, Orlando, FL.

Drews, G., Peters, J., and Dierstein, R. (1983) *Ann. Microbiol. (Inst. Pasteur)* 134B, 151–158.

Feher, G., and Okamura, M.Y. (1978) in *The Photosynthetic Bacteria* (Clayton, R.K., and Sistrom, W.R., eds.), pp. 349–382, Plenum Press, New York.

Fornani, C.S., and Kaplan, S. (1983) *Gene* 25, 291–299.

Gabellini, N., Bowyer, J.R., Hurt, G., Melandri, A., and Hauska, G. (1982) *Eur. J. Biochem.* 126, 105–111.

Gabellini, N., Harnisch, U., McCarthy, J.E., Hauska, G., and Sebald, W. (1985) *EMBO J.* 2, 549–553.

Gabellini, N., and Sebald, W. (1986) *Eur. J. Biochem.* 154, 569–579.

Gibson, J.L., and Tabita, F.R. (1977a) *J. Bacteriol.* 132, 818–823.

Gibson, J.L., and Tabita, F.R. (1977b) *J. Biol. Chem.* 252, 943–949.

Govindjee, ed. (1982) *Photosynthesis,* vols. II and III, Academic Press, Orlando, FL.

Gray, E.D. (1967) *Biochem. Biophys. Acta* 138, 550–563.

Halliwell. B. (1984) *Chloroplast Metabolism. The Structure and Function of Chloroplasts in Green Leaf Cells,* pp. 18–206, Oxford University Press.

Hearst, J.E. (1986) in *Encyclopedia of Plant Physiology: Photosynthesis III,* vol. 19 (Staehelin, L.A., and Arntzen, C.J., eds.), pp. 382–389, Springer-Verlag, New York.

Hearst, J.E., and Sauer, K. (1984a) in *Advances in Photosynthesis. Proceedings of the VIth International Congress on Photosynthesis,* vol. III (Sybesma, C., ed.), pp. 355–359, Martinus Nijhoff/Dr. W. Junk, Boston.

Hearst, J.E., and Sauer, K. (1984b) *Z. Naturforsch.* 39C, 421–424.

Holschuhk, K., Bottomley, W., and Whitfield, P.R. (1984) *Nucleic Acids Res.* 12, 8819–8834.

Imhoff J.F., Truper, H.G., and Pfennig, N. (1984) *Int. J. Syst. Bacteriol.* 34, 340–343.

Kaplan, S. (1978) in *The Photosynthetic Bacteria* (Clayton, R.K., and Sistrom, W.R., eds.), pp. 809–839, Plenum Press, New York.

Kaplan, S., and Arntzen, C.J. (1982) in *Photosynthesis III* (Govindjee, ed.), pp. 65–151, Academic Press, Orlando, FL.

Kiley, P.J., Donohue, T.J., Havelka, W.A., and Kaplan, S. (1987) *J. Bacteriol.* 169, 742–750.

Kiley, P.J., and Kaplan, S. (1987) *J. Bacteriol.* 169, 3268–3275.

Klug, G., Kaufmann, N., and Drews, G. (1984) *FEBS Lett.* 177, 61–65.

Klug, G., Kaufmann, N., and Drews, G. (1985) *Proc. Natl. Acad. Sci. USA* 82, 6485–6489.

Kranz, R.G., and Haselkorn, R. (1986) *Proc. Natl. Acad. Sci. USA* 83, 6805–6809.

Kyle, D.J., Ohad, I., and Arntzen, C.J. (1984) *Proc. Natl. Acad. Sci. USA* 81, 4070–4074.

Lascelles, J. (1975) *Ann. N.Y. Acad. Sci.* 244, 334–347.

Loach, P.A., Parkes, P.S., Miller, J.F., Hinchigen, S., and Callahan, P.M. (1986) in *Molecular Biology of the Photosynthetic Apparatus* (Steinback, K.E., Bonitz, S., Arntzen, C.J., and Bogorad, L., eds.), pp. 197–209, Cold Spring Harbor Laboratory, Cold Spring Harbor, NY.

Marrs, B. (1974) *Proc. Natl. Acad. Sci. USA* 71, 971–973.

Marrs, B. (1981) *J. Bacteriol.* 146, 1003–1012.

Marrs, B., and Kaplan, S. (1970) *J. Mol. Biol.* 49, 297–317.

Metz, J.G., Pakrasi, H., Arntzen, C.J., and Seibert, M. (1987) in *Proceedings of the VII International Congress of Photosynthesis,* vol. 4 (Biggins, J., ed.), pp. 679–682, Martinus Nijhoff, Boston.

Michel, M., and Deisenhofer, J. (1986) in *Encyclopedia of Plant Physiology: Photosynthesis III,* vol. 19 (Staehelin, A.C., and Arntzen, C.J., eds.), pp. 371–381, Springer-Verlag, New York.

Michel, H., Weyer, K.A., Gruenberg, H., et al. (1986) *EMBO J.* 5, 1149–1158.

Michel, H., Weyer, K.A., and Lottspeich, F. (1985) *EMBO J.* 4, 1667–1672.

Monger, T.G., and Parson, W.W. (1977) *Biochim. Biophys. Acta* 460, 393–407.

Muller, E.D., Chory, J., and Kaplan, S. (1985) *J. Bacteriol.* 161, 469–472.

Nargang, F., McIntosh, L., and Somerville, C. (1984) *Mol. Gen. Genet.* 193, 220–224.

Ohad, I., Kyle, D.J., and Arntzen, C.J. (1984) *J. Cell Biol.* 99, 481–485.

Olson, J.M. (1981) in *Origins and Evolution of Eukaryotic Intracellular Organelles* (Frederick, J.F., ed.), pp. 8–19, New York Academy of Sciences, New York.

Peters, J., Takemoto, J., and Drews, G. (1983) *Biochemistry* 229, 5660–5667.

Pfister, K., Steinback, K.E., Gardner, G., and Arntzen, C.J. (1981) *Proc. Natl. Acad. Sci. USA* 78, 791–795.

Quivey, R.G. Jr., and Tabita, F.R. (1984) *Gene* 31, 91–101.

Rochaix, J.D., Dron, M., Rahire, M., and Malnoe, P. (1984) *Plant Mol. Biol.* 3, 363–370.

Sauer, K., and Austin, L.A. (1978) *Biochem.* 17, 2011–2019.

Schachman, H.K., Pardee, A.B., and Stanier, R.Y. (1982) *Arch. Biochem. Biophys.* 38, 245–260.

Schumacher, A., and Drews, G. (1978) *Biochim. Biophys. Acta* 501, 183–194.

Scolnik, P.A., and Haselkorn, R. (1984) *Nature* 307, 289–292.

Solioz, M., Yen, H.C., and Marrs, B.L. (1975) *J. Bacteriol.* 123, 651–657.

Somerville, C.R., and Somerville, S.C. (1984) *Mol. Gen. Genet.* 193, 214–219.

Sprague, S.G., and Varga, A.R. (1986) in *Encyclopedia of Plant Physiology: Photosynthesis III,* vol. 19 (Staehelin, L.A., and Arntzen, C.J., eds.), pp. 603–619, Springer-Verlag, New York.

Staehelin, L.A., and Arntzen, C.J., eds. (1986) *Encyclopedia of Plant Physiology: Photosynthesis* III, vol. 19, Springer-Verlag, New York.

Steinback, K.E., Bonitz, S., Arntzen, C.J., and Bogorad, L., eds. (1985) *Molecular Biology of the Photosynthetic Apparatus,* Cold Spring Harbor Laboratory, Cold Spring Harbor, NY.

Sybesma, C., ed. (1984) in *Advances in Photosynthesis. Proceedings of the VIth International Congress on Photosynthesis, 1983,* Brussels, Belgium, Martinus Nijhoff/Dr. W. Junk, Boston.

Takemoto, J.Y., Peters, J., and Drews, G. (1982) *FEBS Lett.* 142, 227–230.

Taylor, D.P., Cohen, S.N., Clark, W.G., and Marrs, B.L. (1983) *J. Bacteriol.* 154, 580–590.

Webster, G., Cogdell, R., and Lindsay, J. (1980) *FEBS Lett.* 111, 391–394.

Welte, W., Wacker, T., Leis, M., et al. (1985) *FEBS Lett.* 182, 260–264.

Wiemken, V., and Bachofen, R. (1984) *FEBS Lett.* 166, 155–159.

Williams, J.C., Steiner, L.A., Feher, G., and Simon, M. (1984) *Proc. Natl. Acad. Sci. USA* 81, 7303–7307.

Williams, J.C., Steiner, L.A., Ogden, R.C., Simon, M.I., and Feher, G. (1983) *Proc. Natl. Acad. Sci. USA* 80, 6506–6509.

Worland, S.T., Yamagishi, A., Isaacs, S., Sauer, K., and Hearst, J.E. (1987) *Proc. Natl. Acad. Sci. USA* 84, 1774–1778.

Yamamoto, N., and Droffner, M.L. (1985) *Proc. Natl. Acad. Sci. USA* 82, 2077–2081.

Yen, H.C., Hu, N.T., and Marrs, B.L. (1979) *J. Mol. Biol.* 131, 157–168.

Yen, H.C., and Marrs, B. (1976) *J. Bacteriol.* 126, 619–629.

Youvan, D.C., Alberti, M., Begusch, H., Bylina, E.J., and Hearst, J.E. (1984a) *Proc. Natl. Acad. Sci. USA* 81, 189–192.

Youvan, D.C., Bylina, E.J., Alberti, M., Begusch, H., and Hearst, J.E. (1984b) *Cell* 37, 949–957.

Youvan, D.C., Elder, J.T., Sandlin, D.E., et al. (1982) *J. Mol. Biol.* 162, 17–41.

Youvan, D.C., and Ismail, S. (1985) *Proc. Natl. Acad. Sci. USA* 82, 58–62.

Youvan, D.C., and Marrs, B.L. (1984) *Cell* 39, 1–3.

Youvan, D.C., and Marrs, B.L. (1985) in *Molecular Biology of the Photosynthetic Apparatus* (Steinback, K.E., Bonitz, S., Arntzen, C.J., and Bogorad, L., eds.), pp. 173–181, Cold Spring Harbor Laboratory, Cold Spring Harbor, NY.

Yu, P.-L., Hohn, B., Falk, H., and Drew, G. (1982) *Mol. Gen. Genet.* 188, 392–398.

Zannoni, D., and Baccarini-Melandri, A. (1980) in *Diversity of Bacterial Respiratory Systems,* vol. 2 (Knowles, D., ed.), pp. 183–202. CRC Press, Boca Raton, FL.

Zhu, Y.S., Cook, D.N., Leach, F., et al. (1986a) *J. Bacteriol.* 168, 1180–1188.

Zhu, Y.S., and Hearst, J.E. (1986) *Proc. Natl. Acad. Sci. USA* 83, 7613–7617.

Zhu, Y.S., and Hearst, J.E. (1987) in *Proceedings of the VII International Congress of Photosynthesis,* vol. 4 (Biggins, J., ed.), pp. 717–720, Martinus Nijhoff, Boston.

Zhu, Y.S., and Hearst, J.E. (1988) *Proc. Natl. Acad. Sci. USA* 85, 4209–4213.

Zhu, Y.S., and Kaplan, S. (1985) *J. Bacteriol.* 162, 925–932.

Zhu, Y.S., Killey, P.J., Donohue, T.J., and Kaplan, S. (1986b) *J. Biol. Chem.* 261, 10366–10374.

Zsebo, K.M., and Hearst, J.E. (1984) *Cell* 37, 937–947.

Zsebo, K.M., Wu, F., and Hearst, J.E. (1984) *Plasmid* 11, 182–184.

Zuber, H. (1986) in *Encyclopedia of Plant Physiology: Photosynthesis III,* vol. 19 (Staehelin, L.A., and Arntzen, C.J., eds.), pp. 239–251, Springer-Verlag, New York.

Zuber, H., Sidler, W., Füglistaller, P., Brunisholz, R., and Theiler, R. (1986) in *Molecular Biology of the Photosynthetic Apparatus* (Steinback K.E., Bonitz, S., Arntzen, C.J., and Bogorod, L., eds.), pp. 183–195, Cold Spring Harbor Laboratory, Cold Spring Harbor, NY.

Zurawski, G., Bohnert, H.J., Whitfield, P.K., and Bottomley, W. (1982) *Proc. Natl. Acad. Sci. USA* 79, 7699–7703.

PART III

Prospects for Manipulation of Chloroplast Genomes

CHAPTER

13

Organization and Expression of the *Nicotiana* Chloroplast Genome

Masahiro Sugiura

Photosynthesis is the conversion of light energy to chemical energy in plants and certain bacteria. In eukaryotes, photosynthesis takes place in organelles called chloroplasts. The discovery of non-Mendelian mutants of the chloroplast phenotype at the beginning of this century suggested the existence of a separate genetic system in chloroplasts. Since the demonstration of a unique DNA molecule in chloroplasts over 20 years ago, intensive studies of the structure and expression of chloroplast genomes have been made (Sager and Schlanger 1976).

Chloroplast DNAs of higher plants are circular molecules with a size of 120 to 217 kbp. One of the outstanding features of chloroplast DNAs of most higher plants is the presence of two copies of a large inverted repeat. These sequences are separated by a large and a small single-copy region. Chloroplast DNAs are known to contain all the chloroplast rRNA genes (four genes in higher plants) and tRNA genes (about 35 genes) and probably all the genes for proteins synthesized in the chloroplast (about 100 genes) (Dyer 1984; Gray et al. 1984).

The purpose of this chapter is not to review chloroplast genomes from a wide variety of plants and algae, but to discuss the most recent studies on the *Nicotiana* chloroplast genome. Tobacco plant has been a favored material for studies of inheritance and evolution (Smith 1974). There are many interspecific hybrids, chloroplast mutants, and cell lines with altered chloroplast ribosomes (Hughes 1982). Moreover, tobacco cells provide a model system for studying somatic cell genetics because of the recent technical advances in cell and protoplast cultures and protoplast fusion (Takebe and Nagata 1984; Medgyesy et al. 1985).

Readers interested in chloroplast genomes from various organisms are referred to several recent reviews (Dyer 1984; Crouse et al. 1984 and 1985; Palmer 1985; Shinozaki and Sugiura 1986a).

13.1 CHLOROPLAST DNA

Chloroplast DNA from the genus *Nicotiana* exhibits distinct species differences in restriction endonuclease patterns (Atchison et al. 1976; Vedel et al. 1976; Rhodes et al. 1981). A phylogenetic tree of chloroplast DNA evolution in *Nicotiana* has been constructed based on the restriction pattern of chloroplast DNA (Kung et al. 1982). To gain insight into the molecular organization of chloroplast genomes, a detailed knowledge of the positioning of restriction enzyme sites is important. For this purpose physical maps of chloroplast DNA have been constructed in *Nicotiana tabacum* (Jurgenson and Bourque 1980; Seyer et al. 1981; Fluhr and Edelman 1981; Hildebrand et al. 1985; Sugiura et al. 1986) and other *Nicotiana* species (Shen et al. 1982; Zhu et al. 1982; Tassopulu and Kung 1984; Salts et al. 1984).

It has been difficult to isolate large quantities of chloroplast DNA in pure form. The molecular cloning technique has overcome this difficulty. Well-characterized clone banks of *Nicotiana* chloroplast DNAs have been thus constructed (Zhu et al. 1982; Fluhr et al. 1983; Hildebrand et al. 1985). The clone bank of the entire *N. tabacum* (var. Bright Yellow 4) chloroplast DNA as a set of overlapping restriction endonuclease fragments was constructed in order to sequence the whole DNA (Sugiura et al. 1986). DNA fragments must overlap to cover the entire genome; otherwise, very short restriction fragments are overlooked. Using this clone bank, the complete nucleotide sequence of *N. tabacum* chloroplast DNA has been determined (Shinozaki et al. 1986c).

The *N. tabacum* chloroplast DNA is divided into four regions (LSC, SSC, IR_A, and IR_B) (Sugita et al. 1984). The large (LSC) and small (SSC) single-copy regions are 86,684 bp and 18,482 bp long, respectively. The inverted repeats (IR_A and IR_B) have been sequenced separately and found to be completely identical (25,339 bp). The entire genome is thus 155,844 bp long. The complete DNA sequence has been deposited with the EMBL database and presented in the special chloroplast sequence issue of *Plant Molecular Biology Reporter* (Shinozaki et al. 1986d). Figure 13–1 shows a circular map of the *N. tabacum* chloroplast genome. The DNA strand that codes for the large subunit of ribulose-1,5-bisphosphate carboxylase has been designated A, and the complementary strand, B (Deno et al. 1983). The nomenclature for genes follows the proposals of Hallick and Bottomley (1983).

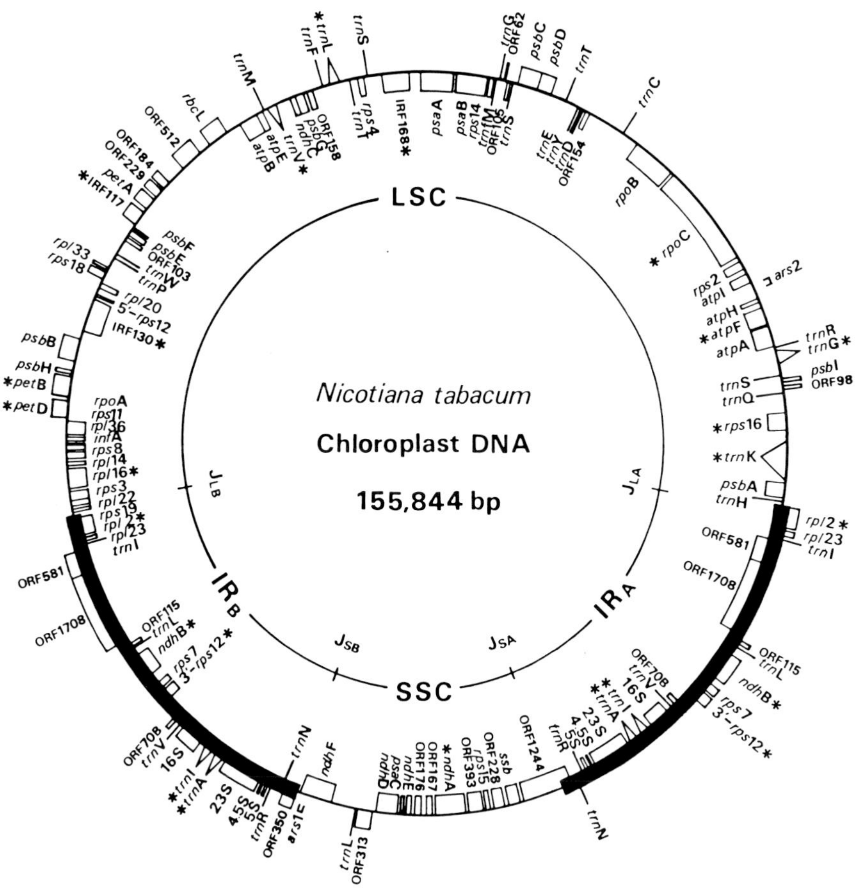

FIGURE 13–1 Circular gene map of the *N. tabacum* chloroplast genome. Genes shown outside the circle are on the A strand and transcribed counterclockwise. Genes shown inside the circle are on the B strand and transcribed clockwise. Asterisks indicate split genes. Major ORFs are included. (With minor modifications, from K. Shinozaki, M. Ohme, M. Tanaka, et al. [1986d]. With permission.)

13.2 GENES FOR rRNAs

Chloroplast ribosomes have a sedimentation coefficient of 70S and are closely related to prokaryotic ribosomes. The rRNA genes were cloned and analyzed first in *N. tabacum* chloroplasts because it is easy to prepare their rRNAs for use as hybridization probes (Sugiura and Kusuda 1979). These genes are arranged as they are in *Escherichia coli* and the cyanobacterium *Anacystis nidulans* (Tomioka et al. 1981), in the order 16S, 23S, 4.5S, and 5S rDNA in both IRs (Takaiwa and Sugiura 1980a, 1982a, and 1982b; Tohdoh and Sugiura 1982) (Figure 13–2). There are consequently two copies of each, or eight rRNA genes per genome. The coding regions for the mature 16S rRNA and 23S rRNA have been determined by S1 mapping, and those for the mature 4.5S rRNA and 5S rRNA by sequencing the mature RNAs (Takaiwa and Sugiura 1980b and 1981).

The 4.5S rRNA that was found associated with the 50S subunits of chloroplasts (Bowman and Dyer 1979) shows 68% and 61% homology with 3′-terminal regions of *A. nidulans* and *E. coli* 23S rRNAs, respectively, (Takaiwa and Sugiura 1982a; Kumano et al. 1983). Chloroplast 4.5S rRNA is likely to be functionally equivalent to the 3′-terminal region of prokaryotic 23S rRNA (Takaiwa and Sugiura 1982a). If this is true, the nature of the 101-bp spacer between the 23S and 4.5S rDNAs is of interest. One possibility is that this sequence is homologous to an intron and that the two split RNA molecules, 23S and 4.5S rRNAs, remain unligated because of incomplete splicing. The fact that the 4.5S rRNA is not phosphorylated at its 5′ end (Takaiwa and Sugiura 1980b) supports this idea, because splicing occurs between 3′-phosphoryl and 5′-hydroxyl ends of RNA chains.

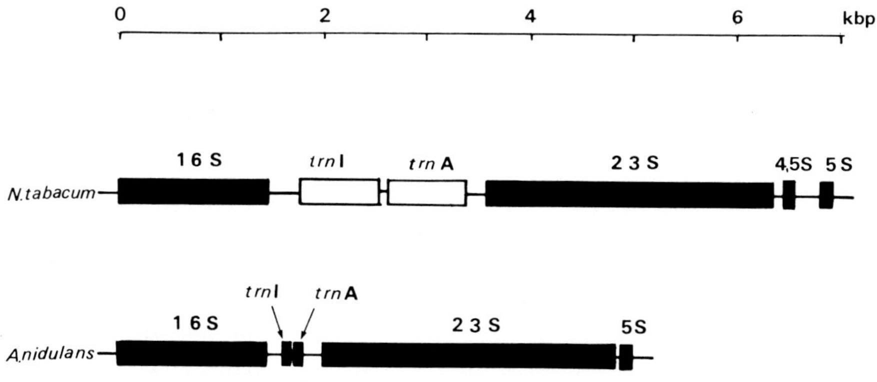

FIGURE 13–2 Organization of rDNA clusters of *N. tabacum* chloroplasts and the cyanobacterium *A. nidulans*. Filled boxes indicate exons, and open boxes, introns. Tobacco *trn*I and *trn*A and *A. nidulans trn*I contain no CCA ends, while *A. nidulans trn*A encodes a CCA end (Takaiwa and Sugiura 1982b; Tomioka and Sugiura 1984).

13.3 GENES FOR tRNAs

All tRNA species involved in chloroplast protein synthesis are believed to be coded in the chloroplast genome. Saturation hybridization of total chloroplast tRNAs to chloroplast DNA indicates that there are 25 to 45 tRNA genes on the chloroplast genome. Using two-dimensional PAGE, 20 tRNAs corresponding to 14 different amino acid–accepting species have been identified in tobacco chloroplasts (Bergmann et al. 1984). Hybridization of individual tRNA species to DNA fragments has permitted the construction of the tRNA gene map of the *N. tabacum* chloroplast genome (Bergmann et al. 1984).

All *N. tabacum* chloroplast DNA fragments that hybridized with the total chloroplast tRNAs have been sequenced. Thirty different tRNA genes have been identified in the DNA sequence (Kato et al. 1981 and 1985; Tohdoh et al. 1981; Deno et al. 1982; Deno and Sugiura 1983 and 1984; Ohme et al. 1984 and 1985; Sugita et al. 1984 and 1985; Yamada et al. 1986; Wakasugi et al. 1986). The complete nucleotide sequence of the *N. tabacum* chloroplast genome has been determined, and no additional tRNA genes could be found using a computer search. Hence these 30 tRNA genes are most likely all of the tRNA genes encoded in the genome (Shinozaki et al. 1986c). Seven of them are located in the IR and therefore the total number of tRNA genes in the genome is 37. Hybridization analysis of the total chloroplast rRNA has revealed that all tRNA genes are expressed. The map position of most of the tRNA genes is consistent with that based on a tRNA/DNA fragment hybridization study (Bergmann et al. 1984). The sequences of the 30 tRNA genes are shown in Table 13–1.

General features of the chloroplast tRNA genes and tRNAs deduced from the DNA sequences are as follows. No tRNA genes code for the 3′-CCA end. Six tRNA genes harbor long single introns (503–2,526 bp) (see Tables 13–1 and 13–2). The presence of introns in chloroplast tRNA genes was first demonstrated in maize *trn*I and *trn*A by Koch et al. (1981). Interestingly, *trn*G-UCC contains a 691-bp intron in the D-stem region (Deno and Sugiura 1984). This seems to be a unique feature of chloroplast genomes. *Trn*K has the longest intron (2,526 bp), and the predicted amino acid sequence of ORF509 found in the *trn*K intron (Sugita et al. 1985) has a local homology with that of the ORF present in the yeast mitochondrial *oxi*3 intron 2. Many of the tRNA genes contain sequences similar to consensus prokaryotic promoters in their upstream regions. A unique feature is the presence of a G_2-C_{72} or C_2-G_{72} base pair in the aminoacyl stem of all the tRNAs. All the tRNAs can form the cloverleaf structure and none has an abnormal structure, as has been reported for some mammalian mitochondrial tRNAs. All the tRNA sequences show higher homology with corresponding bacterial tRNAs than with the corresponding eukaryotic cytoplasmic and mitochondrial tRNAs.

Table 13–2 summarizes the anticodons of *N. tabacum* chloroplast tRNA genes and the codon usage in the sum of 39 protein genes. All possible

TABLE 13–1 Nucleotide Sequences of 30 tRNA Genes from *N. tabacum* Chloroplasts

	Anti codon	Amino stem		D-stem	D-loop	D-stem		Anti stem	Anti loop	Anti stem	Extra arm		TF-stem	TF-loop	TF-stem	Amino stem	
*Ala	(UGC)	GGGGATA	TA	GCTC	AGTT GGT A	GAGC	T	CCGCT	CTTGCAA	GGCGG	ATGT	C	AGCGG	TTCGAGT	CCGCT	TATCTCC	A
Arg	(ACG)	GGGCCTG	TA	GCTC	AGA GGATTA	GAGC	A	CGTGG	CTACGAA	CCACG	GTGT	C	GGGGG	TTCGAAT	CCCTC	CTCGCCC	A
Arg	(UCU)	GCGTCCA	TT	GTCT	AAT GGAT A	GGAC	A	GAGGT	CTTCTAA	ACCTT	TGG	T	ATAGG	TTCAAAT	CCTAT	TGGACGC	A
Asn	(GUU)	TCCTCAG	TA	GCTC	AGT GGT A	GAGC	G	GTCGG	CTGTTAA	CCGAT	TGGT	C	GTAGG	TTCGAAT	CCTAC	TTGGGGA	G
Asp	(GUC)	GGGATTG	TA	GTTC	AATT GGTC A	GAGC	A	CCGCC	CTGTCAA	GGCGG	AAGC	T	GCGGG	TTCGAGC	CCCGT	CAGTCCC	G
Cys	(GCA)	GGCGACA	TG	GCCG	AGT GGT A	AGGC	A	GAGGA	CTGCAAA	TCCTT	TTTT	C	CCCAG	TTCAAAT	CCGGG	TGTCGCC	T
Gln	(UUG)	TGGGGCG	TG	GCCA	AGT GGT A	AGGC	A	ACGGG	TTTTGGT	CCCGC	TATT	C	GGAGG	TTCGAAT	CCTTC	CGTCCCA	G
Glu	(UUC)	GCCCCCA	TC	GTCT	AGT GGTTTA	GGAC	A	TCTCT	CTTTCAA	GGAGG	CAG	C	GGGGA	TTCGAAT	TCCCC	TGGGGGT	A
Gly	(GCC)	GCGGATA	TG	GTCG	AAT GGT A	AAAT	T	TCTCT	TTGCCAA	GGAGA	AGA	T	GCGGG	TTCGATT	CCCGC	TATCCGC	C
*Gly	(UCC)	GCGGGTA	TA	GTTT	AGT GGT A	AAAC	C	CTAGC	CTTCCAA	GCTAA	CGA	T	GCGGG	TTCGATT	CCCGC	TACCCGC	T
His	(GUG)	GGCGGATG	TA	GCCA	AGT GGATCA	AGGC	A	GTGGA	TTGTGAA	TCCAC	CATG	C	GCGGG	TTCAATT	CCCGT	CGTTCGC	C
*Ile	(GAU)	GGGCTAT	TA	GCTC	AGT GGT A	GAGC	G	CGCCC	CTGATAA	GGGCG	AGGT	C	TCTGG	TTCAAGT	CCAGG	ATGGCCC	A
Ile	(CAU)	GCATCCA	TG	GCTG	AAT GGTT A	AAGC	G	CCCAA	CTCATAA	TTGGC	GAATT	C	GTAGG	TTCAATT	CCTAC	TGGATGC	A
*Leu	(UAA)	GGGGATA	TG	GCGA	AATC GGTA G	ACGC	T	ACGGA	CTTAAAA	TCCGT	CGACTTTAAAAATCG	T	GAGGG	TTCAAGT	CCCTC	TATCCCC	A
Leu	(CAA)	GCCTTGG	TG	GTGA	AAT GGTA G	ACAC	G	CGAGA	CTCAAAA	TCTCG	TGCTAAATAGCG	T	GGAGG	TTCGAGT	CCTCT	TCAAGGC	A
Leu	(UAG)	GCCGCTA	TG	GTGA	AATT GGTA G	ACAC	G	CTGCT	CTTAGGA	AGCAG	TGCTAATGCA	T	CTCGG	TTCGAGT	CCGAG	TGGCGGC	A
*Lys	(UUU)	GGGTTGC	TA	ACTC	AAC GGT A	GAGT	A	CTCGG	CTTTTAA	CCGAC	TAGT	T	CCGGG	TTCGAAT	CCCGG	GCAACCC	A
fMet	(CAU)	CGCGGGG	TA	GAGC	AGTTTGGT A	GCTC	G	CAAGG	CTCATAA	CCTTG	AGGT	C	ACGGG	TTCAAAT	CCTGT	CTCCGCA	A
mMet	(CAU)	ACCTACT	TA	ACTC	AGT GGTT A	GAGT	A	CTGCT	TTCATAC	GGCGG	GAGT	C	ATTGG	TTCAAAT	CCAAT	AGTAGGT	A
Phe	(GAA)	GCCGGGA	TA	GCTC	AGTT GGT A	GAGC	A	GAGGA	CTGAAAA	TCCTC	GTGT	C	ACCAG	TTCAAAT	CTGGT	TCCTGGC	A
Pro	(UGG)	AGGGATG	TG	GCGC	AGCTTGGT A	GCGC	G	TTTGT	TTTGGGT	ACAAA	ATGT	C	ACAGG	TTCAAAT	CCTGT	CATCCCT	A
Ser	(GGA)	GGAGAGA	TG	GCCG	AGT GGTTGA	AGGC	G	TAGCA	TTGGAAC	TGCTA	TGTAGGCTTTTGTTTAC	C	GAGGG	TTCGAAT	CCCTC	TCTTTCC	G
Ser	(UGA)	GGAGAGA	TG	GCTG	AGC GGTTGA	TAGC	C	CCGGT	CTTGAAA	ACCGG	TATAGTTTTAACAAAGAACTAT	C	GAGGG	TTCGAAT	CCCTC	TCTCTCC	T
Ser	(GCU)	GGAGAGA	TG	GCTG	AGT GGACTA	AAGC	G	GCGGA	TTGCTAA	TCCGT	TGTACGAGTTAATCGTAC	C	GAGGG	TTCGAAT	CCCTC	TCTTTCC	G
Thr	(GGU)	GCCCTTT	TA	ACTC	AGT GGT A	GAGT	A	ACGCC	ATGGTAA	GGCGT	AAGT	C	ATCGG	TTCAAAT	CCGAT	AAGGGGC	T
Thr	(UGU)	GCCCGCT	TA	GCTC	AGA GGTT A	GAGC	A	TCGCA	TTTGTAA	TGCGA	TGGT	C	ATCGG	TTCGATT	CCGAT	AGCCGGC	T
Trp	(CCA)	GCGCTCT	TA	GTTC	AGTTCGGT A	GAAC	G	TGGGT	CTCCAAA	ACCCG	ATGT	C	GTAGG	TTCAAAT	CCTAC	AGAGCGT	G
Tyr	(GUA)	GGGTCGA	TG	CCCG	AGC GGTTAA	TGGG	G	ACGGA	CTGTAAA	TTCGT	TGGCAATATGTCTA	C	GCTGG	TTCAAAT	CCAGC	TCGGCCC	A
Val	(GAC)	AGGGATA	TA	ACTC	AGC GGT A	GAGT	G	TCACC	TTGACGT	GGTGG	AAGT	C	ATCAG	TTCGAGC	CTGAT	TATCCCT	A
*Val	(UAC)	AGGGCTA	TA	GCTC	AGTT GGT A	GAGC	A	ACTCG	TTTACAC	CGAGA	AGGT	C	TACGG	TTCGAGT	CCGTA	TAGCCCT	A
invariant nucleotides		M	T	RY	AR GG R	RY			YT R			Y	RG	TTCRA Y	CY	M	

* gene containing an intron, R = A or G, Y = T or C, M = G or C

From T. Wakasugi, M. Ohme, K. Shinozaki, and M. Sugiura (1986) *Plant Mol. Biol.* 7, 385–392. With permission.

TABLE 13–2 Anticodons of tRNA Genes and Codon Usage in 39 Protein Genes in *N. tabacum* Chloroplasts

	U			C			A			G			
U	F		321	S		191	Y		259	C		64	U
	F	GAA	161	S	GGA	112	Y	GUA	61	C	GCA	18	C
	L	* UAA	359	S	UGA	95		stop	25		stop	5	A
	L	CAA 2	212	S		48		stop	9	W	CCA	184	G
C	L		202	P		196	H		208	R	ACG 2	190	U
	L		52	P		79	H	GUG	60	R		44	C
	L	UAG	123	P	UGG	118	Q	UUG	285	R		155	A
	L		67	P		65	Q		95	R		42	G
A	I		408	T		228	N		305	S		143	U
	I	* GAU 2	148	T	GGU	130	N	GUU 2	110	S	GCU	45	C
	I	CAU 2	222	T	UGU	152	K	* UUU	373	R	UCU	183	A
	M	CAU 2	231	T		57	K		104	R		65	G
G	V		230	A		342	D		291	G		327	U
	V	GAC 2	73	A		128	D	GUC	85	G	GCC	100	C
	V	* UAC	269	A	* UGC 2	232	E	UUC	405	G	* UCC	311	A
	V		80	A		73	E		120	G		142	G

Boxes, codons recognized by tRNAs (shown as anticodons); *, gene containing an intron; 2, two copies.

From T. Wakasugi, M. Ohme, K. Shinozaki, and M. Sugiura (1986) *Plant Mol. Biol.* 7, 385–392. With permission.

codons are used in the sequences coding for proteins in *N. tabacum* chloroplasts (Wakasugi et al. 1986). The minimum number of tRNA species required for translation of all codons is 32 if normal wobble base pairing is involved in codon-anticodon recognition. Either $tRNA^{Leu}$(UAA) or $tRNA^{Leu}$(CAA) is likely to be dispensable. As shown in Table 13–2, no tRNA that recognizes codons CUU/C(Leu), CCU/C(Pro), GCU/C(Ala), or CGC/A/G(Arg) has been found. If the "two-out-of-three" mechanism can operate in the chloroplast, as has been shown in an in vitro protein-synthesizing system from *E. coli* (Samuelsson et al. 1980), the single $tRNA^{Pro}$(UGG), $tRNA^{Ala}$(UGC), and $tRNA^{Arg}$(ACG) can read all four Pro, Ala, and Arg codons, respectively (GC pairs in the first and second codon-anticodon interaction). There is a gene for $tRNA^{Leu}$(UAG) and, if this tRNA has an unmodified uracil in the first position of the anticodon, it can read all four Leu codons (CUN) by U-N wobble. The bean, spinach, and soybean $tRNAs^{Leu}$(UAG) have unmodified uracil in their anticodons ($UA^{m^7}G$) (Pillay et al. 1984). These 28 tRNAs are therefore likely to be sufficient to read all codons in the *N. tabacum* chloroplast system using the above mechanisms. The possibility that the remaining tRNAs are imported from the cytoplasm or that their genes have unusual structures so as to prevent their detection cannot be excluded.

Seven tRNA genes are present in two copies each because they are localized on IRs. However, there is no apparent relationship between gene copy number and codon usage frequency. From the codon usage, all six tRNA species whose genes contain long introns are frequently used tRNAs. Therefore, the presence of long introns seems not to interfere with effective expression of these tRNA genes (Wakasugi et al. 1986).

13.4 GENES FOR STROMAL POLYPEPTIDES

Ribulose-1,5-bisphosphate carboxylase/oxygenase (Fraction I protein) is the major soluble protein of chloroplasts. This enzyme is composed of large and small subunits, and the large subunit is encoded in chloroplast DNA, while small subunit genes reside in nuclear DNA. The polypeptide composition revealed by isoelectrofocusing shows a unique pattern for each *Nicotiana* species, serving as a genetic marker for chloroplast and nuclear genomes (Kung et al. 1974). The gene for the large subunit (rbc L) has first been cloned and sequenced in maize (McIntosh et al. 1980). In *Nicotiana* the *rbc*L genes of *N. tabacum* (Shinozaki and Sugiura 1982a), *Nicotiana otophora,* and *Nicotiana acuminata* (Lin et al. 1986) have been analyzed. Comparison of the DNA sequences of the three *rbc*L genes revealed a high degree of homology in the coding region. *N. tabacum* is the only plant whose complete amino acid sequence of the large subunit protein has been determined (Amiri et al. 1984). This sequence is in agreement with that deduced from gene sequencing (except for position 284, which is occupied by glycine instead of cysteine).

N. tabacum chloroplast ribosomes are of the 70S type and contain 58 to 62 ribosomal proteins (Capel and Bourque 1982). Analysis of the synthesis of ribosomal proteins in isolated chloroplasts has shown that a chloroplast genome encodes about one third of the ribosomal proteins. The electrophoretic patterns and molecular weight frequency distributions of ribosomal proteins of *N. tabacum* chloroplasts are quite similar to those of *E. coli* (Capel and Bourque 1982). This led to the search for chloroplast ribosomal protein genes through their homology with *E. coli* ribosomal protein genes (Sugita and Sugiura 1983; Shinozaki et al. 1986a; Torazawa et al. 1986; Tanaka et al. 1986). Twenty different sequences coding for polypeptides homologous to *E. coli* ribosomal proteins have been found (Table 13–3). L36 is a new ribosomal protein species recently found associated with the 50S subunit of *E. coli* (Wada and Sako 1987), and the corresponding chloroplast gene is located between *inf*A and *rps*11 (*rpl*36, formerly designated *sec*X). The *rps*19 and *rpl*2 genes have also been sequenced in *N. debneyi* (Zurawski et al. 1984). Among them, *rps*7, *rpl*2, and *rpl*23 are located in IR, and therefore the total number of ribosomal protein genes is 23. The

TABLE 13–3 ***N. tabacum*** **Chloroplast Genes for Ribosomal Proteins**

Genes	*Products*
	30S Ribosomal Proteins
*rps*2	CS2
*rps*3	CS3
*rps*4	CS4
*rps*7	CS7
*rps*8	CS8
*rps*11	CS11
*rps*12	CS12
*rps*14	CS14
*rps*15	CS15
*rps*16	CS16
*rps*18	CS18
*rps*19	CS19
	50S Ribosomal Proteins
*rpl*2	CL2
*rpl*14	CL14
*rpl*16	CL16
*rpl*20	CL20
*rpl*22	CL22
*rpl*23	CL23
*rpl*33	CL33
*rpl*36	CL36

*rps*12, *rps*16, *rpl*2, and *rpl*16 genes contain introns that are 536 to 1,020 bp long.

The most striking feature is the structure of *rps*12. The *rps*12 gene was first identified in *Euglena* chloroplast DNA (Montandon and Stutz 1984). *Euglena rps*12 consists of 125 codons and has no intron. A portion of the *N. tabacum rps*12 gene was then found. This gene was split at least twice (Hallick et al. 1985; Fromm et al. 1986) and its exons (from codon 39 to the last codon 123, 3′-*rps*12) were located in the IR_A and IR_B of the chloroplast DNA (Fromm et al. 1986). The 5′ part of the *rps*12 gene (from the initiation codon to codon 38, 5′-*rps*12) has been found in the LSC of *N. tabacum* (Torazawa et al. 1986). The 5′-*rps*12 is located 28 kbp downstream from 3′-*rps*12 in IR_B on the same strand, or 86 kbp downstream from 3′-*rps*12 in IR_A on the opposite strand. We have designated this gene structure a "divided" gene (Shinozaki et al. 1986c).

The *rpl*23, *rpl*2, *rps*19, *rpl*22, *rps*3, *rpl*16, *rpl*14, *rps*8, *rpl*36, and *rps*11 genes are clustered in this order (*rpl*2 cluster). Surprisingly, this order corresponds to that of the homologous genes in the *E. coli* S10 and *spc* operons (Figure 13–3). This finding raises the interesting possibility that the genes for ribosomal proteins of chloroplast and *E. coli* may have evolved from a common ancestral gene set (Tanaka et al. 1986). The 3′-*rps*12 and *rps*7 genes are arranged as in the *E. coli str* operon and are cotranscribed (Fromm et al. 1986).

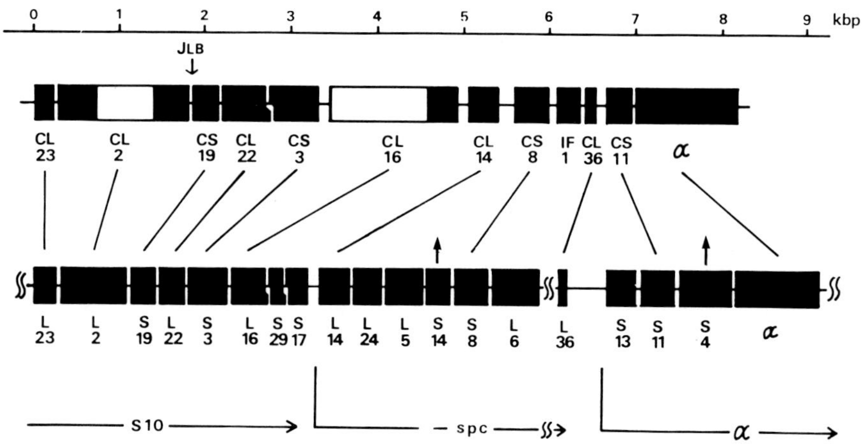

FIGURE 13–3 Comparison of the gene arrangement of the *N. tabacum rpl*2 cluster (top) with those of *E. coli S10* and *spc* operons (bottom). Filled boxes indicate exons, and open boxes, introns. Arrows indicate genes found outside the *rpl*2 cluster on the chloroplast DNA; others are probably located in the nuclear genome (Tanaka et al. 1986; Shinozaki et al. 1986c).

It has been suggested that the chloroplast RNA polymerase in higher plants is nuclear encoded (Lerbs et al. 1985). ORF337 located at the end of the *rpl*2 cluster corresponds to the spinach gene for the α subunit of RNA polymerase (*rpo*A) (Sijben-Müller et al. 1986). ORF1070 has been designated the genes for the β subunit (*rpo*B) (Ohme et al. 1986). A series of ORFs located downstream from *rpo*B shows homology with the gene for the β′ subunit of *E. coli* RNA polymerase. These findings raise the possibility that the chloroplast is an additional site of synthesis of its RNA polymerase subunits (Ohme et al. 1986; Shinozaki et al. 1986c).

13.5 GENES FOR THYLAKOID POLYPEPTIDES

Thylakoid membranes of higher plants have five functionally distinct protein complexes (Herrmann et al. 1985). These are the photosystem I (PSI) complex, the photosystem II (PSII) complex, the light-harvesting chlorophyll protein complex (its proteins are all nuclear coded), the cytochrome *b*/*f* complex, and the coupling factor H^+-ATPase complex.

So far 21 genes for thylakoid polypeptides have been found in the *N. tabacum* chloroplast genome, as listed in Table 13–4 (Sugita and Sugiura 1984; Shinozaki et al. 1986c). These genes have been identified by using the corresponding spinach gene probes (gifts from Dr. R.G. Herrmann) and through their homology with published sequences (Fish et al. 1985; Alt et al. 1984; Morris and Herrmann 1984; Willey et al. 1984; Heinemeyer et al. 1984; Herrmann et al. 1984; Steinmetz et al. 1986). The deduced N-terminal amino acid sequence of ORF73 located between *psb*B and *pet*B matches that reported for the spinach 10-kd phosphoprotein of PSII (Farchaus and Dilley 1986). Therefore this ORF was designated the *N. tabacum* gene for the 10-kd phosphoprotein (*psb*H) (formerly designated *psb*F) (Shinozaki et al. 1986c). ORF52, located between *trn*S and *trn*Q (Deno and Sugiura 1983), corresponds to the gene for the 6-kd component of PSII recently found in the maize genome (*psb*I) (Kato et al. 1987). The gene for the 9-kd polypeptide (an apoprotein for the iron-sulfur centers A and B) of PSI has recently been located in SSC (Hayashida et al. 1987). The *pet*B and *pet*D genes contain single introns of 753 bp and 742 bp, respectively (Tanaka et al. 1987). The *N. debneyi psb*A has also been sequenced (Zurawski et al. 1982).

Of the nine subunits of the H^+-ATPase complex, six are coded for by the chloroplast DNA (subunits α, β, ϵ, I, III, and IV or a). All six *N. tabacum* genes have been identified, as listed in Table 13–4 (Deno et al. 1983 and 1984; Shinozaki et al. 1983 and 1986b). The *atp*F gene contains a 695-bp intron (Shinozaki et al. 1986b).

Mitochondria and chloroplasts are the major sites for energy conversion in the plant cell. The respiratory chain of mitochondria contains a large number of electron transport proteins that act in sequence to transfer electrons from substrate to oxygen. In the chain a reduced nicotinamide adenine

TABLE 13–4 ***N. tabacum*** **Chloroplast Genes for Components of Photosynthetic Apparatus**

Genes	*Products*
*rbc*L	Ribulose-1,5-bisphosphate carboxylase large subunit
	H^+-ATPase
*atp*A	Subunit α
*atp*B	Subunit β
*atp*E	Subunit ϵ
*atp*F	Subunit I
*atp*H	Subunit III
*atp*I	Subunit IV (or a)
	Photosystem I
*psa*A	P700 Apoprotein A1
*psa*B	P700 Apoprotein A2
*psa*C	9-kd Polypeptide
	Photosystem II
*psb*A	D1 protein
*psb*B	51-kd Polypeptide
*psb*C	44-kd Polypeptide
*psb*D	D2 protein
*psb*E	Cytochrome b559, 8-kd subunit
*psb*F	Cytochrome b559, 4-kd subunit
*psb*G	G protein
*psb*H	10-kd Phosphoprotein
*psb*I	I protein
	Cytochrome *b*/*f* complex
*pet*A	Cytochrome *f*
*pet*B	Cytochrome *b*6
*pet*C	Subunit IV

dinucleotide (NADH) dehydrogenase (complex I) accepts electrons from NADH and transfers them to ubiquinone. Evidence for a respiratory-like NADH dehydrogenase and a respiratory chain in the chloroplast of *Chlamydomonas reinhardtii* has been presented (Godde 1982; Bennoun 1982). However, the existence of an NADH dehydrogenase has not been reported in higher plant chloroplasts.

Six *N. tabacum* chloroplast DNA sequences have been found (Shinozaki et al. 1986c; Meng et al. 1986) whose predicted amino acid sequences resemble those of components (ND1, 2, 3, 4, 4L, and 5) of the respiratory chain NADH dehydrogenase from human mitochondria (Chomyn et al. 1985) (Table 13–5). As these sequences are highly expressed in tobacco

TABLE 13–5 Amino Acid Sequence Homologies Between *N. tabacum* Chloroplast NDs and Human Mitochondrial NDs

Subunit	*Tobacco Chloroplast Amino Acids*	*Human Mitochondria Amino Acids*	*Homology*[1]*(%)*
ND1	364	318	31(42)
ND2	387	347	19(30)
ND3	120	115	24(48)
ND4	509	459	21(34)
ND4L	101	98	27(42)
ND5	710	603	22(31)

[1] Homology of identical and similar (in parentheses) amino acids.

chloroplasts, they are most likely to be the genes for components of a chloroplast NADH dehydrogenase (*ndh*A, B, C, D, E, and F) (Matsubayashi et al. 1987). This suggests the existence of a respiratory chain in the chloroplast (chlororespiration) of higher plants, as has been reported in *Chlamydomonas*. A further possibility is that this is an example of transposition in the direction opposite to what has been observed so far (namely, the insertion of chloroplast genes into mitochondrial genomes; Stern and Lonsdale 1982), so that these *ndh*s could be pseudogenes. The *ndh*A and *ndh*B genes contain single introns of 1,148 bp and 679 bp, respectively (Matsubayashi et al. 1987).

13.6 GENE EXPRESSION

Chloroplast genes are transcribed by the chloroplast RNA polymerase. Some of the genes have been shown to be cotranscribed (e.g., *atp*F-*atp*A, *trn*E-*trn*Y-*trn*D, *atp*B-*atp*E, *rpl*2 cluster, 3′-*rps*12-*rps*7, *psb*B cluster, and *rrn*), while others are transcribed individually (e.g., *psb*A, *trn*K, *rps*16, *trn*G-UCC, *trn*V-UAC, and *rbc*L).

Transcriptional initiation sites of the *N. tabacum psb*A, *trn*G-UCC, *trn*EYD, *atp*BE, *psb*B, and *rbc*L genes have been identified by S1 mapping. Upstream of these sites there are sequences highly homologous to bacterial "−10" and "−35" regions. *E. coli* RNA polymerase has been shown to recognize the *N. tabacum rbc*L and *atp*BE promoters and initiate transcription at their authentic initiation sites (Shinozaki and Sugiura 1982b). Expression in *E. coli* and *Bacillus subtilis* of *N. otophora rbc*L has also been demonstrated (Zhu et al. 1984). Identification of *N. tabacum* chloroplast DNA regions active as promoters in *E. coli* has been accomplished using the *gal*K expression plasmid vector (Kong et al. 1984). Recently, essential regions in the spinach *trn*M2, *rbc*L, *atp*BE, and *psb*A promoters have been experimentally identified as similar to the prokaryotic −35 and −10 regions (Gruissem and Zurawski 1985). Many other genes also contain sequences similar to prokaryotic promoters in front of their coding regions and these sequences are most likely their promoters, although they are not yet defined functionally (Crouse et al. 1984; Kung and Lin 1985). Some genes (e.g., *trn*K, *rps*16, *trn*V-UAC, and *rrn*) seem to have multiple promoters. Figure 13–4 lists promoter sequences of several chloroplast genes. The −10 region–like sequences of chloroplast promoters occur three to nine base pairs upstream from the transcriptional initiation sites, while those of *E. coli* promoters are four to ten base pairs upstream from the initiation sites. The distance between the −10 region–like and −35 region–like sequences of chloroplast promoters is 17 to 21 bp, while that of *E. coli* promoters is 14 to 21 bp (Shinozaki and Sugiura 1986a).

Palindromic sequences that can form stem and loop structures in the RNA transcripts are required for rho-factor independent transcription termination of prokaryotic genes. Transcriptional termination sites of the *psb*A, *trn*EYD, *atp*BE, and *rbc*L genes have been identified by S1 mapping.

```
                                 "-35"                              "-10"
rbcL                                                                                  +1
     Tobacco             ATTGGG TTGCGC TATATATATGAAAGAGTA      TACAAT AATG     ATTAT
     Spinach             GTTGGG TTGCGC CATATATATGAAAGAGTA      TACAAT AATG     ATGTAT
     Maize               TTTGGG TTGCGC TATATCTATCAAAGAGTA      TACAAT AATG     ATGGAT
     Chlamydomonas       TGCTAG TTTACA TTATTTTTTATTTCTAAATA    TATAAT ATATTT   AAATGT
psbA                                                                                  +1
     Tobacco             CCTTGG TTGACA CGAGTATATAAGTCATGT      TATACT GTTG     AATAAA
     Spinach             TATTGG TTGACA CGGGCATATAAGGCATGT      TATACT GTTG     AATAAC
     Soy bean            TATTGG TTGACA CTGGTATATAAGTCATGT      TATACT GTTG     AATAAC
     Mustard             CATTGG TTGACA TGGCTATATAAGTCATGT      TATACT GTTC     AATAAC
atpB                                                                                  +1
     Tobacco             ATAGAA TAGATA ATATGGATGGGATTGTC       TATAAT GATA     GACAAA
     Spinach             TTAGTC TTGACA GTGGTATATGTTGTATATGTA   TATCCT AGA      TGTGAA
     Maize               TCTCTG TTGACA GCAATCTATGCTTCACAG      TAGTAT ATA      TTTTGT
psbB                                                                                  +1
     Spinach             CTCCGA TTGCGT ATTGCTACTTATCGAGTA      TAGAAT AGATTT   GTTTCT
rDNA                                                                                  +1
     Maize               GTGGGA TTGACG TGATAGGGTAGGGTTGGC      TATACT GCTGGT   GGCG
     Spinach             GTGGGA TTGACG TGAGGGGGCAGGGATGGC      TATATT TCTG     GGAGCG
     Duckweed            GTGGGA TTGACG TGATAGGGTAGGGATGGC      TATATT GCTGGGAGC C
trn                                                                                   +1
     Tobacco trnV        GTGGCG TTGAGT TTTCTCGACCCTTTGACT      TAGGAT TAGTC    AGTTCT
     Tobacco trnG        CCCTGT TCGACA AAAGTTGCATTTGTA         TACAAT AATCGGG ATTGTA
     Tobacco trnE        ATTATA TTGACA ATTTCAAAAAACTGATCA      TACTAT GATCATA GTATGA

Chloroplast                                                                           +1
conserved sequence          ---g TTGaca --------------------   TAtAaT a-----    a
                                            17 - 21bp                 3 - 9bp

E. coli                                                                               +1
conserved sequence          ---- TTGaca --------------------   TAtAaT ------    a
                                            14 - 21bp                 4 - 10bp
```

FIGURE 13–4 Promoter sequences of chloroplast genes. Transcriptional initiation sites are positioned as +1. In conserved sequences capital letters indicate nucleotides that appear at a frequency of over 80%, and lower case letters, a frequency of 50–80%. (From K. Shinozaki and M. Sugiura [1986] *Jpn. J. Genet.* 61, 371–409. With permission.)

Short IR sequences have been found just before the stop sites (Figure 13–5). This indicates a further prokaryotic feature of the chloroplast genes. One interesting observation is that *atp*BE has two terminators, both of which are located within *trn*M encoded on the opposite strand (see Figure 13–5).

Introns in chloroplast genes were first reported for the 23S rDNA of *Chlamydomonas* (Rochaix and Malnoë 1978). Long introns (0.5–2.5 kbp) have also been found in tRNA genes and protein genes in *Nicotiana* plants. Introns found in chloroplast genes can be classified into three groups on the basis of the compiled intron boundary sequences of chloroplast genes so far reported (Shinozaki et al. 1986a) (Figure 13–6).

1. Group I introns. The intron of *trn*L can be folded into a secondary structure that is similar to the postulated structure of the intron of the self-splicing rRNA precursor of *Tetrahymena* (Michel and Dujon 1983; Bonnard et al. 1984). Chloroplast group I intron sequences may be removed by an autocatalytic reaction, as in the case of the *Tetrahymena* rRNA precursor, or by a "maturase," as in the case of cytochrome *b* and cytochrome oxidase mRNA precursors.
2. Group II introns. Introns of *trn*A and *trn*I can be folded into a secondary structure that is similar to the postulated structure of introns of the genes for maize mitochondrial cytochrome oxidase and yeast cytochrome oxidase and cytochrome *b* (Michel and Dujon 1983). Complex secondary structures may be necessary to excise the group II intron sequences.
3. Group III introns. Introns of *trn*G, *trn*K, *trn*V, *rpl*2, *rps*12, *rps*16, *atp*F, *pet*B, *pet*D, *ndh*A, and *ndh*B have conserved sequences at their boundaries (See Figure 13–6). The conserved boundary sequences of the group III introns are GTGCGNY at the 5′ ends and ATCNRYY(N)YYAY at the 3′ ends (Sugita et al. 1985; Shinozaki et al. 1986a). These conserved boundary sequences resemble those of *Euglena gracilis* chloroplast introns (Hallick et al. 1985); they also resemble the conserved boundary sequences of nuclear gene introns. The group III intron sequences may

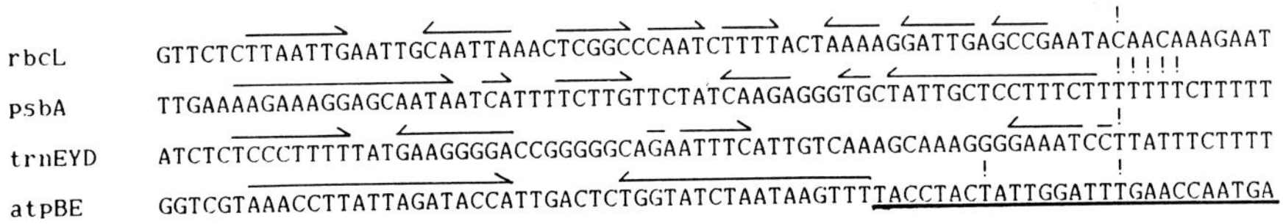

FIGURE 13–5 Terminator sequences of *N. tabacum* genes. Exclamation marks indicate transcriptional termination sites. Arrows show inverted repeated sequences. Sequences underlined are the *trn*M coding region. (From K. Shinozaki and M. Sugiura [1986] *Jpn. J. Genet.* 61, 371–409. With permission.)

			5'EXON	INTRON		3'EXON	INTRON SIZE (bp)
Group I	Tobacco	trnL-UAA	ACGGACTT	AATTGGATTGAGCCTTG---------	---------GAAATTTATCGTAAGAGG	AAAATCCG	503
	Maize	trnL-UAA	ACGGACTT	GATTGTATTGAGCCTTG---------	---------GAAATTTCTAGTAAAAGG	AAAATCCG	458
	Bean	trnL-UAA	ACGGACTT	AATTGTATTGAGCCTTG---------	---------GAAAGTTTTAGTAATCGG	AAAATCCG	451
Group II	Tobacco	trnI-GAU	CCTGATAA	TTGCGTCGTTGTGCCTG---------	---------TGCGATGATTTACTTCAC	GGGGCGAG	707
	Maize	trnI-GAU	CCTGATAA	TT CGTCGTTGTGCCTC---------	---------GGCGATGATTTACTTCAC	GGGGCGAG	949
	Tobacco	trnA-UGC	TCTTGCAA	TTGGGTCGTTGCGATTA---------	---------GGCGGTGGTTTACCCTGC	GGCGGATG	709
	Maize	trnA-UGC	TCTTGCAA	TTGGGTCGTTGCGATTA---------	---------GGCGGTGGTTTACCCTGT	GGCGGATG	806
Group III	Tobacco	trnV-UAC	GTTTACAC	GCGCGCCAATGTTTTTC---------	---------AGTTTGACCTGTT TTAC	CGAGAAGG	571
	Maize	trnV-UAC	GTTTACAC	GTGCGCCAATGCTTTTC---------	---------TGTTTGATCTGTT TTAC	CGAGAAGG	603
	Barley	trnV-UAC	GTTTACAC	GTGCGCCAATGCTTTTC---------	---------CGTTTGATCGTTT TTAC	CGAGAAGG	597
	Tobacco	trnG-UCC	TGGTAAAA	GTGTGATTCGTTCTATT---------	---------ATCGTCGTCGACTATAAC	CCCTAGCC	691
	Wheat	trnG-UCC	TGGTAAAA	GTGTGATTCGTTCTATT---------	---------GAACTTATCTACT CCAT	CCGACTAG	676
	Tobacco	trnK-UUU	GCTTTTAA	GTGCGGCTAGTCTCTTT---------	---------GAACTTATCTACT CCAT	CCGACTAG	2526
	Tobacco	rpl2	ACCTTTGA	GTGCGGTTTGAACTATT---------	---------AGAAGAATCTACT TCAA	CCGATATG	666
	Tobacco	rps16	AAAGCAAC	GTGCGACTTGAAGGACA---------	---------ATCTACATCAATCCCAAT	GAGCCGTC	860
	Tobacco	rps12	GCGTTCTA	GTGCGTTGTAGATTCTT---------	---------TTCGAGATCCACC CTAC	AATATGGG	536
	Tobacco	rpl16	TATGCTTA	GTGTGACTCCGTTGGTT---------	---------ACAACCATCAACTATAAC	CCCAAAAG	1020
	Wheat	atpF	GGGAGTGT	GTGCGAGTTGTCTATTT---------	---------CAAATAATCTACTTTCAT	TAAAAGAT	823
	Tobacco	atpF	GGGAGTGT	GTGCGAGTTGTTTATTT---------	---------AAAATAATCTACTTTCAT	TAAGTGAT	695
	Tobacco	petB	GTATGAGT	GTGTGACTTGTTATAAT---------	---------AGTTCCGCCTATCTCAAT	AAAGTATA	759
	Tobacco	ndhA	ATCTCTAC	GTGTGATTCGGTGAGAC---------	---------GTTATCATCGACTATGAT	TATCTAAC	1148
Group III conserved sequence				GTGCGNY c t g	ATCNRYYNYYAY gc tg a a		
Euglena conserved sequence				GTGYG t c	TARTTNTAY ct ga a a		
Nuclear gene conserved sequence			AAG c	GTGRAGT	YYYYYYNCAG	GG t	

FIGURE 13–6 Exon-intron boundary sequences in chloroplast split genes from higher plants. (From K. Shinozaki and M. Sugiura [1986] *Jpn. J. Genet.* 61, 371–409. With permission.)

be spliced out by a mechanism similar to that used for nuclear mRNA precursors.

Tobacco *rps*12 is divided into one copy of 5′-*rps*12 and two copies of 3′-*rps*12. The 5′-*rps*12 contains exon 1 of 38 codons, and 3′-*rps*12 consists of exon 2 of 78 codons, a 536-bp intron, and exon 3 of 7 codons. Northern blot hybridization revealed that both 5′-*rps*12 and 3′-*rps*12 are transcribed in the chloroplasts. Reverse transcription analysis indicated that trans splicing between the 5′-*rps*12 and 3′-*rps*12 transcripts occurs in vivo (Zaita et al. 1987). Trans-spliced CS12 mRNA has also been observed by electron microscopy using artificially combined exons 1, 2, and 3 DNAs (Koller et al. 1987).

The 3′-flanking sequence of exon 1 has been designated "transon" 1, and the 5′-flanking sequence of exon 2, transon 2 (Zaita et al. 1987). The sequence GTGCGAC at the 5′ end of transon 1 and the sequence GTCAACTTTTCC at the 3′ end of transon 2 fit the conserved boundary sequences of the chloroplast group III introns. A long complementary structure (about 70 bp) can be constructed between transon 1 and transon 2. This structure, if any, may be necessary for, or at least enhance, trans splicing as has been reported in trans splicing in vitro.

13.7 CONCLUSIONS

The sequence and expression analyses have shown both prokaryotic and eukaryotic features of *Nicotiana* chloroplast genes. The genes coding for rRNAs and some proteins (e.g., ribosomal proteins) have substantial sequence homology with their prokaryotic counterparts. The basic regulatory sequences (promoters, terminators, and ribosomal binding sites) are also similar to those in prokaryotic genomes. Some of the gene clusters resemble the corresponding clusters of *E. coli* and cyanobacteria (e.g., *rrn, rpl*2, and *atp* clusters).

Some of the chloroplast genes contain introns similar to those that have been found in eukaryotic genomes. However, introns found in the tRNA genes are very long (up to 2,526 bp), and one intron is located in an unusual position, namely the D-stem region of *trn*G-UCC. The chloroplast splicing mechanisms seem to be more complex than eukaryotic splicing systems. The *rps*12 gene is divided into three parts that are far away from each other and require trans splicing (a divided gene).

The endosymbiotic theory, which proposes that chloroplasts are derived from an ancestral photosynthetic prokaryote related to cyanobacteria, has been supported in part by comparisons between chloroplast and cyanobacterial *rrn* operons. This leads us to speculate that ancestral photosynthetic prokaryotes had introns in their genomes and that existing chloroplast genomes have retained these intron sequences (Tomioka and Sugiura 1984).

REFERENCES

Alt, J., Morris, J., Westhoff, P., and Herrmann, R.G. (1984) *Curr. Genet.* 8, 597–606.

Amiri, I., Salnikow, J., and Vater, J. (1984) *Biochim. Biophys. Acta* 784, 116–123.

Atchison, B.A., Whitfeld, P.R., and Bottomley, W. (1976) *Mol. Gen. Genet.* 148, 263–269.

Bennoun, P. (1982) *Proc. Natl. Acad. Sci. USA* 79, 4352–4356.

Bergmann, P., Seyer, P., Burkard, G., and Weil, J.H. (1984) *Plant Mol. Biol.* 3, 29–36.

Bonnard, G., Michel, F., Weil, J.H., and Steinmetz, A. (1984) *Mol. Gen. Genet.* 194, 330–336.

Bowman, C.M., and Dyer, T.A. (1979) *Biochem. J.* 183, 605–613.

Capel, M.S., and Bourque, D.P. (1982) *J. Biol. Chem.* 257, 7746–7755.

Chomyn, A., Mariottini, P., Cleeter, M.W.J., et al. (1985) *Nature* 314, 592–597.

Crouse, E.J., Bohnert, H.J., and Schmitt, J.M. (1984) in *Chloroplast Biogenesis* (Ellis, R.J., ed.), pp. 83–136, Cambridge University Press, New York.

Crouse, E.J., Schmitt, J.M., and Bohnert, H.J. (1985) *Plant Mol. Biol. Reptr.* 3, 43–89.

Deno, H., Kato, A., Shinozaki, K., and Sugiura, M. (1982) *Nucleic Acids Res.* 10, 7511–7520.

Deno, H., Shinozaki, K., and Sugiura, M. (1983) *Nucleic Acids Res.* 11, 2185–2191.

Deno, H., Shinozaki, K., and Sugiura, M. (1984) *Gene* 32, 195–201.

Deno, H., and Sugiura, M. (1983) *Nucleic Acids Res.* 11, 8407–8414.

Deno, H., and Sugiura, M. (1984) *Proc. Natl. Acad. Sci. USA* 81, 405–408.

Dyer, T.A. (1984) in *Chloroplast Biogenesis* (Baker, N.R., and Barber, J., eds.), pp. 23–69, Elsevier, Amsterdam.

Farchaus, J., and Dilley, R.A. (1986) *Arch. Biochem. Biophys.* 244, 94–101.

Fish, L.E., Kück, U., and Bogorad, L. (1985) *J. Biol. Chem.* 260, 1413–1421.

Fluhr, R., and Edelman, M. (1981) *Mol. Gen. Genet.* 181, 484–490.

Fluhr, R., Fromm, H., and Edelman, M. (1983) *Gene* 25, 271–280.

Fromm, H., Edelman, M., Koller, B., Goloubinoff, P., and Galun, E. (1986) *Nucleic Acids Res.* 14, 883–898.

Godde, D. (1982) *Arch. Microbiol.* 131, 197–202.

Gray, J.C., Phillips, A.L., and Smith, A.G. (1984) in *Chloroplast Biogenesis* (Ellis, R.J., ed.), pp. 137–163, Cambridge University Press, New York.

Gruissem, W., and Zurawski, G. (1985) *EMBO J.* 4, 3375–3383.

Hallick, R.B., and Bottomley, W. (1983) *Plant Mol. Biol. Reptr.* 1, 38–43.

Hallick, R.B., Gingrich, J.C., Johanningmeier, U., and Paasavant, C.W. (1985) in *Molecular Form and Function of the Plant Genome* (von Vloten-Doting, L., Groot, G.S.P., and Hall, T.C., eds.), pp. 211–220, Plenum Publishing, New York.

Hayashida, N., Matsubayashi, T., Shinozaki, K., et al. (1987) *Curr. Genet.* 12, 247–250.

Heinemeyer, W., Alt, J., and Herrmann, R.G. (1984) *Curr. Genet.* 8, 543–549.

Herrmann, R.G., Alt, J., Schiller, B., Widger, W.R., and Cramer, W.A. (1984) *FEBS Lett.* 176, 239–244.

Herrmann, R.G., Westhoff, P., Alt, J., Tittgen, J., and Nelson, N. (1985) in *Molecular Form and Function of the Plant Genome* (von Vloten-Doting, L., Groot, G.S.P., and Hall, T.C., eds.), pp. 233–256, Plenum Publishing, New York.

Hildebrand, M., Jurgenson, J.E., Ramage, R.T., and Bourque, D.P. (1985) *Plasmid* 14, 64–79.

Hughes, K.W., ed. (1982) *Plant Mol. Biol. News.* 3, 46–100.

Jurgenson, J.E., and Bourque, D.P. (1980) *Nucleic Acids Res.* 8, 3505–3516.

Kato, K., Sayer, R.T., and Bogorad, L. (1987) *Proc. Ann. Meet. Jpn. Soc. Plant Physiol. (Urawa),* Abstract 208.

Kato, A., Shimada, H., Kusuda, M., and Sugiura, M. (1981) *Nucleic Acids Res.* 9, 5601–5607.

Kato, A., Takaiwa, F., Shinozaki, K., and Sugiura, M. (1985) *Curr. Genet.* 9, 405–409.

Koch, W., Edwards, K., and Kössel, H. (1981) *Cell* 25, 203–213.

Koller, B., Fromm, H., Galun, E., and Edelman, M. (1987) *Cell* 48, 111–119.

Kong, X.F., Lovett, P.S., and Kung, S.D. (1984) *Gene* 31, 23–30.

Kumano, M., Tomioka, N., and Sugiura, M. (1983) *Gene* 24, 219–225.

Kung, S.D., and Lin, C.M. (1985) *Nucleic Acids Res.* 13, 7543–7549.

Kung, S.D., Sakano, K., and Wildman S.G. (1974) *Biochim. Biophys. Acta* 365, 138–147.

Kung, S.D., Zhu, Y.S., and Shen, G.F. (1982) *Theor. Appl. Genet.* 61, 73–79.

Lerbs, S., Bräutigam, E., and Parthier, B. (1985) *EMBO J.* 4, 1661–1666.

Lin, C.M., Liu, Z.Q., and Kung, S.D. (1986) *Plant Mol. Biol.* 6, 81–87.

Matsubayashi, T., Wakasugi, T., Shinozaki, K., et al. (1987) *Mol. Gen. Genet.* 210, 385–393.

McIntosh, L., Poulsen, C., and Bogorad, L. (1980) *Nature* 288, 556–560.

Medgyesy, P., Fejes, E., and Maliga, P. (1985) *Proc. Natl. Acad. Sci. USA* 82, 6960–6964.

Meng, B.Y., Matsubayashi, T., Wakasugi, T., et al. (1986) *Plant Sci.* 47, 181–184.

Michel, F., and Dujon, B. (1983) *EMBO J.* 2, 33–38.

Montandon, P.E., and Stutz, E. (1984) *Nucleic Acids Res.* 12, 2851–2859.

Morris, J., and Herrmann, R.G. (1984) *Nucleic Acids Res.* 12, 2837–2850.

Ohme, M., Kamogashira, T., Shinozaki, K., and Sugiura, M. (1984) *Nucleic Acids Res.* 12, 6741–6749.

Ohme, M., Kamogashira, T., Shinozaki, K., and Sugiura, M. (1985) *Nucleic Acids Res.* 13, 1045–1056.

Ohme, M., Tanaka, M., Chunwongse, J., Shinozaki, K., and Sugiura, M. (1986) *FEBS Lett.* 200, 87–90.

Palmer, J.D. (1985) *Annu. Rev. Genet.* 19, 325–354.

Pillay, D.T.N., Guillemaut, P., and Weil, J.H. (1984) *Nucleic Acids Res.* 12, 2997–3001.

Rhodes, P.R., Zhu, Y.Z., and Kung, S.D. (1981) *Mol. Gen. Genet.* 182, 106–111.

Rochaix, J.D., and Malnoë, P. (1978) *Cell* 15, 661–670.

Sager, R., and Schlanger, G. (1976) in *Handbook of Genetics,* vol. 5 (King, R.C., ed.), pp. 371–423, Plenum Press, New York.

Salts, Y., Herrmann, R.G., Peleg, N., et al. (1984) *Theor. Appl. Genet.* 69, 1–14.

Samuelsson, T., Elias, P., Lusting, F., et al. (1980) *J. Biol. Chem.* 255, 4583–4588.

Seyer, P., Kowallik, K.V., and Herrmann, R.G. (1981) *Curr. Genet.* 3, 189–204.

Shen, G.F., Chen, K., Wu, M., and Kung, S.D. (1982) *Mol. Gen. Genet.* 187, 12–18.

Shinozaki, K., Deno, H., Kato, A., and Sugiura, M. (1983) *Gene* 24, 147–155.

Shinozaki, K., Deno, H., Sugita, M., Kuramitsu, S., and Sugiura, M. (1986a) *Mol. Gen. Genet.* 202, 1–5.

Shinozaki, K., Deno, H., Wakasugi, T., and Sugiura, M. (1986b) *Curr. Genet.* 10, 421–423.

Shinozaki, K., Ohme, M., Tanaka, M., et al. (1986c) *EMBO J.* 5, 2043–2049.

Shinozaki, K., Ohme, M., Tanaka, M., et al. (1986d) *Plant Mol. Biol. Reptr.* 4, 111–147.

Shinozaki, K., and Sugiura, M. (1982a) *Gene* 20, 91–102.

Shinozaki, K., and Sugiura, M. (1982b) *Nucleic Acids Res.* 10, 4923–4934.

Shinozaki, K., and Sugiura, M. (1986a) *Adv. Biophys.* 21, 57–58.

Shinozaki, K., and Sugiura, M. (1986b) *Jpn. J. Genet.* 61, 371–409.

Sijben-Müller, G., Hallick, R., Alt, J., Westhoff, P., and Herrmann, R.G. (1986) *Nucleic Acids Res.* 14, 1029–1044.

Smith, H.H. (1974) in *Handbook of Genetics,* vol. 2 (King, R.C., ed.), pp. 281–314, Plenum Press, New York.

Steinmetz, A.A., Castroviejo, M., Sayre, R.T., and Bogorad, L. (1986) *J. Biol. Chem.* 261, 2485–2488.

Stern, D.B., and Lonsdale, D.M. (1982) *Nature* 299, 698–702.

Sugita, M., Kato, A., Shimada, H., and Sugiura, M. (1984) *Mol. Gen. Genet.* 194, 200–205.

Sugita, M., Shinozaki, K. and Sugiura, M. (1985) *Proc. Natl. Acad. Sci. USA* 82, 3557–3561.

Sugita, M., and Sugiura, M. (1983) *Nucleic Acids Res.* 11, 1913–1918.

Sugita, M., and Sugiura, M. (1984) *Mol. Gen. Genet.* 195, 308–313.

Sugiura, M., and Kusuda, J. (1979) *Mol. Gen. Genet.* 172, 137–141.

Sugiura, M., Shinozaki, K., Zaita, N., Kusuda, M., and Kumano, M. (1986) *Plant Sci.* 44, 211–216.

Takaiwa, F., and Sugiura, M. (1980a) *Mol. Gen. Genet.* 180, 1–4.

Takaiwa, F., and Sugiura, M. (1980b) *Nucleic Acids Res.* 8, 4125–4129.

Takaiwa, F., and Sugiura, M. (1981) *Mol. Gen. Genet.* 182, 385–389.

Takaiwa, F., and Sugiura, M. (1982a) *Eur. J. Biochem.* 124, 13–19.

Takaiwa, F., and Sugiura, M. (1982b) *Nucleic Acids Res.* 10, 2665–2676.

Takebe, I., and Nagata, T. (1984) in *Cell Culture and Somatic Cell Genetics of Plants* (Vasil, I.K., ed.), pp. 328–339, Academic Press, Orlando, FL.

Tanaka, M., Obokata, J., Chunwongse, J., Shinozaki, K., and Sugiura, M. (1987) *Mol. Gen. Genet.* 209, 427–431.

Tanaka, M., Wakasugi, T., Sugita, M., Shinozaki, K., and Sugiura, M. (1986) *Proc. Natl. Acad. Sci. USA* 83, 6030–6034.

Tassopulu, D., and Kung, S.D. (1984) *Theor. Appl. Genet.* 67, 185–193.

Tohdoh, N., Shinozaki, K., and Sugiura, M. (1981) *Nucleic Acids Res.* 9, 5399–5406.

Tohdoh, N., and Sugiura, M. (1982) *Gene* 17, 213–218.

Tomioka, N., Shinozaki, K., and Sugiura, M. (1981) *Mol. Gen. Genet.* 184, 359–363.

Tomioka, N., and Sugiura, M. (1984) *Mol. Gen. Genet.* 193, 427–430.

Torazawa, K., Hayashida, N., Obokata, J., Shinozaki, K., and Sugiura, M. (1986) *Nucleic Acids Res.* 14, 3143.

Vedel, F., Quetier, F., and Bayen, M. (1976) *Nature* 263, 440–442.

Wada, A., and Sako, T. (1987) *J. Biochem.* 101, 817–820.

Wakasugi, T., Ohme, M., Shinozaki, K., and Sugiura, M. (1986) *Plant Mol. Biol.* 7, 385–392.

Willey, D.L., Auffret, A.D., and Gray, J.C. (1984) *Cell* 36, 555–562.

Yamada, K., Shinozaki, K., and Sugiura, M. (1986) *Plant Mol. Biol.* 6, 193–199.

Zaita, N., Torazawa, K., Shinozaki, K., and Sugiura, M. (1987) *FEBS Lett.* 210, 153–156.

Zhu, Y.S., Duvall, E.F., Lovett, P.S., and Kung, S.D. (1982) *Mol. Gen. Genet.* 187, 61–66.

Zhu, Y.S., Lovett, P.S., Williams, D,M., and Kung, S.D. (1984) *Theor. Appl. Genet.* 67, 333–336.

Zurawski, G., Bohnert, H.J., Whitfeld, P.R., and Bottomley, W. (1982) *Proc. Natl. Acad. Sci. USA* 79, 7699–7703.

Zurawski, G., Bottomley, W., and Whitfeld, P.R. (1984) *Nucleic Acids Res.* 12, 6547–6558.

CHAPTER
14

Genetic Manipulation of the Chloroplast Genome

John C. Gray

Chloroplasts are one of a group of organelles, the plastids, that are vital for the metabolic functioning of plants. Chloroplasts are the site of photosynthesis and many biosynthetic reactions, but other, non-green plastids also carry out essential metabolic processes. Amyloplasts are the plastids responsible for starch synthesis in non-green parts of plants, chromoplasts synthesize large amounts of carotenoids in some fruits and flowers, and a whole host of biochemical reactions, including fatty acid and amino acid synthesis, takes place in colorless plastids throughout the plant. The manipulation of many of these biochemical pathways, ultimately to alter the quality of crop yields, is the aim of the plant breeder. The alteration of the protein, or enzyme, complement of plastids by using genetic manipulation methods requires a large amount of background information on the location and expression of genes for plastid proteins, as well as means for introducing genes into selected genomes in the plant cell.

I am extremely grateful to Richard Blyden and Chris Howe for valuable discussions, and to the Science and Engineering Research Council and the Agricultural and Food Research Council for supporting research in my laboratory.

Genes for plastid proteins are located either in the nuclear genome, in which case the proteins are synthesized in the cytoplasm and targeted to the plastids, or in the plastid genome. The plastid genome is of limited size and codes for relatively few proteins (see below), the vast majority of plastid proteins being coded by nuclear genes. The alteration of existing genes for plastid proteins, perhaps by homologous recombination with genes manipulated in vitro, requires a knowledge of the subcellular location of the genes. The introduction of novel proteins, or enzymes, into plastids can be accomplished by two routes: insertion into the nuclear genome of a gene construct containing the gene of interest fused to sequences coding for a plastid transit peptide plus appropriate transcriptional control signals, or insertion of the gene directly into the chloroplast for expression by the chloroplast genetic apparatus. The first route, the introduction of manipulated genes into the nuclear genome, has been successful using *Agrobacterium*–Ti plasmid–derived vectors for the introduction of aminoglycoside phosphotransferase activity into chloroplasts (Van den Broeck et al. 1985; Schreier et al. 1985). This approach is currently the method of choice for the introduction of proteins into plastids. The second route, the introduction of genes into plastids, that is, the genetic transformation of plastids, is the subject of this chapter. To date, there has been only one report of the successful transformation of chloroplasts (De Block et al. 1985); in the main, this chapter will consider necessary background information and future possibilities for genetic manipulation. First, however, the reported successful transformation of tobacco chloroplasts must be considered.

14.1 CHLOROPLAST TRANSFORMATION BY *AGROBACTERIUM TUMEFACIENS*

The only reported successful transformation of chloroplasts was achieved using an *Agrobacterium tumefaciens*–Ti plasmid construct containing a chimeric gene consisting of the promoter region of the nopaline synthase gene (*nos*) fused to the coding sequence of the chloramphenicol acetyltransferase (CAT) gene (*cat*) (De Block et al. 1985). The structure of the construct is shown in Figure 14–1. Also included in the construct between the border sequences of T-DNA were complete genes for nopaline synthase, aminoglycoside phosphotransferase type II (*aph*II), and β-lactamase. The construct was introduced into tobacco leaf protoplasts by cocultivation with *A. tumefaciens* containing the hybrid Ti plasmid. Chloramphenicol-resistant (Cm^R) calli were selected on media containing 10 μg/ml chloramphenicol. Only a small proportion (8%) of the resistant calli contained nopaline, suggesting that the *nos* gene had been inactivated or lost (nos^-) from the region of DNA containing the *cat* gene. Of twenty plantlets regenerated from Cm^R-nos^- calli only eight (40%) were able to root on chloramphenicol-containing medium, and CAT activity was lost from the shoots of plants

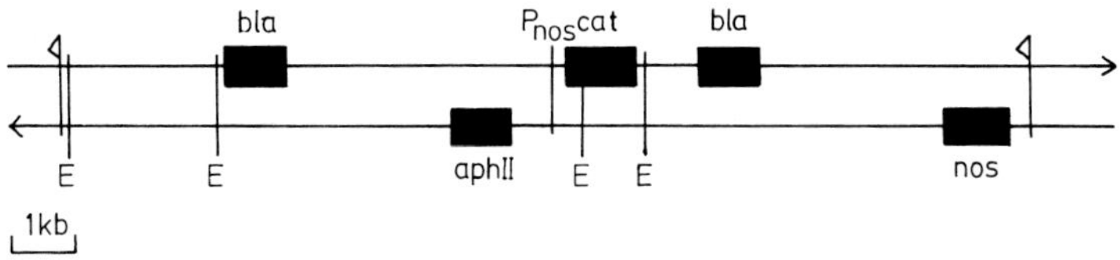

FIGURE 14–1 Structure of the T-DNA construct used for chloroplast transformation. The location and direction of transcription of the chimeric *nos-cat* gene and genes for aminoglycoside phosphotransferase (*aph*II), β-lactamase (*bla*), and nopaline synthase (*nos*) are shown between the T-DNA border sequences (flags). *Eco*RI restriction sites (E) in the construct are also shown. (Adapted from De Block et al. [1985].)

grown in the absence of chloramphenicol. Cell fractionation studies on leaves of a Cm^R plant (rGV3002) grown in the presence of chloramphenicol indicated that the CAT activity was localized in the chloroplast fraction. The chloroplast fraction also contained aminoglycoside phosphotransferase activity with kanamycin as the substrate. The mode of inheritance of chloramphenicol resistance was examined using the flowering Cm^R plant rGV3002. The results of reciprocal crosses with wild-type *Nicotiana tabacum* indicated that the *cat* gene was not transmitted through the pollen, but was maternally inherited in the progeny, with marked variation in relative CAT activity in different seedlings. Hybridization of a 900-bp *Eco*RI fragment containing a large portion of the *cat* gene to Southern blots of restriction digests of nuclear, mitochondrial, and chloroplast DNA indicated a chloroplast location of the *cat* gene. However, the *cat* gene probe hybridized to two fragments of 12.0 kbp and 5.4 kbp with different intensities in *Xho*I digests of different samples of DNA. As there were no *Xho*I sites in the *cat* gene construct, this suggested that the DNA integrated into the chloroplast DNA. The two hybridizing fragments may correspond either to two independently integrated fragments or to fragments of which one is a rearranged version of the other. Unfortunately these *Xho*I fragments were not recloned to allow the characterization of the integrated fragments and their sites of insertion in the plastid DNA. Hybridization with other probes indicated that these *Xho*I fragments also contained sequences homologous to the *aph*II gene.

The loss of CAT activity from plants grown in the absence of chloramphenicol suggests that the *cat* gene was not stably maintained in the chloroplast DNA. This suggests that the *cat* gene construct integrated into only a small proportion of the chloroplast DNA molecules, possibly at a site or sites that disrupt normal chloroplast gene function. In the presence of chloramphenicol the *cat* gene would be maintained, possibly with the normal chloroplast gene function being provided by chloroplast DNA molecules without *cat* gene inserts. In the absence of selection pressure the *cat* gene inserts would be lost, probably by recombination with wild-type chloroplast DNA molecules. However, in the absence of information on the site of integration of the *cat* gene, the possible fate of the gene constructs must remain speculative.

The expression of the *cat* gene in chloroplasts also raised questions concerning the transcription of the gene. The construct introduced into the tobacco protoplasts consisted of the promoter region of the *nos* gene fused to the *cat* gene. The *nos* promoter would normally be recognized by RNA polymerase II in the plant nucleus and would not be expected to be recognized by chloroplast RNA polymerase. De Block et al. (1985) suggested that the *nos* promoter region contained the sequence blocks TTG and TAAATT that could provide prokaryotic transcription signals for chloroplast RNA polymerase. However, neither the initiation site nor the sizes of the transcripts of the *cat* genes in chloroplasts were determined experimentally. It is therefore possible that the *cat* genes were transcribed as part of one of the large transcriptional units found in chloroplast DNA.

The reported transformation of tobacco chloroplasts with the *cat* gene has therefore raised a number of questions that will be answered only when more transformed plants are available for analysis. Unfortunately there have been no further reports of chloroplast transformation, and it must be suspected that chloroplast transformation using *Agrobacterium*–Ti plasmid-derived vectors is not a fully reproducible technique.

14.2 THE CHLOROPLAST GENOME

Any considered attempts at the genetic transformation of plastids require a knowledge of the organization of the chloroplast genome and the regulation of expression of its genes. The complete nucleotide sequences of chloroplast DNA from the liverwort, *Marchantia polymorpha* (Ohyama et al. 1986) and from tobacco (Shinozaki et al. 1986) have recently been determined and provide a major source of information on the organization of chloroplast genes. The study of the regulation of expression of chloroplast genes is still in its infancy, but it is clear that the transcription of chloroplast genes and the processing of transcripts are not simple events.

Only a few salient points on the organization and expression of chloroplast genes need to be emphasized here. The first is the small size of the chloroplast genome (within the range of 120–217 kbp for land plants, with the majority approximately 150 kbp; Palmer 1985). This small size obviously limits the amount of potential genetic information, or number of genes, in chloroplast DNA. Chloroplast DNAs from *Marchantia* and tobacco have been shown to contain the genetic information for a full complement of rRNAs and tRNAs and for a limited number of proteins. The proteins identified to date can be divided into two groups: those involved in the chloroplast transcription and translation apparatus, and those associated with the photosynthetic apparatus. In addition a large number of open-reading frames, or potential genes, have been identified by nucleotide sequencing, but their functions are not known. A map showing the location of these genes and open-reading frames in the tobacco chloroplast genome is shown in Figure 14–2. The organization of genes in other chloroplast genomes can be derived from this arrangement by a series of simple inversions (Palmer 1985). Although the tobacco chloroplast genome appears to be tightly packed with genes, in places there are quite large gaps between the genes. The largest gap, 2.4 kbp, occurs between the gene for cysteine-tRNA (*trn*C) and ORF 154, but numerous other gaps in excess of 1 kbp occur at various places around the genome. These gaps may be suitable sites for the integration of foreign genes.

Studies of transcription of chloroplast genes indicate that in many respects the features are typical of prokaryotic genes (Whitfeld and Bottomley 1983; Hanley-Bowdoin and Chua 1987). The promoter structure appears to be identical to the structure of the *Escherichia coli* promoter, with a −35 sequence resembling the *E. coli* consensus TTGACA and a −10 sequence resembling the *E. coli* consensus TATAAT. Studies using a chloroplast transcription system in vitro with promoter constructs mixing −35 and −10 sequences from several chloroplast genes indicate that maximum rates of transcription are obtained with the *E. coli* −35 and −10 consensus sequences separated by 17 bp (Gruissem and Zurawski 1985). The promoter for the *psb*A gene for the Q_B-binding protein of the photosystem II reaction center provides the closest match to this optimal promoter structure. However, transcriptional activity of chloroplast genes does not depend solely on the promoter structure, but on the organization of genes and their promoters in chloroplast DNA as well. The promoters for the *rbc*L and *atp*B genes, which are separated by about 150 bp, do not function independently, and RNA polymerase binding at the *rbc*L promoter interferes with transcription of the *atp*B gene, possibly by sterically hindering or by changing the DNA conformation. Deletion of the *rbc*L promoter leads to enhanced *atp*B transcription (Bradley and Gatenby 1985), and these effects on *atp*B transcription are eliminated by increasing the distance between the promoters (Hanley-Bowdoin and Chua 1987). These studies indicate that the inclusion of a strong promoter sequence in a construct for chloroplast transformation is not the only consideration, and that the arrangement of this promoter with respect to other promoters in the construct, or after integration into chloroplast DNA, is important.

Termination of transcription of chloroplast genes has not been studied in any great detail, but stable stem-loop structures may be predicted to exist based on the sequences near the 3′ ends of transcripts (Whitfeld and Bottomley 1983). This suggests that transcription termination in chloroplasts may be similar to that in *E. coli* (Rosenberg and Court 1979), but the

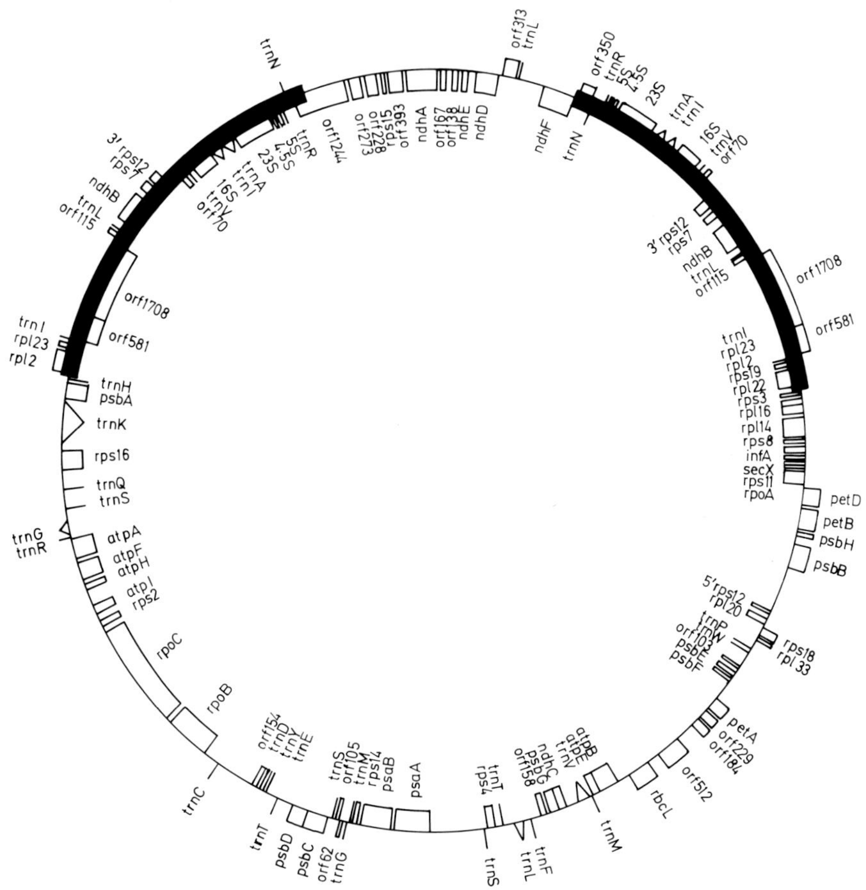

FIGURE 14–2 Map of the 156-kbp tobacco chloroplast genome showing identified genes and open-reading frames. Genes on the inner circle are transcribed clockwise, genes on the outer circle are transcribed counterclockwise. The inverted repeat sequences are boxed. Gene designations are as follows: *trn,* transfer RNA; *rrn,* ribosomal RNA; *rpl* and *rps,* ribosomal proteins of large and small subunits; *inf,* initiation factor; *rpo,* RNA polymerase; *psa,* photosystem I; *psb,* photosystem II; *pet,* cytochrome *b-f* complex; *atp,* ATP synthase; *rbc,* ribulose bisphosphate carboxylase; *ndh,* NADH dehydrogenase. (Adapted from Shinozaki et al. [1986].)

involvement of a rho-like termination factor in chloroplasts has not been established. Large stem-loop structures predicted to exist between *pet*D and *rpo*A (Phillips and Gray 1984) and between *pet*A and *psb*EF (Willey, unpublished observations) may be transcription terminators for these two pairs of convergently transcribed genes.

Many chloroplast genes are cotranscribed and arranged in transcriptional units that have been conserved during evolutionary changes in chloroplast gene organization (Palmer 1985). These transcriptional units appear to be transcribed from promoters upstream from the first gene in the unit and terminate downstream from the last gene. However, in most instances complex patterns of transcripts from these transcriptional units have been detected, suggesting that there is extensive processing of the primary transcripts. The largest transcriptional units studied so far are the 16S rDNA, *trn*I, *trn*A, 23S rDNA, 4.5S rDNA, 5S rDNA region (Strittmatter et al. 1985); the *psb*B, *psb*H, *pet*B, *pet*D region (Morris and Herrmann 1984; Hird et al. 1986; Westhoff et al. 1986); and the *atp*I, *atp*H, *atp*F, *atp*A region (Bird et al. 1985; Cozens et al. 1986; Hennig and Herrmann 1986). In each case a large transcript covering all the genes in each unit has been observed, but processing, including the removal of intron sequences, appears to be responsible for the generation of smaller gene-specific transcripts. Many other chloroplast genes, including *atp*B and *atp*E (Zurawski et al. 1982), *psb*E and *psb*F (Herrmann et al. 1984), *psa*A and *psa*B (Fish et al. 1985), *psb*D and *psb*C (Holschuh et al. 1984), and *trn*E, *trn*Y, and *trn*D (Ohme et al. 1985) are cotranscribed, and it is important that foreign genes integrated into the chloroplast genome do not disrupt these transcriptional units.

The translation of transcripts of protein-coding genes by chloroplast ribosomes appears to be very similar to translation by bacterial ribosomes. The mRNAs from most highly expressed genes contain a putative ribosome-binding site complementary to a sequence at the 3′ end of 16S rRNA, just upstream from the translation initiation site. This sequence, for example GGAGG or a variant thereof, resembles the bacterial ribosome-binding site recognized by Shine and Dalgarno (1974). However there have been no functional studies on these putative ribosome-binding sequences in chloroplasts. The presence of overlapping or very closely spaced genes in transcriptional units implies that translational coupling may occur, but this has not yet been demonstrated in chloroplast extracts. However, if translational coupling of adjacent genes occurs, the integration of foreign genes in these regions may disrupt chloroplast gene expression.

The regulation of chloroplast gene expression during plant development is not well studied, but ideas that regulation occurs primarily at the level of transcription may be too simplistic. It is now clear that the expression of a number of chloroplast genes is regulated posttranscriptionally. That the synthesis of chlorophyll-binding proteins of photosystems I and II is regulated at the level of translation in barley was shown by Klein and Mullet (1986). This study showed that transcripts of the *psa*A and *psa*B genes were present in dark-grown plants, but that synthesis of the proteins was blocked at the level of translation in the dark. In *Spirodela,* the synthesis of the *psb*A gene product was regulated at the level of translation and by the degradation of the protein (Fromm et al. 1985). The degradation of unassembled *rbc*S gene product in *Chlamydomonas* chloroplasts has been observed (Schmidt and Mishkind 1983), indicating that chloroplasts have efficient systems for the removal of unwanted or damaged proteins. These systems may cause problems with the expression of foreign genes in chloroplasts. Our lack of any real knowledge of the mechanisms for regulating the expression of chloroplast genes in nongreen plastids hinders our study of the expression of foreign genes in tissues other than leaves.

14.3 INTRODUCTION OF DNA INTO PLASTIDS

A number of possible methods for the introduction of DNA into plastids may be considered. In general preference might be given to natural means of entry of genetic information, but there appear to be no natural agents such as viruses or bacteria that obligatorily transfer genetic information into plastids. Similarly there is no information on whether chloroplasts are able to take up DNA across their envelope membranes. The current dogma is that there is no transfer of RNA or DNA across the chloroplast envelope during chloroplast biogenesis, but there has been no direct experimental examination of this idea. In the apparent absence of clear-cut natural mechanisms of DNA uptake, a number of artificial means of achieving uptake may be considered.

14.3.1 DNA Uptake by Chloroplasts In Vivo

The possibility that chloroplasts possess a DNA uptake system that is a relic from their free-living prokaryotic ancestors must be considered. A number of bacterial species, especially *Bacillus, Streptococcus,* and *Haemophilus* species, are transformable by naked DNA in vivo (Smith et al. 1981) and possess special mechanisms for DNA uptake. The reported transformation of chloroplasts following the introduction into plant cells of *Agrobacterium*–Ti plasmid–derived vectors (De Block et al. 1985) suggests that chloroplasts may be able to take up exogenous DNA, although nothing is known about the mechanism of DNA uptake. If chloroplasts are able to take up DNA, then introduction of DNA into the cytoplasm of plant cells may lead to uptake by chloroplasts. Direct gene transfer, including electroporation, appears to be an efficient means of introducing foreign DNA into protoplasts (Shillito et al. 1985), and it may be expected that in a small proportion of these protoplasts DNA may be taken up by chloroplasts. If this DNA were to be stably maintained in the chloroplasts it should be manifested by non-Mendelian maternal inheritance in offspring from crosses of plants regenerated from the protoplasts. However, the absence so far of any clear-cut results from direct gene transfer experiments (Potrykus et al. 1985; Hain et al. 1985) cannot be taken to indicate that the chloroplasts do not have a DNA uptake system. It is highly likely that the constructs tested so far would be unlikely to operate in the chloroplasts and that no consideration was given to their maintenance in the chloroplasts. Further experimentation needs to be carried out to assess DNA uptake by chloroplasts in vivo.

The possibility of engineering a DNA uptake system into the chloroplast envelope may be considered in the future when something is known about targeting proteins to the chloroplast envelope. A DNA uptake system similar to that present in *Haemophilus influenzae* (Goodgal 1982) would appear to offer considerable advantages. In this transformation system a specific 11-bp sequence in the DNA is recognized by a cell-surface receptor, leading to internalization of intact double-stranded DNA. Many details of this process are not clear, and it is not known how many genes are required to specify the DNA uptake apparatus (Goodgal 1982). However, when this system, or others like it, is fully defined genetically, it may be possible to transfer gene constructs to the nuclear genome of plants and target the proteins to the chloroplast envelope. DNA introduced into the cell cytoplasm would then be specifically recognized by the chloroplast uptake system.

14.3.2 Microinjection of DNA into Chloroplasts In Vivo

Microinjection of DNA directly into chloroplasts appears to be possible, although as yet there are no reports that it has been successful. Microinjection of *Chlamydomonas* chloroplasts has been suggested as a possible route to chloroplast transformation because of the large size of the single chloroplast, which occupies over 40% of the cell volume (Rochaix et al. 1984). Microinjection of DNA into nuclei in protoplasts, leading to transformation, has been achieved (Crossway et al. 1986), suggesting that there should be few technical problems with microinjection of chloroplasts in protoplasts of higher plants. The number of chloroplasts injected per cell will need to be considered, because little is known of the fate of chloroplasts and chloroplast DNA during the various stages of plant regeneration from protoplasts.

14.3.3 DNA Uptake by Chloroplasts In Vitro

Some of the problems encountered in attempting to introduce DNA into chloroplasts in vivo, such as the numbers of chloroplast treated and the low concentrations of DNA in the vicinity of the chloroplasts, may be overcome by attempting to introduce DNA into chloroplasts in vitro. However even if in vitro methods were successful, the manipulated chloroplasts would then have to be returned to plant cells, and there is currently no successful method for the introduction and stable maintenance of isolated chloroplasts into plant cells. There have been a number of reports of the successful uptake of isolated chloroplasts by plant protoplasts (Potrykus 1973; Bonnett and Eriksson 1974; Bonnett 1976; Davey et al. 1976; Landgren and Bonnett 1979; Uchimiya 1979; Giles et al. 1980), but stable incorporation of the chloroplasts and subsequent regeneration of transformed plants has not been achieved. It has been demonstrated that polyethylene glycol–induced uptake of chloroplasts by protoplasts results in the loss of the chloroplast envelope by fusion with the plasma membrane of the protoplasts (Davey et al. 1976). Liposome-mediated transfer of chloroplasts has been suggested as a means of transferring large numbers of chloroplasts into protoplasts (Giles et al. 1980; Whitaker and Evans 1986), but the morphological characteristics of the transferred chloroplasts have not been examined by electron microscopy.

There are several possible means for achieving DNA uptake by chloroplasts in vitro. As discussed previously, chloroplasts may have a DNA uptake system; this would probably operate most efficiently in vitro, where high concentrations of DNA could be supplied for the chloroplasts. However, other means such as electroporation may be required to allow access of externally added DNA to the chloroplast stroma. There are no reports of the electroporation of isolated chloroplasts, and it is not clear what effects electroporation will have on the chloroplast envelope membrane. Because of the high permeability of the outer envelope membrane, it seems that the low-ionic-strength media used in electroporation would penetrate the intermembrane space, and electroporation would affect only the inner envelope membrane. However high-molecular-weight DNA might not penetrate the outer membrane and hence might not have access to the pores.

Similar problems due to the dual membrane nature of the chloroplast envelope may also be encountered with attempts to deliver DNA in liposomes (Lurquin 1979; Ostro et al. 1980). Simple liposomes fusing with the outer envelope membrane would merely deliver their contents into the intermembrane space. More complex liposomes would be required to deliver DNA to the stroma.

It appears from the above that there are no easy routes for the introduction of foreign DNA into chloroplasts. However if methods are successfully developed then the next question concerns the maintenance and replication of the foreign DNA sequences in the chloroplast. Two strategies require consideration. The first is the integration of foreign sequences into chloroplast DNA molecules so that the foreign sequences will be replicated and inherited together with the native chloroplast genes. The second is the introduction of foreign sequences in autonomously replicating plasmids that may replicate independently of chloroplast DNA molecules. These strategies are discussed below.

14.4 INTEGRATION OF FOREIGN GENES INTO CHLOROPLAST DNA

The results of the hybridization experiments reported by De Block et al. (1985) suggest that the *cat* gene construct was integrated, perhaps at two sites, in tobacco chloroplast DNA. The mechanism of integration is not known but the enzymes of the chloroplast recombination system may have been involved. Although little is known about the chloroplast recombination system, the evidence for recombination in chloroplasts is clear-cut. Recombination between genetic markers in the *Chlamydomonas* chloroplast genome and in the chloroplast genome of a somatic cell hybrid of *N. tabacum* and *Nicotiana plumbaginifolia* has been demonstrated (Sager and Ramanis 1970; Medgyesy et al. 1985). Recombination across the large inverted repeat sequences, found in the chloroplast genomes of most plants, is probably responsible for the two types of chloroplast DNA molecules having the single copy sequences in different relative orientations (Palmer 1983). Intermolecular recombination may be responsible for the generation of circular dimers of chloroplast DNA observed by electron microscopy (Kolodner and Tewari 1979). In all these examples recombination is likely to have taken place across homologous sequences, whereas the integration of the *cat* gene into tobacco chloroplast DNA most likely took place at nonhomologous sequences. Homologous recombination is likely to take place at a much higher frequency than nonhomologous recombination, and this may be one reason for the low numbers of recombinants obtained by De Block et al. (1985). It would seem advisable when attempting to obtain integration of foreign DNA into chloroplast DNA to include chloroplast DNA sequences in the foreign gene construct to promote homologous recombination. This should lead to the insertion of foreign genes at specified sites in the chloroplast DNA, so that the disruption of essential genes or transcriptional units can be avoided. Although the organization of genes in chloroplast DNA from *N. tabacum* and *M. polymorpha* may be predicted from the complete nucleotide sequences of the chloroplast genomes (Ohyama et al. 1986; Shinozaki et al. 1986), our current knowledge of the transcription of these genes is far from complete, so it may be difficult to choose a nontranscribed region of chloroplast DNA to be the site of insertion of foreign genes (see Section 14.2).

Regions of wheat chloroplast DNA that may be potentially useful for the integration of foreign DNA have been identified by Howe (1985). The endpoints of an inversion in wheat chloroplast DNA were shown to be associated with copies of a 70-bp repeated sequence, part of which was highly homologous to a 15-bp sequence (5′-GCTTTTTATACTAA-3′) in the core region of the bacteriophage lambda attachment site. Site-specific recombination across this 15-bp sequence in bacteriophage lambda DNA and in *E. coli* chromosome DNA results in the integration of the bacteriophage lambda genome into that of *E. coli*. Integration of bacteriophage DNA into a fragment of wheat chloroplast DNA containing two copies of the 70-bp repeat sequence was demonstrated in an *E. coli* strain lacking the primary attachment site for bacteriophage lambda (Howe 1985). This indicates that these chloroplast sequences are recombinogenic in *E. coli*, and have probably acted as such in chloroplasts during evolution to generate the inversion found in chloroplast DNA of wheat, barley, maize, and other cereals. This inversion appears not to have affected the expression of genes in the vicinity of the endpoints of the inversion, and these sequences may therefore represent potentially useful sites for the integration of foreign DNA. Unfortunately it is not possible to estimate the rate of recombination across the 70-bp repeated sequences in wheat chloroplasts because a subsequent inversion event has rearranged the repeat sequences so that they are now direct repeats. Recombination across these direct repeats would result in the elimination of an approximately 20-kbp fragment from the chloroplast DNA molecule.

Integration of bacteriophage lambda DNA into the *E. coli* chromosome requires the phage-encoded *int* protein as well as a host-encoded protein, integration host factor (IHF). These proteins recognize the 15-bp core and some flanking sequences of the phage and bacterial attachment sites (*att*) and lead to the formation of a large recombinogenic protein-DNA complex. The inclusion of genes for the *int* protein and the IHF in constructs designed for integration into wheat chloroplast DNA via *att* sequences may be seriously considered. The lambda *int* gene lies just to the right of *att*P (Sanger et al. 1982) and the genes for the α and β subunits of IHF have recently been characterized (Flamm and Weisberg 1985).

14.5 AUTONOMOUSLY REPLICATING PLASMIDS

The construction of autonomously replicating plasmids for the transformation of chloroplasts requires sequences that will act as origins of replication in chloroplasts. Unfortunately such sequences have not been unequivocally identified in higher plant chloroplast DNA. Origins of replication have been identified by electron microscopy in *Chlamydomonas* (Waddell et al. 1984) and *Euglena* (Manning and Richards 1972; Koller and Delius 1982; Ravel-Chapuis et al. 1982), and nucleotide sequences of these regions have been determined (Schlunegger and Stutz 1984; Wu et al. 1986). Both sequences are AT rich and may be folded into complex stem-loop structures, but there is no direct homology between the sequences. The lack of homology between the *Euglena* and *Chlamydomonas* sequences may be related to the different modes of replication in these organisms. In *Euglena* a single replication loop is observed by electron microscopy and it appears that DNA replication proceeds unidirectionally away from the origin of replication for about 5 kbp before proceeding in both directions around the circular chloroplast genome (Koller and Delius 1982). In *Chlamydomonas* two replication loops located about 6.5 kbp apart are observed by electron microscopy. At each origin, replication initiates with the formation of a displacement (D) loop resulting from the synthesis of one daughter strand. After the synthesis of 350 ± 130 bp of DNA, synthesis proceeds in both directions (Waddell et al. 1984).

Two loops have also been observed in pea and maize chloroplast DNA and have been identified as origins of replication (Kolodner and Tewari

1975a and 1975b). The D loops are located about 7 kbp apart in both plants. Kolodner and Tewari (1975b) suggested that replication is initiated by the formation of these two D loops, with the displacing daughter strands complementary to opposite parental strands. The D loops are extended toward each other and initiate the formation of a Cairns theta structure. Replication then proceeds bidirectionally around the circular genome, terminating at a site 180 degrees from the initiation sites. This initial round of replication is then used to start a rolling circle round of replication. The rolling circle replication mechanism may be a means of ensuring rapid synthesis of chloroplast DNA. This suggested mechanism of chloroplast DNA replication involving theta and rolling circle structures is similar to that found for bacteriophage lambda in *E. coli*. A scheme for the replication of chloroplast DNA is shown in Figure 14–3. The DNA seqences involved in the initiation of replication of higher plant chloroplast DNA have not been identified.

Attempts to identify chloroplast origins of replication by their ability to promote autonomous replication in yeast (*Saccharomyces cerevisiae*) appear to be misguided. Vallet and Rochaix (1985) showed that *Chlamydomonas* chloroplast sequences promoting autonomous replication in *Chlamydomonas* (ARC) and autonomous replication in yeast (ARS) are distinct from the origin of chloroplast DNA replication. ARS sequences have been located and characterized in petunia (Overbeeke et al. 1984; de Haas et al. 1986) and tobacco (Uchimiya et al. 1983; Ohtani et al. 1984). The location of these sequences is shown in Figure 14–4. Extensive homology between an ARS element located in the small single-copy region near the 23S rRNA gene in petunia and tobacco and a region of *Euglena* chloroplast DNA containing the origin of replication has been detected (de Haas et al. 1986). However the region of petunia chloroplast DNA showing the reported homology is highly AT rich (84%), and gaps have to be introduced into the petunia and *Euglena* sequences for the alignment of five sequence blocks. At the present time there is no other corroborative evidence suggesting that this region of petunia chloroplast DNA represents an origin of replication.

An attempt to identify the chloroplast origin of replication in maize chloroplast DNA using cloned restriction fragments as templates for DNA synthesis in vitro by a partially purified preparation of chloroplast DNA polymerase has been described (Gold et al. 1987). This study identified a region of approximately 9 kbp that is especially active as a template (marked as "ori?" in Figure 14–4). The smallest of the strongly active fragments has been sequenced and shows homology with the region of *Chlamydomonas* chloroplast DNA that has been identified as an origin of replication (Wu et al. 1986). However the highest sequence homology is in a region that could code for a protein homologous to *E. coli* ribosomal protein L16, and it is not clear if the sequence similarities are due to similar origins of replication or to the conserved ribosomal protein gene. This region of maize chloroplast DNA did not contain sequences that promoted autonomous replication in yeast (Gold et al. 1987).

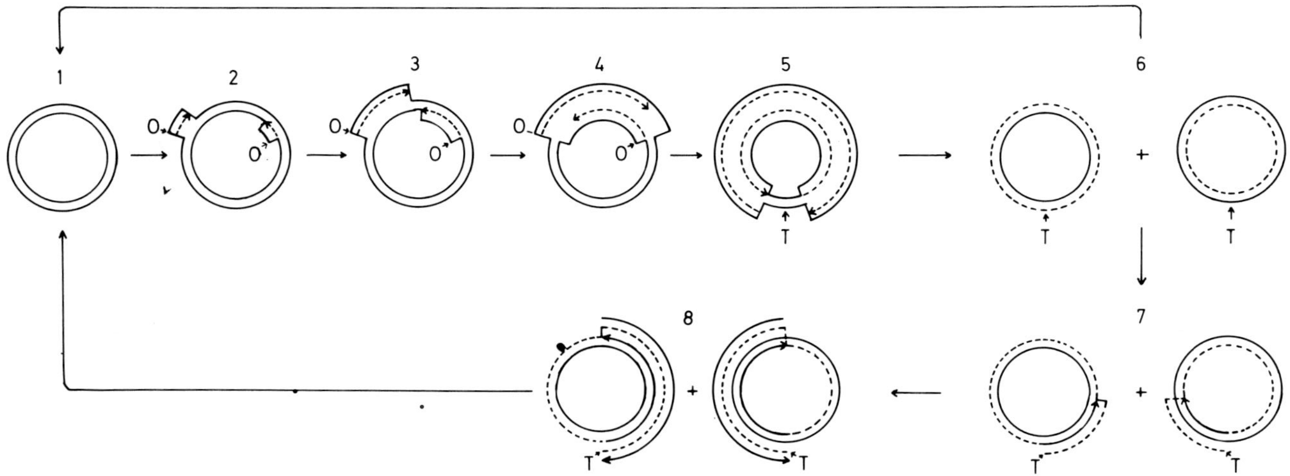

FIGURE 14–3 Model for the replication of higher plant chloroplast DNA. 1, closed circular parental molecule; 2, molecule with two D loops; 3, molecule with expanded D loops; 4 and 5, Cairns' theta structure intermediates; 6, nicked progeny molecules; 7 and 8, rolling circle intermediates; 0, origins of D-loop synthesis; T, terminal of the Cairns replication. Daughter strands are shown as arrows. (Adapted from Kolodner and Tewari [1975a].)

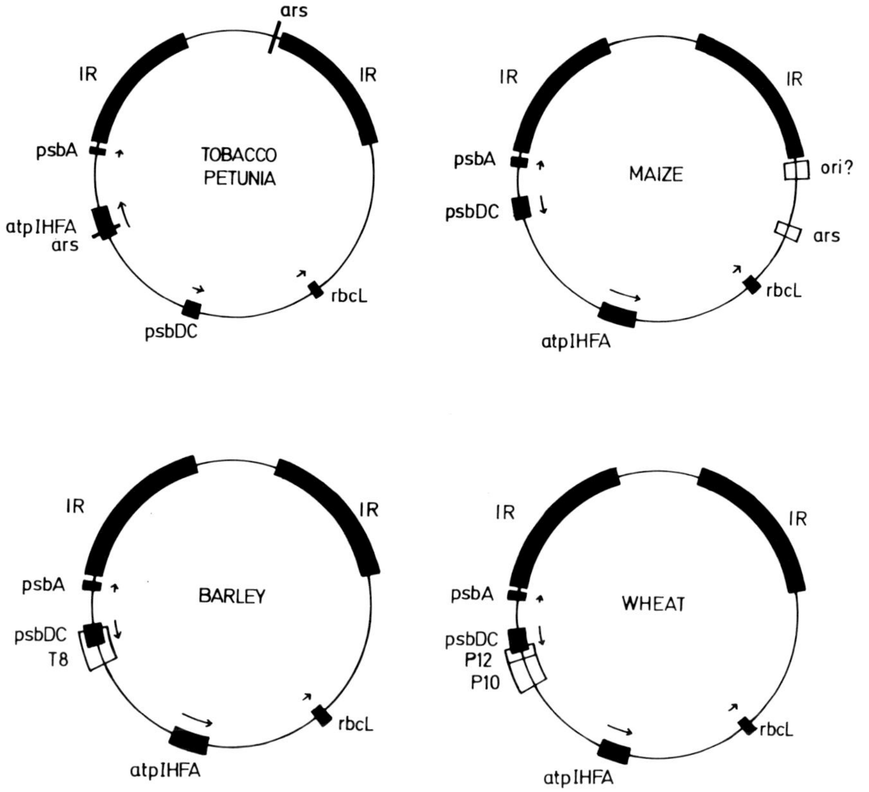

FIGURE 14–4 Putative origins of replication of higher plant chloroplast genomes. Regions identified by ARS activity (ars), template activity in vitro (ori), or retention in deleted plastid DNA molecules (T8, P10, P12) are shown on circular maps of chloroplast DNA. The inverted repeat (IR) sequences and *psb*A, *psb*DC, *atp*IHFA, and *rbc*L loci are shown to indicate the organization of the chloroplast genomes in the different plants. Putative origins of replication are shown for petunia (de Haas et al. 1986), tobacco (Shinozaki et al. 1986), maize (Gold et al. 1987), barley (Day and Ellis 1985), and wheat (Day and Ellis 1984).

Perhaps the clearest, albeit indirect, evidence for the location of a chloroplast origin of replication was obtained from studies on the deleted forms of plastid DNA in albino plants produced from anther culture of wheat and barley (Day and Ellis 1984 and 1985). Individual wheat and barley plants regenerated from separate pollen calli contained plastid DNA that had suffered major deletions (over 80% in one case). Examination of these deleted plastid DNA molecules by Southern hybridization indicated that in wheat only two *Pst*I fragments [5.2 kbp (P10) and 1.2 kbp (P12)] were retained unaltered in size in all plants (Day and Ellis 1984), and in barley a

4.5 kbp *Sst*I fragment (T8) was retained in all albino plants (Day and Ellis 1985). The location of these restriction fragments is shown in Figure 14–4. It was suggested that these restriction fragments may delimit a region of the chloroplast genome that contains an origin of replication. The wheat *Pst*I fragments and the barley *Sst*I fragment map to a similar region of the chloroplast genome. This region appears to be the best candidate for a chloroplast origin of replication, but it must be considered that other sequences containing origins of replication may have been lost by deletion in these plants.

If chloroplast DNA is replicated by a rolling circle mechanism as suggested by Kolodner and Tewari (1975a), then knowledge of the sequences involved in the termination of the Cairns'-type replication and the initiation of the rolling circle round of replication will be essential for the construction of an autonomously replicating plasmid for chloroplasts. However the apparent absence of a constant, retained sequence 180 degrees from the postulated origin of replication in the deleted plastid DNAs suggests there may not be any sequence specificity required for the termination or the initiation of rolling circle replication.

These studies highlight some of the difficulties likely to be encountered in attempts to construct an autonomously replicating plasmid for chloroplast transformation studies. In addition to an origin of replication it will probably be necessary to include a dominant selectable marker for the identification of transformed plants. The most-suitable selectable markers for chloroplast transformation are discussed below.

14.6 SELECTABLE MARKERS FOR CHLOROPLAST TRANSFORMATION

Unless an extremely efficient method of chloroplast transformation is discovered, it will be necessary to include a dominant selectable marker gene in the constructs produced either for integration or for autonomous replication. Several of the dominant selectable markers used in plant transformation confer resistance to antibiotics that act on chloroplast protein synthesis and appear to be suitable selectable markers for chloroplast transformation. The bacterial *aph*II gene from transposon Tn5, which confers resistance to the related aminoglycoside antibiotics, neomycin, kanamycin, and G418, has been widely used as a dominant selectable marker for plant transformation (Bevan et al. 1983; Fraley et al. 1983; Herrera-Estrella et al. 1983a). These antibiotics cause misreading during protein synthesis on 70S ribosomes, and thereby inhibit chloroplast protein synthesis. The bacterial *cat* gene from transposon Tn9, which confers resistance to chloramphenicol, another inhibitor of protein synthesis on 70S ribosomes, has been shown to be a selectable marker for chloroplast transformation (De Block et al. 1985) and for plant transformation (Herrera-Estrella et al.

1983b). The other selectable markers used for plant transformation do not appear to be suitable for chloroplast transformation. Expression of the bacterial genes for the methotrexate-insensitive dihydrofolate reductase (Herrera-Estrella et al. 1983a) and the hygromycin phosphotransferase (Lloyd et al. 1986) in chloroplasts would probably not protect the cells from the toxic effects of methotrexate and hygromycin, which exert their major effects outside plastids.

The inclusion of *aph*II or *cat* genes in constructs for chloroplast transformation would require that suitable sequences for the initiation and termination of transcription and for initiation of translation be provided as described in Section 14.2.

14.7 PROSPECTS

The prospects for efficient, reproducible chloroplast transformation do not look bright in the short term. There appear to be no rational means of transformation that can guarantee any reasonable degree of success. *Agrobacterium*–Ti plasmid–mediated gene transfer does not appear to be an efficient and reproducible method for chloroplast transformation. However a number of methods for the introduction of DNA into chloroplasts have been considered (see Section 14.3) and these offer some hope of success. Even if efficient methods for the introduction of DNA into chloroplasts are developed, considerable problems concerning the integration or autonomous replication of foreign genes still exist. However, work in progress in a large number of laboratories can be expected to provide the basic knowledge of the transcription and replication of chloroplast DNA required for chloroplast transformation. The expression of foreign genes in chloroplasts must be considered a possibility in the long term, but in the short term the introduction of gene constructs into the nuclear genome and targeting of the protein product to the chloroplast would appear to be the method of choice for altering the protein complement of chloroplasts. Nuclear transformation and protein targeting to plastids would appear to be the method of choice for alteration of the protein complement of nongreen plastids also.

REFERENCES

Bevan, M., Flavell, R.B., and Chilton, M-D. (1983) *Nature* 304, 184–187.

Bird, C.R., Koller, B., Auffret, A.D., et al. (1985) *EMBO J.* 4, 1381–1388.

Bonnett, H.T. (1976) *Planta* 131, 229–233.

Bonnett, H.T., and Eriksson, T. (1974) *Planta* 120, 71–79.

Bradley, D., and Gatenby, A. (1985) *EMBO J.* 4, 3641–3648.

Cozens, A.L., Walker, J.E., Phillips, A.L., Huttly, A.K., and Gray, J.C. (1986) *EMBO J.* 5, 217–222.

Crossway, A., Oakes, J.V., Irvine, J.M., et al. (1986) *Mol. Gen. Genet.* 202, 179–185.

Davey, M.R., Frearson, E.M., and Power, J.B. (1976) *Plant Sci. Lett.* 7, 7–16.

Day, A., and Ellis, T.H.N. (1984) *Cell* 39, 359–368.

Day, A., and Ellis, T.H.N. (1985) *Curr. Genet.* 9, 671–678.

De Block, M., Schell, J., and Van Montagu, M. (1985) *EMBO J.* 4, 1367–1372.

de Haas, J.M., Boot, K.J.M., Haring, M.A., Kool, A.J., and Nijkamp, H.J.J. (1986) *Mol. Gen. Genet.* 202, 48–54.

Fish, L.E., Kuck, U., and Bogorad, L. (1985) *J. Biol. Chem.* 243, 1413–1421.

Flamm, E.L., and Weisberg, R.A. (1985) *J. Mol. Biol.* 183, 117–128.

Fraley, R.T., Rogers, S.G., Horsch, R.B., et al. (1983) *Proc. Natl. Acad. Sci. USA* 80, 4803–4807.

Fromm, H., Devic, M., Fluhr, R., and Edelman, M. (1985) *EMBO J.* 4, 291–295.

Giles, K.L., Vaughn, V., Ranch, J.P., and Emery, J. (1980) *In Vitro* 16, 581–584.

Gold, B., Carrillo, N., Tewari, K.K., and Bogorad, L. (1987) *Proc. Natl. Acad. Sci. USA* 84, 194–198.

Goodgal, S.H. (1982) *Annu. Rev. Genet.* 16, 169–192.

Gruissem, W., and Zurawski, G. (1985) *EMBO J.* 4, 3375–3383.

Hain, R., Stabel, P., Czernilofsky, A.P., et al. (1985) *Mol. Gen. Genet.* 199, 161–168.

Hanley-Bowdoin, L., and Chua, N.-H. (1987) *Trends Biochem. Sci.* 12, 67–70.

Hennig, H., and Herrmann, R.G. (1986) *Mol. Gen. Genet.* 203, 117–128.

Herrera-Estrella, L., De Block, M., Messens, E., et al. (1983a) *EMBO J.* 2, 987–995.

Herrera-Estrella, L., Depicker, A., Van Montagu, M., and Schell, J. (1983b) *Nature* 303, 209–213.

Herrmann, R.G., Alt, J., Schiller, B., Widger, W.R., and Cramer, W.A. (1984) *FEBS Lett.* 176, 239–244.

Hird, S.M., Willey, D.L., Dyer, T.A., and Gray, J.C. (1986) *Mol. Gen. Genet.* 203, 95–100.

Holschuh, K., Bottomley, W., and Whitfeld, P.R. (1984) *Nucleic Acids Res.* 12, 8819–8834.

Howe, C.J. (1985) *Curr. Genet.* 10, 139–145.

Klein, R.R., and Mullet, J.E. (1986) *J. Biol. Chem.* 261, 11138–11145.

Koller, B., and Delius, H. (1982) *EMBO J.* 1, 995–998.

Kolodner, R., and Tewari, K.K. (1975a) *J. Biol. Chem.* 250, 8840–8847.

Kolodner, R., and Tewari, K.K. (1975b) *Nature* 256, 708–711.

Kolodner, R., and Tewari, K.K. (1979) *Proc. Natl. Acad. Sci. USA* 76, 41–45.

Landgren, C.R., and Bonnett, H.T. (1979) *Plant Sci. Lett.* 16, 15–22.

Lloyd, A.M., Barnason, A.R., Rogers, S.G., et al. (1986) *Science* 234, 464–466.

Lurquin, P.F. (1979) *Nucleic Acids Res.* 6, 3773–3784.

Manning, J.E., and Richards, O.C. (1972) *Biochemistry* 11, 2036–2043.

Medgyesy, P., Fejes, E., and Maliga, P. (1985) *Proc. Natl. Acad. Sci. USA* 82, 6960–6964.

Morris, J., and Herrmann, R.G. (1984) *Nucleic Acids Res.* 6, 2837–2850.

Ohme, M., Kamogashira, T., Shinozaki, K., and Sugiura, M. (1985) *Nucleic Acids Res.* 13, 1045–1056.

Ohtani, T., Uchimiya, H., Kato, A., et al. (1984) *Mol. Gen. Genet.* 195, 1–4.

Ohyama, K., Fukuzawa, H., Kohchi, T., et al. (1986) *Nature* 322, 572–574.

Ostro, M., Lavelle, D., Paxton, W., Matthew, B., and Giacomoni, D. (1980) *Arch. Biochem. Biophys.* 201, 392–402.

Overbeeke, N., Haring, M.A., Nijkamp, H.J.J., and Kool, A.J. (1984) *Plant Mol. Biol.* 3, 235–241.

Palmer, J.D. (1983) *Nature* 301, 92–93.

Palmer, J.D. (1985) *Annu. Rev. Genet.* 19, 325–354.

Phillips, A.L., and Gray, J.C. (1984) *Mol. Gen. Genet.* 194, 477–484.

Potrykus, I. (1973) *Z. Pflanzenphysiol.* 70, 364–366.

Potrykus, I., Paszkowski, J., Saul, M.W., Petruska, J., and Shillito, R.D. (1985) *Mol. Gen. Genet.* 199, 169–177.

Ravel-Chapuis, P., Heizman, P., and Nigon, V. (1982) *Nature* 300, 78–81.

Rochaix, J.C., Erickson, J., Goldschmidt-Clermont, M., Schneider, M., and Vallet, J.M. (1984) in *Gene Manipulation in Plant Improvement* (Gustafson, J.P., ed.), pp. 577–603, Plenum Press, New York.

Rosenberg, M., and Court, D. (1979) *Annu. Rev. Genet.* 13, 319–353.

Sager, R., and Ramanis, Z. (1970) *Proc. Natl. Acad. Sci. USA* 65, 593–600.

Sanger, F., Coulson, A.R., Hong, G.F., Hill, D.F., and Petersen, G.B. (1982) *J. Mol. Biol.* 162, 729–773.

Schlunegger, B., and Stutz, E. (1984) *Curr. Genet.* 8, 629–634.

Schmidt, G.W., and Mishkind, M. (1983) *Proc. Natl. Acad. Sci. USA* 80, 2632–2636.

Schreier, P.H., Seftor, E.A., Schell, J., and Bohnert, H.J. (1985) *EMBO J.* 4, 25–32.

Shillito, R.D., Saul, M.W., Paszkowski, J., Muller, M., and Potrykus, I. (1985) *Biotechnol.* 3, 1099–1103.

Shine, J., and Dalgarno, L. (1974) *Proc. Natl. Acad. Sci. USA* 71, 1342–1346.

Shinozaki, K., Ohme, M., Tanaka, T., et al. (1986) *EMBO J.* 5, 2043–2049.

Smith, H.O., Danner, D.B., and Deich, R.A. (1981) *Annu. Rev. Biochem.* 50, 41–68.

Strittmatter, G., Gozdzicka-Jozefiak, A., and Kossel, H. (1985) *EMBO J.* 4, 599–604.

Uchimiya, H. (1979) *Naturwissenschaften* 66, 314–315.

Uchimiya, H., Ohtani, T., Ohgawara, T., et al. (1983) *Mol. Gen. Genet.* 192, 1–4.

Vallet, J.-M., and Rochaix, J.-D. (1985) *Curr. Genet.* 9, 321–324.

Van den Broeck, G., Timko, M.P., Kausch, A.P., et al. (1985) *Nature* 313, 358–363.

Waddell, J., Wang, X.-M., and Wu, M. (1984) *Nucleic Acids Res.* 9, 3843–3856.

Westhoff, P., Farchaus, J.W., and Herrmann, R.G. (186) *Curr. Genet.* 11, 165–169.

Whitaker, R.J., and Evans, D.A. (1986) in *Handbook of Plant Cell Culture,* vol. 4 (Evans, D.A., Sharp, W.R., and Ammirato, P., eds.), pp. 172–196, Macmillan, New York.

Whitfeld, P.R., and Bottomley, W. (1983) *Annu. Rev. Plant Physiol.* 34, 279–310.

Wu, M., Lou, J.K., Chang, D.Y., Chang, C.H., and Nie, Z.Q. (1986) *Proc. Natl. Acad. Sci. USA* 83, 6761–6765.

Zurawski, G., Bottomley, W., and Whitfeld, P.R. (1982) *Proc. Natl. Acad. Sci. USA* 79, 6260–6264.

CHAPTER 15

A Perspective on the Biotechnology of Ribulose Bisphosphate Carboxylase/Oxygenase

Harry Roy

The biotechnology of the photosynthetic enzyme ribulose bisphosphate carboxylase/oxygenase (E.C.4.1.1.39) (Rubisco) is of special interest because it relates to the possibility of improving photosynthetic yields by suppressing the process of photorespiration in plants.

Photorespiration is manifested as a light-dependent *consumption* of oxygen and *evolution* of CO_2 by plants (Figure 15–1). It is generally accepted that the oxygen consumption step is catalyzed by Rubisco. Paradoxically, this enzyme is best known for its key role in the Calvin cycle of photosynthesis, in which it catalyzes the carboxylation of ribulose bisphosphate by CO_2 (*consumption* of CO_2), leading to the formation of two molecules of D-3-phosphoglycerate (Figure 15–2). Concomitantly oxygen is *evolved* by the plants because of the extraction of protons and electrons from water that are needed to reduce CO_2 to carbohydrate. The carboxylation of ribulose bisphosphate leads to a net gain of fixed carbon, but the oxygenation of ribulose bisphosphate not only undoes a carbon-carbon bond formed in

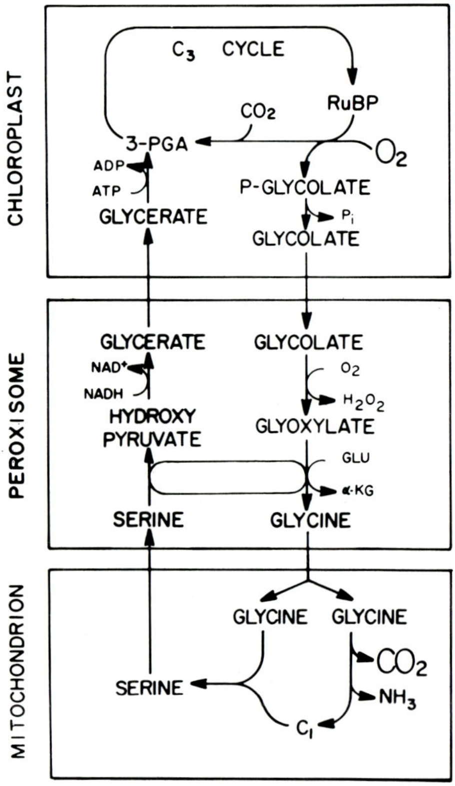

FIGURE 15–1 Photorespiration in plants. Oxygen is consumed by reaction with ribulose bisphosphate (RuBP) at the active site of Rubisco in the chloroplast. The product phosphoglycolate is dephosphorylated and metabolized further in peroxisomes and mitochondria, liberating CO_2. Some of the carbon of phosphoglycolate is returned to the Calvin cycle in the chloroplast, after an energy-consuming process. (From J.K.Hoober [1984] *Chloroplasts,* Plenum Press, New York. With permission.)

a previous turn of the Calvin cycle, but also leads to actual loss of fixed carbon as CO_2. This is due to the cleavage of glycine derived ultimately from phosphoglycolate, an immediate product of the oxygenation of ribulose bisphosphate. Some of the released CO_2 may be refixed, but this has a cost. The remaining organic carbon from the decarboxylation of glycine eventually finds its way back to the Calvin cycle in the form of triose phosphate, which represents some salvage. Under normal conditions, oxygen (rather than CO_2) reacts with three out of ten ribulose bisphosphate molecules at the active site of Rubisco. Despite attempts to rationalize the phenomenon of photorespiration, most investigators have concluded that the oxygenation

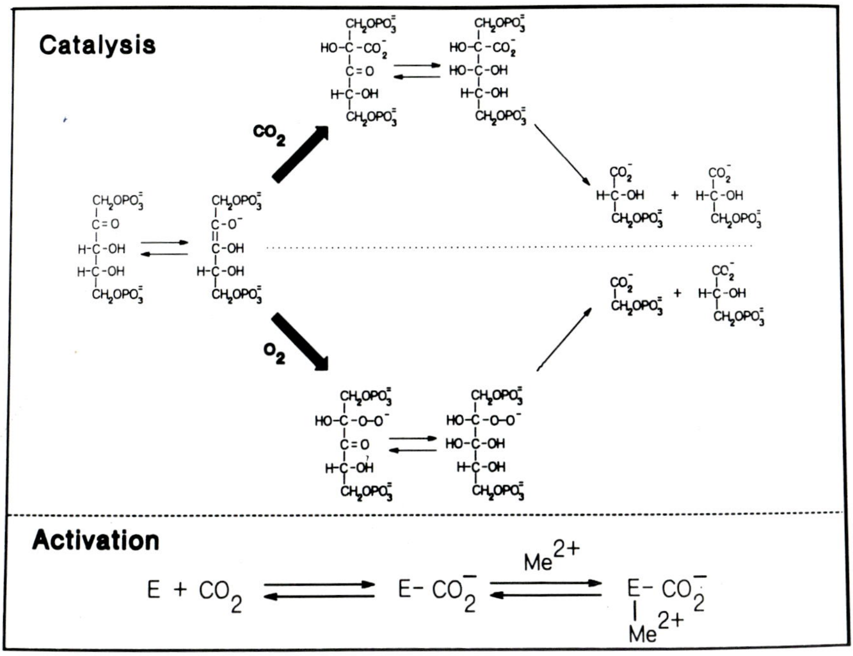

FIGURE 15–2 Biochemical mechanism of action of Rubisco. Rubisco is activated in vitro by the formation of a carbamate, which can bind Mg^{2+}. After binding of ribulose bisphosphate, the enzyme mediates the formation of an enediol intermediate. This can react with either CO_2 (heavy arrow, upper) or oxygen (heavy arrow, lower). In the first case, two molecules of D-3-phosphoglycerate are produced. In the second, however, one of the molecules produced is 2-phosphoglycolate. (From J. Pierce [1986] *Plant Physiol.* 81, 943–945. With permission.)

of ribulose bisphosphate is just a chemical consequence of the existence of oxygen in the atmosphere. The remaining steps in the photorespiratory pathway offer no advantage to the plant other than salvage of the products of this oxygenation reaction (Ogren 1984). Consequently it appears that if photorespiration could be done away with, photosynthetic yields might be increased by at least 20–30%. Empirical evidence suggests that the enhancement might be even greater than this (Hardy et al. 1978) (see Section 15.1).

15.1 PROPERTIES OF RUBISCO

Rubisco is found in chemoautotrophic bacteria, photosynthetic bacteria, cyanobacteria, algae, and plants. In the bacterium *Rhodospirillum rubrum,* the enzyme contains two identical subunits of 50.5 kd whose amino acid sequence has been deduced from nucleic acid sequence data (Nargang et al. 1984). In *Rhodopseudomonas sphaeroides,* there is evidence for two forms (I and II) of the enzyme. The form II enzyme contains six catalytic subunits. The form I enzyme contains eight catalytic (large) subunits and eight small subunits of approximately 14 kd. The latter hexadecameric structure is the only one found in cyanobacteria, algae, and higher plants (Miziorko and Lorimer 1983). In most eukaryotes that contain the enzyme, the large subunits are encoded in chloroplast DNA, while the small subunits are encoded in nuclear DNA. The small subunits are synthesized in the cytosol, whence they are taken up by the chloroplasts, cleaved, and assembled with large subunits into holoenzyme (Chua and Schmidt 1978). The nucleotide sequences of large and small subunit genes have been determined in a wide variety of plants and algae (Miziorko and Lorimer 1983). Because of the dominance of plants in the environment and the high concentration of Rubisco in chloroplasts Rubisco is believed to be the most abundant protein on the planet (Ellis 1979). Rubisco has been a favorite object of study in plant biochemistry and development; more recently it has become a major target of biotechnological manipulation (Ogren 1984; Somerville 1984; Miziorko and Lorimer 1983; Jensen et al. 1987).

About ten years ago, Hardy et al. (1978) summarized a number of features of Rubisco that appeared to need fixing. This may have been a response to the questions of molecular biologists, who asked what the biochemists wanted to change in this molecule. Some of Hardy's comments even seemed amusing—for example, that Rubisco's molecular weight was too high—but many seemed to foreshadow current developments. In particular, Hardy's field data, indicating that low atmospheric CO_2 limits productivity of crop plants, still provide a strong practical reason to concentrate on understanding the biochemistry, synthesis, and assembly of this enzyme.

Since 1978, a great deal has been learned about all these aspects of Rubisco, and we are probably close to seeing some practical attempts at engineering this enzyme in higher plants or cyanobacteria. What sort of results may we expect?

15.2 FIXING RUBISCO

The most popular objective in research aimed at engineering Rubisco has been to compensate for the fact that three out of ten ribulose bisphosphate molecules get oxygenated instead of carboxylated at the active site of the enzyme in C_3 plants. The attendant waste of energy and subsequent loss of fixed carbon make Rubisco an obvious target for biotechnological intervention. This characteristic of Rubisco is summarized mathematically as the specificity value, V_cK_o/V_oK_c, where V_c and V_o are the V_{max} values for carboxylation and oxygenation of ribulose bisphosphate, and K_c and K_o are the K_m values for these two reactions. There is evidence that the specificity value is subject to evolutionary variation (Jordan and Ogren 1983), ranging from 9 to 15 for certain photosynthetic bacteria, to about 50 for cyanobacteria, and up to about 80 for many higher plants (reviewed by Ogren 1984). Thus it is reasonable to suppose that an optimal chemical structure at the active site that would maximize this value might be designed or discovered. To design such a structure it is necessary to know what the active site looks like. Considerable progress in this regard has been made: Schneider et al. (1986) have determined the structure of the *R. rubrum* enzyme, which contains only two large subunits, at 2.9-Å resolution. The active site can be identified by difference Fourier analysis and by analysis of active site residues whose positions are known in the primary sequence (Lorimer 1981; Herndon et al. 1982; Fraij and Hartman 1982). Rational in vitro mutagenesis studies with this enzyme, now under way, will be significantly aided by the new structural information now emerging (see below).

Culminating progress over the years (Baker et al. 1977; Holzenburg et al. 1987), structural analysis of hexadecameric bacterial or plant Rubisco appears now to be achieving the high resolution reported for the dimeric *R. rubrum* enzyme (Chapman et al. 1988). There is uncertainty over whether all hexadecameric Rubiscos are structurally alike; the model for the tobacco enzyme (Chapman et al. 1988) appears to conflict with that for the *Alcaligenes eutrophus* enzyme (Holzenburg et al. 1987). It appears that the quaternary structure may be quite sensitive to ligands both for the bacterial (Holzenburg et al. 1987) and the higher-plant enzyme (Johal et al. 1985), but it is not yet clear whether the conformational changes that occur are of the magnitude suggested for the bacterial enzyme.

Even without detailed structural information for hexadecameric Rubisco, one can nevertheless pursue the biotechnological Holy Grail through the use of genetic analysis and manipulation. For example, it should be possible to create a large number of mutations in Rubisco genes in bacteria or cyanobacteria, screen the mutants for increased phostosynthetic efficiency, and analyze the structure of the altered Rubisco genes in the most-efficient variants (Pierce et al. 1988). One can also screen photosynthetic mutants of higher plants for alterations in Rubisco. An example of such a mutant is the sigma mutant in *Oenothera*, which appears to have a nonsense codon leading to premature termination of the large subunit (Gordon et al. 1982; Winter and Herrmann, submitted for publication). Alternatively, one can look for mutants of the Rubisco gene after mutagenesis of cells and characterize them (Spreitzer and Mets 1980; Beasley and Mets 1986). Dron et al. (1983) characterized one such mutation that leads to complete inactivation of Rubisco in *Chlamydomonas*. This mutation changed Gly171 to aspartate, located near Lys175, identified as an essential amino acid (reviewed by Miziorko and Lorimer 1983). By screening for revertants, additional information can be gained. A true revertant has been found (Spreitzer et al. 1986b), and other revertants have been detected. These reversions have been due to intergenic suppression, because they segregate away from the original mutant gene subsequently (Spreitzer et al. 1986a; Spreitzer and Chastain 1986 and 1987). This leaves open the possibility of recovering revertants altered in the Rubisco gene itself, which might indicate which amino acid changes would impinge positively upon kinetic aspects of the enzyme. Readers are referred to earlier reviews of photosynthetic mutations by Somerville (1984 and 1986).

15.3 MOLECULAR ANALYSIS OF RUBISCO FUNCTION

Considerable progress has been made in the effort to use the tools of biotechnology to study the molecular basis for Rubisco function. The *R. rubrum* enzyme, for example, which contains only large subunits, can be cloned as a fusion protein in active form in *Escherichia coli* (Somerville and Somerville 1984). Recently this clone has been reconstructed to produce an active enzyme with the correct amino terminal (Larimer et al. 1986). These developments make the *R. rubrum* gene the vehicle of choice for those who wish to examine the effect of amino acid substitutions on catalysis. The overall catalytic mechanism is illustrated in Figure 15–2.

Terzaghi et al. (1986) sought to make a relatively innocuous change near Lys329, which had been identified earlier as an active site residue. They accordingly replaced Met330 with a leucine residue. The kinetic properties of the enzyme were altered significantly, and evidence was found for a weaker binding of carboxyl-arabinitol bisphosphate (CABP), a transition-state analogue.

Estelle et al. (1986) sought to test the hypothesis that activation of the enzyme by carbamate formation at Lys191 might be imitated by substituting a glutamate residue at this position. They found that the altered enzyme had an activity at least six orders of magnitude lower than that of the wild-type. They found that the quaternary complex containing enzyme, $^{14}CO_2$, Mg^{2+}, and CABP was considerably less stable in the mutant enzyme than in the wild-type. Since Mg^{2+} is thought to interact with the carbamate normally, these authors inferred that the mutant enzyme was unable to bind Mg^{2+}, and they calculated that the glutamate carboxyl oxygens were about 3.6 Å closer to the polypeptide backbone in the mutant than were the oxygens of the carbamate in the wild-type. They concluded that this was probably too far to permit appropriate binding of Mg^{2+} to the enzyme. According to current models of catalysis (van Dyk and Schloss 1986; Pierce 1986), the carboxylated transition state intermediate is stabilized by binding to Mg^{2+}.

Gutteridge (1986) reported changing Asp198 to glutamate, on the theory that this would be a relatively innocuous change that might preserve activity. He reported a 30% reduction in the rate of both the carboxylation and the oxygenation reaction; rates of activation of the enzyme by CO_2 and Mg^{2+}, and K_ms for substrates, were unchanged. Thus the changes in rates were attributed to altered values for k_{cat}. Electron paramagnetic resonance (EPR) measurements indicated that the binding of metal was affected by the introduction of the extra methylene group. The data suggest a role for metal binding for this residue, but the relationship could be indirect, considering the rather mild effect of the substitution.

Niyogi et al. (1986) replaced a putative active site residue, His291, with alanine. This resulted in an enzyme about 25% as active as the wild-type. On this basis, they concluded that His291 (corresponding to His298 of the spinach enzyme) is not essential for catalysis. It could be playing a helpful role, of course, and the possible mechanism of this role would depend on the amino acid's exact location as deduced from the crystallographic data.

Hartman et al. (1987) substituted arginine, serine, histidine, glutamine, cysteine, alanine, or glycine for Lys166. The substitution of arginine reduced activity to less than 0.02% of wild-type; substitution of serine reduced activity to less than 0.2% of wild-type; and substitution of the others reduced activity to less than 0.1% of wild-type. On this basis, and considering the pK of this residue and the pH dependence of catalysis, the authors concluded that Lys166 probably is crucial to catalysis, most likely by removing the proton at C3 in ribulose bisphosphate, which results in formation of the enediol intermediate, which is believed to react with substrate CO_2. These experiments required the synthesis of several different oligonucleotide primers in order to permit each of the different substitutions to take place. Saturation mutagenesis, the elegant and powerful technique of substituting up to 19 amino acids at selected sites in a gene (Bylina et al. 1986), could be applied to subunits of Rubisco. This would reduce the labor involved in obtaining the mutants.

Larimer et al. (1987) reported complementation between engineered mutants altered in critical residues in the carboxyl terminal domain (Glu48) and amino terminal domain (Lys166) of *R. rubrum* Rubisco. These data, in conjunction with structural information, suggest that the active site may be shared between the carboxyl terminal domain of one subunit and the amino terminal domain of the other subunit in this dimeric enzyme.

In all the above-mentioned studies, the enzyme is a poorer catalyst when engineered, so what is the basis for our hope of improvement? Clearly, these pioneering studies are providing information that, together with the structural model now available for this enzyme, should allow the formation of a definitive description of the catalytic mechanism. With this basis of understanding, it is not unreasonable to expect that someone may be able to propose a rational change that may actually improve the specificity value. Of course, given the great structural difference between the dimeric *R. rubrum* enzyme and the hexadecameric enzyme of higher plants, and the former's low specificity value (Ogren, 1984), the *R. rubrum* enzyme possibly is not itself a good tool for engineering improved photosynthesis in higher plants. For this and other reasons a number of investigators are trying to set up similar analytical systems for the study of the hexadecameric enzymes characteristic of most other autotrophic organisms, as described below.

15.4 CLONING AND EXPRESSION OF HEXADECAMERIC RUBISCO

The *Anabaena* Rubisco, which contains both large and small subunits, can be cloned and expressed in *E. coli,* although the amount of active enzyme assembled is well below the amount of large subunits synthesized, most of which remain insoluble (Gurevitz et al. 1985). The *Anacystis nidulans* enzyme also can be synthesized in active form in *E. coli* (Tabita and Small 1985). The *Chromatium* enzyme, which contains small subunits, can be synthesized and expressed at high levels (15% of soluble protein) in *E. coli* (Viale et al. 1985). The *Synechococcus* large subunit can assemble with small subunits from higher plants, both in vitro (Andrews and Lorimer 1985) and in *E. coli* (Van der Vies et al. 1986). The Rubisco genes of the hydrogen bacterium *A. eutrophus* can be expressed as active enzyme in *Pseudomonas aeruginosa* (Andersen et al. 1986). These results offer considerable promise for future analysis of the type being carried out on the *R. rubrum* enzyme.

Despite the success with prokaryotic enzymes, all attempts to obtain active cloned Rubisco using large subunit genes from eukaryotic organisms have been unsuccessful so far (Musgrove and Ellis 1986; Newman and Cattolico 1987; Bradley et al. 1986). This is a curious result, since there are few amino acid sequence differences between the large subunits of higher plants and those of cyanobacteria (Curtis and Haselkorn 1983) and cyanobacterial large subunits can assemble with higher-plant small subunits in *E. coli* (Van der Vies et al. 1986). However, these differences are often not conservative with respect to charge or hydrophobicity, and so they may affect binding sites or the solubility properties of the polypeptides. According to this interpretation, the Rubisco large subunit binding protein discovered by Barraclough and Ellis (1980) in pea chloroplasts may be of considerable importance. In reactions mediated by ATP, large subunits derived from complexes with this binding protein have been shown to assemble into Rubisco in vitro (Roy et al. 1982; Bloom et al. 1983; Milos and Roy 1984 and 1985). In the presence of antibodies directed against this binding protein, this assembly is inhibited (Cannon et al. 1986) (Figure 15–3). It thus appears that all assembly-competent large subunits of Rubisco in pea chloroplast extracts are associated at least transiently with this binding protein. It is possible that the binding protein compensates for the amino acid differences between cyanobacterial and higher-plant large subunits, and keeps large subunits soluble in the chloroplasts until they can interact with small subunits. One reason for this might be that small subunit supplies from the cytoplasm of higher plants may be less tightly regulated than those from cyanobacteria. If this is the case, it may be necessary to design engineered

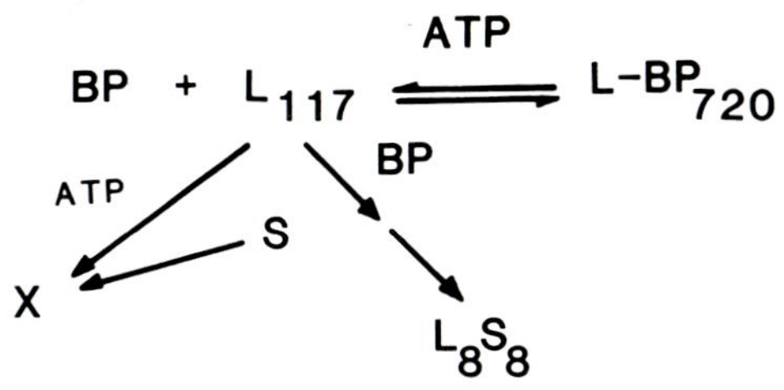

FIGURE 15–3 Model for assembly of higher plant Rubisco. In this model, large subunits occur in complexes of approximately 117 kd (based on apparent sedimentation coefficient of 7S and gel filtration data) and greater than 700 kd (based on nondenaturing PAGE, sucrose gradient sedimentation, and gel filtration). These are indicated as L_{117} and L-BP_{720}, respectively. ATP can mediate the dissociation of the latter high-molecular-weight form into binding protein (BP) monomers and large subunits, which apparently leads to the formation of L_{117} complexes. Since antibodies to BP inhibit assembly and radioactive large subunits sediment at 7S, and since under no circumstances have we observed direct losses of radioactivity from the 29S complex into 18S Rubisco (L_8S_8), assembly of Rubisco is depicted as a small subunit (S) reaction with L_{117}, followed by more than one unknown step. The binding protein either is part of the L_{117} complex or associates with participating subunits later during the assembly pathway (multiple arrows). There is evidence for a competing ATP-dependent step (arrows pointing to X, designating unknown products) that inhibits the participation of small subunits or L_{117} in the assembly process. (Adapted from S. Cannon, P. Wang, and H. Roy [1986] *J. Cell. Biol.* 103, 1327–1335. With permission.)

large subunits in such a way that they can interact with the binding protein, in order to fit into the plant's regulatory scheme for Rubisco assembly. Or it may be desirable to supply cyanobacterial small subunits to the plant at the same time. In order to study the behavior of plant large subunits in *E. coli,* it may be necessary to co-clone genes for the binding protein (of which there are two subunits, α and β) (Musgrove et al. 1987) or to compensate in some way for their supposed absence from this often-used host (perhaps by using a different host). Another possibility is to transform Rubisco-deficient mutant plants with engineered Rubisco genes, using electroporation or Ti-based plasmids as vehicles (Winter 1986) (see Section 15.5).

Attention should be paid to the possibility that an analysis of the amino acid sequence differences between cyanobacterial large subunits and those of higher plants may be especially informative not only about the assembly capacity of the subunits, but also about the factors responsible for increased specificity values in plants as compared with cyanobacteria (Ogren 1984). For example, Phe50 in *Anabaena* is replaced by proline in *Chlamydomonas,* spinach, and maize; Gln213 by methionine; Thr227 by tyrosine; and Pro246 by glycine. These residues are not within active site regions, but some of them are close enough to affect the conformation of residues at the active site (Curtis and Haselkorn 1983). It might be thought that such a comparison would be vitiated by the substantial differences between cyanobacterial and higher-plant small subunit sequences (Nierzwicki-Bauer et al. 1984); but at least in one case, a hybrid Rubisco containing cyanobacterial large subunits and higher-plant small subunits has the same specificity value as the cyanobacterial enzyme (Andrews and Lorimer 1985). It is possible that changes such as those listed above include one or two that were critical in evolution from the low-efficiency cyanobacterial-type enzyme to the current relatively efficient forms of Rubisco found in higher plants.

15.5 FIXING PLANTS

Moving an altered gene into a higher-plant chloroplast is a challenging problem. Difficulties include the facts that the chloroplast genome of higher plants is polyploid and that many plants are not susceptible to the stable integration of recombinant DNA carried by the most commonly used vectors, which are based on the properties of *Agrobacterium* Ti plasmids. It is possible that the technique of electroporation (Paszkowski et al. 1984; Potrykus et al. 1985) may be used to facilitate transfer of DNA into cells of species that are not hospitable to *Agrobacterium* DNA. An alternative strategy is to integrate a modified large subunit gene into nuclear DNA, arranging for transport of the protein product to the chloroplast by attaching to the gene the sequence encoding the transit peptide that mediates the transfer of another nuclear-coded chloroplast protein into chloroplasts (Van den Broeck et al. 1985; Wasmann et al. 1986; Cheung et al. 1988). Very recently this approach has been taken by Winter (1986), who reports having cured the sigma mutation in the primrose *Oenothera* by transferring an altered large subunit gene into nuclear DNA using a Ti-based vector. This very interesting experiment suggests a new way to study the possibility of altering Rubisco for improved photosynthetic productivity.

15.6 SOME CRITICAL RESERVATIONS

Some species of plants appear able to dispense with photorespiration (the C_4 species especially). C_3 plants appear to be quite healthy when grown under optimal conditions at high CO_2 concentrations, which suppress photorespiration (Hardy et al. 1978; Ogren 1984). These considerations argue that photorespiration is an adaptive salvage mechanism that only partly compensates for the oxygenation of ribulose bisphosphate under present atmospheric conditions. According to this view, oxygenation is simply a consequence of the chemistry of ribulose bisphosphate at the active site of Rubisco (Lorimer and Andrews 1973). This view, while superficially pessimistic, nevertheless permits the hope that phostorespiration may somehow be suppressed by altering the chemistry at the active site of Rubisco. A minority view is that attempts to diminish photorespiration by altering carboxylase might run up against regulatory mechanisms that are adapted to maintaining a certain metabolic flux through this pathway. This might have the function of protecting the plants against the consequences of drought (discussed by Heber and Krause 1980; Ogren 1984). In wheat, Evans (1986) and Evans and Austin (1986) have studied high-specific-activity variants of Rubisco that are transmitted maternally, and find that even though these variant enzymes have very different specific activities in vitro, the plants harboring them have the same photosynthetic characteristics as the normal variants, that is, they are not different with respect to leaf gas-exchange properties. Hanson and Peterson (1986) have proposed an additional source of photorespiratory CO_2 loss at high temperature, besides the decarboxylation of glycine. If their idea is correct, it is possible that alteration of the unknown enzymes controlling this loss might have some effect on crop yields. Such enzymes might have little to do with Rubisco per se, and their functions might not be affected by attempts to engineer photosynthesis through an altered Rubisco. Thus it should be kept in mind that our knowledge of plant biochemistry and metabolism may be insufficient for us to predict accurately the consequences of a given genetic engineering experiment.

15.7 ALTERNATIVE BIOLOGICAL STRATEGIES FOR ENHANCING PHOTOSYNTHESIS

The strategy of altering the specificity value of Rubisco, apparently followed by precursors of today's plants and algae, is not the only conceivable means of reducing photorespiration. For example, the C_4 metabolic variants of photosynthesis are carried out by a number of species of plants that are quite widely separated in evolution and that therefore appear to represent recent adaptations. In these plants, the first step in CO_2 fixation is carried out in mesophyll cell chloroplasts by phosphoenolpyruvate carboxylase, resulting in the formation of four-carbon acids that are transported to an entirely novel cell type containing modified chloroplasts. There the dicarboxylic acids are decarboxylated, and the released CO_2 is only then fixed into organic form by Rubisco. Apparently the increased CO_2 concentration in these chloroplasts suppresses the oxygenase reaction (Ogren 1984). This trick is very elaborate, in that it requires a complex new pathway of differentiation of chloroplasts in two cell types, leading to both biochemical and structural differences that no genetic engineer would dare to invent if there were no natural model to contemplate. One might think that if there were an easier way to compensate for photorespiration (such as altering Rubisco), the plants would have used that rather than evolve such a complex solution to the problem. However, natural systems are not obliged to obey our rules.

Algae have a still different strategy: They concentrate inorganic carbon in the cell (Moroney et al. 1986) and employ carbonic anhydrase to facilitate equilibration between HCO_3^- and CO_2. Mutants of *Chlamydomonas* deficient in this mechanism excrete large amounts of glycolate, presumably produced by dephosphorylation of the phosphoglycolate produced by the oxygenase reaction (reviewed by Ogren 1984). Recently, it has been found that there are at least three genetic complementation groups concerned with this mechanism in *Chlamydomonas* (Moroney et al. 1986). It is not clear what would happen if such a system were engineered into higher-plant chloroplasts. Would it adversely affect the cytosol by raising its pH? Would it place a load on the pH gradient–generating mechanism in the thylakoid membranes of the chloroplast? Or would it raise the CO_2 concentration in the stroma of the chloroplast enough to improve net photosynthesis? The technology for answering these questions appears to exist. The answers would be of intrinsic scientific interest and possibly of profound practical importance in the long run.

Another important point, which was not understood ten years ago, is that in vivo not all the Rubisco molecules are in an activated state (Perchorowicz et al. 1981). It has been found that the activation state of the enzyme is regulated by light, but not in such a way as to be explained by changes in CO_2, Mg^{2+}, or pH in the stroma that are known to occur and to be qualitatively in the right direction for activation of the enzyme. A clue to the mystery of this regulation was provided by the discovery of a mutant in *Arabidopsis* that is defective with respect to activation of Rubisco (Somerville et al. 1982). It now appears that the activation of Rubisco in vivo is mediated by an enzyme, activase, that is itself probably activated by light in vivo (Salvucci et al. 1986). This protein has been purified (Salvucci et al. 1986b), and a cDNA has been cloned and sequenced (Wernecke et al. 1988a). Perhaps activase could be made more efficient, leading indirectly to an increased proportion of active Rubisco in vivo. Whether this would lead to increased photosynthetic efficiency can be questioned, but the data of Hardy et al. (1978) suggest that any increase in the frequency of collision between CO_2 and active Rubisco should yield more biomass.

Another interesting point is the apparent correlation between nuclear ploidy, Rubisco content, and increased photosynthesis. Dean and Leech (1982) reported that in a polyploid series of wheat, Rubisco content increased. In this study, the amount of Rubisco correlated very closely with nuclear DNA content. In the hexaploid variety, there was a decrease in plastid area and plastid DNA, indicating that the actual concentration of Rubisco probably increased in the chloroplasts of this strain as compared with the tetraploid or diploid strains. Similar studies have been carried out by Schrader and his colleagues (Meyers et al. 1982; Molin et al. 1982) and Randall and his colleagues (Joseph et al. 1981). Their data indicated an effect of ploidy not only on Rubisco content, but also on net photosynthesis. The data indicate that the abundance of Rubisco is under nuclear genetic control.

To develop information on what particular genes mediate this control, Jellings et al. (1983) carried out a survey of a complete monosomic series in hexaploid wheat. They found that the deletion of group 4 chromosomes significantly reduced Rubisco amounts. They also found that deletions of certain other specific chromosomes led to significant increases in Rubisco amount. These studies have exposed a complex genetic control system that adjusts Rubisco quantity. It would appear to be possible to determine which genes known to be involved in Rubisco function are located on these chromosomes, and in this way to bridge the gap between knowledge of cytological and knowledge of molecular biological phenomena. The underlying practical idea has been that simply increasing the concentration of Rubisco (a strategy the plants clearly have already employed) should increase photosynthetic rates, and that small subunit genes and perhaps other genes probably offer a route to doing this. However, it may not be possible to increase Rubisco amounts without increasing chloroplast volume or number in plant cells (Pyke and Leech 1987).

A recent comment by Pickersgill (1986) may be relevant in this regard: he calculated, based on the structure of crystalline higher-plant Rubisco, that there is a physical upper limit to the concentration of active sites of Rubisco. This limit is about 10 mM, which can be compared with the 4-mM level thought by many to exist in vivo. Considering the need for aqueous channels in the chloroplast to permit the diffusion of substrates and macromolecules on their way to (and from? see Hattori and Margulies 1986) the thylakoid membrane for assembly, it seems likely that the physiological upper limit for Rubisco must be between 4 and 10 mM in active sites. Thus, increased quantities of Rubisco in leaves would appear to require increased total chloroplast volume above this limit. It would therefore be of interest to know whether the Rubisco in the hexaploid lines of wheat studied by Leech et al. approaches a crystalline state in the chloroplasts. If there is room for more carboxylase in the chloroplast compartment of the cell, the prospects for a biotechnological approach could be quite exciting. It is well known that small subunit genes can be engineered into transgenic plants (see Poulsen et al. 1986 for a recent example) and that these respond to endogenous phytochrome-mediated and blue light receptor–mediated environmental signals, leading to light-controlled increases in the abundance of small subunit mRNA, which leads in turn to the light-dependent synthesis and assembly of heterologous small subunits with the host Rubisco large subunits. Very active promoters, such as that for a CaMV gene (Paszkowski et al. 1984), could be spliced onto these genes, leading to high rates of transcription. Presumably the rates could be placed under light control by means of alterations in the CaMV promoter sequence. One could ask whether, in a transformed plant containing a superactive small subunit gene, the abundance of Rubisco increases. Alternatively, one could transform large numbers of plants with cloned nuclear genes and ask if any of these turn out large amounts of Rubisco. Difficulties here are related to the large size of plant genomes and the cumbersome nature of the screening procedures that would be necessary. Perhaps this is one of the problems that could best be approached using *Arabidopsis,* which has a small nuclear genome (Estelle and Somerville 1986), or algae, which can be manipulated with microbiological techniques. Studies of this type would be of intense theoretical interest as well as practical importance, since we really have little evidence on how the nucleus controls chloroplast formation. Notwithstanding the above considerations, recent data indicate that in some lines of amphidiploid wheat, Rubisco concentrations are not correlated with chromosome number, as judged by comparison with the corresponding diploid lines (Tingey and Andersen 1986). In this case, however, the data were expressed on a fresh weight basis, and so it is not easy to compare them with the other work cited above.

In summary, it appears that biotechnology is working now to help us understand the biochemical mechanism of Rubisco activity and the regulation of its synthesis and assembly. An important goal of research on the biotechnological application of this knowledge is to overcome the block to progress imposed by the inability of higher-plant Rubisco to assemble into active enzyme in *E. coli*. This may require expression of the Rubisco large subunit binding protein in this organism, or it may require the use of a different host organism as a testing system to see if an improved higher-plant Rubisco can be made.

ADDENDUM

The Rubisco subunit binding protein has recently been found to be homologous to gp groEL, an *E. coli* host protein which is needed for growth and for the assembly of bacteriophages T4, T5, and λ (Hemmingsen et al. 1988).

REFERENCES

Andersen, K., Wilke-Douglas, M., and Caton, J. (1986) *Biochem. Soc. Trans.* 14, 29–31.

Andrews, T.J., and Lorimer, G. (1985) *J. Biol. Chem.* 260, 4632–4636.

Baker, T.S., Eisenberg, D., and Eiserling, F. (1977) *Science* 196, 293.

Barraclough, R., and Ellis, R.J. (1980) *Biochim. Biophys. Acta* 608, 19–31.

Beasley, E.M., and Mets, L.J. (1986) *Plant Physiol.* 80(suppl.), 53.

Bloom, M., Milos, P., and Roy, H. (1983) *Proc. Natl. Acad. Sci. USA* 80, 1013–1017.

Bradley, D., Van der Vies, S., and Gatenby, A.A. (1986) *Philos. Trans. R. Soc. Lond.* [*Biol.*] 313, 447–458.

Bylina, E.J., Ismail, S., and Youvan, D.C. (1987) in *Proceedings of the VII International Congress on Photosynthesis* (Biggins, J., ed.), p. 307, Brown University, Providence, RI, August 10–15, 1986, Martinus Nijhoff, Dordrecht, The Netherlands.

Cannon, S., Wang, P., and Roy, H. (1986) *J. Cell Biol.* 103, 1327–1335.

Chapman, M.S., Suh, S.W., Kurmi, P.M.G., et al. (1988) *Science* 241, 71–74.

Cheung, A.Y., Bogorad, L., Van Montagu, M., and Schell, J. (1988) *Proc. Natl. Acad. Sci. USA* 85, 391–395.

Chua, N.H., and Schmidt, G.W. (1978) *Proc. Natl. Acad. Sci. USA* 75, 6110–6114.

Curtis, S.E., and Haselkorn, R. (1983) *Proc. Natl. Acad. Sci. USA* 80, 1835–1839.

Dean, C., and Leech, R. (1982) *Plant Physiol.* 70, 1605–1608.

Dron, M., Rahire, M., Rochaix, J.D., and Mets, L. (1983) *Plasmid* 9, 321–324.

Ellis, R.J. (1977) *Biochim. Biophys. Acta* 463, 185–215.

Ellis, R.J. (1979) *Trends in Biochem. Sci.* 4, 241–244.

Estelle, M., Hanks, J., McIntosh, L., and Somerville, C.R. (1986) *J. Biol. Chem.* 260, 9523–9526.

Estelle, M.A., and Somerville, C.R. (1986) *Trends Genet.* April, 89–93.

Evans J.R. (1986) *Planta* 167, 351–358.

Evans, J.R., and Austin, R.B. (1986) *Planta* 167, 344–350.

Fraij, B., and Hartman, F.C. (1982) *J. Biol. Chem.* 257, 3501–3505.

Gordon, K.H.J., Crouse, E.J., Bohnert, H.J., and Herrmann, R.G. (1982) *Theor. Appl. Genet.* 61, 373–384.

Gurevitz, M., Somerville, C.R., and McIntosh, L. (1985) *Proc. Natl. Acad. Sci. USA* 82, 6546–6550.

Gutteridge, S. (1986) *Biochem. Soc. Trans.* 14, 28–29.

Hanson, K.R., and Peterson, R.B. (1986) *Arch. Biochem. Biophys.* 246, 332–346.

Hardy, R.W.F., Havelka, U.D., and Quebedeaux, B. (1978) in *Photosynthetic Carbon Assimilation* (Siegelman, H.W., and Hind, G., eds.), pp. 265–278, Plenum Press, New York.

Hartman, F.C., Soper, T.S., Niyogi, S.K., et al. (1987) *J. Biol. Chem.* 262, 3496–5301.

Hattori, T., and Margulies, M.M. (1986) *Arch. Biochem. Biophys.* 244, 630–640.

Heber, U., and Krause, G.H. (1980) *Trends Biochem. Sci.* 5, 32–34.

Hemmingsen, S.M., Woolford, C., van der Vies, S.M., et al. (1988) *Nature* 333, 330–334.

Herndon, C.S., Norton, I.C., and Hartman, F.C. (1982) *Biochemistry* 21, 1380–1385.

Holzenburg, A., Mayer, F., Harauz, G., et al. (1987) *Nature* 325, 730–732.

Hoober, J.K. (1984) *Chloroplasts,* Plenum Press, New York.

Jellings, A.J., Leese, B.M., and Leech, R.M. (1983) *Mol. Gen. Genet.* 192, 272–274.

Johal, S., Partridge, B.E., and Chollet, R. (1985) *J. Biol. Chem.* 260, 9894–9904.

Jordan, D.B., and Ogren, W.L. (1983) *Arch. Biochem. Biophys.* 227, 425–433.

Joseph, M.C., Randall, D.D., and Nelson, C.J. (1981) *Plant Physiol.* 68, 894–898.

Larimer, F.W., Machanoff, R., and Hartman, F.C. (1986) *Gene* 41, 113–120.

Larimer, F.W., Lee, E.H., Mural, R.J., Super, T.S., and Hartman, F.C. (1987) *J. Biol. Chem.* 262, 15327–15329.

Lorimer, G.H. (1981) *Annu. Rev. Plant Physiol.* 32, 349–383.

Lorimer, G.H., and Andrews, T.J. (1973) *Nature* 243, 359.

Meyers, S.P., Nichols, S.L., Baer, G.R., Molin, W.T., and Schrader, L.E. (1982) *Plant Physiol.* 70, 1710–1714.

Milos, P., and Roy, H. (1984) *J. Cell. Biochem.* 24, 153–162.

Milos, P., and Roy, H. (1985) in *Molecular Biology of the Photosynthetic Apparatus* (Steinback, K.E., Arntzen, C., Bogorad, L., and Bonitz, S., eds.), Cold Spring Harbor Laboratory, Cold Spring Harbor, NY.

Miziorko, H., and Lorimer, G.H. (1983) *Annu. Rev. Biochem.* 52, 507–535.

Molin, W.T., Meyers, S.P., Baer, G.R., and Schrader, L.E. (1982) *Plant Physiol.* 70, 1710–1714.

Moroney, J.V., Tolbert, N.E., and Sears, B.B. (1986) *Mol. Gen. Genet.* 204, 199–203.

Musgrove, J.E., and Ellis, R.J. (1986) *Philos. Trans. R. Soc. Lond.* [*Biol.*] 313, 419–428.

Musgrove, J.E., Johnson, R.A., and Ellis, R.J. (1987) *Eur. J. Biochem.* 163, 529–534.

Nargang, R., McIntosh, L., and Somerville, C. (1984) *Mol. Gen. Genet.* 193, 220–224.

Newman, S., and Cattolico, R.A. (1987) in *Proceedings of the VII International Congress on Photosynthesis Abstracts* (Biggins, J., ed.), Abstract 210-A, Brown University, Providence, RI, August 10–15, 1986, Martinus Nijhoff, Dordrecht, The Netherlands.

Nierzwicki-Bauer, S.A., Curtis, S.E., and Haselkorn, R. (1984) *Proc. Natl. Acad. Sci. USA* 81, 5961–5965.

Niyogi, S.K., Foote, R.S., Mural, R.J., et al. (1986) *J. Biol. Chem.* 261, 10087–10092.

Ogren, W.L. (1984) *Annu. Rev. Plant Physiol.* 35, 415–442.

Paszkowski, J., Shillito, R.D., Saul, M., et al. (1984) *EMBO J.* 3, 2717–2722.

Perchorowicz, J.T., Raynes, D.A., and Jensen, R.G. (1981) *Proc. Natl. Acad. Sci. USA* 78, 2985–2989.

Pickersgill, R.W. (1986) *Biochem. J.* 236, 311.

Pierce, J. (1986) *Plant Physiol.* 81, 943–945.

Pierce, J., Carlson, T.J., and Williams, J.G.K. (1988) *Proc. Natl. Acad. Sci. USA* (in press).

Potrykus, I., Paszkowski, J., Saul, M.W., Petruska, J., and Shillito, R.D. (1985) *Mol. Gen. Genet.* 199, 169–177.

Poulsen, C., Fluhr, R., Kauffman, J.M., Boutry, M., and Chua, N.H. (1986) *Mol. Gen. Genet.* 205, 193–200.

Pyke, K.A., and Leech, R.M. (1987) *J. Exp. Bot.* 38, 1949–1956.

Roy, H., Bloom, M., Milos, P., and Monroe, M. (1982) *J. Cell. Biol.* 94, 20–27.

Salvucci, M.E., Portis, A.R., and Ogren, W.L. (1985) *Photosyn. Res.* 7, 193–201.

Salvucci, M.E., Portis, A.R., and Ogren, W.L. (1986a) *Anal. Biochem.* 153, 97–101.

Salvucci, M.E., Portis, A.R., and Ogren, W.L. (1986b) *Plant Physiol.* 80, 655–659.

Schneider, G., Lindquist, Y., Branden, C.I., and Lorimer, G. (1986) *EMBO J.* 5, 3409.

Somerville, C.R. (1984) *Oxford Surv. Plant Mol. Cell Biol.* 1, 103–131.

Somerville, C.R. (1986) *Annu. Rev. Plant Physiol.* 37, 467–507.

Somerville, C.R., Portis, A.R., and Ogren, W.L. (1982) *Plant Physiol.* 70, 381–387.
Somerville, C.R., and Somerville, S.C. (1984) *Mol. Gen. Genet.* 193, 214–219.
Spreitzer, R.J., and Chastain, C.J. (1986) *Plant Physiol.* 80(suppl.), 53.
Spreitzer, R.J., and Chastain, C.J. (1987) *Curr. Genet.* 11, 611–616.
Spreitzer, R.J., Chastain, C.J., and Ogren, W.L. (1986a) *Curr. Genet.* 9, 83–90.
Spreitzer, R.J., and Mets, L.J. (1980) *Nature* 285, 114–115.
Spreitzer, R.J., Rahire, M., and Rochiaix, J.D. (1986b) *Curr. Genet.* 9, 229–232.
Tabita, F.R., and Small, C.L. (1985) *Proc. Natl. Acad. Sci. USA* 82, 6100–6103.
Terzaghi, B.E., Laing, W.A., Christeller, J.T., Petersen, G.B., and Hill, D.F. (1986) *Biochem. J.* 235, 839–846.
Tingey, S., and Andersen, W.R. (1986) *J. Exp. Bot.* 37, 625–632.
Van den Broeck, G., Timko, M.P., Kausch, A.P., et al. (1985) *Nature* 313, 358–363.
Van der Vies, S.M., Bradley, D., and Gatenby, A.A. (1986) *EMBO J.* 5, 2439–2444.
Van Dyk, D.E., and Schloss, J.V. (1986) *Biochemistry* 25, 5145–5156.
Viale, A.M., Kobayashi, H., Takabe, T., and Akazawa, T. (1985) *FEBS Lett.* 192, 283–288.
Wasmann, C.C., Reiss, B., Bartlett, S.G., and Bohnert, H. (1986) *Mol. Gen. Genet.* 205, 446–453.
Wernecke, J.M., Zielinski, R.E., and Ogren, W.L. (1988) *Proc. Natl. Acad. Sci. USA* 85, 787–791.
Winter, P. (1986) Ph.D. dissertation, p. 156, University of Düsseldorf, Düsseldorf.

CHAPTER

16

Applications of Nucleic Acid Electron Microscopy and In Situ Hybridization Techniques in the Study of Plant Genomes

Madeline Wu

The basic protein film technique developed by Kleinschmidt and Zahn (Kleinschmidt and Zahn 1959; Kleinschmidt 1968) initiated nucleic acid electron microscopy. This technique allows the electron microscopic visualization of double-stranded DNA molecules that have been incorporated and extended into a monolayer film of denatured basic protein. Subsequent modifications in the technique, such as the use of formamide to destabilize random base interactions and to extend single-stranded DNA (Westmoreland et al. 1969; Davis et al. 1971), have increased its range of applications. Since then, a variety of related techniques have been developed for solving specific problems in the nucleic acid field. The methods and applications have been described in several comprehensive reviews (Younghusband and Inman 1974; Chow et al. 1975; Ferguson and Davis 1978, Brack 1980). Recently, the rapid advance of recombinant DNA technologies has demanded various effective methods to characterize cloned DNA molecules. Because of its simplicity and reproducibility, nucleic acid electron microscopy has frequently been used to obtain information that is difficult or time consuming to obtain by other experimental approaches. In this chapter, several examples of its application in characterizing the structural and functional organization of plant genomes have been chosen to illustrate the scope of its utilization.

A wealth of gene mapping studies within a genome have been achieved by in situ hybridization of labeled nucleic acids to their complementary sequences on chromosomes. This technique was first established by Gall and Pardue (1971). Further modifications have increased the sensitivity of this method, and it has extensively been used for mapping cloned genetic sequences on mammalian metaphase chromosomes (Tereba et al. 1979; Robins et al. 1981; Harper and Saunders 1981; Gerhard et al. 1981; Wu and Davison 1981; Li et al. 1986). Recently, this technique has been successfully adapted for gene mapping study of some plant genomes.

The application of both techniques in the investigation of functionally important sequences in various plant genomes will be described here. The chapter is not intended to be a comprehensive literature survey, but rather an introduction to a few experimental approaches.

16.1 ELECTRON MICROSCOPY OF ORGANELLE GENOMES

A plant cell contains interdependent genetic systems in the nucleus, in plastids, and in mitochondria. In recent years, there has been renewed interest in studying the roles of cytoplasmic genes in plant development and crop improvement. This is not surprising, since the mitochondrion and chloroplast are the major sites of energy conversion and presumably play a vital role in determining the vigor and productivity of a plant. The small size of an organelle genome makes it suitable for electron microscopic study. Consequently, a substantial amount of detailed information on the organelle genetic systems has been acquired by using nucleic acid electron microscopy.

The initial observation of intact chloroplast DNA molecules was made by Manning et al. (1971). Using electron microscopy, they detected circular DNA molecules with a contour length of 40 μm in lysates of *Euglena gracilis* chloroplasts. Large circular chloroplast DNA molecules have since been observed in several species of higher plants (Kolodner and Tewari 1972; Manning et al. 1972, Herrmann et al. 1975; Kolodner and Tewari 1975a) and also in the green algae *Chlamydomonas reinhardtii* and liverwort *Sphaerocarpos castellanii* (Behn and Herrmann 1977). In all cases the chloroplast DNA molecules appeared to have a homogeneous contour length that differed in length from species to species. Electron microscopy is an accurate and reproducible method for determining the size of DNA molecules if suitable DNA molecules are included in each experiment as an internal length standard.

The observation of a supercoiled form of chloroplast DNA by electron microscopy (Manning et al. 1972; Kolodner and Tewari 1972) led to the development of a popular isolation procedure for intact chloroplast DNAs utilizing cesium chloride–ethidium bromide density gradients (Kolodner et al. 1976). Convincing evidence for the homogeneity of the circular chloroplast DNA molecules was initially obtained by using a denaturation mapping technique developed by Inman and Schnos (1970). Using electron microscopy, Kolodner and Tewari (1975b) demonstrated that all of the chloroplast DNA molecules isolated from *Pisum sativum* had the same denaturation pattern and therefore had an identical base sequence. This conclusion was confirmed by restriction site mapping.

When circular chloroplast DNA molecules of many species were nicked, denatured, briefly renatured, and examined under the electron microscope, a large inverted repeat was visualized as a double-stranded region flanked by single-stranded regions. The sizes of the inverted repeat and the spacer regions between the inverted sequences of the chloroplast DNAs from *Spinacia oleracea, Lactuca sativa,* and *Zea mays* were determined by electron microscopy (Kolodner et al. 1975). In two species of tobacco, *Nicotiana acuminata* and *Nicotiana tabacum,* the sizes of the inverted repeat measured by electron microscopy were 22.67 ± 0.78 kbp and 19.28 ± 0.61 kbp, respectively. Figure 16–1 shows that in *N. tabacum* chloroplast DNA, three short inverted repeats adjacent to the long inverted repeat were detected by electron microscopy (Shen et al. 1982). This structural detail could have easily been overlooked by restriction site mapping using a few restriction enzymes.

The mitochondrial genome of higher plants appear to be much larger and more complex than those of animals or fungi. Kinetic-complexity measurements of a range of plant mitochondrial DNAs gave values between 7.3×10^7 and 1.6×10^9 with little evidence for extensive sequence reiteration (Ward et al. 1981). Digestion of a variety of plant mitochondrial DNAs with different restriction enzymes yields a surprisingly complex pattern of fragments in differing molar ratios (Quetier and Vedel 1977). Electron microscopy was used to visualize DNA molecules isolated from purified maize mitochondria. A heterogeneous population of circular molecules ranging from 0.5 μm to 30 μm in contour length and linear molecules with molecular weights up to 120 megadaltons were detected (Leving et al. 1979).

Electron microscopic studies on mitochondrial DNAs prepared from the normal fertile and the S-male and T-male sterile cytoplasms of maize showed that the circular DNA molecules fell into specific size classes and that these size classes were unique to each cytoplasm (Levings et al. 1979). Circular mitochondrial DNA molecules that can be identified as supercoiled form in cesium chloride–ethidium bromide gradients appear to be found in significantly higher concentrations in mitochondrial DNA preparations from

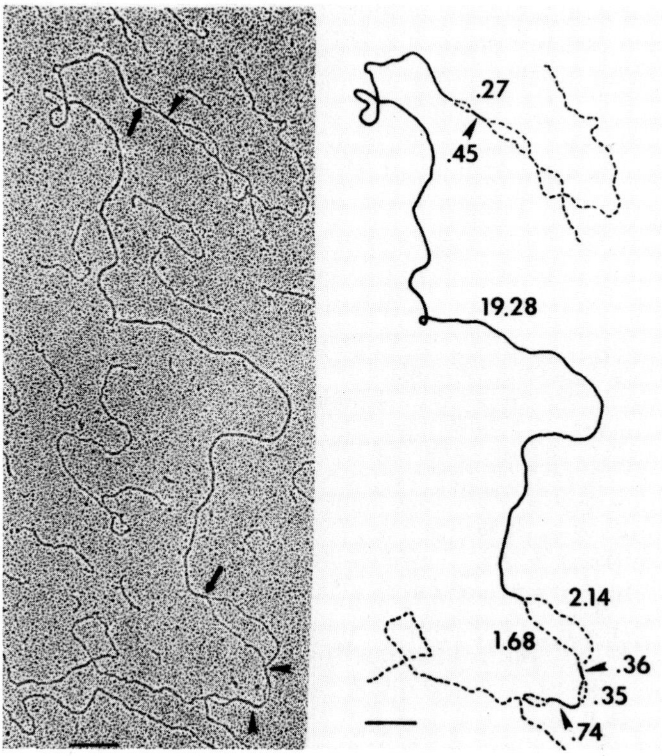

FIGURE 16–1 An electron micrograph and its explanatory tracing show the long inverted repeat region of the chloroplast DNA prepared from *N. tabacum*. Arrows point to the junctions. Several adjacent short inverted repeats are marked with arrowheads. The average length of each region was measured on more than five molecules and is indicated in kilobases in the explanatory tracing. The bar represents 1 kb. (From G.F. Shen, K. Chen, M. Wu, and S.D. Kung [1982] *Mol. Gen. Genet.* 187, 12–18. With permission.)

tissue culture cells (Quetier and Vedel 1977; Sparks and Dale 1980; Brennicke and Blanz 1982). Several size classes of supercoiled mitochondrial DNAs were isolated from cells in a suspension culture derived from *N. tabacum* Wisconsin 38. Three of the smallest size classes—10.1 kbp, 20.2 kbp, and 30.3 kbp, respectively—were characterized through denaturation, heteroduplex analyses using electron microscopy. The 20.2-kbp molecule was found to be a head-to-tail dimer of the 10.1-kbp size class. The 30.3-kbp size class was found to contain two kinds of molecules, a head-to-tail trimer of the 10.1-kbp size class (X_3) and a second molecule (ABC) that had a completely different sequence composition (Dale et al. 1983). Figure 16–2 shows that in a particular denaturation condition, X_3 shows three symmetrical denaturation loops, while ABC is extensively denatured. This result was also confirmed by restriction mapping of cloned molecules (Dale et al. 1983).

In summary, nucleic acid electron microscopy is effective for the study of organelle genomes. It can assist, complement, and guide restriction mapping.

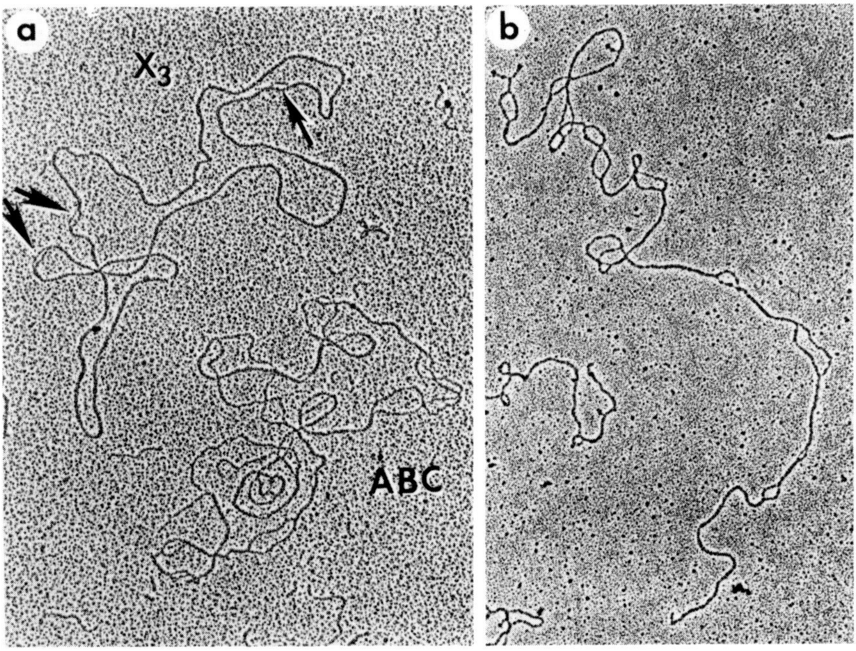

FIGURE 16–2 Electron micrographs show the denaturation profile of (A) a mixture of two types of molecules designated X_3 and **ABC**. X_3 and **ABC** were copurified in a 0.4% agarose gel from mitochondrial DNA isolated from the suspension culture of *N. tabacum* Wisconsin 38. The denaturation profile as well as the restriction map (Dale et al. 1983) shows that X_3 is the trimer of a 10.1-kb size class, and ABC consists of three smaller size classes. The three symmetrical denaturation loops in X_3 are marked with arrows. ABC is extensively denatured under these conditions. The spreading solution contained 0.1 M Tris (pH 8.5), 0.01 M Na_3EDTA, and 80% formamide. The hypophase contained 0.01 M Tris (pH 8.5), 1 mM Na_3EDTA, and 50% formamide. (B) A subunit of **ABC** in a moderately denatured condition. The spreading solution contained 0.1 M Tris (pH 8.5), 0.01 M Na_3EDTA, and 75% formamide. The hypophase contained 0.01 M Tris (pH 8.5), 1 mM Na_3EDTA, and 45% formamide.

16.2 DNA-DNA HETERODUPLEX ANALYSIS

Heteroduplex analysis is a powerful and widely used method for mapping regions of sequence homology and sequence nonhomology between related nucleic acids. For example, if a mixture of two related double-stranded DNAs, AA′ and BB′, is denatured and renatured, the reaction mixture will consist of the homoduplexes AA′ and BB′, the heteroduplexes AB′ and A′B, and unrenatured single strands. The heteroduplexes are the molecules of interest. Segments along a heteroduplex that are double-stranded represent regions of homology where AA′ and BB′ have identical or very similar sequences; segments of the heteroduplex that are single stranded are regions where the two nucleic acids differ markedly in sequence. This technique has been used extensively to characterize cloned DNA fragments. A few examples of the application of this technique will be described.

Sequence homologies between the whole rRNA operon of *Vicia faba* chloroplast genome and a cloned rRNA operon of *Escherichia coli* were studied by heteroduplex analysis (Delius and Koller 1980). Circular chloroplast DNA was hybridized to the restriction fragment BamHl B from the DNA of the transducing phage λrif^{d} 18, which carries the rRNA operon rrn B (Kiss et al. 1978). Analysis of the heteroduplexes showed homologies in the 16S and 23S rRNA regions, but none in the spacer region. Basic protein film spreadings of analogous heteroduplexes that were prepared from two restriction fragments in increasing concentration of formamide revealed regions of incomplete homology in the 23S rRNA region (Delius and Koller 1980). Heteroduplexes between restriction fragments of wheat and maize chloroplast DNAs that contain the gene for the large subunit of Rubisco (rbc L) and its flanking sequences were prepared. Examination of such heteroduplexes clearly showed that sequence divergence occurred most rapidly in spacer regions between genes. The coding sequences for proteins were more conserved (Koller et al. 1982).

In both higher plants and the unicellular green algae *Chlamydomonas* the replication of chloroplast DNA has been shown to initiate with the formation of a displacement loop (D loop) at specific sites (Kolodner and Tewari 1975b; Waddell et al. 1984). In *Chlamydomonas,* functionally important DNA sequences adjacent to one D-loop site were delimited by heteroduplex study. The 5.5-kbp EcoRl fragment containing one D-loop site was isolated from *C. reinhardtii* and cloned. The cloned sequence was used as a probe to identify the equivalent EcoRl fragment from a different species of *Chlamydomonas*. Heteroduplexes between the two EcoRl fragments were prepared and observed in the electron microscope. Five homologous regions—A, B, C, D, E—with respective lengths of 0.51 kbp, 0.35 kbp, 0.42 kbp, 0.28 kbp, and 0.88 kbp were detected (Figure 16–3). A D-loop site was mapped within the homologous region C (Waddell et al. 1984; Wang et al. 1984). Vallet and Rochaix (1985) determined that the 0.35-kbp conserved region B could promote autonomous replication in *Chlamydomonas,* and

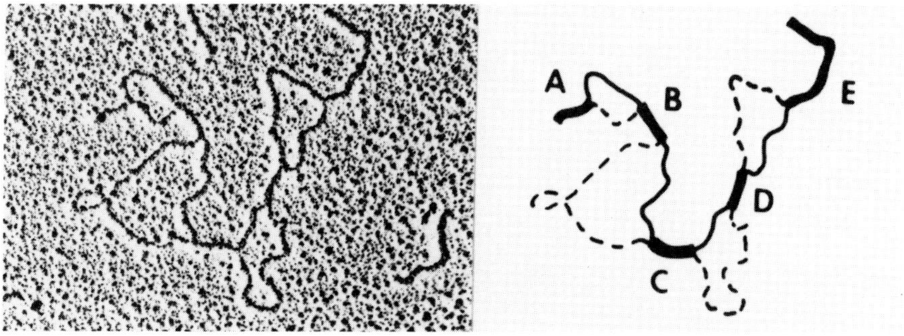

FIGURE 16–3 Electron micrograph and explanatory tracing of a heteroduplex molecule prepared from the EcoR1 inserts of clones CR-13 and CI-10. CR-13 and CI-10 were cloned from the chloroplast DNAs isolated from two different species of *Chlamydomonas, C. reinhardtii* and *Chlamydomonas* sp. strain WXM, respectively. Each EcoR1 insert contains a displacement loop region where the replication of chloroplast DNA initiates (Waddell et al. 1984). The formamide concentration used in the spreading solution was 50%, and in the hypophase, 20%. The homologous regions between those two EcoR1 fragments are labeled A, B, C, D, and E consecutively in the explanatory tracings. A D loop was located in region C. The bar represents 1 kbp of double-stranded DNA.

that the 0.88-kbp conserved region E contained a yeast autonomously replicating sequence (ARS) (Vallet and Rochaix 1985).

Heteroduplex analysis has also been used to characterize cloned nuclear genes. In soybean, lectin gene structure in a genetic line that accumulates normal levels of seed lectin (Le^+ line) and in a mutant line that produces no detectable seed lectin (Le^- line) was investigated. Each genetic line contains two lectin genes, L1 and L2; L1 encodes seed lectin, while the function of L2 is unknown. The L1 lectin gene of both genetic lines has been cloned. Heteroduplexes of L1 lectin gene regions isolated from these two genetic lines showed that the Le^--line gene contained a 3.4-kbp insertion (Goldberg et al. 1983a).

After its initial isolation, the activator (Ac) element of maize has been identified and visualized by using heteroduplex studies (Fedoroff et al. 1983). Genomic DNA sequences homologous to a waxy (Wx) cDNA plasmid (Shure et al. 1983) were cloned from strains with (Ac wx-m9) and without (Wx 9-rl) Ac insertion at the locus. Heteroduplexes prepared between the two recombinant clones showed that the maize DNA inserts were homologous except for a region of approximately 4 kbp near one end. A heteroduplex prepared between the cloned AC wx-m9 Bgl II fragment and a cloned Wx EcoRl fragment (Shure et al. 1983) also showed that these two fragments differed by a 4-kbp sequence. Therefore, the 4-kbp insertion detected in strain Ac wx-m9 corresponded to the Ac element. Detailed restriction endonuclease mapping also confirmed this conclusion (Fedoroff et al. 1983). Extensive similarity between the sequence of the AC element and the sequence of the dissociation (Ds) element was also revealed by heteroduplex study using subclones containing the AC 9 and Ds 9 insertion sequences (Fedoroff et al. 1983).

16.3 ELECTRON MICROSCOPY OF DNA-RNA HYBRIDS

The thermal stabilities of DNA-RNA hybrids are substantially greater than those of DNA-DNA duplexes in aqueous electrolyte solutions containing high concentrations of formamide (Casey and Davison 1979). Under appropriate conditions, RNA will hybridize with double-stranded DNA and displace the noncomplementary DNA strand. The structure that forms has been termed an R loop and is easily recognized under the electron microscope (Kaback et al. 1979). In several laboratories, this technique has been used to study plant genes.

The presence of the rRNA genes on the inverted repeat of chloroplast DNA in *Z. mays* and *S. oleracea* was confirmed by electron microscopy. Denatured chloroplast DNA was annealed to rRNA under conditions where RNA-DNA hybrids are more stable than DNA-DNA hybrids. The temperature of the hybridization reaction was then lowered to allow intramolecular renaturation of the inverted repeat sequences. Electron microscopy of these molecules showed that the rRNA hybridization resulted in the formation of R loops in the duplex region formed by the intramolecular hybridization of the inverted DNA sequences (Bedbrook et al. 1977; Kolodner and Tewari 1979). In the chloroplast genome of *C. reinhardtii,* R-loop analysis was used to demonstrate that each of the 23S rRNA genes is interrupted by a 940-bp intervening sequence at a site 270 bp from the 5′ end (Rochaix and Malnoe 1978).

The major storage proteins of maize seeds compose a group of alcohol-soluble polypeptides that are collectively known as zein. The genes encoding zein proteins occur in several different linkage groups (Valentini et al. 1979) that are developmentally and coordinately regulated. An EcoRl library of maize DNA was constructed in the lamda vector Charon 4A. A genomic clone, λZg99, that cross hybridizes with zein cDNAs was isolated. R loops of λZg99 and zein mRNAs were prepared for electron microscopy. Only a single R loop was formed. This suggested the absence of any large intervening sequence in the coding region. It also confirmed that the genes were not clustered and there was only one zein gene present in the 13.2-kbp genomic clone (Petersen et al. 1982). Structure and expression of several seed protein genes from soybean have been investigated. Hybridization studies show that lectin and kunitz trypsin inhibitor proteins are encoded by small gene families containing approximately two and five genes each, respectively (Goldberg et al. 1981). R-loop studies again failed to reveal detectable introns in any lectin or kunitz trypsin inhibitor gene investigated to date (Goldberg et al. 1983b).

Several different soybean inbred lines have been identified that are defective in the expression of lectin gene and express a lectinless phenotype (Pueppke and Hymowitz 1982). The relative levels of lectin mRNA sequence in Le^+ and Le^- embryo polysomes were compared by RNA blot analysis. Le^- embryos contain reduced levels of lectin gene transcripts. R-loop analyses showed that the 3.4kb insertion element detected by the heteroduplex study that was described in Section 16.2 is actually inserted within the lectin coding region in several Le^- lines. The presence of this insertion element is correlated with a reduced level of lectin gene transcripts (Goldberg et al. 1983a).

The enzyme phosphoenolpyruvate (P-ePrv) carboxylase plays an important role in C_4 and crassulacean acid metabolism photosynthesis. P-ePrv carboxylase is responsible for the fixation of atmospheric CO_2 (Edwards and Walker 1983). One cDNA clone for maize P-ePrv carboxylase has been isolated and used for the selection of genomic clones from a maize genomic library (Hudspeth et al. 1986). R loops prepared from one genomic clone and the total mRNA showed that at least five introns were detected in the coding gene (Figure 16–4).

16.4 LOCALIZATION OF DNA REPLICATION INITIATION SITES BY ELECTRON MICROSCOPY

DNA replication is one of the most fundamental cellular processes. Electron microscopy of replication intermediates has been used extensively to map origins of DNA replication. The replication of corn and pea chloroplast

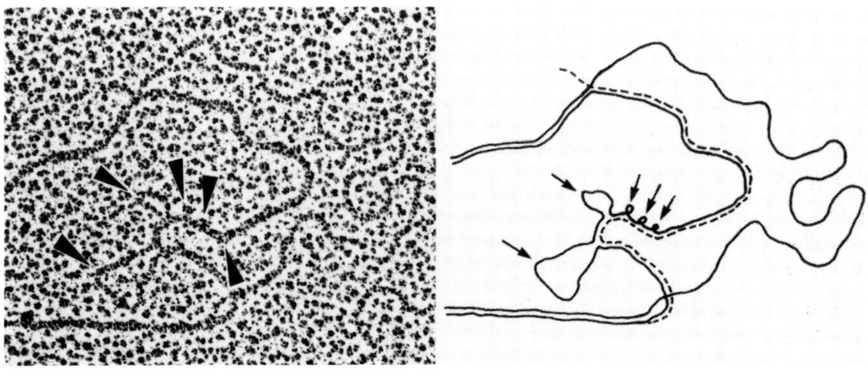

FIGURE 16–4 Electron micrograph and explanatory tracing of an R-loop molecule prepared from a genomic clone containing the coding region of P-ePrv carboxylase and total mRNApoly (A). The five introns are marked in the tracing (arrows) and in the micrograph (arrowheads).

DNAs has been shown to initiate with the formation of two D loops located 7.1 kbp apart (Kolodner and Tewari 1975a and 1975b). Two independent electron microscopic studies revealed that, in *E. gracilis,* chloroplast DNA replication starts from one unique position located upstream from the 5′ end of a supplementary 16S rRNA gene (Ravel-Chapuis et al. 1982; Koller and Delieus 1982).

In *C. reinhardtii,* chloroplast DNA replication origins were mapped using electron microscopy. Chloroplast DNA, isolated from cells of a synchronized culture harvested at various stages, was digested with either EcoRl or BamHl and examined under the electron microscope. Restriction fragments containing D loops that represented early replication intermediates were photographed and measured to determine the position of replicated sequences in relation to the restriction enzyme sites. An example of an EcoRl fragment containing either a D loop or a replication fork is shown in Figure 16–5. Statistical analyses identified that D loops were predominately located at definite sites in two EcoRl fragments and one BamHl fragment. Using the restriction map constructed by Rochaix (1978), D loops were located at two positions on the physical map of chloroplast DNA (Waddell et al. 1984). One replication origin was mapped about 10 kbp upstream of the 5′ end of a 16S rRNA gene. The second origin was spaced 6.5 kbp apart from the first origin and was about 16.5 kbp upstream of the same 16S rRNA. Electron microscopic observation also revealed that initiations at those two sites were not always synchronized. Replication initiated with the formation of a D loop resulting from the synthesis of one daughter strand. After a short initial lag phase, corresponding to the syn-

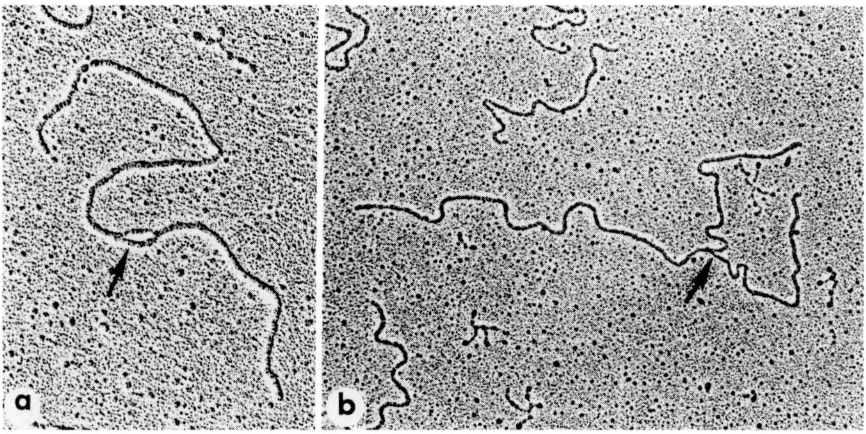

FIGURE 16–5 Electron micrograph of a chloroplast DNA EcoR1 fragment containing (A) a displacement loop and (B) a replication fork. Arrows point to the replicated regions.

thesis of 350 ± 130 bp of one daughter strand, DNA synthesis then proceeded in both directions. Both D-loop regions were preferred binding sites of undetermined protein complexes (Figure 16–6). One D-loop site was first cloned in the *E. coli* plasmid pBR325 to yield the clone R-13. The D-loop region was subsequently subcloned in pBR322 to yield the clone SC3-1. Electron microscopy of the closed circular DNA of both recombinant plasmids showed preferred denaturation in the D-loop region (Figure 16–7). This data suggested that the D-loop region was AT rich (Wang et al. 1984). This result was confirmed by DNA sequence analysis (Wu et al. 1986). A functional role for the D-loop region was also supported by biochemical studies (Wu et al. 1986; Gold et al. 1987).

16.5 CHROMOSOMAL LOCALIZATION OF CLONED GENES BY IN SITU HYBRIDIZATION

Natural gene vectors, such as Ti and Ri plasmids and transposable elements, that could achieve the transfer and stable maintenance of a defined DNA sequence into the nuclear genome of transformed plant cells have been

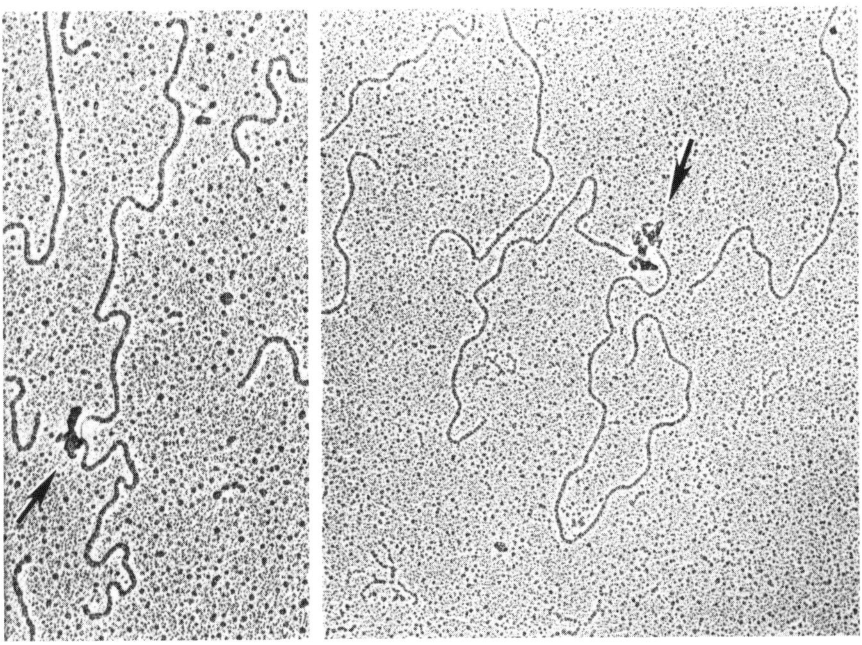

FIGURE 16–6 Electron micrographs show large protein complexes on the EcoR1 fragment of chloroplast DNA of *C. reinhardtii*. The protein binding site is adjacent to the D-loop site on each fragment. Arrows point to the protein complexes.

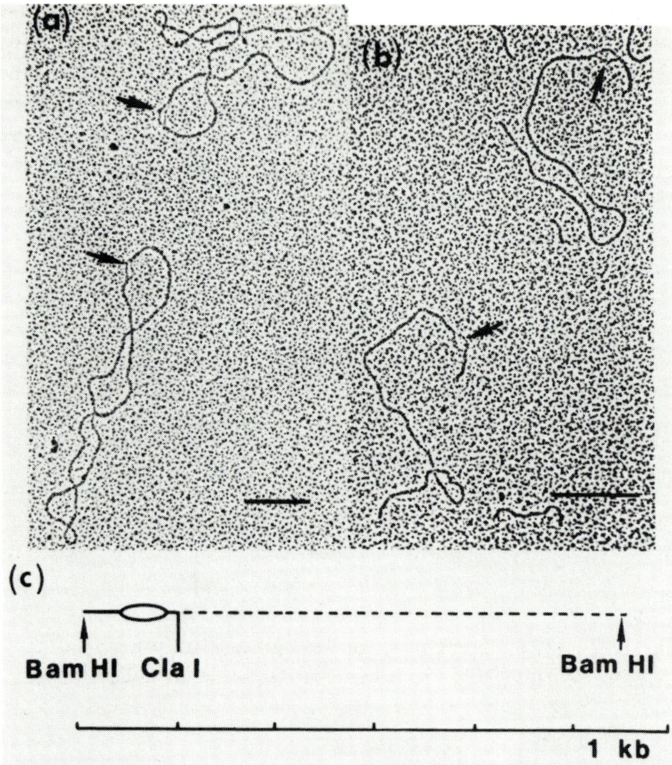

FIGURE 16–7 (A) An electron micrograph of closed circular DNA of CR-13. The DNA sample was prepared using 50% formamide in the spreading solution and 20% formamide in the hypophase; arrows point to the small denaturation loops. (B) Electron micrograph of BamHl digest of SC3-1. The closed circular DNA of SC3-1 was treated with psoralen and glyoxal sequentially to fix the denatured region prior to BamHl digestion. Arrows point to the small denaturation loops. Bars represent 1 kb. (C) A diagram shows position of the preferentially denatured region relative to the position of the restriction sites of BamHl and Clal in SC3-1. Sequence of pBR322, the cloning vector, is represented with a dotted line.

investigated extensively (Bevan and Chilton 1982a and 1982b; Burr and Burr 1982; Couragė-Tebge et al. 1983; Fedoroff 1983; Nester et al. 1984). Recently, many transgenic plants have been successfully constructed using Ti or Ri plasmids (Murai et al. 1983; Broglie et al. 1983; Ambros et al. 1986a and 1986b).

Extensive studies of the chromosomal locations and the expressions of oncogenes in several animal tumor cell lines strongly suggest that chromosomal rearrangements could result in the transfer of one gene from a region that is transcriptionally silent to a region that is actively transcribed (Neel et al. 1982; Taub et al. 1982). Therefore the optimal expression of the transferred gene will not only depend upon the promoter utilized during gene splicing, but will also depend upon the chromosomal insertion site of the transferred gene. The chromosomal localization of genes by a sensitive in situ hybridization technique will assist the characterization of transgenic plants and facilitate the studies of gene regulation and gene expression.

Since the original technique was developed, in situ hybridization has been used extensively to map repeated genes in polytene and metaphase chromosomes at the light-microscope level (for review see Eckhardt 1976). Recent refinements that enhance the efficiency of hybridization and the sensitivity of hybrid detection have made it possible to identify and localize specific cellular or viral DNA and RNA sequences even when present in low copy number (Tereba et al. 1979; Gall and Pardue 1971). Furthermore, by using radioactive probes with high specific activity and/or by including dextran sulfate in the hybridization mixture for network formation, unique sequences have been mapped to metaphase chromosomes (Tereba et al. 1979; Gerhard et al. 1981; Harper and Saunders 1981). C-onc genes (Neel et al. 1982), two genetic probes within the human major histocompatibility complex (Morton et al. 1984), N-myc gene (Li et al. 1986), and many other genes have been mapped on various human chromosomes by using the sensitive in situ hybridization technique. In several systems, the chromosomal in situ hybridization technique has been successfully extended to the electron microscope level (Wu and Davidson 1981; Hutchison et al. 1982; Li et al. 1986).

Using biotinylated probes and in situ hybridization, the presence of a 17-kbp T-DNA in *Crepis capillaris* (2N = 6) metaphase chromosome was detected. The T-DNA was introduced into the genome by infecting *Crepis* stem segments with *Agrobacterium rhizogenes* (Ambros et al. 1986a). The same technique was used to determine the chromosomal insertion sites of *A. rhizogenes* Ri T-DNA in transformed roots of *C. capillaris*. The results indicated that T-DNA was present in a different chromosomal location in each root line and that each chromosome had been a target for T-DNA insertion at least once (Ambros et al. 1986b).

Among higher plants, maize has the most extensive genetic information. The ten meiotic parchytene chromosomes in maize sporocytes can be distinguished by their lengths, the length ratios of the long arm to the short arm, and the knobs at specific regions (McClintock 1959). The structural details of each extended pachytene chromosome makes it feasible to conduct high-resolution gene mapping using in situ hybridization. The highly repeated gene coding for 5S rRNA has been located on maize parchytene chromosome 2 by in situ hybridization. The DNA sequences homologous to 5S rRNA were located in the long arm at 88% of the distance out from the centromere (Mascia et al. 1981).

Recently, a model system has been developed for mapping small single-copy genes on maize pachytene chromosome in my laboratory using cloned Wx locus as a probe. The Wx locus in maize determines the amylose content of endosperm tissue and pollen. There is evidence that it encodes a starch granule–bound uracil diphosphate–glucose starch transferase (Tsai 1974). A 10.8-kbp EcoRl fragment containing most or all of the sequence encoding the Wx mRNA was cloned and was kindly provided to me by Dr. Nina Fedoroff. Southern analysis indicated that the sequence is present as a single-copy gene (Shure et al. 1983). Using nick-translated probes and a hybridization condition favoring the network formation in the in situ hybridization mixture, the 10.8-kbp Wx sequence was located exclusively at a unique site in the short arm of chromosome 9 (Figure 16–8). A subsequent mapping experiment using a 4.5-kbp subclone of the 10.8-kbp genomic clone also gave a positive signal at the same site (Shen and Wu 1987). Therefore,

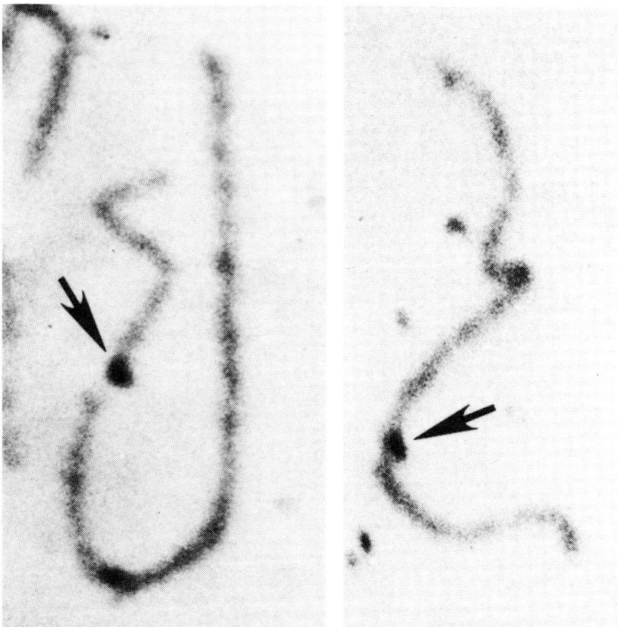

FIGURE 16–8 Light micrographs of two chromosome 9s prepared from maize microsporocytes at the midpachytene stage. The chromosomes were hybridized with a ^{3}H-labeled 10.8-kbp EcoR1 fragment containing the Wx locus. Arrows point to the silver grains resulting from autoradiography.

this conventional in situ hybridization technique has been adapted for mapping small, single-copy genes on maize pachytene chromosomes.

REFERENCES

Ambros, P.F., Matzke, M.A., and Matzke, A.J.M. (1986a) *Chromosoma* 94, 11–18.

Ambros, P.F., Matzke, A.J.M., and Matzke, M.A. (1986b) *EMBO J.* 5, 1073–1077.

Bedbrook, J.R., Kolodner, R., and Bogorad, L. (1977) *Cell* 11, 739–749.

Behn, W., and Herrmann, R.G. (1977) *Mol. Gen. Genet.* 157, 25–30.

Bevan, M.W., and Chilton, M. (1982a) *Annu. Rev. Genet.* 16, 357–384.

Bevan, M.W., and Chilton, M. (1982b) *J. Mol. Appl. Genet.* 1, 539–546.

Brack, C. (1980) *CRC Crit. Rev. Biochem.* 10, 113–169.

Brennicke, A., and Blanz, P. (1982) *Mol. Gen. Genet.* 187, 461–466.

Broglie, R., Coruzzi, G., Lamppa, G., Keith, B., and Chau, N.-H. (1983) *Biotechnol.* 1, 55–61.

Burr, B., and Burr, F. (1982) *Cell* 29, 977–986.

Casey, J., and Davidson, N. (1979) *Nucleic Acids Res.* 4, 1539–1552.

Chow, L.T., Scott, J.M., and Broker, T.R. (1975) *The Electron Microscopy of Nucleic Acids—Laboratory Manual,* Cold Spring Harbor Laboratory, Cold Spring Harbor, NY.

Courage-Tebge, U., Doring, H.-P., Fedoroff, N., and Starlinger, P. (1983) *Cell* 34, 388–393.

Dale, R.M.K., Wu, M., and Kiernan, M.C.C. (1983) *Nucleic Acids Res.* 11, 1673–1685.

Davis, R.W., Simon, M., and Davidson, N. (1971) in *Methods in Enzymology,* vol. XXI (Grossman, L., and Moldave, K., eds.), pp. 413–428, Academic Press, Orlando, FL.

Delius, H., and Koller, B. (1980) *J. Mol. Biol.* 142, 247–261.

Eckhardt, R.C. (1976) in *Handbook of Genetics,* vol. 5, *Molecular Genetics* (King, R.C., ed.), pp. 31–53, Plenum Press, New York.

Edwards, G., and Walker, D.A. (1983) *C3, C4: Mechanisms, and Cellular and Environmental Regulation, of Photosynthesis*, pp. 299–353, University of California Press, Berkeley.

Fedoroff, N. (1983) in *Mobile Genetic Elements,* (Shapiro, J., ed.), pp. 1–63, Academic Press, Orlando, FL.

Fedoroff, N., Wessler, S., and Shure, M. (1983) *Cell* 35, 235–242.

Ferguson, J., and Davis, R.W. (1978) in *Advanced Techniques in Biological Electron Microscopy II* (Koehler, J.K., ed.), pp. 123–171, Springer-Verlag, New York.

Gall, J.G., and Pardue, M.L. (1971) in *Methods in Enzymology,* vol. XXI (Grossman, L., and Moldave, K., eds.), pp. 470–480, Academic Press, Orlando, FL.

Gerhard, D.S., Kawasaki, E.S., Bancroft, F.C., and Szabo, P. (1981) *Proc. Natl. Acad. Sci. USA* 78, 3755–3759.

Gold, B., Tewari, K.K., and Bogorad, L. (1987) *Proc. Natl. Acad. Sci. USA* 84, 194–198.

Goldberg, R.B., Hoschek, G., Ditta, G.S., and Breidenbach, R.W. (1981) *Dev. Biol.* 83, 218–231.

Goldberg, R.B., Hoschek, G., and Vodkin, L.O. (1983a) *Cell* 33, 465–475.

Goldberg, R.B., Jofuku, D., and Vodkin, L.O. (1983b) *J. Cell. Biochem.* 7B(suppl.), 1181.

Harper, M.E., and Saunders, G.F. (1981) *Chromosoma* 83, 439.

Herrmann, R.G., Bohner, H.J., Kowallik, K.V., and Schmitt, J.M. (1975) *Biochem. Biophys. Acta* 378, 305–317.

Hudspeth, R.L., Glackin, C.A., Bonner, J., and Grula, J.W. (1986) *Proc. Natl. Acad. Sci. USA* 83, 2884–2888.

Hutchison N.J., Langer-Safer P.R., Ward D.C., and Hamkalo B.A. (1982) *J. Cell. Biol.* 95, 609–618.

Inman, R.B., Schnos, M. (1970) *J. Mol. Biol.* 49, 93–98.

Kaback, D.B., Angerer, L.M., and Davidson, N. (1979) *Nucleic Acids Res.* 6, 2499–2517.

Kiss, A., Sain, B., Kiss, I., et al. (1978) *Gene* 4, 137–152.

Kleinschmidt, A.K. (1968) in *Methods in Enzymology,* vol. XIIB (Colowick, S.P., and Kaplan, N.O., eds.), pp. 361–377, Academic Press, Orlando, FL.

Kleinschmidt, A.K., and Zahn, R.K. (1959) *Z. Naturforsch.* B124, 770–775.

Koller, B., and Delius, H. (1982) *EMBO J.* 1, 995–998.

Koller, B., Delius, H., and Dyer, T.A. (1982) *Eur. J. Biochem.* 122, 17–23.

Kolodner, R., and Tewari, K.K. (1972) *J. Biol. Chem.* 247, 6355–6364.

Kolodner, R., and Tewari, K.K. (1975a) *Biochim. Biophys. Acta* 402, 372–390.

Kolodner, R., and Tewari, K.K. (1975b) *J. Biol. Chem.* 250, 4888–4895.

Kolodner, R., and Tewari, K.K. (1979) *Proc. Natl. Acad. Sci. USA* 76, 41–45.

Kolodner, R., Tewari, K.K., and Warner, R.C. (1976) *Biochim. Biophys. Acta* 447, 144–155.

Kolodner, R., Warner, R.C., and Tewari, K.K. (1975) *J. Biol. Chem.* 250, 7020–7026.

Levings, C.S. III, Shah, D.M., Hu, W.W.L., Pring, D.R., and Timothy, D.H. (1979) in *Extrachromosomal DNA* (Cummings, D.J., Borst, P., David, I., Weissman, S.M., and Vox, C.F., eds.), pp. 63–73, Academic Press, Orlando, FL.

Li, C., Wu, M., Margitich, I.S., and Davidson, N. (1986) *Chromosoma* 93, 305–312.

Manning, J.E., Wolstenholme, D.R., and Richards, O.C. (1972) *J. Cell. Biol.* 53, 594–601.

Manning, J.E., Wolstenholme, D.R., Ryan, R.S., Hunter, J.A., and Richards, O.C. (1971) *Proc. Natl. Acad. Sci. USA* 68, 1169–1173.

Mascia, P.N., Rubenstein, I., Phillips, R.L., Wang, A.S., Lu, Z.X. (1981) *Gene* 15, 7–20.

McClintock, B. (1959) *Carnegie Inst. of Washington Year Book* 58, 452.

Murai, N., Sutton, D.W., Murray, M.G., et al. (1983) *Science* 222, 476–482.

Neel, B.G., Jhanwar, S.C., Changanti, R.S.K., and Hayward, W.S. (1982) *Proc. Natl. Acad. Sci. USA* 79, 7842–7849.

Nester, E.W., Gordon, M.P., Amasino, R.M., and Yanofsky, M.F. (1984) *Annu. Rev. Plant Physiol.* 35, 387–413.

Petersen, K., Devereux, J., Wilson, D.R., Sheldon, E., and Larkins, B.A. (1982) *Cell* 19, 1015–1026.

Pueppke, S.G., and Hymowitz, T. (1982) *Crop Sci.* 22, 558–560.

Quetier, F., and Vedel, F. (1977) *Nature* 268, 365.

Ravel-Chapuis, P., Heizmann, P., and Nigon, V. (1982) *Nature* 300, 78–81.

Robins, D.M., Ripley, S., Henderson, A.S., and Axel, R. (1981) *Cell* 23, 29–39.
Rochaix, J.-D. (1978) *J. Mol. Biol.* 126, 597–617.
Rochaix, J.-D., and Malnoe, P. (1978) *Cell* 15, 670–681.
Shen, D., Wang, Z.F., and Wu, M. (1987) *Chromosoma* 95, 311–314.
Shen, G.F., Chen, K., Wu, M., and Kung, S.D. (1982) *Mol. Gen. Genet.* 187, 12–18.
Shure, M., Wessler, S., and Fedoroff, N. (1983) *Cell* 35, 225–233.
Sparks, R.B. Jr., and Dale, R.M.K. (1980) *Mol. Gen. Genet.* 180, 351–355.
Taub, R., Kirsch, I., Morton, C., et al. (1982) *Proc. Natl. Acad. Sci. USA* 79, 7837–7841.
Tereba, A., Lai, M.M.C., and Murti, K.G. (1979) *Proc. Natl. Acad. Sci. USA* 76, 6486–6490.
Tsai, C.Y. (1974) *Biochem. Genet.* 11, 83–96.
Valentini, G., Soave, C., and Ottaviano, E. (1979) *Heredity* 42, 33–40.
Vallet, J.-M., and Rochaix, J.-D. (1985) *Curr. Genet.* 9, 321–324.
Waddell, J., Wang, X.-M., and Wu, M. (1984) *Nucleic Acids Res.* 12, 3843–3856.
Wang, X.-M., Chang, C.H., Waddell, J., and Wu, M. (1984) *Nucleic Acids Res.* 12, 3857–3872.
Ward, B.L., Anderson, R.S., and Bendich, A.J. (1981) *Cell* 25, 793.
Westmoreland, B.C., Szybalski, W., Ris, H. (1969) *Science* 163, 1343–1348.
Wu, M., and Davidson, N. (1981) *Proc. Natl. Acad. Sci. USA* 78, 7059–7063.
Wu, M., Lou, J.K., Chang, D.Y., Chang, C.H., and Nie, Z.Q. (1986) *Proc. Natl. Acad. Sci. USA* 83, 6761–6765.
Younghusband, H.B., and Inman, R.B. (1974) *Annu. Rev. Biochem.* 43, 605–619.

CHAPTER 17

Molecular Evolution of *Nicotiana* Chloroplast Genomes

Shain-dow Kung

Recent advances in biotechnology have provided numerous new opportunities to study the fundamental mechanisms of life at the molecular level, and have thus far refined and expedited the approaches in the search of the secret of life. One of the many successful applications of biotechnology has been in the study of plant molecular biological phenomena. By employing recombinant DNA technology, rapid progress has been made in the understanding of many molecular processes in plants. The purpose of this chapter is to describe our present knowledge of the molecular evolution of the *Nicotiana* chloroplast genome gained through the uses of biotechnology in recent years. The *Nicotiana* chloroplast genome is one of several higher plant chloroplast genomes that have been widely studied and well characterized. Others are those of maize (Bogorad et al. 1984), spinach (Whitfeld and Bottomley 1983), bean, pea (Chu et al. 1981; Palmer et al. 1983), wheat (Howe et al. 1982), and *Oenothera* (Gordon et al. 1981).

There are many advantages to using the *Nicotiana* system to study molecular evolution. This genus consists of 64 species, many of which can be crossed to produce viable hybrid plants. Thus, the mode of inheritance of many cytoplasmic components, including organelles, can be determined. In addition, excellent records on taxonomy, cytology, and evolution of

Nicotiana species are available (Goodspeed 1954). Furthermore, the biochemistry and genetics of many chloroplast genome–coded proteins from many species of *Nicotiana* have been well documented (Kung 1984). In recognizing and appreciating these properties, my laboratory has focused for many years on the study of *Nicotiana* chloroplast proteins and recently its chloroplast genomes. The results obtained from studying this genus through the application of recombinant DNA technology reflect those properties unique to *Nicotiana* as well as those general properties of other plants.

17.1 CHLOROPLAST GENOMES

17.1.1 Historical Notes

The study of higher plant chloroplast genomes goes back some 30 years when Feulgen-positive material was first detected in chloroplasts (Chiba 1951). This discovery created excitement as well as controversy. The confusion that ensued ended in the early 1960s when more-direct biochemical and biophysical approaches confirmed the existence of the chloroplast genome (Kirk 1971; Kung 1977). Electron microscopy and CsCl density gradient centrifugation were the major techniques used in identifying the chloroplast genome, though results obtained from CsCl density gradients were not without contradiction. The density value at or above 1.705 g/cm^3 initially assigned to the chloroplast genome was proved to be erroneous (Kirk 1971). This once-accepted value was ultimately corrected to 1.696 g/cm^3, but no definitive reason for this misidentification was ever discovered.

After the existence of the chloroplast genome was firmly established, attention was quickly directed to more-quantitative and more-qualitative analyses. It immediately became clear that the chloroplast genome was organized into a circular form that existed in multiple copies per chloroplast and was of a size sufficient to encode about 200 average-sized proteins (Kung 1977). Thus, subsequent efforts were focused on the identification of the structure and function of chloroplast genes.

The first function assigned to chloroplast genes was to code for chloroplast rRNAs (Tewari and Wildman 1970) and some chloroplast proteins (Wildman et al. 1973b). The use by Wildman's group at the University of California, Los Angeles, in the early 1970s of a combination of genetic and molecular approaches to the *Nicotiana* system in order to localize on the chloroplast genome the coding site of the large subunit of ribulose-1,5-bisphosphate carboxylase/oxygenase (Rubisco) further stimulated interest in the study of functions of the chloroplast genome (Wildman et al. 1973a). By the middle of the 1970s this interest was amplified and expanded by Bogorad's group at Harvard; they quickly captured the newly developed recombinant DNA technology and applied it to the study of the molecular biological phenomena of the maize chloroplast genome (Coen et al. 1977). It was the introduction of the recombinant DNA technology into the study of chloroplast genomes that made it possible to study the molecular evolution described here. Today, the complete nucleotide sequence of the *Nicotiana tabacum* chloroplast genome has been elegantly determined by Sugiura's group (Shinozaki et al. 1986).

17.1.2 The *N. tabacum* Chloroplast Genome

Chloroplast genomes from over 35 *Nicotiana* species have been isolated and studied (Atchison et al. 1976; Rhodes et al. 1981; Salts et al. 1984). The well-characterized chloroplast genome from *N. tabacum* has been mapped in detail (Jurgenson and Bourque 1980; Seger et al. 1981; Fluhr and Edelman 1981; Tassopulu and Kung 1984; Hildebrand et al. 1985; Shinozaki et al. 1986) using clone banks of *Nicotiana* chloroplast genomes constructed from at least two species, *N. tabacum* (Fluhr et al. 1983) and *Nicotiana otophora* (Zhu et al. 1982).

The entire genome is 155,844 bp long and consists of four well-divided regions: the large and small single copy (LSC, SSC) regions separated by the inverted repeats (IR_A, IR_B). The LSC and SSC regions have 86,684 and 18,428 bp, respectively. The IR_A and IR_B regions are identical, having 26,399 bp each. The genome is double stranded and circular.

The *N. tabacum* chloroplast genome contains two copies of the 23S, 16S, 5S, and 4.5S rRNA genes in the inverted repeats; 30 tRNA genes distributed throughout the genome; 20 genes or reading frames for ribosomal proteins; 21 genes or reading frames for thylakoid proteins; and 6 genes for NADH dehydrogenase components. Chloroplast genes are transcribed by the chloroplast RNA polymerase polycistronically or monocistronically. They possess the prokaryotic promoters and terminators as well as the eukaryotic introns (Kung and Lin 1985; Shinozaki et al. 1986).

17.2 COMMONALITY AND DIVERSITY OF *NICOTIANA* CHLOROPLAST GENOMES

17.2.1 Restriction Maps

Recognition sequences generated by various restriction enzymes from the chloroplast genome provide ideal markers for the construction of physical maps. The size of most higher plant chloroplast genomes is well within the range of size amenable to direct restriction enzyme mapping techniques. Chloroplast genomes prepared by the standard method are also pure enough for mapping and cloning experiments. Different restriction enzymes usually cleave the DNA at different frequencies, and the one that cuts the DNA under study infrequently provides the simplest tool for obtaining a map covering the entire genome. Among the ten restriction enzymes (EcoRI, BamHI, HindIII, XhoI, SmaI, KpnI, pstI, SalI, PvuII, and BglI) tested on the *Nicotiana* chloroplast genome, six produced 10 to 20 fragments suitable for mapping work. The most frequently used restriction enzymes for mapping the *Nicotiana* chloroplast genome are SalI, SmaI (Jurgenson and Bourque 1980), PvUII, XhoI, BglI (Seyer et al. 1981), and BamHI (Zhu et al. 1982). Complete, detailed physical maps of the *N. tabacum* chloroplast genome based on digests of several restriction enzymes have been constructed (Sugiura 1988).

Figure 17–1 is the SmaI restriction map of chloroplast genomes from eight *Nicotiana* species. It illustrates the variations in the SmaI sites and the genome sizes among the eight species. For example, the total length of the genome of *Nicotiana acuminata* is considerably longer than that of other species (Shen et al. 1982). On the other hand, the genome size of *Nicotiana knightiana* is slightly shorter than others (Tassopulu and Kung 1984). Deletions, insertions, or inversions occur frequently during the course of evolution. They occur in relatively defined regions on the genome (Salts et al. 1984). Most restriction site variability is in the LSC region adjacent to the inverted repeats, and the two major deletion sites are only 5–6 kb apart (see Figure 17-1). These areas are termed the *hot spots* (Tassopulu and Kung 1984).

17.2.2 Commonality and Diversity Revealed by the Restriction Fragment Patterns

Commonality among the chloroplast genomes of higher plants has long been expected. This expectation originated from the observation that the chloroplast genome is highly conserved. There is considerable biochemical and functional evidence to support this. For example, the chloroplast genome exists in multiple copies per chloroplast, contains similar coding information, and exhibits uniform physicochemical properties. Therefore, the discovery that the restriction fragment patterns of chloroplast genomes from over 35 *Nicotiana* species were all different was unexpected. The identical restriction fragment patterns of *N. tabacum* and *Nicotiana sylvestris*, the only exception to this widespread diversity, reflect maternal inheritance of chloroplast genomes (Kung 1984), in that phylogenetically, *N. sylvestris* was the female parent of *N. tabacum* (Gray et al. 1974).

When the *Nicotiana* chloroplast genome was digested with a number of restriction enzymes, the results showed clearly that EcoRI and BamHI enzymes produced far more fragments and exhibited a much higher resolving power to uncover sequence differences. The 40-odd fragments generated by EcoRI enzyme can be arranged into six groups (Kung et al. 1982). The first two groups consist of 10 to 13 fragments that are extremely variable. In the remaining fragments in the other four groups there are more commonalities than diversities. It is such a combination of commonalities among groups and diversities among individuals in restriction fragment patterns that forms the basis of species specificity and genus commonality among *Nicotiana* chloroplast genomes.

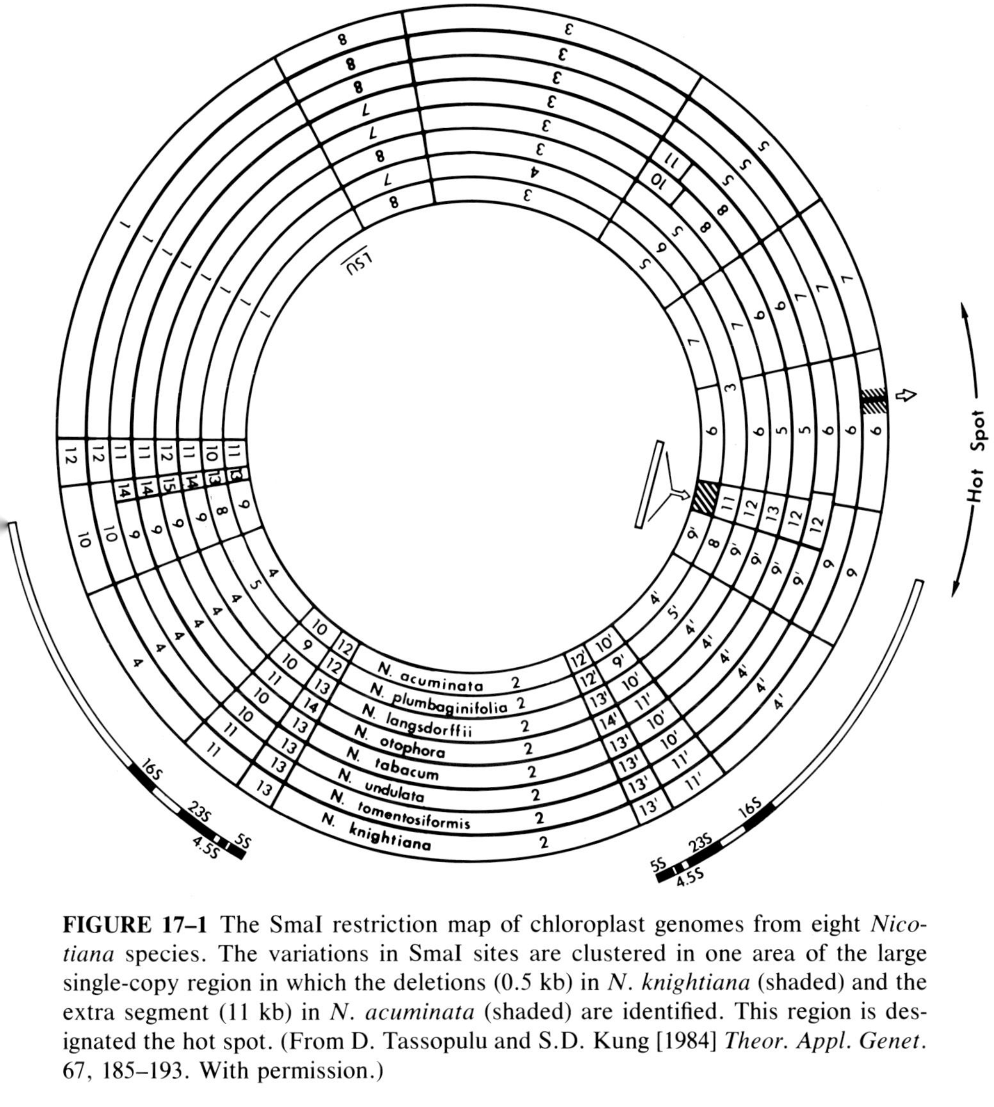

FIGURE 17–1 The SmaI restriction map of chloroplast genomes from eight *Nicotiana* species. The variations in SmaI sites are clustered in one area of the large single-copy region in which the deletions (0.5 kb) in *N. knightiana* (shaded) and the extra segment (11 kb) in *N. acuminata* (shaded) are identified. This region is designated the hot spot. (From D. Tassopulu and S.D. Kung [1984] *Theor. Appl. Genet.* 67, 185–193. With permission.)

17.3 MOLECULAR EVOLUTION OF *NICOTIANA* CHLOROPLAST GENOMES

Alteration of DNA is the main force of evolution. It is evident from the wide diversity of *Nicotiana* chloroplast genomes that there have been considerable changes in their DNA throughout the course of evolution. It has been demonstrated in *Nicotiana* that the mechanism of chloroplast genome alteration involves point mutation, inversion, deletion, duplication, and possible recombination (Shen et al. 1982; Tassopulu and Kung 1984; Kung et al. 1981 and 1982). Point mutations are primarily responsible for the observed gain and loss of many restriction sites. They occur frequently in relation to other mutations, and are clustered in hot spots (Tassopulu and Kung 1984). Large as well as small deletions have frequently been detected primarily in the LSC region (Salts et al. 1984). The largest deletion so far detected is about 11 kb long. Whether this segment, which accounts for 7% of the *Nicotiana* chloroplast genome, contains any structural genes is not known. The smallest deletion so far detected involves only about 50–300 bp (Salts et al. 1984). Duplication mutations are confirmed by the existence of inverted and small tandem repeats.

The molecular evolution of the *Nicotiana* chloroplast genome can be studied at both the restriction fragment pattern and the nucleotide sequence levels. At the restriction fragment pattern level the commonality and diversity among *Nicotiana* species are remarkable. The species specificity and genus generality are unique in the plant kingdom. At the nucleotide sequence level, the homology among rbcL genes is extremely high, exceeding 99.5%. When the nucleotide sequence of a particular segment within a gene is compared among species there is total homology.

17.3.1 At the Restriction Fragment Pattern Level

17.3.1.1 The Genomes. *Nicotiana* chloroplast genomes possess many distinct properties and unique features. The fragment patterns generated by restriction enzymes exhibit a high degree of diversity as well as a general commonality (Rhodes et al. 1981; Kung et al. 1981): The individual restriction pattern is species specific for any given species, while the overall configuration is characteristic of the genus *Nicotiana* (Kung et al. 1981). In many species, even variation of a fragment contains sufficient information to indicate its taxonomic position or to reflect its evolutionary relationship. Therefore, restriction patterns of *Nicotiana* chloroplast genomes can be used in this context to identify species in a manner similar to the taxonomic, cytogenetic (Goodspeed 1954), and biochemical (Chen et al. 1976) classifications. This is due mainly to the existence of the unusual taxonomic and evolutionary relationships among *Nicotiana* species. Many of the 64 *Nicotiana* species originated through interspecific hybridization, followed by doubling of the chromosomes (Goodspeed 1954). In each interspecific hybridization the chloroplast genome was generally inherited maternally (Wildman et al. 1973b). Thus, the present-day species with 24 paired chromosomes are more advanced than those possessing 12 paired chromosomes. Consequently, if the chloroplast genome of a species with 12 paired chromosomes shares a fragment pattern identical to that of a 24-pair species, a close phylogenetic relationship between them is implied. It is likely that the former is an ancestor (female) of the latter. The best-studied phylogenetic relationship is that between *N. tabacum* (N = 24) and *N. sylvestris* (N = 12), in which *N. sylvestris* was the female parent in the original cross (Gray et al. 1974) and both species possess identical chloroplast genomes (Rhodes et al. 1981), as directed by the mode of maternal inheritance.

The following description is aimed at demonstrating the molecular evolution of chloroplast genomes from many closely related *Nicotiana* species with different levels of ploidy. The sequence of chloroplast genome evolution in *Nicotiana* is determined based on the restriction patterns from a single restriction enzyme without radioisotope labeling of the fragments.

The restriction enzyme used in this evolutionary study is SmaI, which generates many species-specific fragment patterns in *Nicotiana* chloroplast genomes. Among the 35 *Nicotiana* species examined there are a total of 21 distinct SmaI fragments (Table 17–1). Eight are present in every species, and 13 are variable and scattered. On the average there are 13 to 15 bands in each species. Fragments 17 and 20 are doublets in all species, whereas fragment 15 exists in duplicate in most species, but is missing in *Nicotiana tomentosa, Nicotiana tomentosiformis, N. knightiana*, and *Nicotiana paniculata* (see Table 17–1). Only in *N. tabacum, N. sylvestris*, and *Nicotiana longiflora* are there two copies of fragment 18. Since the molecular weight of each fragment is known, it is possible to trace the origin and conversion of each variable fragment in each species. It is thus deduced that the chloroplast genome of *Nicotiana langsdorffii*, a more-advanced species, evolved from that of *Nicotiana plumbaginifolia* by the introduction of a single SmaI site into fragment 3 of the chloroplast genome of *N. plumbaginifolia*. Likewise, a gain of a single SmaI site in fragment 6 of *N. langsdorffii*'s chloroplast genome at different positions generated the chloroplast genome of *N. sylvestris* and that of *N. otophora*. The two new SmaI fragments generated in each case are fragments 12 and 18 for *N. sylvestris*, and fragments 13 and 16 for *N. otophora* (Figure 17–2). This is confirmed by nick translation and hybridization experiments.

In *N. tomentosa, N. tomentosiformis, N. knightiana*, and *N. paniculata*, loss instead of gain of the SmaI site seems to be the mechanism of the evolution of their chloroplast genome. The point mutations result in gains as well as losses of restriction sites.

Comparing the two species *N. tomentosa* and *N. paniculata* one finds that all fragments except one are equal (see Table 17–1). Thus, the reduction of the size of fragment 7 (*N. tomentosa*) to the size of fragment 8 (*N. paniculata*) may indicate a deletion. To date, no polymorphism for a restriction site has been observed in this study, although such intraspecific variability was reported in *Nicotiana debneyi* (Scowcroft 1979). On the basis of this information a sequence of evolution of chloroplast genomes from several species in *Nicotiana* has been proposed. Figure 17–2 is a diagram-

matic representation of this evolution indicating the sequential addition of three, and the deletion of two, SmaI sites during the course of chloroplast genome evolution.

Figure 17–3 is a phylogenetic tree of chloroplast genome evolution within this genus. Goodspeed (1954) assumed that the ancestral stocks having five pairs of chromosomes diverged into several branches. It appears that the present-day species *N. plumbaginifolia* represents one of them at the 12-pair level. In this proposed scheme of molecular evolution of the *Nicotiana* chloroplast genome, *N. plumbaginifolia* may have served as the ancestral stock from which all the Australian and some of the American species were derived. It is proposed here that the chloroplast genome of *N. langsdorffii* originated directly from *N. plumbaginifolia*. There is adequate cytogenetic evidence to support this claim (Goodspeed 1954). *N. langsdorffii* may in turn have given rise to *Nicotiana noctiflora, N. longiflora*, and *Nicotiana alata*. The evidence that *N. noctiflora* includes *N. langsdorffii* elements is available (Goodspeed 1954). At this point of evolution, the ancestor of *N. noctiflora* had migrated to Australia and formed the current Australian population.

Chloroplast genomes from *N. tomentosa, N. tomentosiformis, N. knightiana*, and *N. paniculata* are clearly separated from those of the other species studied here. Their immediate ancestor is different from but probably similar to *N. plumbaginifolia*. The close relationship among this group is strongly supported by other evidence inlcuding cytogenetic evidence (Goodspeed 1954). Furthermore, the evidence obtained from amino acid analysis and peptide mapping of the large subunit of Rubisco (Kung et al. 1977) and the analysis of restriction fragments of chloroplast genomes (Kung et al. 1982) divide these species into two subgroups: *N. tomentosa* and *N. tomentosiformis* make up one, and *N. knightiana* and *N. paniculata* make up the other. This is in agreement with the results obtained previously (Rhodes et al. 1981).

17.3.1.2 The Inverted Repeats. One distinct feature of the chloroplast genome is the existence of inverted repeats that contain genes for all rRNAs, some tRNAs, and other RNAs. Already, there have been suggestions that these inverted repeats may protect the genome from structural alterations (Bedbrook and Bogorad 1976). However, there is no explanation as to how the inverted repeats evolved or why they evolved in some but not in all chloroplast genomes. Based on the results obtained from restriction mapping and electron microscopic observation of *Nicotiana* chloroplast genomes, the molecular evolution of the inverted repeat was proposed (Tassopulu and Kung 1984). This proposal states that the present chloroplast genomes may have evolved from a much larger one that consisted of two identical copies of chloroplast genomes joined in a head-to-head and tail-to-tail fashion: a total genome inverted repeat. As existed in procaryotic

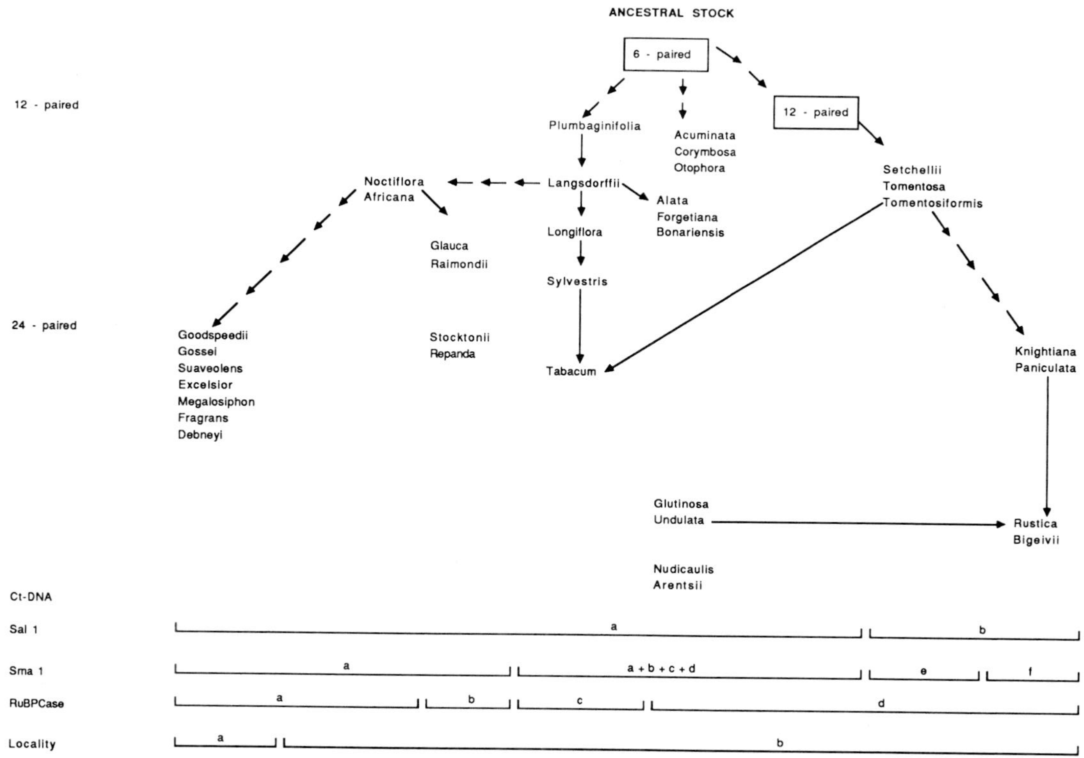

FIGURE 17–3 A phylogenetic tree of chloroplast genome evolution in *Nicotiana* based on the restriction pattern of chloroplast genomes and supported by other biochemical (Rubisco) and geographical distribution data. Short arrows indicate the possible sequence of evolution. *N. tabacum* originated from a cross between *N. sylvestris* and *N. tomentosiformis*, whereas *N. rustica* was a product of *N. paniculata* and *N. undulata*, as indicated by long arrows. Differences among chloroplast genomes are indicated at bottom of figure and designated **a** to **f**. (Expanded from S.D. Kung, Y.S. Zhu, and G.F. Shen [1982] *Theor. Appl. Genet.* 61, 73–79. With permission.)

systems, this structure would have had many advantages over the linear form. During the course of evolution, the size of the chloroplast genome may have been reduced by frequent deletions in some defined regions, resulting in some defined configuration such as the current forms of chloroplast genomes with and without inverted repeats (Figure 17–4). These inverted repeats may be a vestige of what once was two full identical chloroplast genomes. A complete deletion of one (or equivalent) of the two original copies of a chloroplast genome would generate a much smaller genome size with no inverted repeats (see Figure 17–4).

If this proposal is correct, then one would expect to find a correlation between the size of the inverted repeats and total genome size. Indeed, there is evidence to support this. Kolodner and Tewari (1979) observed that

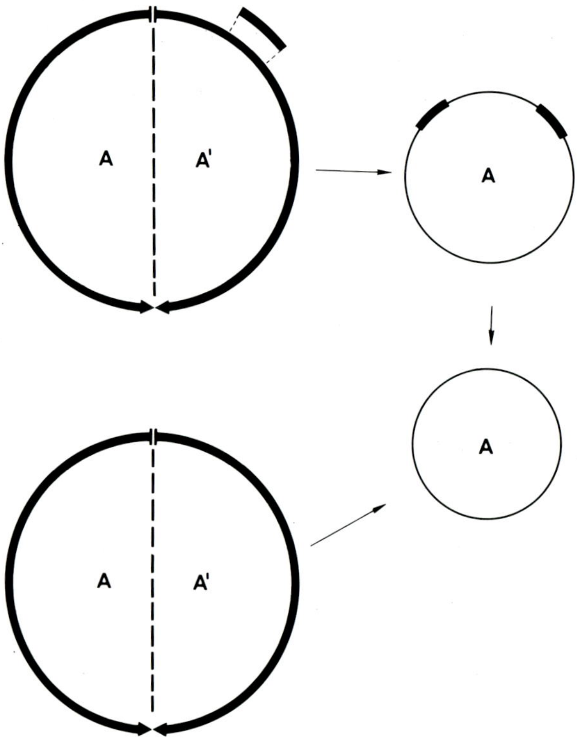

FIGURE 17–4 A proposed evolutionary pathway of the chloroplast genome. A and A′ represent two identical chloroplast genomes joined in a head-to-head and tail-to-tail fashion, creating a total genome inverted repeat. When the most part of A′ was deleted, leaving only a small portion as marked, a genome containing the inverted repeats was generated (upper right). If the A′ was totally deleted, a genome without the inverted repeat was created (lower right).

the inverted repeats represent a constant portion of the entire chloroplast genome. Regardless of the variation in genome size, the inverted repeats always constitute approximately 16% of the total length of the chloroplast genome (Tassopulo and Kung 1984). For example, spinach and corn chloroplast genomes are 153.5 and 136.1 kb, respectively, as measured by electron microscopy. In both cases, about 16% of the total genome is the inverted repeat. In *N. acuminata*, as well as in duckweed, the size of the inverted repeats and their corresponding chloroplast genomes are proportionally larger. Likewise, when one of the inverted repeats is absent from a species, the genome size is reduced by approximately the size of one copy of an inverted repeat (20–25 kb) to about 120 kb, such as occurs in pea, chickpea, and broad bean (Koller and Delius 1980; Palmer and Thompson 1982; Chu and Tewari 1982).

In a recent review of the evolutionary trends in chloroplast genome organization, Palmer (1985) pointed out that many size variations found among angiosperm chloroplast genomes can be accounted for by changes in the size of inverted repeats. By spreading or shrinking the inverted repeats, the genome size can be increased or decreased. At one extreme, the geranium chloroplast genome is the largest determined so far (217 kb) among higher plants; its size can be accounted for by the inverted repeat enlarged from 25 to 76 kb (Palmer 1985). This enlargement produces many duplicate genes that would exist otherwise as single copies. At the other extreme, some legumes (pea and broad bean) have no inverted repeats. The absence of inverted repeats is probably due to deletion of the entire segment equivalent to the inverted repeat. These examples provide strong evidence in support of the origin or molecular evolution of the inverted repeats as proposed by Tassopulu and Kung (1984).

17.3.2 At the Nucleotide Sequence Level

17.3.2.1 The rbcL Genes. Rapid advances in biotechnology have greatly simplified the sequencing of many genes, and every gene sequenced brings us one step closer to understanding the molecular evolution of that gene. Recently, the entire chloroplast genome from *N. tabacum* has been sequenced by Sugiura's group (Shinozaki et al. 1986). Whenever a gene from one plant is sequenced it is always compared with the available sequences from other plants. Such a comparison invariably adds valuable information as to the molecular evolution of that gene.

The rbcL genes from three species of *Nicotiana* (Shinozaki and Sugiura 1982; Lin et al. 1986), maize (McIntosh et al. 1980), spinach (Zurawski et al. 1981), and other plants have been completely sequenced. The nucleotide sequences of maize and spinach diverge by 16% and 10%, respectively, from those of *Nicotiana*. The overall sequence is thus highly conserved. Because of the redundancy of the codons, the overall amino acid homology is 90% and 92% for maize and spinach, respectively. The amino acid sequences at the active sites are totally conserved. The sequences around the N-terminals of rbcL are also highly conserved, although the C-terminals exhibit some variability.

The degree of conservation is extremely high if the nucleotide sequences of the rbcL genes of different *Nicotiana* species are compared. The rbcL genes from *N. tabacum* (Shinozaki et al. 1986), *N. otophora*, and *N. acuminata* (Lin et al. 1986) all contain 1,431 bp coding for 477 amino acid residues. The overall nucleotide sequences of the rbcL genes from *N. otophora* and *N. acuminata* are very similar to that of *N. tabacum* (Shinozaki and Sugiura 1982). There are only eight base changes between *N. tabacum* and *N. otophora* and between *N. otophora* and *N. acuminata*, resulting in two amino acid substitutions each. There are only four base changes between *N. acuminata* and *N. tabacum*, generating two amino acid replacements. There is one change among the charged residues, and there are no unusual features in the resulting polypeptides.

This represents the first comparative sequence analysis of rbcL genes from three different species within a genus. There are only ten base alterations in the coding region, comprising 1,431 bp among three species. Six of these changes occurred at identical positions. The degree of sequence homology is in excess of 99.5%. This is much higher than the degree of sequence homology of the rbcL genes found among many plant species, which ranges from 80 to 90%. There are 141 base changes between *Nicotiana* and spinach, and 224 between *Nicotiana* and maize.

17.3.2.2 Promoters of rbcL Genes. As do procaryotes, chloroplast promoters contain two conserved hexamers separated by a short stretch of about 17 bp. The first hexamer is TTGACA, resembling the −35 region of procaryotic promoter sequences (Rosenberg and Court 1979) in which the trimer TTG is highly conserved. In the second hexamer, the conserved sequence TATAAT is identical to the −10 region (Rosenberg and Court 1979). Within this hexamer, three bases (TA---T) are highly conserved. The first TA pair has been present in all but a few cases examined so far. Similar to the procaryotic promoters, the last "T" in this hexamer is also highly conserved, appearing in all but a few of the 60 promoters compiled (Kung and Lin 1985). Overall, the distribution of bases in each position in these chloroplast promoters is remarkably similar to that of procaryotic promoters (Table 17–2).

Recently, many chloroplast genes and their 5′ flanking regions have been sequenced and studied (Kung and Lin 1985). In most cases, there are conserved promoter sequences in front of the chloroplast genes. The promoter sequences of rbcL genes from different *Nicotiana* species are identical (see Table 17–2). Furthermore, the −35 and −10 regions of the rbcL promoters from several plant species so far studied are also identical (see

TABLE 17–2 The psbA and rbcL Promoters from Different Plant Species

Plant Species	*psbA Promoter*	*rbcL Promoter*
	−35 ... −10	
S. nigrum	TTGACACGAGTATATAAGTCATGTTATACT	
Mustard	TTGACATGGCTATATAAGTCATGTTATACT	
Soybean	TTGACACTGGTATATAAGTCATGTTATACT	
N. debneyi	TTGACACGAGTATATAAGTCATGTTATACT	−35 ... −10
N. tabacum	TTGACACGAGTATATAAGTCATGTTATACT	TTGCGCTATATATATGAAAGAGTATACAAT
Spinach	TTGACACGGGCATATAAGGCATGTTATACT	TTGCGCCATATATATGAAAGAGTATACAAT
N. otophora		TTGCGCTATATATATGAAAGAGTATACAAT
Broad bean		TTGCGCCATACATATGAAACAGTATACAAT
Pea		TTGCGCCATACATATGAAAGAGTATACAAT
Maize		TTGCGCTATATCTATCAAAGAGTATACAAT
Wheat		TTGCGCTATATCTATCAAAGAGTATACAAT

From S.D. Kung, S. Akada, S. Mongkolsuk, et al. (1986) in *The Chrondriome* (S.H. Mantel, G.P. Chapman, and P.F. Street, eds.), pp. 142–163, Longman, UK. With permission.

Table 17–2). This high degree of conservation of nucleotide sequences among the rbcL promoters may have important biological implications and thus deserves further investigation. The only other promoter that resembles the procaryotic types and is highly conserved is that of psbA (see Table 17–2).

Available evidence suggests that both rbcL and psbA promoters are strong signals. Both contain highly conserved sequences: the TTG in the −35 region, and the first TA pair and the last T in the −10 region, with a near-optimal distance of 18 bp. In fact, the psbA promoter is only a single base away from being "perfect." Both promoters are totally conserved in the region from −35 to −10 among plant species, being identical in all seven plant species sequenced to date for rbcL and psbA. There is no such feature found in any other promoters of any other chloroplast genes (see Chapter 13). Both rbcL and psbA promoters are AT rich (70%) and possess a TATA-like element between the −35 and −10 regions. This eukaryotic element is also functional in chloroplasts (Link and Langridge 1984). In addition to the unique structural features, there are some obvious functional properties that support the notion that they are strong promoters. For example, rbcL and psbA mRNAs are the most abundant mRNA species in chloroplasts (Mattoo et al. 1984). Both proteins are synthesized in large amounts in chloroplasts. It is conceivable that the structural features are responsible for this functional consequence.

17.4 CONCLUDING REMARKS

Nicotiana chloroplast genomes are the best-analyzed higher plant organelle genomes to date. The complete nucleotide sequence has been determined for the chloroplast genome of *N. tabacum*, the first chloroplast genome so characterized. This would have been impossible without the use of recombinant DNA technology. Over 35 species of *Nicotiana* chloroplast genomes have been examined at least at the restriction pattern level. This represents the largest number of species studied in any genus. The data accumulated so far provide the most-comprehensive information for understanding the molecular evolution of chloroplast genomes. This includes the evidence obtained at different molecular levels: the entire genome, the inverted repeat, and the rbcL gene and its promoters. There is commonality as well as diversity at the genomic level. The evolution of the inverted repeat presents an interesting theory as to its origin. The difference among nucleotide sequences of the rbcL genes within the genus *Nicotiana* is less than 0.6%. The rbcL promoters have identical sequences in all higher plants examined so far.

It should be stressed that the limited differences in the restriction patterns generated by SmaI provide excellent molecular markers for evolutionary studies on chloroplast genomes as well as on the speciation of

Nicotiana. The use of a single restriction enzyme makes this system more convenient than any others available today for identifying steps in chloroplast genome evolution. Patterns of sequential losses and gains of SmaI recognition sites among *Nicotiana* chloroplast genomes during the course of evolution can be demonstrated. This can be best illustrated using the following species: *N. plumbaginifolia, N. langsdorffii, N. sylvestris, N. tabacum, N. tomentosa*, and *N. paniculata*. The chloroplast genomes of *N. sylvestris* and *N. tabacum* were evolved, not necessarily directly, from *N. langsdorffii*, which was in turn evolved from *N. plumbaginifolia*. This is in agreement with the direction of origin or evolution of *Nicotiana* species. For example, *N. tabacum*, which has a somatic chromosome number of 48, is a more-modern *Nicotiana* species than *N. plumbaginifolia*, which has 20 chromosomes. The larger chromosome number of the modern species is, of course, the result of chromosome doubling after interspecific hybridization.

The most-interesting evidence related to the molecular evolution of chloroplast genomes is their size variation. The chloroplast genome from *N. acuminata* is 7% larger than all other 34 *Nicotiana* chloroplast genomes so far examined. The extra fragment is located in the LSC region on the right-hand border of one inverted repeat. The addition of this segment results in an enlargement of the inverted repeats and consequently the size of the genome. It is likely the expansion or reduction of the inverted repeat that alters the size of chloroplast genomes (Palmer 1985). This is certainly interesting in the evolutionary sense. The proposed route of evolution of the inverted repeat involves many deletions. Chloroplast genes not essential for its survival could be deleted. On the other hand genes that code for the major enzymes remain highly conserved. Finally, the identical nucleotide sequence of the rbcL promoters analyzed here may reflect the structural requirement needed for efficient function. This is parallel to the total conservation of sequences at the active site of Rubisco in all plant species sequenced to date. It is entirely conceivable that there is a correlation between functional requirements and structural features. If this is so, the strength of a promoter may be predicted to a certain degree from its structural features.

REFERENCES

Atchison, B.A., Whitfield, P.R., and Bottomley, W. (1976) *Mol. Gen. Genet.* 148, 263–269.

Bedbrook, J.R., and Bogorad, L. (1976) *Proc. Natl. Acad. Sci. USA* 73, 4309–4313.

Bogorad, L., Crossland, L.D., Fish, L.E., Krebbers, E.T., and Kuck, U. (1984) in *Biochemistry of Photosynthetic Apparatus* (Thornber, J.P., Stachelin, L.A., and Hallick, R.B., eds.), pp. 257–272, Alan R. Liss, New York.

Chen, K., Johal, S., and Wildman, S.G. (1976) in *Genetics and Biogenesis of Chloroplast and Mitochondria* (Bucker, T.H., Neupert, W., Sebald, W., and Werner, S., eds.), pp. 3–11, Elsevier, Amsterdam.

Chiba, Y. (1951) *Cytologia* 16, 259–264.

Chu, N.M., Oishi, K.K., and Tewari, K.K. (1981) *Plasmid* 6, 279–292.

Chu, N.M., and Tewari, K.K. (1982) *Mol. Gen. Genet.* 186, 23–32.

Coen, D.M., Bedbrook, J.R., Bogorad, L., and Rick, A. (1977) *Proc. Natl. Acad. Sci. USA* 74, 5487–5491.

Deno, H., Shinozaki, K., and Sugiura, M. (1983) *Nucleic Acids Res.* 11, 2185–2191.

Fluhr, R., and Edelman, M. (1981) *Mol. Gen. Genet.* 181, 484–490.

Fluhr, R., Fromm, H., and Edelman, M. (1983) *Gene* 25, 271–280.

Goodspeed, T.H. (1954) *The Genus* Nicotiana, *Chronica Botanica Co.*, Waltham, MA.

Gordon, K.H.J., Crouse, E.J., Bohnert, H.J., and Herrmann, R.G. (1981) *Theor. Appl. Genet.* 61, 373–384.

Gray, J.C., Kung, S.D., Wildman, S.G., and Sheen, S.J. (1974) *Nature* 252, 226–227.

Hildebrand, M., Jurgenson, J.S., Ramage, R.T., and Bourque, D.P. (1985) *Plasmid* 14, 64–79.

Howe, C.J., Bowman, C.M., Dyer, T.A., and Gray, J.C. (1982) *Mol. Gen. Genet.* 186, 525–530.

Jurgenson, J.E., and Bourque, D.P. (1980) *Nucleic Acids Res.* 8, 3505–3516.

Kirk, J.T.O. (1971) in *Autonomy and Biogenesis of Mitochondria and Chloroplasts* (Boardman, N.K., Linnane, A.W., and Smillie, R.M., eds.), pp. 267–276, Elsevier, Amsterdam.

Koller, B., and Delius, H. (1980) *Mol. Gen. Genet.* 188, 305–308.

Kolodner, R., and Tewari, K.K. (1979) *Proc. Natl. Acad. Sci. USA* 76, 41–45.

Kung, S.D. (1977) *Annu. Rev. Plant Physiol.* 28, 401–437.

Kung, S.D. (1984) *CRC Crit. Rev. Plant Sci.* 1, 227–267.

Kung, S.D., Akada, S., Mongkolsuk, S., et al. (1986) in *The Chrondriome* (Mantel, S.H., Chapman, G.P., and Street, P.F., eds.), pp. 142–163, Longman, UK.

Kung, S.D., and Lin, C.M. (1985) *Nucleic Acids Res.* 13, 7543–7549.

Kung, S.D., Zhu, Y.S., Chen, K., Shen, G.F., and Sisson, V.A. (1981) *Mol. Gen. Genet.* 183, 20–24.

Kung, S.D., Zhu, Y.S., and Shen, G.F. (1982) *Theor. Appl. Genet.* 61, 73–79.

Lin, C.M., Liu, Z.Q., and Kung, S.D. (1986) *Plant Mol. Biol.* 6, 81–87.

Link, G., and Langridge, U. (1984) *Nucleic Acids Res.* 12, 945–958.

McIntosh, L., Poulsen, C., and Bogorad, L. (1980) *Nature* 288, 556–560.

Mattoo, A., Foffman-Falk, H., Marder, J.B., and Edelman, M. (1984) *Proc. Natl. Acad. Sci. USA* 81, 1380–1384.

Palmer, J.D. (1985) *Annu. Rev. Genet.* 19, 325–354.

Palmer, J.D., Singh, G.P., and Pillay, D.T.N. (1983) *Mon. Gen. Genet.* 190, 13–19.

Palmer, J.D., and Thompson, W.F. (1982) *Cell* 29, 537–550.

Rhodes, P.R., Zhu, Y.S., and Kung, S.D. (1981) *Mol. Gen. Genet.* 182, 106–111.

Rosenberg, M., and Court, D. (1979) *Annu. Rev. Genet.* 13, 319–353.

Salts, Y., Herrmann, R.G., Peleg, N., et al. (1984) *Theor. Appl. Genet.* 69, 1–14.

Scowcroft, W.R. (1979) *Theor. Appl. Genet.* 172, 134–141.

Seyer, P., Kowallik, K.V., and Herrmann, R.G. (1981) *Curr. Genet.* 3, 189–204.

Shen, G.F., Chen, K., Wu, M., and Kung, S.D. (1982) *Mol. Gen. Genet.* 187, 12–18.

Shinozaki, K., Ohme, M., Tanaka, M., et al. (1986) *Plant Mol. Biol. Reptr.* 4, 111–147.

Shinozaki, K., and Sugiura, M. (1982) *Gene* 20, 91–102.

Tassopulu, D., and Kung, S.D. (1984) *Theor. Appl. Genet.* 67, 185–193.

Tewari, K.K., and Wildman, S.G. (1970) in *Control of Organelle Development* (Miller, P.L., ed.), pp. 147–179, Academic Press, Orlando, FL.

Whitfield, P.R., and Bottomley, W. (1983) *Annu. Rev. Plant Physiol.* 34, 279–310.

Wildman, S.G., Kawashima, N., Bourque, D.P., et al. (1973a) in *The Biochemistry of Gene Expression in Higher Organisms* (Pollak, J.K., and Lee, J.W., eds.), pp. 443–456, D. Reidel, Holland.

Wildman, S.G., Lu-Liao, C., and Wong-Staal, F. (1973b) *Planta* 113, 293–312.

Zhu, Y.S., Duvall, E.F., Lovett, P.S., and Kung, S.D. (1982) *Mol. Gen. Genet.* 187, 61–66.

Zurawski, G., Perrat, B., Bottomsley, W., and Whitfeld, P.R. (1981) *Nucleic Acids Res.* 9, 3259–3269.

PART IV

Applications of Biotechnology in Plant Systems

CHAPTER 18

Genetic Engineering for Crop Improvement

Robert T. Fraley

Since their development in 1983, the availability of gene transfer systems has already led to several important insights into the regulation of gene expression and protein function in plants. With the increasing efforts to refine and develop vector systems and the progress that has been made in the identification and isolation of plant genes, the next several years will bring about a dramatic expansion in our understanding of gene structure and function at the molecular level.

In many cases, our technical ability to isolate and transfer genes has surpassed the level of understanding of their biochemical and cell biological properties required for their rational manipulation. However, the wealth of interesting genetic, physiological, environmental, and agronomic problems available for study ensures that there will be a rapid transition from basic research to agricultural applications. In this chapter, the development of genetic transformation systems will be reviewed and discussed. Special emphasis will be placed on research areas where advances have permitted the application of transformation technology to agronomic crops. The progress that has been made in the identification of agronomic traits that confer herbicide, viral disease, and insect tolerance will be reviewed with an emphasis on research programs developed by me and my colleagues.

18.1 PLANT TRANSFORMATION

The rapid progress that has been made in the development of gene transfer systems for higher plants has surprised even the most optimistic researchers in the field. Today, nearly two dozen species of crop plants can be routinely manipulated using available *Agrobacterium tumefaciens* transformation systems. Within the next two to three years it is likely that all major dicotyledonous crop species will be accessible to improvement using transformation technology. The major technical advances that have permitted the extension of gene transfer technology to crop plants have been (1) the construction of fascile vectors based on the *A. tumefaciens*–Ti plasmid system, (2) the availability of a broad spectrum of selectable and scoreable transformation marker genes, and (3) the development of explant-based regeneration systems that efficiently couple the transformation and regeneration processes. Each of these areas is discussed in detail below.

18.1.1 Ti Plasmid Vectors

The fundamental features of the *A. tumefaciens*–Ti plasmid system have been previously reviewed (Bevan and Chilton 1982; Schell and van Montagu 1983; Fraley et al. 1986). To harness this natural gene transfer system, derivative Ti plasmids that lack the phytohormone genes responsible for tumor formation have been constructed. Examples of such plasmids include pGV3850 (Zambryski et al. 1983) and pTiB6S3-SE (Fraley et al. 1985); these plasmids retain a functional *vir* region, *Agrobacterium* origin of replication and optionally, T-DNA border sequences and DNA segments that facilitate homologous recombination with intermediate vectors. There are basically two methods for transferring specific genes into *Agrobacterium* using intermediate vectors: the first method utilizes integration by recombination to combine the intermediate vector with a disarmed Ti plasmid to reconstruct a disarmed T-DNA segment. Although this method works satisfactorily, a disadvantage is the low efficiency of the recombination event.

A second method takes advantage of the fact that *vir* region functions can operate in trans relative to the T-DNA, allowing the disarmed Ti plasmid and an intermediate plasmid containing a border sequence to exist independently in *Agrobacterium*. Such binary, or trans, vectors contain an *Agrobacterium* origin of replication and are represented by Bin 19 (Bevan 1984) or pMON505 (Fraley et al. 1986).

18.1.2 Selectable and Scoreable Markers

Aminoglycoside resistance conferred by various neomycin phosphotransferases (NPTs) was the first widely used selectable marker in plant transformation studies (Herrera-Estrella et al. 1983; Fraley et al. 1983; Bevan et al. 1983) and it remains a marker of choice for many plant systems. In the last five years, many additional selectable markers have been developed that complement NPT and extend the useful range of plants in which selection of transformants can be readily achieved (Table 18–1).

The availability of additional selectable marker genes allows for multiple transformation of selected lines and for optimization of the selection process for different species. For example, NPT genes function efficiently in solanaceous plants, but we find that hygromycin resistance is superior in *Arabidopsis* (Lloyd et al. 1986) and that gentamicin resistance functions better in certain legumes, including alfalfa (unpublished observation).

The nopaline and octopine synthase genes represent the first generation of useful scoreable marker genes; however, these genes suffer from lack of sensitivity, flexibility, and quantitation. Significant progress has been made in the identification and construction of other scoreable marker genes (Table 18–2). The new markers are not only useful in analyzing gene expression and monitoring inheritance of foreign DNA inserts, but are also valuable in demonstrating the targeting of transformation to particular cell types in tissue explants.

18.1.3 Regeneration Systems

Early transformation studies utilized protoplast-based methods; however, the technical difficulties and species limitations associated with the use of protoplasts represent significant problems. The use of tissue explants such

TABLE 18–1 Selectable Markers for Plant Transformation

Selectable Marker	*Reference*
Neomycin phosphotransferase (type II)	Bevan et al. 1983; Fraley et al. 1983; Herrera-Estrella et al. 1983
Neomycin phosphotransferase (type I)	Fraley et al. 1983
Hygromycin phosphotransferase	Lloyd et al. 1986; Waldron et al. 1985; Watrud et al. 1985
Bacterial dihydrofolate reductase	De Block et al. 1984
Mammalian dihydrofolate reductase	Eichholtz et al. 1987
Gentamicin acetyltransferase	Hayford et al. 1988
Streptomycin phosphotransferase	Jones, personal communication
5-enolpyruvylshikimate-3-phosphate synthase	Shah et al. 1986
Acetolactate synthase	Haughn et al. 1988
Bromoxynil nitrilase	Stalker, personal communication
Phosphinothricin acetyltransferase	De Block et al. 1987

TABLE 18–2 Scoreable Marker Genes for Plant Transformation

Scoreable Markers	*Reference*
Nopaline synthase	Depicker et al. 1982
Octopine synthase	De Greve et al. 1983
β-galactosidase	Helmer et al. 1984
Chloramphenicol acetyltransferase	De Block et al. 1984
Firefly luciferase	Ow et al. 1986
β-glucuronidase	Jefferson et al. 1988

TABLE 18–3 Species in Which Transgenic Plants Have Been Produced

Tomato	Alfalfa	White clover	Sugar beet
Potato	Peas	Flax	Lotus
Petunia	Lettuce	Cotton	Cucumber
Tobacco	Sunflower	Cabbage	Asparagus
Carrot	Oil seed rape	Pear	Celery
Poplar	*Arabidopsis*		

as leaves (Horsch et al. 1985), stems (Lloyd et al. 1986), or cotyledons (McCormick et al. 1986) effectively combine ease of regeneration with the unique ability of *Agrobacterium* to transform intact plant tissues. Other approaches involve transformation of cultured cells or establishment of transformed cell lines prior to selection and regeneration. Since most plants can be regenerated from tissue explants, this powerful approach should extend transformation technology to all dicotyledonous crop species. A key area of current research is the optimization of strains, vectors, and selectable markers with particular regeneration systems. The success of this approach is evident from the rapidly growing list of transgenic plant species (Table 18–3).

18.2 CROP IMPROVEMENT

Plant breeding has played a major role in modern agriculture by providing plants with superior disease resistance and increased yields. Genetic engineering promises to have a significant impact on crop improvement by accelerating breeding efforts and by dramatically expanding the available gene pool. While efforts to identify and transfer genes of agronomic importance into plants are just beginning, some degree of success has already been achieved in engineering selective tolerance to herbicides, viral diseases, and insect pests. We present here a summary of the results of these studies with an emphasis on our research programs.

18.2.1 Herbicide-Tolerant Plants

Advances have been particularly dramatic in the engineering of selective herbicide tolerance because existing knowledge of herbicide mode of action and metabolism has permitted rapid identification of key target genes. It seems clear that within the next five to ten years, plants with commercial-level selective tolerance toward many major herbicides—including Roundup^{R}, sulfonylureas and imidazolinones—as well as several newly introduced products will be available. The availability of crops with selective resistance to many of today's potent broad-spectrum herbicides will exert a dramatic effect on weed control practices by accelerating the trend toward commercializing fewer more-effective, less-costly, and environmentally more-acceptable weed control products.

Glyphosate, the active ingredient in Roundup herbicide, is a broad-spectrum nonselective herbicide used to control annual and biennial sedges, grasses, and broadleafed weed species. It is rapidly absorbed by foliar tissue and is quickly translocated to various plant organs. It is rapidly metabolized in the soil and has no residual activity. The shikimate pathway enzyme, 5-enolpyruvylshikimate-3-phosphate synthase (EPSPS) involved in aromatic amino acid biosynthesis has been identified as the specific target of glyphosate in bacteria (Amrhein 1984). Subsequent studies have shown that glyphosate also inhibits this enzyme in higher plants (Mousdale and Coggins 1984; Rubin et al. 1984; Nafziger et al. 1984).

Our researchers (Shah et al. 1986) utilized high-level expression of a plant EPSPS gene to engineer glyphosate-resistant plants. A full-length cDNA clone for EPSPS was isolated from a glyphosate-resistant suspension cell line of *Petunia hybrida*. The amino acid sequence predicted from the nucleotide sequence indicated that the enzyme was synthesized as a precursor polypeptide with an amino-terminal transit peptide sequence of 72 amino acids. The transit peptide is responsible for posttranslational targeting of the precursor enzyme to the chloroplast (della-Cioppa et al. 1986). The wild-type petunia EPSPS cDNA was placed under control of the promotor for the 35S transcript of CaMV, and the resulting chimeric gene was transferred to a binary vector. Introduction of the chimeric EPSPS genes into petunia cells led to the growth of callus at concentrations of glyphosate sufficient to inhibit completely the proliferation of wild-type callus. Transformed petunia plants were regenerated and these plants were tolerant to formulated glyphosate at 0.9 kg/ha, approximately four times the quantity necessary to kill the control plants. However, the growth of these plants was reduced relative to the growth of unsprayed controls.

More recently, experiments have been carried out with a mutant EPSPS gene that encodes an enzyme that is six thousand-fold less sensitive to glyphosate inhibition than the wild-type petunia enzyme. Transformed tobacco plants expressing the glyphosate-resistant EPSPS gene were found to be significantly more tolerant to glyphosate than plants overexpressing the wild-type EPSPS gene. Tobacco plants expressing the mutant gene displayed no visible injury when sprayed with 0.9 kg/ha of glyphosate; the treated plants flowered normally and set seed at levels identical to those for unsprayed controls.

In 1987, we carried out field evaluations of genetically engineered tomato plants containing the mutant EPSPS gene. The site chosen for the field test was a farm located three miles outside the farming community of Jerseyville, Illinois. The test plot was well drained and was not located near a flood plain or tomato production areas. The site was fenced to protect the plants from damage by animals. The local community and state officials were briefed in advance; as expected, there was strong local support for the tests. Over 9,000 tomato transplants were planted between the first and third weeks of June; agronomic conditions that closely paralleled commercial practices were chosen.

The Roundup-tolerant tomatoes containing the mutant petunia EPSPS gene yielded comparably to controls in the absence of herbicide treatment, indicating that EPSPS levels have no effect on plant growth. Twenty-two independently derived transgenic tomato lines were evaluated in the test; the best plants survived Roundup applications of over 1 lb/acre; in contrast, control plants were rapidly killed at this concentration. The treated transgenic plants displayed excellent vegetative tolerance, but showed reduced and delayed flowering relative to unsprayed controls. Given the extreme sensitivity of tomatoes to damage by Roundup exposure, the dramatic increase in tolerance observed in the field results is extremely encouraging and indicates that tolerance to commercial concentration (0.5–1 lb/acre) should be achievable.

18.2.2 Virus-Tolerant Plants

An important application of genetic engineering technology is in the area of producing resistance to pathogens. At present, very little is known about genetic mechanisms of disease resistance in plants. Observations made 40 to 50 years ago pointed to a possible mechanism for increasing host resistance to viral pathogens. In 1929 McKinney observed that tobacco infected with one strain of a virus resisted infection by a second related strain (McKinney 1929). This phenomenon, named cross-protection, could be demonstrated with a number of viruses for which distinct strains could be found (reviewed by Hamilton 1980; Sequeira 1983; Palukaitis and Zaitlin 1984; Fulton 1986). Although the mechanisms of cross-protection are not fully understood, several investigators have suggested that protection from viral disease could be achieved with the help of gene transfer technology (Hamilton 1980; Sequeira 1984; Palukaitis and Zaitlin 1984; Beachy et al. 1985; Bevan et al. 1985; Sanford and Johnston 1985).

Powell Abel et al. (1986) constructed an expression vector containing the 35S promoter from CaMV, a cDNA encoding the coat protein (CP) gene of the U1 strain of tobacco mosaic virus (TMV), and a polyadenylation signal from the nopaline synthase gene. This vector was introduced into tobacco and tomato cells and plants were regenerated. Progeny of self-fertilized transgenic plants expressing high levels of the CP gene (0.05–0.1% of total soluble cell protein) were inoculated with TMV. These plants either did not develop an infection or disease symptoms developed more slowly than in the control plants after inoculation with the U1 strain of TMV. The plants expressing the CP gene were also delayed in symptom development after inoculation with a severe TMV strain, PV230, which is immunologically related to the U1 strain (Nelson et al. 1987). The numbers of chlorotic and necrotic local lesions on the inoculated leaves of transgenic *Xanthi* and *Xanthi* nc plants infected with PV230 or U1 were 70% and 90% lower, respectively, than the numbers of lesions observed on the control plants. As observed in classical cross-protection, the decrease in the numbers of chlorotic and necrotic lesions could be at least partially overcome by inoculation with viral RNA. Thus the presence of CP on the challenge virus was necessary for maximum protection (Nelson et al. 1987).

The phenomenon of genetically engineered cross-protection was extended to another virus, alfalfa mosaic virus (AlMV) (Tumer et al. 1987). AlMV differs from TMV in many respects. TMV is a rigid rod-shaped virus that encapsidates a single RNA molecule. It encodes at least four proteins in three open reading frames (Hirth and Richards 1981). In contrast, AlMV is a bacilliform virus that has a tripartite genome consisting of RNA molecules 1, 2, and 3. RNA 4, a subgenomic RNA of RNA 3, encodes the CP (Jaspers 1985). Unlike infection with TMV RNA, which does not require CP, infection with AlMV RNA requires addition of RNA 4 or CP (Bol et al. 1971; Houwing and Jaspers 1978). Thus, TMV and AlMV are clearly distinguished by their morphological features, genome structure, strategies of viral gene expression, early steps in replication, and modes of transmission (AlMV is transmitted mechanically and by aphids; TMV is primarily mechanically transmitted).

Tumer et al. (1987) have cloned RNA 4 encoding the CP of AlMV and engineered it into an expression vector downstream from the CaMV 35S promoter. The cDNA is flanked at the 3′ end by the nopaline synthase polyadenylation site. Leaf disks of tobacco and tomato were transformed with *A. tumefaciens* containing this construct, selected for kanamycin resistance, and regenerated into plants (Horsch et al. 1985; McCormick et al. 1986). The amount of AlMV CP expressed in different transgenic tobacco plants varied between 0.1% and 0.4% of total extractable leaf protein. The amount of AlMV CP expressed in transgenic tomato plants was between

0.1% and 0.8% of the total extractable leaf protein. The segregation ratio of the CP gene was determined in the seedling progeny of the self-fertilized transgenic plants. Progeny from a single transgenic tobacco plant were then inoculated with different concentrations of AlMV (20 μg/ml, 10 μg/ml, and 5 μg/ml) and symptoms were monitored in a growth chamber. Typical symptoms of AlMV infection appeared in three to four days on the inoculated leaves of the control seedlings. Of the seedlings that expressed the AlMV CP, 80% of the plants infected with 20 μg/ml of AlMV and 90% of the plants infected with 5 and 10 μg/ml of AlMV did not develop symptoms by the end of two weeks. As expected, decreasing the inoculum concentration decreased the number of CP-positive and CP-negative plants that showed symptoms on the inoculated leaves and increased the amount of time it took to produce symptoms by one to twelve days. Similar results were obtained for transgenic tomato plants. Recently, cross-protection against PVX and cytomegalovirus has also been obtained in our laboratory using an identical strategy.

An important question is whether the protection observed in transgenic plants is the same as that found in plants that are cross-protected against virus infection. Although the answer is not known, the two types of protection show similarities. Both types of protection manifest as a delay in disease development. In the case of the transgenic plants, the delay is the result of a 90–100% reduction in the numbers of lesions on the plant. Some plants escape infection completely and if infection is established, spread of virus is slowed down in the transgenic plants. In the case of TMV, protection could be partially overcome both in the transgenic and cross-protected plants by increasing the concentration of the virus in the challenge inoculum. In the case of AlMV, the protection could not be overcome at the highest concentration of inoculum (50 μg/ml). The protection observed against TMV could also be overcome in both tomato and tobacco plants if the challenge inoculum contained the virus RNA rather than virions (Nelson et al. 1987). Regardless of the precise mechanism involved in the virus disease resistance in transgenic plants, genetically engineered cross-protection provides a generally applicable way of producing virus-resistant plants.

Transgenic tomatoes carrying the TMV CP gene were evaluated in field tests during the summer. The TMV-tolerant plants were hand inoculated with 10–40 μg/ml of TMV (U1 strain) at different intervals after transplanting. In all cases, 100% of the control plants exhibited virus symptoms and tested positive for the presence of TMV. In contrast, less than 5% of the transgenic tomatoes containing the chimeric CP gene displayed symptoms. Where symptoms did appear on the transgenic plants they were quite mild and were restricted to the inoculated leaves. Biochemical tests confirmed the lack of systemic spread of the virus in the transgenic plants. In the absence of virus pressure, the transgenic tomato plants expressing TMV CP yielded comparable to controls, indicating that the presence of CP did not affect plant growth. Interestingly the transgenic plants showed no yield reduction after virus infection, whereas the control plants suffered 23–33% yield losses. This result was somewhat surprising, since the U1 TMV strain caused only mild symptoms on the control plants. These results may indicate, as many plant pathologists suspect, that subclinical levels of viral diseases in plants may cause significant but generally unobserved yield reductions.

18.2.3 Insect-Tolerant Plants

Another application of genetic engineering with important implications for crop improvement has been in the production of insect-resistant plants. Progress in engineering insect resistance in transgenic plants has been achieved through the expression in plants of the insect toxin gene of *Bacillus thuringiensis (B.t.) B. thuringiensis* is an entomocidal bacterium that produces a parasporal protein crystal. Most strains of *B.t.* are toxic to lepidopteran larvae (Dulmage 1981) although some strains with toxicity to dipteran (Goldberg and Margalit 1977) or coleopteran (Kreig et al. 1983; Herrnstadt et al. 1986) larvae have also been described. The insect toxicity of *B.t.* resides in the parasporal protein crystal that in the case of lepidopteran-active strains is composed of toxin protein subunits of approximately 130,000 kDa (Bulla et al. 1981). Genes encoding the lepidopteran-specific toxins from several strains of *B.t.* have been cloned and sequenced (Adang et al. 1985; Schnepf and Whiteley 1985; Shibano et al. 1986; Thorne et al. 1985; Hofte et al. 1986; Wabiko et al. 1986; Fischhoff et al., unpublished observations). These genes are largely similar; however, there are regions of significant variability at both the nucleotide sequence and amino acid sequence levels. Deletion variants of several of these genes have been constructed in vitro and tested for toxicity after expression in *Escherichia coli* (Adang et al. 1985; Schnepf and Whiteley 1985; Shibano et al. 1985; Hofte et al. 1986; Wabiko et al. 1986; Fischhoff et al. 1987). Taken together these experiments indicate that the region essential for toxicity resides in the N-terminal portion of the protein, extending approximately from amino acid 29 to amino acid 610.

Fischhoff et al. (1987) have engineered and expressed in tomato plants a lepidopteran-specific toxin gene from *B.t.* subsp. *kurstaki* HD-1 (Watrud et al. 1985). DNA sequence analysis has shown that this gene is very similar to genes reported from *B.t.* subsp. *berliner 1715* (Hofte et al. 1986; Wabiko et al. 1986). A fragment of this gene encoding amino acids 1 through 725 was isolated and shown to encode a functional toxin by expression in *E. coli*. A chimeric truncated toxin gene was constructed in which the *B.t.* toxin gene was flanked by the CaMV 35S promoter and the 3′ end of the nopaline synthase gene.

Tomato plants transformed with this chimeric gene were recovered by the method of McCormick et al. (1986). Plants were assayed for expression of the *B.t.* toxin gene by Northern analysis, and a polyadenylated RNA of the size expected of a full-length transcript was detected. The plants were also analyzed for toxicity to *Manduca sexta* (tobacco hornworm), an insect that is sensitive to the *B.t.* toxin and known to feed on tomato leaves. All of the larvae applied to the transgenic plants were killed within a few days, and the plants showed very little evidence of feeding damage. Larvae applied to control plants survived and eventually consumed the plants completely. These experiments demonstrate the feasibility of producing transgenic plants resistant to some lepidopteran insects through the expression of the *B.t.* toxin gene.

Transgenic tomato plants containing the *B.t.* gene were field tested during the summer. The plants were infested with egg masses of tobacco hornworm and *Heliothis zea* (tomato fruitworm). The level of insect control observed in the field tests was generally superior to that observed in growth chamber experiments. Under conditions where control plants were totally defoliated by the hornworm infestation, tomato plants containing the *B.t.* gene suffered no agronomic damage. Similar results were obtained with control and *B.t.*–containing tomato plants following natural or inoculated infestations by tomato fruitworm. Fruit damage on control plots was 17–23%; whereas fruit damage on *B.t.* plants was only 4–9%. The excellent insect control observed under field conditions may be a result of using egg mass inoculations instead of young larvae, as in the growth chamber experiments; this may have allowed more effective control at early stages in insect development. It is also possible that the feeding deterrent effects associated with the *B.t.* insecticidal protein were more pronounced under natural field conditions.

Future prospects for genetically engineering insect tolerant plants will likely continue to involve the protein insect toxins from *B.t.* for several reasons. The *B.t.* toxins show a high degree of specificity. Individual toxins are active only against a single order of insects (e.g., Lepidoptera or Coleoptera), and within a given order not all insects are equally sensitive. In spite of this high degree of specificity, the *B.t.* toxins are active against some insect pests of major agronomic importance such as *H. zea* (corn earworm/cotton bollworm), in the case of lepidopteran-specific toxins, and *Leptinotarsa decemlineata* (Colorado potato beetle), in the case of the coleopteran-specific toxins. The lepidopteran-specific toxins, which have been intensively studied, are considered safe; they have no known activity against mammals, fish, or nontarget invertebrates.

18.3 CONCLUSIONS

The results described above indicate the potential for using gene transfer technology for crop improvement. However with all the attention focused on agricultural applications, it is important to emphasize the impact this powerful tool will have on advancing basic research in all aspects of the plant sciences. Precise localization of the regulatory sequences for tissue-specific and environmentally modulated expression of genes has just begun for a small number of plant genes. The availability of specific molecular probes is already extending research in plant physiology, biochemistry, and cell biology.

The field testing of genetically engineered plants containing herbicide-, virus-, and insect-tolerance genes represents an important step in the commercialization of plant biotechnology research. In view of recent research breakthroughs, increased public awareness, and the existence of new regulatory policies, such tests have undoubtedly taken on greater significance than deserved. It is important to remember that many of the first tests have not been conducted with commercial cultivars and that several years of plant breeding efforts, field evaluations, and scale-up lie ahead before improved crops will be marketed. At the same time though, the technology is developing quickly and already issues such as regulatory costs and registration time lines are becoming key concerns of companies attempting to develop improved genetically engineered crops.

While the commercial significance of the current year's field tests will be determined only after much additional evaluation, several important observations can be noted:

- Field experiments were carried out with local community support.
- There were no adverse impacts of engineered plants on the test site environment.
- Introduced traits did not affect plant growth or reduce yield.
- Field performance was as good as or better than performance in greenhouse tests.

It is important that the investigators evaluating field tests of genetically engineered plants recognize and respond quickly to the need for testing of additional plants at multiple locations. It will also be important in subsequent evaluations to follow normal agronomic practices in field tests, including letting crops complete their reproduction cycles, and to test in normal production areas. Finally it is critical that regulations dealing with the commercialization of genetically engineered plants recognize the inherent low risk of the technology and not draw undue attention to the particular biotechnique used to improve plants.

REFERENCES

Adang, M.J., Staver, M.J., Rochleau, T.A., et al. (1985) *Gene* 36, 289–300.

Baulcombe, D.C., Saunders, G.R., Bevan, M.W., Mayo, M.A., and Harrison, B.D. (1986) *Nature* 321, 446–449.

Beachy, R.N., Powell Abel, P., Oliver, M.J., et al. (1985) *Biotechnology in Plant Science* (Zaitlin, M., Day, P., and Hollaender, A., eds.), pp. 265–275, Academic Press, Orlando, FL.

Bevan, M., and Chilton, M.D. (1982) *Annu. Rev. Genet.* 16, 357–384.

Bevan, M. (1984) *Nucleic Acids Res.* 12, 8711–8718.

Bevan, M., Flavell, R.B., and Chilton, M.-D. (1983) *Nature* 304, 184–187.

Bevan, M.W., Mason, S.E., and Goelet, P. (1985) *EMBO J.* 4, 1921–1926.

Bol, J.F., Van Vloten-Doting, L., and Jaspers, E.M.J. (1971) *Virology* 46, 73–85.

Bulla, L.A., Kramer, K.J., Cox, D.J., et al. (1981) *J. Biol. Chem.* 256, 3000–3004.

Comai, L., Facciotti, D., Hiatt, W.R., et al. (1985) *Nature* 317, 741–744.

De Block, M., Herrera-Estrella, L., Van Montagu, M., Schell, J., and Zambryski, P. (1984) *EMBO J.* 3, 1681–1689.

De Block, M., Botterman, J., Vandewiele, M., Dockx, J., and Thoen, C. (1987) *EMBO J.* 6, 2513–2518.

De Greve, H., Dhaese, P., Seurinck, J., et al. (1983) *J. Mol. Appl. Genet.* 1, 499–511.

della-Cioppa G., Bauer, S.C., Klein, B.K., et al. (1986) *Proc. Natl. Acad. Sci. USA* 83, 6873–6877.

Depicker, A., Stachel, S., Dhaese, P., Zambryski, P., and Goodman, H. (1982) *J. Mol. Appl. Genet.* 1, 561–573.

Dulmage, H.T. (1981) *Microbial Control of Pests and Plant Diseases 1970–1980* (Burges, H.D., ed.), pp. 193–222, Academic Press, Orlando, FL.

Eichholtz, D., Rogers, S., Horsch, R., et al. (1987) *Somatic Cell Mol. Genet.* 13, 67–76.

Fischhoff, D.A., Bowdish, K.S., Perlak, F.J., et al. (1987) *Bio/Technol.* 5, 807–813.

Fraley, R.T., Rogers, S.G., and Horsch, R.B. (1986) *CRC Crit. Rev. Plant Sci.* 4, 1–46.

Fraley, R.T., Rogers, S.G., Horsch, R.B., et al. (1983) *Proc. Natl. Acad. Sci. USA* 80, 4803–4807.

Fraley, R.T., Rogers, S.G., Horsch, R.B., et al. (1985) *Bio/Technol.* 3, 629–637.

Fromm, M.E., Taylor, L.P., and Walbot, V. (1986) *Nature* 319, 791–793.

Fulton, R.W. (1986) *Annu. Rev. Phytopathol.* 24, 67–81.

Goldberg, L.J., and Margalit, J. (1977) *Mosquito News* 37, 353–358.

Hamilton, R.I. (1980) *Plant Disease,* vol. 5 (Horsfall, J.G., and Cowling, E.B., eds.), pp. 279–303, Academic Press, Orlando, FL.

Haughn, G., Smith, J., Mazur, B., and Somerville, C. (1988) *Mol. Gen. Genet.* (in press).

Hayford, M., Medford, J., Hoffmann, N., Rogers, S., and Klee, H. (1988) *Plant Physiol.* (in press).

Helmer, G., Casadaban, M., Bevan, M., Kayes, L., and Chilton, M.-D. (1984) *Bio/Technol.* 2, 520–527.

Herrera-Estrella, L., Deblock, M., Van Montagu, M., and Schell, J. (1983) *EMBO J.* 2, 987–994.

Herrnstadt, C., Soares, G.G., Wilcox, E.R., and Edwards, D.L. (1986) *Bio/Technol.* 4, 305–308.

Hirth, L., and Richards, K.E. (1981) *Adv. Virus Res.* 26, 145–199.

Hofte, H., De Greve, H., Seurinck, J., et al. (1986) *Eur. J. Biochem.* 161, 273–280.

Horsch, R.B., Fry, J.B., Hoffmann, N.L., et al. (1985) *Science* 227, 1229–1231.

Houwing, C.J., and Jaspers, E.M.J. (1978) *Biochemistry* 17, 2927–2933.

Jaspers, E.M.J. (1985) *Molecular Plant Virology,* vol. 1 (Davies, J.W., ed.), pp. 155–221, CRC Press, Boca Raton, FL.

Jefferson, R., Kavanaugh, T., and Bevan, M. (1988) *EMBO J.* 6, 3901–3908.

Krieg, A., Huger, A.M., Langenbruch, G.A., and Schnetter, W. (1983) *Z Angew. Entomol.* 96, 500–508.

Lamppa, G.K., Morelli, G., and Chua, N.-H. (1985) *Mol. Cell. Biol.* 5, 1370–1378.

Lloyd, A.M., Barnason, A.R., Rogers, S.G., et al. (1986) *Science* 234, 464–466.

McCormick, S., Niedermeyer, J., Fry, J., et al. (1986) *Plant Cell Rep.* 5, 81–84.

McKinney, H.H. (1929) *J. Agricult. Res.* 39, 557–578.

Mousdale, D.M., and Coggins, J.R. (1984) *Planta* 160, 78–83.

Nafziger, E.D., Widholm, J.M., Steinrucken, H.C., and Kilmer, J.L. (1984) *Plant Physiol.* 76, 571–574.

Nelson, R.S., Powell Abel, P., and Beachy, R. (1987) *Virology* (in press).

Ow, D., Wood, K., Deluca, M., et al. (1986) *Science* 234, 856–859.

Palukaitis, P., and Zaitlin, M. (1984) *Plant-Microbe Interactions. Molecular and Genetic Perspectives*, vol. 1 (Kosuge, T., Nester, E.N., eds.), pp. 420–429, Macmillan, New York.

Powell Abel, P., Nelson, R.S., De, B., et al. (1986) *Science* 232, 738–743.

Rubin, J.L., Gaines, C., and Jansen, R.A. (1984) *Plant Physiol.* 75, 839–845.

Sanford, J.C., and Johnston, S.A. (1985) *J. Theor. Biol.* 113, 395–405.

Sequeira, L. (1984) *Trends Biotechnol.* 2, 25–29.

Schell, J., and Van Montagu, M. (1983) *Bio/Technol.* 1, 175–180.

Schnepf, H.E., and Whiteley, H.R. (1985) *J. Biol. Chem.* 260, 6273–6280

Shah, D.M., Horsch, R.B., Klee, H.J., et al. (1986) *Science* 233, 478–481.

Shibano, Y., Yamagata, A., Nakamura, N., et al. (1985) *Gene* 34, 243–251.

Steinrucken, H.C., and Amrhein, N. (1984) *Biochem. Biophys. Res. Commun.* 94, 1207–1212.

Thorne, L., Garduno, F., Thompson, T., et al. (1986) *J. Bacteriol.* 166, 801–811.

Tumer, N.E., O'Connell, K.M., Nelson, R.S., et al. (1987) *EMBO J.* 6, 1181–1188.

Wabiko, H., Raymond, K.C., and Bulla, L.A. (1986) DNA 5, 305–314.

Waldron, C., Murphy, E., Roberts, J., et al. (1985) *Plant Mol. Biol.* 5, 103–108.

Watrud, L.S., Perlak, F.J., Tran, M.-T. et al. (1985) *Engineered Organisms in the Environment* (Halvorson, H.O., Pramer, D., and Rogul, M., eds.), pp. 40–46, American Society for Microbiology, Washington, DC.

Van Den Elzen, P., Townsend, J., Lee, K., and Bedbrook, J. (1985) *Plant Mol. Biol.* 5, 299–302.

Zambryski, P., Joos, H., Maes, M., et al. (1983) *EMBO J.* 12, 2143–2150.

INDEX